SECOND EDITION

Switching Power
Converters
MEDIUM AND HIGH POWER

SECOND EDITION

Switching Power
Converters
MEDIUM AND HIGH POWER

Dorin O. Neacsu

CRC Press
Taylor & Francis Group
Boca Raton   London   New York

CRC Press is an imprint of the
Taylor & Francis Group, an **informa** business

CRC Press
Taylor & Francis Group
6000 Broken Sound Parkway NW, Suite 300
Boca Raton, FL 33487-2742

First issued in paperback 2017

© 2014 by Taylor & Francis Group, LLC
CRC Press is an imprint of Taylor & Francis Group, an Informa business

No claim to original U.S. Government works
Version Date: 20131023

ISBN 13: 978-1-138-07571-9 (pbk)
ISBN 13: 978-1-4665-9192-9 (hbk)

Visit the Taylor & Francis Web site at
http://www.taylorandfrancis.com

and the CRC Press Web site at
http://www.crcpress.com

Contents

PART I   Conventional Power Converters

PART II   *Other Topologies*

Preface

Power electronics represents a branch of electronics dedicated to the controlled conversion of electrical energy. This conversion includes adaptation of power to diverse applications such as voltage or current power sources, electrical drives, active filtering in power systems, distributed generation and smart grid, electrochemical processes, inductive heating, lighting and cooking control, distributed generation, and naval or automotive electronics. This very broad range of applications has stimulated research and development, and new control methods of power hardware are suggested each day. The medium- and high-power converter systems require multidisciplinary knowledge of basic power electronics, digital control and hardware, sensors, analog preprocessing of signals, thermal management, reliability, protection devices and fault management, or mathematical calculus.

Because of this great number of technical solutions with many variations of the same concepts, it is somewhat difficult for the practicing engineer or for a student to keep track of new developments or to find the most appropriate solution in the given time. It is therefore easier to develop a reasoning based on system-level understanding of the problem rather than aiming at an encyclopedia collection of solutions. This naturally moves the question from "*how to do it?*" to "*what is better to do?*" Therefore, a good engineer involved in industrial activities needs to also understand technology evolution, market timing, component availability and technology cycling, social requirements for environment and reliability, all of these in addition to the classical now circuit design.

Libraries and bookstores offer a great number of books on power electronics, mostly academic textbooks of a theoretical nature without getting too deeply into the practical aspects of technology development.

Other publications offer a multitude of design solutions grouped into an encyclopedia style handbook. Given the need for multidisciplinary knowledge at the edge between academic and industrial preoccupations, as well as the large variety of applications, the technical information often stretches beyond the offer of a handbook.

Conversely, this book offers a technology review rather than a collection of design procedures. Readers are offered information about the history of important achievements, current performance expectations, the technology evolution from radical to incremental solutions, *S*-curve and the cycle of performance achievement in technology development, and modern requirements for either standards or design for reliability. All of these distinguish this book on the library shelf as a very unique view of the technology of power electronics systems. This way, the design and development decisions are not made solely by circuit investigation, and a multitude of other technology-related criteria can be considered.

This book can also be seen as a digest of cutting-edge results in the field of medium- and high-power converters presented in a precise manner, with a fair amount of examples and references. From the numerous papers, patents, and research notes published throughout the world during the last 25 years, those methods mostly relevant to the industry have been selected as samples of the technology evolution.

The position of each topic in the history of power electronics technology and its contribution to performance improvement is highlighted and justified. The most incisive focus of this book is dedicated to the PWM algorithms and it is hoped that this book presents this concept at its best.

The presentation flows from simple facts to advanced research topics and readers are required to have only a minimal background in electrical engineering or power electronics. Chapters in the first part of the book end with problems to help readers improve their learning. This combination of theory and examples is the result of the author's many years of teaching at different universities as well as his vast industrial "hands-on" experience.

The book begins with an overview of industrial power converters and power semiconductors dedicated to medium- and high-power operations, and includes aspects about the market. After a brief technology review of power semiconductors in Chapter 2, Chapters 3 through 5 define the basics of operating a conventional three-phase inverter with pulse width modulation. Chapters 6 through 10 are dedicated to the practical aspects of implementation with many examples from known industrial platforms. Chapters 11 through 17 are dedicated to other special three-phase topologies and their control. Chapter 14 introduces a solution that has been used more frequently during the past few years to achieve higher power from conventional lower-power converters. The parallel or interleaved operation of conventional three-phase inverters helps increase the power capacity by the addition of multiple low-power units already available on the market.

Finally, Chapter 18 features a future-looking research topic related to conceiving novel converter topologies using *Intelligent Power Modules* as building bricks, under the *Network of Switches* concept, with benefits in reliability improvement and loss reduction. It is the author's belief that the *Network of Switches* concept can represent for contemporary power electronics what the transition from the bipolar transistor to the integrated circuit meant for analog electronics in 1970s.

This book covers modern topics pertaining to medium- and high-power converters used in three-phase DC/AC or AC/DC conversion, and it can serve as an advanced textbook for graduate students or as a reference book for engineers working in the industry.

Dorin O. Neacsu
Iasi, Romania
and
Westford, Massachusetts

MATLAB® and Simulink® are registered trademarks of The MathWorks, Inc. For product information, please contact:

The MathWorks, Inc.
3 Apple Hill Drive
Natick, MA 01760-2098 USA
Tel: 508 647 7000
Fax: 508-647-7001
E-mail: info@mathworks.com
Web: www.mathworks.com

Acknowledgments

I thank all the professors, managers, and colleagues who helped my personal development as an engineer and facilitated acquiring the knowledge shared within this book. Their leadership and vision in power electronics have aided in depicting the cutting-edge trends in modern high power switching converters, and I hope this book will make a positive impact in the formation of a new generation of power electronics engineers. It is the role of leaders to find other leaders and to unlock for them the possibility of making a positive impact.

I am entirely grateful to Professors Mihai Lucanu and Dimitrie Alexa who encouraged and supervised my beginnings at the Technical University of Iasi, Romania. Many of the research results published in this book stem from educational programs attended under their direction. Special thanks to Professors Venkatachari Rajagopalan (Canada) and Frede Blaabjerg (Denmark) who introduced me to the IEEE and the world of highly competitive modern technologies.

I thank all my colleagues from US industry including Kaushik Rajashekara, James Walters, T.V. Sriram, Fani Gunawan, and Balarama Murty of Delphi Automotive; Toshio Takahashi, David Tam, Brian Pelly, HuaHuu Nguyen, and Eric Person of International Rectifier; Ted Lesster, Bogdan Borowy, William Bonnice, Geoffrey Lansberry, and Evgeny Humansky of SatCon Technology Corporation; and James Worden, Viggo Selchau-Hansen, Lance Haines, Beat Arnet, Lu Jiang, and Don Lucas of Azure Dynamics/Solectria.

Author

Dorin O. Neacsu, PhD (IEEE M'95, SM'00) received MS and PhD degrees in electronics from the Technical University of Iasi, Iasi, Romania, in 1988 and 1994, respectively. Dr. Neacsu also holds an MSc degree in engineering management from the prestigious Gordon Institute of Tufts University, Medford, Massachusetts, USA.

Dr. Neacsu spent his mandatory industrial probation with TAGCM-SUT Iasi, Romania from 1988 to 1990, and was an associate professor in the Department of Electronics, at the Technical University of Iasi, between 1990 and 1999. During this time, he held visiting positions at Universite du Quebec a Trois Rivieres, Canada, and General Motors/Delphi, Indianapolis, Indiana, USA. Dr. Neacsu held different positions (consultant, engineer, product manager) within US industry from August 1999 to June 2006, at International Rectifier, SatCon Technology, and Azure Dynamics/Solectria. This led to innovative work in advanced gate drivers, isochronous hot-swap paralleling of three-phase power converters, and interleaved power converters. After August 2006, Dr. Neacsu held visiting or temporary appointments at diverse US academic institutions including the University of New Orleans, Massachusetts Institute of Technology, and the United Technologies Research Center. Dr. Neacsu is currently an associate professor at the Technical University of Iasi, Romania.

Dr. Neacsu has published more than 80 papers and research notes in IEEE transactions, conferences, and other international journals, on all continents, and has presented seven tutorials at IEEE conferences. He holds three US patents and has co-written several university textbooks in Canada and Romania, and a book on simulation modeling of power converters. Dr. Neacsu is a senior member of the IEEE, has served as a reviewer for several IEEE transactions, and has been a member of the technical program committees or organizing committees at various IEEE conferences. His research activities are in static power converters, power semiconductor devices, PWM algorithms, microprocessor control, and modeling and simulation of power converters.

1 Introduction to Medium- and High-Power Switching Converters

1.1 MARKET FOR MEDIUM- AND HIGH-POWER CONVERTERS

1.1.1 TECHNOLOGY STATUS

Power electronic converters have been one of the fastest-growing market sectors in the electronics industry over the last 40 years [1]. Power electronic devices are at the heart of many modern industrial and consumer applications and account for $18 billion per year in direct sales, with an estimated $570 billion through sales of other products that include power electronic modules.

The main application areas for power electronics are in power quality and protection, switch-mode power conversion, batteries, and portable power sources, automotive electronics, solar energy technology, communications power, and motion control (classification similar to a *Darnell Group* market report). The technology behind most products within these markets is on the saturation side of the performance's *S*-curve. The industry's efforts are concentrated in optimization of production and cost efficiency. The Organization of Electronics Manufacturers (OEM) has shown a clear trend for the power supply sector to stay away from custom-designed products and to optimize the standard, modified standard, and modular configurable products.

In power electronics, technology has developed under the pressure of the industry's needs, and there are many excellent papers written both by industry and university peers over more than 30 years. In synchronous with the technology status, current academic efforts target organization of information, book and tutorial writing, as well as improving the educational means. Moreover, there are new emerging regions of the world, and an ever-increasing number of new students in engineering in new places. This moves the focus of large corporations from achieving the technological leadership toward global market supremacy by production volume, diversity, and global and regional coverage.

Such an impressive count of sources of information may be overwhelming. However, each publication has its own goals, from basic student education textbooks, to industrial design handbooks, or niche tutorials. It is the intention of the present book to understand current technology within a business perspective and to present the existing engineering *hands-on* knowledge in an organized manner. This book focuses on medium- and high-power converters and the main applications at this power level are

- High-voltage DC transmission lines
- Locomotives

- Ship propulsion
- Large- or medium-sized uninterruptible power supply (UPS) systems
- Motor control from horsepower range to multi-MVA
- Propulsion of electric or hybrid vehicles
- Servo-drives, robot, or welding machine systems
- Elevator systems
- Distributed generation for renewable energy sources
- Appliances, air conditioners, refrigerators, microwave ovens, and washing machines
- Automobile electronics, power steering, power windows, doors, or seats
- Switch-mode power supply for industrial applications
- Consumer electronics, power supplies for VCR, TV sets, and radio
- Distribution systems for computers

Since the book deals with intimate details of designing and working with power electronic converters at medium- and high-power levels, without too much details at the application level, this introductory chapter briefly discusses the most attractive and emerging applications.

The introduction to the first edition has insisted on a series of market realities and numbers since the beginning of the twenty-first century which quest for technological leadership corporations still have. Such commitment for technological performance has generally favored a mathematical approach, a competition based on quantities explicitly shown both in market and technological achievements. Currently, we are witnessing a shift from the interest for quantitative expression of success toward the interest in global coverage and image. The newest financial annual reports of many corporations are less rich in numeric data and more informative on the geopolitical plans of the corporation. Sensitive to this trend, the introduction of the current edition of this book will review more the major technological achievements and less the market numerical data.

The most advanced efforts in power electronics are covering the following activities:

- Semiconductors
  - Application development and assimilation of SiC/GaN devices
  - New generations of power ICs, taking advantage of new IC technology platforms
- Low-power converters
  - Digital power supplies, especially those used for server/computer applications
    - Processor power controller with Intel VR10/11, or AMD VID support
      - Digital power supply with communication, and variable voltage controller, multiple operation modes
    - Generate and/or meet new standards, at the cross-disciplinary field between power supplies and servers/computers
  - Lower-voltage output, for newer generations of processors (like 100 A at 1 V)
  - Lower-voltage input for energy harvesting devices such as thermo-electric generators

- Conventional low-voltage applications
  - Improved power density and efficiency (sustaining efforts at system level, including thermal management)
  - Use of new materials in passive components (magnetics and capacitors), with redesign at the converter level to accommodate their peculiar performance
  - Generate and/or meet standards derived from the saturation of performance
  - Application development
    - Light systems, including multiple LEDs
    - Energy management in automotive systems, with networks of multiple motor drives
  - Given the existing production lines operating at high volume, we have efforts in reliability and protection—models, calculations, physics of failure
- Medium-voltage converters
  - Improved power density and efficiency (sustaining efforts at system level, including thermal management)
  - New topologies and afferent control for better use of energy while taking advantage of the existing saturation limits of performance *if sustaining would not do it, need to cross fields—you need to be really good to see it at the system level*
  - Application development
    - Smart grid, including communications along the transmission of energy, for better energy management
      - New products, for energy metering, sensing, production, storage, and transfer
      - New algorithms for software calculation of various performance indices
    - Application development and assimilation of power converter technology (motor drives) within HVAC and refrigeration systems (fairly new product applications)
      - This may require novel and appropriate control algorithms
    - Integration of renewable energy sources
      - Photovoltaics
      - Wind, sun, and water—this topic slowly moves into sustaining mode
  - Given the existing production lines operating at high volume, we have efforts in reliability and protection—models, calculations, and physics of failure
- High-voltage converters
  - Introduction/design of new power semiconductor devices
  - More power electronic control of energy, including active filters, STATCOM devices, power quality controllers, and so on
  - Inventive protection devices for high-voltage environment

- Brand new applications in experimental physics and medical equipment
  - HV pulse power, plasma science, and scanning microscope systems
  - Laser diode, LED, or other projection lamps

1.1.2 TRANSPORTATION ELECTRIFICATION SYSTEMS

The most important current application for power electronic systems lies within the electrification of the transportation systems.

1.1.2.1 Automotive

With a continuously evolving market and a continuous demand for new vehicles, the automotive sector embraces more and more electronic-based features. They range from entertainment systems to propulsion systems. This market is expected to double its growth rate in the coming years (from 2005 to 2015). A study [2] states an annual growth rate of 15.5% for the automotive sector, that is the strongest growth market.

These are new divisions for the power electronics market, but they must develop quickly due to the increased demand for efficiency, comfort, and safety. Another study has counted about 80 small-power drives, including two modern cars, in a middle-class American family's household. The power electronic products used in home applications are designed for low voltage and low power. Low-power servo-drives are described in this book.

Propulsion systems for advanced electric and hybrid vehicles are another emerging application field, with numerous electric and hybrid vehicles already released on the market.

The contemporary efforts are targeting

- Extended energy storage capabilities
- Fast and contactless chargers
- Advanced control algorithms
- Improved reliability to sustain the lifetime expectancy in the automotive products
- New designs for permanent magnet motors

1.1.2.2 Aviation

Another transportation-related effort is related to the More Electric Aircraft concept [3,4]. The large aircraft systems feature electrical power systems in the range of 1 MW. The power installed within the electrical power systems is increasing with the introduction of more and more power electronics processing to replace the conventional hydraulic systems. Such a large power needs to be processed by power electronics systems with advantages such as "power on demand," increased efficiency, small form factor, and so on.

Several modules are well-known examples of success stories:

- Integrated drive generator
- Variable speed constant frequency converter driven by engine
- Auxiliary power unit (115 V, 400 Hz)

- Emergency power source driven by the Ram Air Turbine (RAT system)
- High-voltage DC (HVDC) power systems and conversion to low voltage
- Flight control actuation, brake systems, and doors
- Air conditioning and heating systems

Challenges are related to harmonics, reduction of electromagnetic inference (EMI), and increasing power density. All of these fuel an impressive contemporary R&D effort for More Electric Aircraft based on power electronics concepts.

1.1.2.3   Railways

An emerging application for medium-voltage motor drives consists of rail propulsion systems in the multi-MW range [5–6]. Coming a long way from the beginning of the twentieth century, four major railway power supply systems are the most extended nowadays:

- DC 1.5 kV (6.5% of market in 2003 [5])
- DC 3 kV (30.3% in 2003 [5])
- AC 50 Hz, 25 kV (44.8% in 2003 [5])
- AC 16.7 Hz, 15 kV (13.6% in 2003 [5])

Additionally, subway systems mostly use DC 600–800 V.

The development, especially in Europe and Japan, of power electronics used in locomotive propulsion has encouraged replacement of gate turn-off thyristors (GTOs) switches by their modern insular gate bipolar transistor (IGBT) counter-parts. Traditional GTO solutions [6] were in use since early 1990 s, in the 6.4 MW EuroSprinter locomotive built by Krauss-Maffei and Siemens. Other examples are the locomotives RENFE8252 in Spain and CPLE5600 in Portugal. Other impor-tant corporations playing a role in railway electrification are Alstom, Bombardier Transportation (which absorbed the former ABB/ADTranz), or new comers such as the Swiss' Stadler Rail in cooperation with ABB.

Contemporary R&D efforts target:

- Power electronics for high-speed trains
- Expansion of onboard energy storage
- New concepts in permanent magnet synchronous motors
- Electronic transformer, especially for the 16.7 Hz applications
- Improvements in flux-orientated control algorithms
- Optimization of system level issues for the transmission and distribution of energy

1.1.2.4   Marine Power Systems

The ship marine power systems represent contemporary effort for market success. Over the last 5 years (2008–2013), numerous large corporations, or small business endeavors have released an impressive number of new devices and systems to this sector.

Electric propulsion has redeemed itself as the proper choice for large cruise ships and is already accepted more and more for warships. Unfortunately, simple operating profiles of some low-power vessels or commercial pressures make the all-electric solution not generally attractive. There exist many types of ships between these two extremes in which an all-electric solution can be successful. This solution provides potential for safer, more flexible, and sustainable vessels in the future as well as increased effectiveness in war and reduced life-cycle cost within the warship fleet.

In the past, the U.S. Navy acquired and implemented all-electric ship-propulsion systems for warships and submarines. Since the late 1990s, Eaton NCD (currently DRS Technologies) has already delivered a 2.2 MW brushless DC motor drive for submarine propulsion [7], and such efforts are continuing within the defense industry around the world.

Current efforts are mainly targeting expansion to smaller vessels, offering more and more electrical equipment for propulsion, services, auxiliary power, and navigation for either military, commercial, or entertainment vessels. Integration of small renewable energy sources, improved onboard energy storage, power management, and actuator supply are trying to cope up with the ever-increasing power demand.

1.1.3 TRADITIONAL INDUSTRIAL APPLICATIONS

1.1.3.1 Motor Drives

Despite the decreasing number of production facilities and the reduction of new facility development, the industrial sector is still asking for motor drives. The market share for the motor drives has developed steadily during the last 40 years [1,10–12]. This market opportunity has been followed with a strong R&D effort leading to a continuous technology development. However, advanced knowledge has allowed complete automation of the production lines, which has soon led to excess capacity and which, in turn, has resulted in a decrease in the revenue growth rate from 16.6% in 1970 to 5.5% in 2000 [8]. The resulting price erosion has been overcome by introducing new semiconductor devices and improving control algorithms and motor designs to reduce the cost, improve the efficiency, and increase the applicability to a large number of uses. Moreover, the motor drive market has in time a larger share of the nonindustrial products' market, in contrast to the trends of the last 20 year, when the end-market has been industrial. Home appliance power electronics and motor drives are good examples in this sense.

1.1.3.2 Grid-Tied Power Supplies

Another large business profile for electronic power converters is the UPS or grid-related applications. In 2001, the total worldwide market for UPS alone was at $5.3 billion. A derivative from this market is distributed generation, which is probably (since 2002) the most dynamic R&D sector in power electronics in the United States. The combination of a grid power supply and a nonconventional power source such as a diesel generator, a fuel-cell, or a wind turbine requires power electronics conditioning and protection. The appropriate power converters do not really bring anything new in their structure or packaging, but their control is a challenge yet to be solved.

Other related consumer markets include the AC/DC power supply, the PC and work-station power supply markets, and the communication power market. As these markets use only low-voltage systems, they are therefore not the direct focus of this book. However, these emergently ask for more power within the distribution system that is controlled with electronics.

1.1.3.3 Medium Voltage

Another new market segment deals with medium-voltage motor drives, in the range of 3300 V and 2000 A. High-voltage IGBTs have been introduced recently, and they take more and more the role of the GTOs. The complete picture here is filled with new devices, such as the integrated gate commutated thyristor (IGCT), a traditional IGBT device with the gate driver colocated with the power semiconductor. Motor drives delivering 19,000 HP are nowadays built by companies such as Siemens Corporation.

Different solutions for multilevel inverters are of interest for medium- and high-voltage applications allowing operation of up to 25 kV. A special approach consists of a stack of connected single-phase inverters, which is being extensively analyzed in the ABB and Daimler laboratories.

Harmonic performance is limited, however, due to limited switching frequency. New device materials such as silicon-carbide (SiC) may make the dream of high-frequency switching come true for medium- or high-voltage applications.

1.2 BOOK COVERAGE

The biggest difference from the other fields of electronics is the *power* coordinate. If students learn about a circuit or method for any other class of applications of electronics, they can easily manage to debug or put into service versions of that circuit from different manufacturers or within different applications. Power electronic circuits and their applications, however, are very different. The topology of a three-phase rectifier equipped with thyristors can be used for a 500 MVA HVDC transmission line or for a 1 kW welding machine. We can understand the basic operation of the three-phase phase-controlled rectifier from a college textbook, but the two systems are extremely different in reality. Each *thyristor* circuit explained in the textbook has a different implementation in practice, ranging from a half-inch T0–220 package to a building of six floors. The protection circuits are also very different, and range from no protection at all to sets of computer-controlled panels and automatic hot-swap replacement units. Finally, the cooling system could range from environmental air to complex systems of pumps or fans that by themselves have large installed power, often controlled with variable frequency through power electronics equipment.

Given this diversity of power levels and applications, different power semiconductor switches are more suitable for each case. Figure 1.1 stretches over the whole range of possible switching frequencies and installed power achievable with a single device. For larger power levels, multiple converters can be hardware connected in parallel. Modern power semiconductor devices, especially those of high power, require a good knowledge and control of their dv/dt and di/dt variations. These can be achieved through gate control as well as through circuit design, as shown

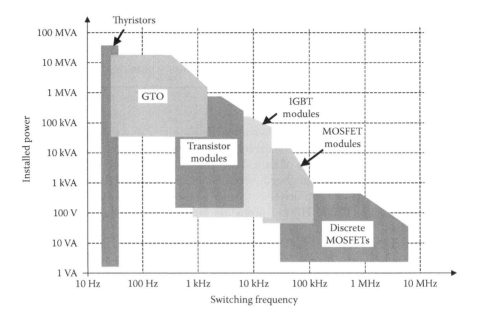

FIGURE 1.1   Power switches availability.

in Chapter 2, which is dedicated to understanding the operation and parameters of diverse power semiconductor devices.

The core of any power electronic converter is the control algorithm, and multiple options for pulse width modulation (PWM) algorithms are detailed in the book. The implementation possibilities for these control methods are shown from a historical perspective to allow the reader to understand that the semiconductor technology embedded in microcontrollers influence continuously the way we think about the PWM algorithms.

Each class of converters is dedicated a separate chapter, including topologies that are in demand by industry as well as topologies that lost somewhat their appeal while they still offer a huge academic and educational benefit.

The final chapter presents advanced concepts considered by the author as possible to find success in certain niche applications where conventional products still leave room for improvement with advanced concepts.

1.3 ADJUSTABLE SPEED DRIVES

A three-phase adjustable speed drive (ASD) comprises not just the power converter; it is a whole system that includes the power converter.

Figure 1.2 shows a complete ASD system consisting of:

- A three-phase rectifier system able to convert the grid three-phase system into a DC voltage
- An intermediate DC circuit usually composed of a large capacitor bank

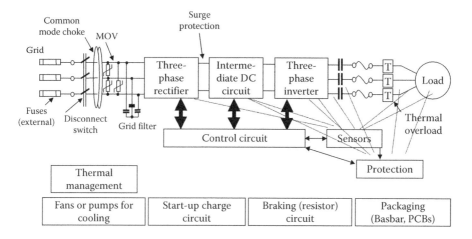

FIGURE 1.2   Global view on a three-phase ASD system.

- A three-phase inverter able to generate variable frequency, and variable voltage in the three-phase system
- A control circuit built with a digital signal processor (DSP), microcontroller, or programmable logic circuit (PLC) device
- Sensors and analog-signal preprocessing
- Connect/disconnect power switches, fuses, or protection circuitry
- Thermal-management system based on heatsinks or coldplates and a cooling system
- Start-up circuit with charging of the DC bus capacitor
- Braking resistor circuit

1.3.1   AC/DC CONVERTER

The input stage, called the three-phase rectifier, is built in many applications with rectifier diodes. The DC bus voltage is therefore quasi-constant at 1.35 times the line-to-line voltage (VLL). For a system with a VLL of 460 V, the DC voltage equals 620 V. Rarely, this power converter is made with silicon-controlled rectifier (SCR) devices in order to control the DC voltage with a method called phase control. Both solutions introduce very large harmonics of the current on the grid. These can become bothersome at large power levels, pollute the grid, and create problems for other users.

Another solution used often during the last few years consists of active front-end rectifiers built with controllable devices such as IGBTs or MOSFETs. Such three-phase power converters can process power directly, or they can be used as active filters to deliver the difference between the square-wave current produced by the diode rectifier and an ideal sinusoidal waveform. Thus, they result rated at a lower power level. If a direct three-phase active converter is used, the DC bus voltage can be higher due to the boost operation of that converter. This is advantageous, as it is easier to manipulate high-power levels from a high-voltage source.

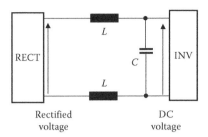

FIGURE 1.3   Intermediate DC circuit.

1.3.2 INTERMEDIATE CIRCUIT

The intermediate circuit is also called the *DC Link,* as it really is a DC link between the input rectifier and the output inverter. It serves as a power storage device. It is composed of a reactor inductor and a capacitor bank. Inductors filter the current through the capacitor in order to limit losses and heating. These two components are the bulkiest parts in the converter system panel (Figure 1.3).

Many manufacturers of ASD use a very large capacitance on the bus in order to ensure a power ride by enabling the motor to continue to operate when grid power is interrupted. Because of this large capacitance, however, it takes a longer time for these capacitors to discharge once power is turned off.

1.3.3   DC CAPACITOR BANK

The main functions of the DC capacitor bank are to

- Filter the harmonic ripple produced by the switching devices to produce clean sine-in, sine-out waveforms.
- Provide a stable voltage to ensure the control system's stability.
- Store energy useful for quick transients in the output.
- Work together with the brake resistor to limit DC voltage during regeneration of the "inverter" power stage.
- Limit overvoltages (clamp) before the system protection takes over and shutdown the power devices or start other auxiliary protection.

If the load is unbalanced or nonlinear, an alternative current circulates through the DC bus at twice the fundamental frequency. Depending on the value of the DC capacitor, this current can produce an oscillation of the DC bus voltage. Additional capacitive kVA in the DC link seems mandatory for inverters that feed unbalanced or nonlinear loads. This implies:

- Increased weight, volume, cost
- Selection of DC link to satisfy the maximum expected imbalance or worst-case nonlinear
- Increased losses and reduced reliability of the DC link components

Different active filtering solutions are considered to solve this problem.

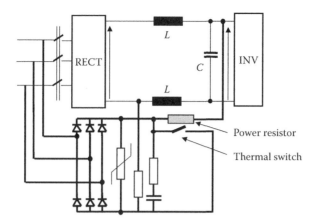

FIGURE 1.4   Principle of a soft-charge circuit.

1.3.4 Soft-Charge Circuit

ASDs at power levels above 30 HP (22.5 kW) use a soft-charge circuit for powering up the drive. Without this circuit, the in-rush current will be very large at power-on due to the extremely small impedance of the discharged DC bus capacitor. This large in-rush current would blow the grid fuses if not damage the rectifier semiconductors.

Figure 1.4 presents a possible soft-charge circuit. It basically adds a power resistor in the path of capacitor charging. This power resistor is also protected with a thermal switch able to disconnect above a certain temperature. After the voltage on the capacitor is larger than a minimum value, the power converter is disconnected through the grid disconnect switch. Due to the cooling requirements for the power resistor, the ASD can start only after 1 or 2 min.

1.3.5 DC Reactor

The other important part of the intermediate circuit consists of the DC reactor. This is also called *choke* or *DC coils*. It has two basic functions:

- Reduce the harmonics of the current by about 40%, with advantages in the power source or grid current.
- Help reduce power interruptions to avoid numerous nuisance shut-downs.

1.3.6 Brake Circuit

The intermediate circuit may also contain a brake circuit that takes the power from the DC bus when the drive is decelerating or stopping (Figure 1.5). Its operation is very simple: when the voltage across the capacitor bank increases above a certain level, the IGBT is turned-on and the power resistor is connected across the DC bus, at the inverter input. The inverter current now feeds a parallel R–C circuit. A large part of this current circulates through the resistor along with the discharge current from

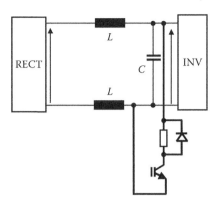

FIGURE 1.5   Brake circuit.

the capacitor. Usually, the brake circuit is part of the ASD, and the brake resistor is something the user adds depending on his requirements for a specific application.

One alternative to using the brake circuit is to transfer the excess power back to the grid through a power converter. This is called *regeneration* due to its efficiency advantages. However, one drawback is that it produces harmonics on the grid voltage affecting the incoming power to the converter. Finally, another option is to transfer the power excess to another drive's DC bus capacitor. This is sometimes called *load sharing*.

1.3.7   Three-Phase Inverter

The third major component of the system is the three-phase inverter. This is used for conversion of energy from DC voltage in an AC three-phase system with variable frequency and variable voltage. Typically, the topology is based on six IGBTs connected in a bridge; this will be discussed later in Chapter 3. Control of the three-phase inverter for this purpose is called pulse width modulation (PWM). Different PWM methods will be introduced in Chapters 4 and 5. Other topologies for DC/AC conversion are also presented in this book in Chapters 6 and 7.

1.3.8   Protection Circuits

A very important function for the whole ASD system is represented by the protection circuitry. We have protection for each power semiconductor device at over-voltage, over-current, over-temperature, or at problems within the gate drivers. The appropriate protection circuits will be presented in Chapter 3. More protection at the system level includes input or output fuses.

1.3.9   Sensors

Voltage on the DC bus and of the output currents is monitored through sensors. Some manufacturers use two current sensors at the output of the inverter while others use three sensors, one for each phase.

1.3.10   MOTOR CONNECTION

Large-power ASDs include motor coils that allow the operation of the motor far from the ASD system. For instance, the standard distance for a *Danfoss* drive is up to 300 m (1000 ft) for unshielded (unscreened) cable and 150 m (500 ft) for shielded (screened) cable [9]. If these coils are not used, the standard distance from the drive to the motor is as low as 50 m (160 ft) [9].

1.3.11   CONTROLLER

All of these blocks are supervised, monitored, and controlled from a central controller module. This is usually implemented on a digital circuit built around a micro-controller, DSP, PLC, or field programmable gate array (FPGA), application-specific integrated circuits (ASIC).

There are several functions that mandatorily must be included in the system:

- System command
  - System initialization
  - Run auto-test program
  - Define start/stop functions and check their operation
  - Define acceleration/deceleration of the system
  - Define sense of rotation or direction of displacement
  - Interfaces
    - Display data
    - User-interface
    - Communication with upper hierarchical level
- Control and regulation
  - Control algorithm
  - Data acquisition and digital processing
  - Regulation
  - Limits of control variables
  - Nonlinear characteristics
- Rectifier control when it is not built with only diodes
  - Synchronization
  - Command angle generation
  - Harmonic control
  - Power factor control
  - Gate control
  - Inverter control
  - Three-phase system generation
  - PWM generation
  - Minimum pulse control
  - Change of voltage and frequency
  - Limit of the operation range
- Supervision
- Protection

- Diagnosis
- Data storage
- Report to upper level through communication interface

Chapters 6 and 7 will provide details about the experimental aspects of implementing these functions in modern microcontrollers.

Power converters used for ASD applications generally need to satisfy some requirements or standards. Typical requirements are next presented.

1.4   GRID INTERFACES OR DISTRIBUTED GENERATION

Power electronics has been used for controlling and monitoring power transfer through HVDC links, especially in countries such as Canada and Brazil with isolated or local power systems. The back-to-back connection of controlled rectifier bridges on both ends of a DC transmission line allows control of up to 150 MW after the AC/DC/AC conversion [10]. However, these systems are rather rare, and the extensive use of power electronics in power systems is increasing as either active filters or grid interfaces. Many utility companies are providing solutions for power quality at the facility level on the utility side of the power meter. This multi-MW equipment is expensive and not likely to find success in the market.

A separate class of applications deals with nonconventional power sources, such as fuel cells, solar power, micro-turbines, or wind power. These projects with distributed energy sources manage local power generation in the range of 1 kW to 1 MW. For instance, one of the largest fuel-cell-based equipment is installed in Anchorage, Alaska, and accounts for 1 MW [11,12].

Special features are included in power converter controls in order to transfer energy from any of these energy sources or conventional batteries to grid [13]. At higher-power levels, this energy is exchanged on three-phase systems. Two operation modes are typical for these applications:

- *Grid parallel*: power converters that synchronize with the grid while exchanging energy from or to the grid.
- *Stand-alone*: that maintains three-phase voltage generation while the grid is disconnected.

Definitely, the control system must be able to switch between these modes any time the grid is lost or re-appears suddenly. Such requirements are also present in a conventional UPS system. The distributed generation system can also combine power delivery from the grid and the alternate source of energy.

The power electronic system maintains many of the protection and connection features presented for the ASD case. Let us take a closer look at the requirements of the grid interface.

The switching nature of operating power converters has led to various concerns about the quality of the grid at the point where the power converter is connected. Many standards have been elaborated in this respect. Some of these follow general requirements for inverters, some are specific for the grid connection. Any new power electronic equipment dedicated to a grid interface must obey regulations. Unfortunately, there are different grid voltage systems in the world, and grid requirements are different from country to country. Constraints to low-voltage grid applications around the world are presented next. Appropriate standards are next quoted, and they can be consulted for larger grid voltage systems:

- Nominal voltage ratings and operating tolerances for 60 Hz electric power systems from 100 V through 230 kV [14]
- Voltage sags analysis and methods of reporting sag characteristic graphically and statistically [15]
- Guidelines and limits for current and voltage distortion levels on transmission and distribution circuits [16]
- Powering and grounding sensitive electronic equipment [17]
- Monitoring of single-phase and polyphase AC power systems [18]
- Incompatibility of modern electronic equipment with a normal power system [19]
- Distributed resources interconnected with electric power systems [20]

1.4.1 GRID HARMONICS

Most European countries require compliance with EN61000-3-2. It lays down absolute limits for each individual harmonics. Japan's regulations are also derived from EN61000-3-2. Australia, the United States, and the United Kingdom set relative limits with a Total Harmonic Distortion (THD) of the current of 5% maximum and maximum values for each individual harmonic. Methods for minimizing those grid harmonics are presented in Chapter 9.

Power converters are also subject to harmonics from grid. The harmonics of the mains (grid) voltage a converter can cope with are given in the European standard EN60146-1-1 [13,14].

1.4.2 POWER FACTOR

A power factor of 1.00 is considered the best case, while anything higher than 0.8 is acceptable. If these levels cannot be achieved with the power system itself, additional units are used for power factor correction. This is the case of large inductive loads on the grid or on silicon-controlled rectifiers.

The high-frequency components of the input currents can be further reduced with chokes on the mains or on the DC link. DC link chokes also prevent resonance with the grid impedance. The incorporation of DC chokes on the power converter structure reduces the harmonic currents by up to 40% [13,14].

1.4.3 DC CURRENT INJECTION

It is very important not to inject DC components on the grid. Many countries avoid transformerless connection of switching converters to the grid. The operation of the power-switching converter must be symmetrical, so as not to produce DC components. The amount of DC current accepted by different countries is very different. A maximum of 0.5 mA is allowed in the United Kingdom; Australia's regulations allow a maximum of 0.5% of the power converter's rated current or 5 mA, whichever is greater; the U.S. regulations limit DC to a maximum of 0.5% of rated current; Japan allows a maximum 1% of rated current; and Germany a maximum 1 A per power converter connected to grid [13,14].

1.4.4 Electromagnetic Compatibility and Electromagnetic Inference

Step-switching waveforms of up to 15 V/ns or 5 A/ns generate EMI in both conducted and radiated forms. The conducted EMI is generated in differential (symmetrical) mode or common (asymmetrical) mode. Symmetrical mode EMI is generated when currents flow into the connection lines due to the power semiconductor variation of current (di/dt). The common mode EMI is produced due to the high (dv/dt) and parasitic capacitances to ground or connecting lines.

The radio or EMI interference produced by power converters depends on a number of factors:

- Switching frequency of the converter
- Slope of current and voltage at switching
- Impedance of the mains power supply
- Length of cables from grid and to the motor

Standards have been defined for previous applications of power converters, and they are re-applied to these grid interfaces. The most used standards for EMI are the German standard VDE or the Europe standard EN55011 (Figure 1.6). Appliances are covered by Europe standard EN55014, while power converter products are covered by EN61800-3.

The interference conducted to grid is usually reduced with a filter composed of coils and capacitors. If the power converter is not built with this filter, it can be purchased separately: "class A" for industrial applications and "class B" for household

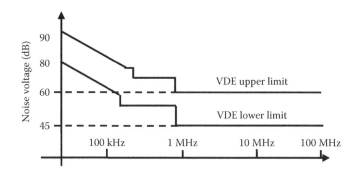

FIGURE 1.6 Example of EMI standard requirements.

applications. Moreover, using screened or armored cables limits the interference generated from power converter to the switching motor.

A new trend in EMI protection is the use of converter methods to reduce the common mode voltages; this will be analyzed in detail in Chapter 6.

1.4.5 FREQUENCY AND VOLTAGE VARIATIONS

It is accepted that power converters connected to grid can operate only within certain voltage and frequency windows. The system is considered stable within these windows. Along with voltage or frequency limits, a maximum allowable run-on time is also defined, and it varies considerably from country to country. Table 1.1 shows these limits [20].

1.4.6 MAXIMUM POWER CONNECTED AT LOW-VOLTAGE GRID

The maximum power installed in a power converter used as a grid interface is not always regulated by standards. Single-phase converters can be connected to low-voltage systems if their power is below 4.6 kW in Germany or Austria, 5 kW in the United Kingdom or Italy, and 10 kW in Australia. Three-phase converters can be connected to a low-voltage grid if their power is below 25 kW in Mexico, 30 kW in Australia, and 100 kW in Portugal. Obviously, higher-power converters can be connected to three-phase systems with higher voltages.

TABLE 1.1
Voltage and Frequency Variations

Country	Voltage			Frequency		
	Max V	Min V	Run-On Time (s)	Max Hz	Min Hz	Run-On Time (s)
Australia	270	200	2	50–52	48–50	2
Austria	253	195	0.2	50.2	49.8	0.2
Denmark	253	195	0.2	50.5	49.5	0.2
Germany	253	195	0.2	50.2	49.8	0.2
Italy	264	184	0.1 (@ 264 V) 0.15 (@ 184 V)	50.3	49.7	0.1
Japan	120	80	0.5–2	51.5	48.5	0.5–2
Mexico	132	108	2	61	59	—
The Netherlands	244	207	0.1	52	48	2
Portugal	264	195	0.1–1	50.25	49.75	0.1
Switzerland	264	195	0.2	51	49	0.2
The United Kingdom	253	207	Disconnect	50.5	47	Disconnect
The United States	164	64	0.022–0.100	60.6	59.3	0.1

Source: Data compiled from Panhuber, C. 2001. PV System Installation and Grid Interconnection Guidelines in Selected IEA countries, Raport IEA-PVPS, T5, April.

1.5   MULTICONVERTER POWER ELECTRONIC SYSTEMS

The advent of power electronic applications in industry changed the focus from issues related to building the power converter to issues related to system development and interaction between different power converters. Many modern industrial systems are composed of several ASDs connected to the same DC bus in *multidrive or multimodule* configurations. Modular design in multiconverter applications is based on the knowledge gained by individual analysis of each power converter. Power quality, efficiency, and system stability are affected by the interdependency between power stages.

Examples of multiconverter applications are:

- Industrial multidrive systems
- Parallel operation achieving higher-power levels
- Electric or hybrid electric vehicles
- Aircraft power electronic systems
- Ship power electronic systems
- Space electronic systems

Figure 1.7 shows a schematic of a modern power electronic system. The power source can be the industrial AC grid followed by an AC/DC power conversion, or the main power source can be a nonconventional power source, such as solar, wind, or thermal energy. After the appropriate conversion, the whole power resides on the DC bus. This bus supplies several motor drives. Some of them can dynamically be on the motoring mode, some on regeneration. The important thing is to manage the power on the DC bus so that the voltage is kept within two certain limits. This raises new problems, such as the stability of the DC bus at different loads. If one of the ASDs is working at constant torque with its speed regulated, its power can be considered constant. A load with constant power presents negative dynamic impedance that is a source of instability on the bus. Chapter 9 makes an extensive analysis of multiconverter power electronic systems.

Multidrive or multiconverter systems have several advantages:

- *Modularity*: quick and easy to integrate in panels and cabinets
- Scalability

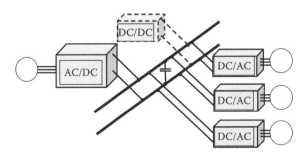

FIGURE 1.7   Complex power electronics system comprised of several drives.

- Redundancy
- *Reliability*: easy to replace a faulty module with a new one *(hot-swap possible)*
- Electronic gearing
- *Flexibility*: modules can be customized to any application
- Use of the same control cards and software for a large number of applications
- The same personnel training requirements across a wide power range
- Reduced-size library of *AutoCad* drawings, easy to integrate in a new design
- Lead-time reduction and money savings by minimizing spare requirements
- Same packaging and power density across the whole power range
- Technical advantages of using a single, high-power DC bus structure
- Optimized cooling system

1.6   CONCLUSION

Power electronics has emerged as a well-established technology with a broad range of applications. This chapter has shown the application range for medium and high-power converters, and it has focused on ASDs and grid interfaces. Constraints and standards to be met by different power converters within these applications are briefly listed. Equipment involving power converters are being increasingly used in all domains of our lives. Most of this energy is processed at medium and high power through power converters. The following chapters take an in-depth look at the theory of three-phase power converters, giving details of their problems and providing many solutions that can be implemented.

REFERENCES

1. Neacsu, D. 2005. Business plan for R&D operations in power electronics, Graduation Project, Tufts University, Gordon Institute, April 26, Scientific Advisers: Professors Arthur Winston and Mary Viola.
2. Anon, 2006. Market and Technology Study Automotive Power Electronics 2015, Arthur D. Little White Paper.
3. Wheeler, P. 2009. The more electric aircraft why aerospace needs power electronics, *European Power Electronics and Applications Conference, EPE'09*, pp.1–30.
4. Said, W. 2011. Aircraft electric power system—A power conversion perspective, *IEEE Power and Energy Magazine*, August.
5. Steimel, A. 2010. Power-electronics issues of modern electric railway systems, *10th International Conference on Development and Application Systems*, Suceava, Romania, pp. 1–6.
6. Bochetti, G., Bordignon, P., Perna, M., and Venanzo, P. 1993. 3MW converter for high power universal locomotive based on deionized water cooled GTO module—Improvements and type tests, *Proceeding of 5th European Power Electronics Conf. (EPE'93)*, Brighton, UK, 1993, pp. 241–246.
7. Divan, D. and Brumsickle, W.E. 1999. Powering the next millennium with power electronics, *Proceedings of the IEEE 1999 International Conference on Power Electronics and Drive Systems IEEE PEDS*, 7–17, Hong Kong, 1999, pp. 7–10 vol. 1.
8. Anon, 2001. Electronic motor drive market projected to top $19 billion by 2005. Appliance Design Magazine (www.appliancedesign.com), August 14, 2001.

9. Drives 101, Danfoss lessons, Danfoss Internet Documentation, www.danfoss.com.

10. Baker, M.H. and Bruges, R.P. 1991. Design and experience of a back-back HVDC link in western Canada, in *Proceedings of the IEE Conference on Advances in Power Systems Control Operation and Management*, Hong Kong, pp. 686–693.

11. Gilbert, S. 2001. The nation's largest fuel cell project: A 1 MW fuel cell power plant deployed as a distributed generation resource, Project Dedication August 9, 2000, *IEEE Rural Electric Power Conference*, Anchorage, Alaska, 2001, pp. A4/1–A8/1.

12. Bartos, F.J. and Gulalo, G. 1998. Power modules and devices advance motor controls. *Control Eng. J.*, 45(6), 91–101.

13. IEEE P1547 Std Draft 10 Standard for Distributed Resources Interconnected with Electric Power Systems, 2003.

14. ANSI C84.1–1989 American National Standards for Electric Power Systems and Equipment Ratings (60 Hertz).

15. IEEE Std 493–1900 IEEE Recommended Practice for Design of Reliable Industrial and Commercial Power Systems (IEEE Gold Book).

16. IEEE Std 519–1992 IEEE Recommended Practice and Requirements for Harmonic Control in Electric Power Systems, 1992.

17. IEEE Std 1100–1992 IEEE Recommended Practice for Powering and Grounding Sensitive Electronic Equipment (IEEE Emerald Book), 1992.

18. IEEE Std 1159–1995 IEEE Recommended Practice for Monitoring Electric Power Quality, 1995.

19. IEEE Std 1250–1995 IEEE Guide for Service to Equipment Sensitive to Momentary Voltage Disturbances, 1995.

20. Panhuber, C. 2001. PV System Installation and Grid Interconnection Guidelines in Selected IEA countries, Raport IEA-PVPS, T5, April.

Part I

Conventional Power Converters

2 High-Power Semiconductor Devices

2.1 A VIEW ON THE POWER SEMICONDUCTOR MARKET

Power semiconductor components are at the core of any power electronic converter. They have a history of more than 50 years, reach in technology development and market success. Since the technology behind these devices is not new, the differences between the newly released components and the role of these changes are not always easy to understand for a student. For this reason, a brief market survey is herein presented with the goal of outlining why efforts are made for certain performance indices of the power semiconductor devices. It is also important to understand the specifics of the semiconductor industry. Since production is based on large capital equipment, the technology development is done in a cyclical manner.

Power semiconductor devices are at the heart of many modern industrial and consumer end-use applications and come in different size and ratings. The application objectives are ranging from low power supplies of tens of watt to 4 MW locomotives or 10 MW steel rollers.

The power discrete market was estimated at $12.9 billion in 2011, having grown by just over 2% from 2010 [1]. It is worthnoting the continued high demand for discrete IGBTs that accounted for nearly all the growth. This was fuelled by new products for domestic appliances such as room air-conditioning and washing machines targeting especially the Asian markets. Sales of standard power MOSFETs and thyristors declined slightly in 2011 given the limited new development within the economic crisis.

According to IMS Research, the market for power semiconductor modules grew faster than that for discrete power semiconductors in 2010, increasing by 32% to $4.6 billion. The power module market growth was continuing in 2012 despite slowing demand.

The power semiconductor devices most related to our book topic are MOSFET, IGBT, and diverse modern variations of thyristors (SCR). The last 20 years have seen spectacular improvement in technology and performance. The technological S-curves related to the power device capacity are shown in Figure 2.1 [4].

The power MOSFET device was introduced in early 1980s with starting parameters of 3–5 A for the drain current, up to 400 V breakdown voltage and turn-off time in the range of 1.2 ms. Technology development allowed improvement of ratings to different sets of 9 A/600 V or 100 A/50 V and decrease of the turn-off time to 600 ns. The most recent technology advances include the CoolMOS devices that are able to switch 20 A/600 V with a turn-off time of around 100 ns. Among all sorts

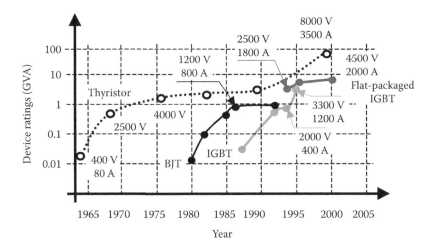

FIGURE 2.1   Technology *S*-curves with maximum device ratings as parameter.

of MOSFET devices, the largest market increase is now seen in the high current applications where new devices are released continuously.

IGBT devices combine the advantages of bipolar and MOSFET transistors into a device dedicated to power-switching converters operated under high current and high voltage. These devices are the most useful for the class of converters presented in this book, and we will dedicate more space to the presentation of IGBT's parameters.

The history of IGBTs starts also in early 1980s, but the real technological advent was in late 1980s and early 1990s when several generations of IGBT devices have been developed by a series of companies. Snapshots of performance evolution are next included:

- 1986: starting parameters 50 A/600 V/3 ms
- 1990: commonly from 50 A to 400 A/1000 V/1.8 ms
- 1995: commonly from 50 A to 400 A/1200 V/1.3 ms
- 1996: 800 A/1600 V/1.6 ms
- 1997: 1200 A/3,3 kV/2.2 ms
- 2000: 50 A to 400 A/1200 V/0.4 ms to 0.8 ms
- 2000: 1000 A/3,3 kV are available in smaller series
- 2004: commonly 1000 A/1700 V/1.2 ms, in small series up to 1200 A/6 kV

Given their application to high-power converters, the focus was on the improvement of parameters that relate to the power conversion. During the last 20 years, we have seen technology evolution with effects in

- Current handling capability—increased four times since 1982
- Voltage handling capabilities—increased four times

- Turn-off time dropped 20 times, to around 100 ns today
- Switching frequencies from 2 kHz in early 1980s to 150 kHz in 1999 and 200 kHz nowadays

The evolution of the IGBT market was also impressive over the last 10 years. The 1995 world market for IGBT market was estimated at $200 million, the European market taking the largest share (approximately 45%). The global market increased to $800 million in 2003, and it topped $1 billion in 2005.

All the above performance-related information refers to the limits of the IGBT technology. It is worthlooking also into the market depth. Most of the power converter applications are in lower power range (around kW) rather than the multi-100 kW power range. Also, the production volumes are higher in low kW power range. For instance, there are more inverterized A/C units than locomotives in the world. This explains why the number of vendors and the number of product variations are higher in the lower power range. Figure 2.2 shows a snapshot of all the IGBT product offering of a distributor (digikey) for North American market, in November 2011. It is clearly shown that the most IGBT product types are for 600 V and <50 Amp ratings. Knowing the peculiar aspects of protection and control for these devices covers more applications than focusing on the multi-kW IGBTs. It is true that the companies with products in higher power range also work on direct supplier agreements rather than with distributors.

The success of power semiconductor devices in existing applications and the appearance of new applications encouraged the development of new concepts. Today, emerging high-frequency power semiconductor devices (Example 1–10 kW switched at 100 s kHz) are a very hot R&D topic.

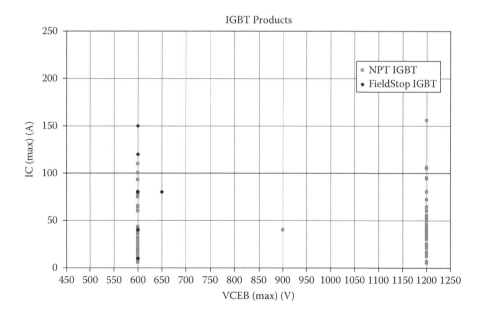

FIGURE 2.2 Cumulative view on the digikey offer for IGBT, on November 2011.

Currently, the discrete power semiconductor devices target the following markets [2]:

- IT and consumer for 33.9%
- Automotive for 12.5%
- Industrial equipment 23.8%
- Consumer appliances 29.8%

The first two categories use mostly lower power devices, and the last two are more related to our book. However, the advent of hybrid electric vehicle (HEV) power electronics is expected to increase the market share of high-voltage automotive applications.

A special market segment refers to the integrated circuits dedicated to power management and motor control. This sector is very dynamic with large investments over the last years.

2.2 POWER MOSFETs

2.2.1 OPERATION

Power MOSFET devices are faster than bipolar transistors, as they do not have excess minority carrier that should be moved during turn-on and turn-off. A positive voltage is applied at turn-on on the gate circuit. The equivalent gate capacitance is charged through an external gate resistor. When this gate voltage rises above the $V_{GS(th)}$, a current starts circulating in the drain circuit with a (di/dt) determined by both the internal semiconductor structure and the external circuit. During this time interval, charge is stored within both C_{ds} (drain-source) and C_{gs} (gate-source). This state ends when drain current reaches the level of the current determined by the external circuit (the current is clamped at the load current). As no variation of the current is possible, the voltage across the gate-source circuit remains constant at a level depending on the load-circuit current. This level is called the Miller plateau.

During this state, the gate-source capacitance has a constant voltage, and all the gate current charges the gate-drain capacitance. This determines the trip of the drain source voltage towards the ground. When this voltage reaches a low level, the gate-source voltage increases to the level of the control voltage.

During the first two states of the turn-on transient, electrical charges are moved through the stray capacitances or depletion-layer capacitances, and the equivalent circuit model for transient analysis in cut-off and active regions are shown in Figure 2.3.

The last state shown in Figure 2.4 corresponds to a drain-source voltage $V_{DS} < V_{GS} - V_{GSth}$, when the MOSFET device enters the ohmic region. In power-switching converters, $V_{GS} \gg V_{GSth}$ (typically, 15 V > 4 V), and the boundary for the ohmic region is sometimes approximated with $V_{DS} < V_{GS}$, the equivalent circuit model for the ohmic region (Figure 2.5). The drain-source resistance corresponds to the conduction loss, mostly arising from the drain-drift region. This is the most important performance index for MOSFET devices. Modern MOSFETs go as low as 5 mV $R_{DS(on)}$.

The capacitances C_{gd} and C_{gs} are not constant during the transient. A better model can be defined with values varying with the voltage across them. The capacitance

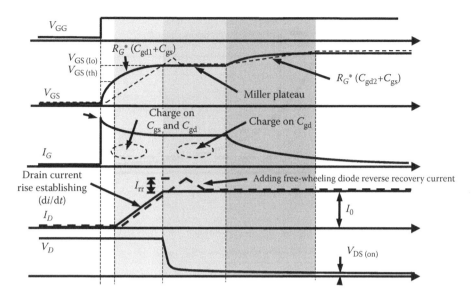

FIGURE 2.3 Generic turn-on waveforms for an IGBT/MOSFET power device.

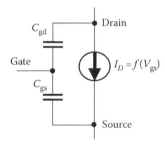

FIGURE 2.4 Model for the transient analysis in cut-off and active regions.

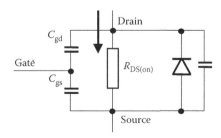

FIGURE 2.5 Model for the analysis of the ohmic region of a MOSFET device.

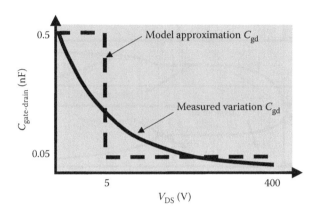

FIGURE 2.6 Variation of the gate-drain capacitance.

C_{gd} shows a substantial change that can be approximated with a two-step variation (Figure 2.6).

The gate-source capacitance is constant on the first interval, increases with voltage on the second interval because of the gate oxide capacitance of drain overlap, and it is constant during and after the third interval. The final value is three to four times higher than the initial value (both values are in the range of few nF).

MOSFET datasheets provide values of C_{ISS}, C_{RSS}, and C_{OSS}. The following relationships help relate these parameters to inter-junction parasitic capacitances $C_{gd} = C_{RSS}$, $C_{gs} = C_{ISS} - C_{RSS}$, $C_{ds} = C_{OSS} - C_{RSS}$.

The switching speed is not only determined within the input capacitance and gate resistor circuit, but the Miller threshold level and the device transconductance also matter. This is illustrated in Figure 2.7. The first slope (from zero to the Miller threshold) is determined by the input capacitance, which is higher for the second device. However, the second device has a higher transconductance and, therefore, requires less voltage at its gate for a given amount of collector current. The device with the smaller input capacitance is not always faster.

At turn-off, the gate voltage goes to zero and the gate's equivalent capacitance starts to discharge through the gate resistance (Figure 2.8). Both C_{gd} and C_{gs} are discharged at the first interval. When the gate voltage reaches the Miller plateau, it

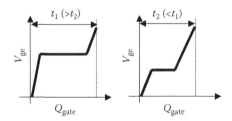

FIGURE 2.7 Gate switching characteristics for two devices.

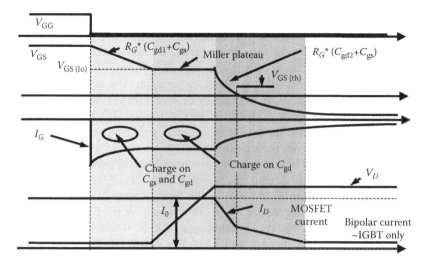

FIGURE 2.8 Generic turn-off waveforms for an IGBT/MOSFET power device.

is clamped until the drain voltage increases to the bus voltage. During this interval, charge is changed with the C_{gd} capacitance only. Finally, the current decreases to zero at the last interval, whereas the drain-source voltage remains at the bus voltage level. The device can be considered turned-off when the gate voltage goes below the threshold voltage. Figure 2.8 also presents the turn-off characteristic of an IGBT device. IGBT devices will be described in the next section. Notice the difference between the MOSFET turn-off with the current tail due to bipolar effect and the use of a bipolar voltage for the gate control.

The MOSFET's switching characteristics also depend on the external circuit. Switching currents on inductive loads imposes special precautions, including a free-wheeling diode for the reactive current (Figure 2.9). As this diode is not an ideal switch, its reverse recovery current has an important influence on the switching characteristics. The dotted lines in both Figures 2.3 and 2.8 illustrate how the reverse recovery current of the free-wheeling diode influences the switching of the power MOSFET.

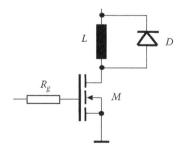

FIGURE 2.9 Using free-wheeling diodes for inductive switching.

There is also another reason for the use of the antiparallel free-wheeling diode. The internal structure of the MOSFET device features a p–n junction between the source and the drain. Under certain conditions, a negative current may free-wheel through this parasitic diode, which may happen especially within an inverter leg when a MOSFET turns off and the other one turns on. The conduction of the parasitic diode becomes a problem because of its slow turn-off (or long reverse recovery time) when the opposing MOSFET tries to turn-on. If the body diode of one MOSFET conducts when the opposing device is switched on, then a short circuit occurs similar to the shoot-through condition.

The historical solution to this problem consists in using two additional diodes for each MOSFET. A fast diode (can be a Schottky diode) is connected in series with the MOSFET source preventing the body diode from turning-on. A second fast diode is used in parallel with the MOSFET to allow a path for the free-wheeling current. Schottky diodes are nowadays available up to 200 V, whereas other fast recovery diodes are available at higher voltages. Moreover, MOSFET devices are mainly sold with the fast diode integrated within the same package for ease of use.

Numerous modern MOSFET devices eliminate this problem by creating a fast body diode. For instance, International Rectifier has introduced a 500 V HEXFET in the power MOSFET family, with fast body-diode characteristics that eliminate the need for additional Schottky and high-voltage diodes, reducing component count, cost, and layout space. The maximum reverse recovery time for the body diodes in the L-Series HEXFET devices is <250 ns, and even shorter for lower-current devices.

Note that the MOSFET semiconductor structure has a parasitic bipolar transistor formed with the body region of the MOSFET as the base, the source as the bipolar emitter, and the drain as the bipolar collector. The base of such transistors should be kept at a low voltage, which can otherwise cause negative effects.

- The MOSFET breakdown voltage will be reduced to the collector–emitter voltage of this transistor.
- The bipolar transistor can turn on accidentally without any possibility of being turned off by control. (This is called MOSFET latch-up.)
- A fast turn-off of the MOSFET would produce the turn-on of the parasitic bipolar transistor through the portion of the gate-drain capacitance that would connect the base to the collector. This can be prevented with series diodes on each drain.

Modern technology avoids the presence of this parasitic bipolar transistor (Figure 2.10). For instance, modern MOSFET devices have (dv/dt) larger than 10,000 V/μs.

What concerns the evolution of the MOSFET technology, we have witnessed several generations of successful components:

- Planar MOSFET
- Quasi-planar junction
- Superjunction MOSFET

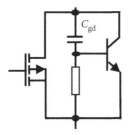

FIGURE 2.10   Parasitic bipolar transistor.

Their performance for representative Fuji devices in the 600 V class is next illustrated:

- Technology evolution with time (Figure 2.11)
- Evolution of the figure of merit "$R_{on} * Q_{gd}$" over time (Figure 2.12) [3]
- Evolution of figure of merit "$R_{on} * Area$" over time (Figure 2.13) [3]

Since different technologies are more suitable for low voltage power switches (60 V) that mostly target automotive applications, the performance of products in this category is presented separately. This is presented for information only, as these devices are not directly used in medium and high-power converters (that is the topic of this book). Criteria for illustration of results are

- Technology evolution with time (Figure 2.14)
- Evolution of the figure of merit "$R_{on} * Q_{gd}$" over time (Figure 2.15) [3]
- Evolution of figure of merit "$R_{on} * Area$" over time (Figure 2.16) [3]

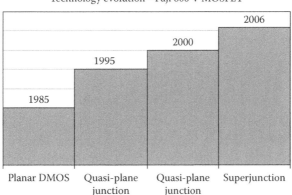

FIGURE 2.11   Technology evolution and years of product release.

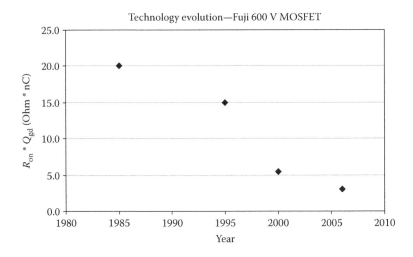

FIGURE 2.12   Evolution of the figure of merit "$R_{on} * Q_{gd}$" over time.

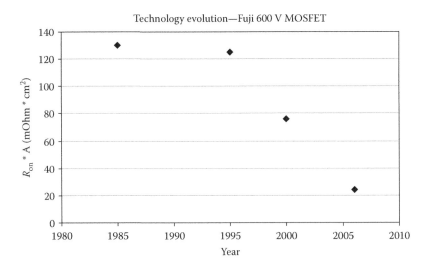

FIGURE 2.13   Evolution of figure of merit "$R_{on} * Area$" over time.

2.2.2   CONTROL

Design of gate drivers depends on the switching characteristics. The switching times given in datasheets as electrical characteristics are for resistive load switching. The performance curves are for half-bridge inductive load, as they are the most prevalent application of IGBTs.

Circuits used to control power MOSFET devices are called gate drivers. A MOSFET gate driver has the simple task of providing a voltage for the gate control,

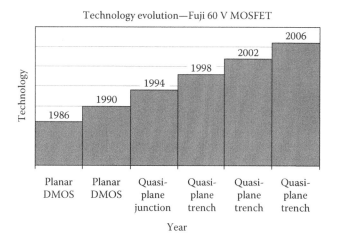

FIGURE 2.14 Technology evolution and years of product release.

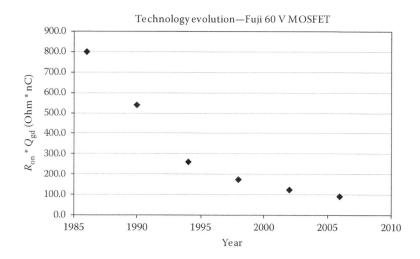

FIGURE 2.15 Evolution of the figure of merit "$R_{on} * Q_{gd}$" over time.

and it does not require a large amount of current. The gate current is large at the beginning and limited by the resistance at the gate circuit. Depending on the level of the gate threshold voltage, gate control is usually performed with voltages at logic level (5 V) or from complementary metal oxide semiconductor buffers (15–20 V). Many integrated gate driver circuits are available for both situations, along with protection circuits for fast shutdown (e.g., TPS2812).

Given the small amount of capacitance to be charged, MOSFET gate drivers should ensure a fast variation of the control voltage with slopes below 20 ns.

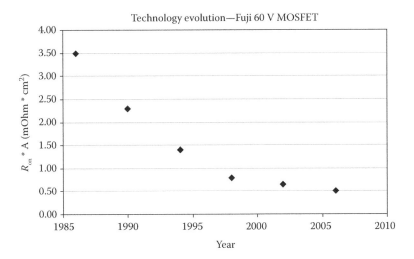

FIGURE 2.16   Evolution of figure of merit "R_{on} * Area" over time.

The gate-source voltage should not be higher than a datasheet parameter, $V_{GS(max)}$. This is determined by the requirement that the gate oxide not be broken down by a large electric field. Another consideration is the paralleling of MOSFETs, which is presented in Chapter 12.

2.3   INSULATED GATE BIPOLAR TRANSISTORS

2.3.1   OPERATION

IGBTs combine the advantages of bipolar transistors, such as low conduction losses, with the merits of MOSFETs, such as shorter switching times. For this reason, the switching behavior of IGBTs can be analyzed based on MOSFET models described earlier. The conduction interval is better modeled with the characteristics of a saturated bipolar transistor. Because of the smaller voltage drop at conduction, IGBT devices are used at higher voltages than the MOSFET devices (Figure 2.17).

Without entering into the details of the semiconductor structure, let us first focus on the IGBT model presented in Figure 2.18. It considers the IGBT formed as a Darlington combination of a main MOSFET device and a pnp transistor. Unlike the conventional Darlington, the MOSFET device carries most of the current. The parasitic npn transistor has the same origin and effect as the parasitic transistor from the MOSFET structure.

Switching characteristics of IGBT devices are highly similar to those of the MOSFET devices. The major difference consists in the bipolar effect at turn-off, when a tail current still persists for a certain amount of time. Because of this tail current, the IGBT devices are not very suitable for use within zero-voltage switching (ZVS) applications and generally introduce additional switching loss.

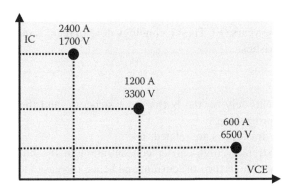

FIGURE 2.17   Present limits of the IGBT technology (example from EUPEC product line).

The tail current is also the source of an interesting design trade-off. This current exists due to the charge stored in the drift region. As the MOSFET section is OFF and there is no reverse voltage applied to the device to generate a negative drain current, there is no possibility of removing the stored charge. This can finally be removed by recombination within the IGBT. Here comes the trade-off. The excess carrier lifetime is required to be large for a small voltage drop in the conduction state. This would determine a slow recombination and a long existence of the tail current.

This is the most-used method to minimize the magnitude of the tail current, or the magnitude of the bipolar current within the IGBT device. The device is designed to have 90% or more drain current passing through the MOSFET structure and only a small amount of current through the bipolar transistor, which can be achieved with a low beta of the pnp transistor.

An alternative technological solution to this problem is the so-called punch-through (PT) technology. The PT IGBT minimizes current-tailing by shortening the duration of the tailing time, with one n+ layer acting like a sink for the excess holes. This buffer layer allows the drift region to be smaller than that of the non-punch through (NPT) IGBTs, resulting, consequently, in a reduced voltage drop in the conduction state.

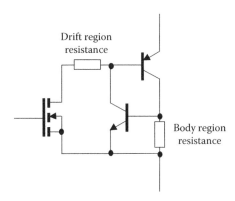

FIGURE 2.18   Equivalent model of an IGBT device.

The characteristics of both NPT IGBT and punch through (PT) IGBT have been improved over the years (F). Their technology development is marked by the following top characteristics:

PT IGBT:

- Improvements rely on the buffer layer structure and the carrier lifetime reduction processing.
- The major drawbacks are related to
  - The extremely high-carrier concentration in the buffer layer causes undesired high turn-off current and hence increased losses.
  - The processing to reduce carrier lifetime results in an increased forward voltage drop.

NPT IGBT:

- Improvements rely on geometry optimization allowing reduction of the wafer layer thickness.
- The major drawback relates to the thick n-drift region that leads to higher static and dynamic losses. This thick n-drift region is needed for voltage blocking.

The newer Field-Stop (FS) IGBT structures are overcoming the drawbacks of the two structures by vertically shrinking the NPT IGBT to a structure with a thin n-drift region, while inserting a low doped field-stop layer instead of the high doped buffer layer used in the PT IGBT (Figure 2.19). Therefore, the Field-Stop IGBT exhibits reduced overall losses and better high-voltage performance.

The weakly doped field-stop layer in the FS IGBT gives a trapezoidal electric field distribution under forward blocking (similar to the PT IGBTs), which is more desirable than the triangular electric field distribution of NPT IGBTs. This further provides a reduction of the drift region thickness for the same blocking voltage.

The advantages of the NPT IGBT devices (low-efficiency emitter and the high-carrier lifetime) are still maintained because the field-stop layer only pins the electric field under forward blocking without reducing the p-emitter injection efficiency, unlike the injection efficiency reduction from the highly doped buffer layer in the PT structure.

Variations of the Field-Stop IGBT technology are improving further these main properties (like the Advanced Field Stop trench IGBT, AFS-IGBT).

The actual performance of these three technologies (NPT IGBT, PT IGBT, and FS IGBT) can be briefly observed within the following figures, compiled from various sources featuring Fuji devices [4–7]. While different manufacturers report merits and demerits of their own technologies, sampling the product performance of a single manufacturer illustrates somewhat the state of the technology at a given time. Criteria illustrated in these figures are

- Technology timeline (Figure 2.20)
- Maximum current for each chip on the respective technology (Figure 2.21)
- Evolution of inverter loss over time (Figure 2.22)

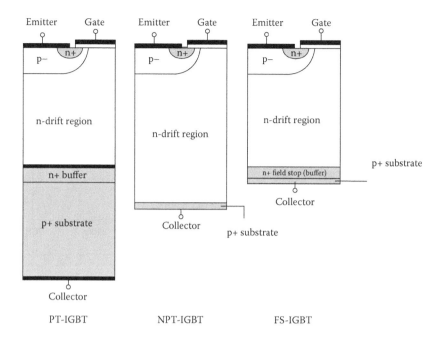

FIGURE 2.19 Principle of the three IGBT technologies.

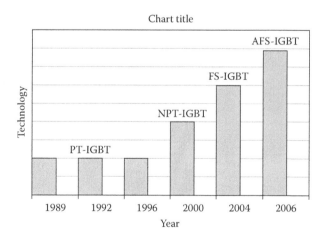

FIGURE 2.20 Approximate launching year for each technology.

- Chip area (size) over time (Figure 2.23)
- Semiconductor design rule over time (Figure 2.24)

Very similar results are reported in [8] for the technology evolution for IGBT devices made by Infineon and in [9] for IGBT devices made by Semikron. The performance of the FS IGBT technology is also investigated in [10,11].

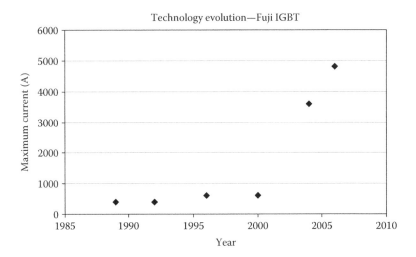

FIGURE 2.21   Maximum current achievable with each technology.

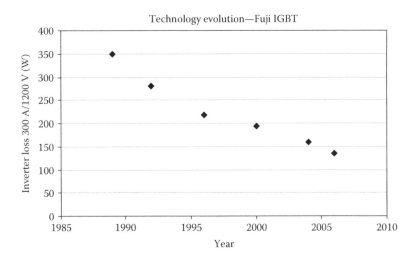

FIGURE 2.22   Evolution of inverter loss over time.

The latest generation of IGBT devices, after the emergence of FS IGBT, featured the *trench gate*. Trench gate technology developed in parallel with field-stop, and it was incorporated in both MOSFET and IGBT devices, independent of the application of the field-stop features. The advantages of trench gate devices consist of the use of a vertical channel that requires less area compared to the horizontal channel of planar structure. This provides a greater cell density, greater channel width/unit area, and a lower $R_{DS(on)}$. Moreover, there is no parasitic JFET between adjacent cells, and this also helps a greater cell density and lower $R_{DS(on)}$. The advantage of a smaller

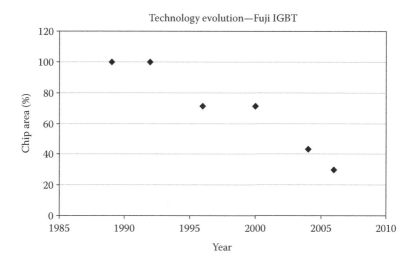

FIGURE 2.23 Evolution of chip area over time.

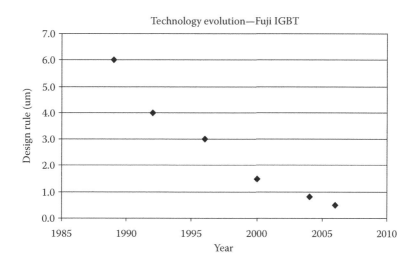

FIGURE 2.24 Evolution of semiconductor design rule over time.

ON-resistance is even more prominent at lower voltage ratings, where the channel resistance represents a more important contribution to the ON-resistance.

2.3.2 Control, Gate Drivers

2.3.2.1 Requirements

The major requirements for the IGBT gate driver are highly similar to those of a MOSFET device. However, IGBT devices usually require a negative gate voltage for turning-off, with the exception of a class of IGBT devices designed for operation with unipolar voltage. Negative OFF-state control voltage and appropriate gate resistance can prevent cross-conduction. When the high-side IGBT turns on within an inverter leg, the voltage across the low-side IGBT increases with a high (dv/dt).

This induces a current in the gate of the IGBT that may produce turn-on of the low side device short of the DC bus. A negative gate voltage prevents this by providing a different path for the gate current. The same possible turn-on due to (dv/dt) can take place within a MOSFET as well, but the input capacitance is different and the chances of cross-conduction are minimal. This can be understood by looking at the ratio of reverse transfer capacitance to the input capacitance, which is larger for IGBTs (C_{res}/C_{ies}). This produces an increased Miller effect, and a larger noise is coupled from collector to gate. However, certain low power IGBTs do not need negative gate voltage for turn-off, as their design minimizes C_{res} (reverse transfer). Another reason for the negative gate voltage at IGBTs is of the operation at higher voltages with increased (dv/dt) coupling of noise.

Figure 2.25 shows the minimal requirements of the gate driver circuit: a power supply able to ensure enough gate current, a gate-driving circuit, and a gate resistor. As the IGBT can float with respect to ground at the power stage, both the power supply and the gate circuitry should be isolated from the inverter ground. This gives room to a limited number of gate-driver configurations [12,13].

- Gate drivers with potential separation
- Gate driver with inductive transfer of power (power supply of up to 1 MHz intermediate frequency) and a direct information transfer
- Gate driver with inductive transfer of energy (power supply of up to 20 kHz intermediate frequency) and optocoupler transfer of information
- Gate drivers without potential separation
- Gate driver with bootstrap for power supply of high-side and level shifter for sending switching control information

In all these designs, a series resistor is employed at both turn-on and turn-off (Figure 2.25) that is usually implemented with a passive resistor. Advanced gate driver design requires different resistors for turn-on and turn-off. The value of the

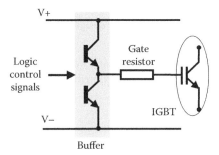

FIGURE 2.25 Gate driver concept.

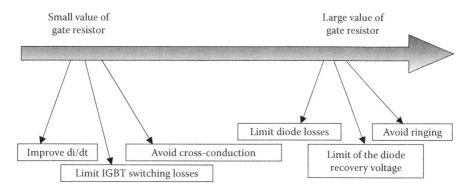

FIGURE 2.26 Effect of the gate resistor.

gate resistor within the range of values suggested by the IGBT/MOSFET manufacturer influences different aspects of the switching process. Figure 2.26 illustrates this graphically [14].

The gate current and the appropriate power of the voltage supply depend on the operating frequency, bias control voltages, and total gate charge. The total gate charge is published in IGBT/MOSFET datasheets, depending on gate-control voltage. The gate charge necessary for switching is very important to establish the switching performance of a MOSFET or IGBT. The lower the charge, the lower is the gate-drive current needed for a given switching time [15].

The average gate current can be calculated as $i_s = Q \cdot$ freq. The total gate power can be estimated as $P = i_s \cdot (V_G + - V_G -)$. Therefore, the power requirements for the gate circuit are reduced to a small-power level and a high-peak gate current. This peak current can be roughly estimated as $I_{G(Peak)} = (V_G + - V_G -)/R$.

2.3.2.2 Optimal Design of the Gate Resistor

As shown in Figure 2.26, the selection of the gate resistor influences performance of the converter. Considering an advanced gate driver with different gate resistors for turn-on and turn-off [16] allows us to perform an optimal selection of the gate resistor in order to control (dv/dt) and (di/dt) during switchings. The converter can thus be designed to operate snubberless [17,18].

Design constraints for optimization include:

* Minimum $(dv/dt)_{on}$ at highest current
* Maximum allowable peak voltage at turn-off
* Minimum $(dv/dt)_{off}$ for any current
* Avoiding cross-conduction through the (dv/dt)-induced current

The following calculations refer to Figure 2.25.

2.3.2.2.1 dv/dt at Turn-On

Let us observe Figure 2.3. Immediately after the turn-on signal has been applied, the V_{GE} voltage rises from V_{Gmin} to $V_{GE(th)}$ due to the currents flowing through the input capacitances (C_{ge} and C_{gc}). The rate of rise is almost linear across the equivalent input capacitance (as approximation of a part of an exponential curve with a time constant of $\tau_1 = R_{GON} * C_{ies}$).

After the gate voltage passes $V_{GE(th)}$, the collector current begins to increase with a rate given by the current/voltage characteristics:

$$\left(\frac{di_C}{dt}\right) = g_{fe} \cdot \left(\frac{dv_{GE}}{dt}\right) \approx g_{fe} \cdot \frac{I_g}{C_{ies}} \tag{2.1}$$

The current within the gate circuit can be described with the equation:

$$C_{ies} \cdot \frac{dv_{C_{ies}}}{dt} \approx \frac{V_G^+ - V_{GE}\left(I_L + I_{RM}\right)}{R_{Gon}} \tag{2.2}$$

where $V_{GE}(I_L + I_{RM})$ represents the Miller voltage, and V_G^+ is the positive gate supply voltage.

The voltage equation for the gate circuit should include the parasitic inductance that is composed of IGBT package inductance and external connection inductance. This value is usually provided in the device datasheet as L_E.

$$V_G^+ = R_G \cdot I_g + L_E \cdot \frac{di_C}{dt} + V_{GE}\left(I_C\right) \tag{2.3}$$

It yields:

$$\left(\frac{di_C}{dt}\right)_{on} = \frac{V_G^+ - V_{Miller}}{\left(\dfrac{C_{ies} \cdot R_{Gon}}{g_{fe}}\right) + L_E} = \frac{V_G^+ - V_{GE}\left(I_L + I_{RM}\right)}{\left(\dfrac{C_{ies} \cdot R_{Gon}}{g_{fe}}\right) + L_E} \tag{2.4}$$

After the gate voltage reaches the Miller plateau, the entire gate current is flowing through the gate-collector capacitance, causing the collector voltage to drop at a rate of

$$\left(\frac{dv}{dt}\right)_{on} = \frac{dv_{CE}}{dt} = \frac{i_G}{C_{gc}} = \frac{V_G^+ - V_{GE}\left(I_C\right)}{R_{Gon} \cdot C_{gc}} \tag{2.5}$$

with the maximum value at $I_C = I_{Load}$.

If we want to obtain a (dv/dt) higher (faster) than a specified minimum value, the gate resistor should be less than:

$$R_{Gon} \leq R_{Gon,max} = \frac{V_G^+ - V_{GE}(I_L)}{C_{gc} \cdot \left(\dfrac{dv}{dt}\right)_{on}^{req}} \tag{2.6}$$

2.3.2.2.2   dv/dt at Turn-Off

After the turn-off control signal has been applied, and the IGBT crosses the active region, the V_{GE} is clamped to a constant value that is the voltage needed to maintain the I_L load current.

$$V_{GE}(I_L) = V_{GE(th)} + \frac{1}{2} \cdot \frac{I_L}{g_{fe}} \tag{2.7}$$

The entire gate current flows through C_{gc} and causes the voltage variation:

$$\left(\frac{dv_{GC}}{dt}\right) = \frac{dv_{CE}}{dt} = \frac{i_G}{C_{gc}} = \frac{-V_G^+ + V_{GE}(I_L)}{R_{Goff} \cdot C_{gc}} = \frac{V_{GE(th)} + \frac{1}{2} \cdot \frac{I_L}{g_{fe}} - V_G^+}{R_{Goff} \cdot C_{gc}} \tag{2.8}$$

If we want to obtain a (dv/dt) higher (faster) than a specified minimum value, the gate resistor should be less than:

$$R_{Goff} = \frac{V_{GE}(I_L)}{C_{gc} \cdot \left(\dfrac{dv}{dt}\right)^{req}_{off}} \tag{2.9}$$

2.3.2.2.3   Peak Voltage at Diode Turn-Off

The overvoltage across the inverter leg should be limited to a pre-established value (for instance 5%). This over-voltage occurs at any fast current variation (switching) in the inverter leg, and it is most important at diode turn-off when the recovery current is superimposed to the conventional switching of the load current. This occurs when IGBT turns on. The design value for the gate resistance yields from Equation 2.4:

$$\left(\frac{C_{ies} \cdot R_{Gon}}{g_{fe}}\right) + L_E = \frac{V_G^+ - V_{GE}(I_L + I_{RM})}{\left(\dfrac{di_C}{dt}\right)} \Rightarrow$$

$$R_{Gon} \geq R_{Gon,min} = \frac{g_{fe}}{C_{ies}} \cdot \left[\frac{V_G^+ - V_{GE}(I_L + I_{RM})}{\left(\dfrac{di_C}{dt}\right)^{req}_{on}} - L_E\right] \tag{2.10}$$

This design equation can be expressed also directly from the peak voltage:

$$R_{Gon} \geq R_{Gon,min} = \frac{g_{fe}}{C_{ies}} \cdot \left[\frac{V_G^+ - V_{GE}(I_L + I_{RM})}{\dfrac{|V_{DC} - V_{Peak}|}{2 \cdot L_{st}}} - L_E\right] \tag{2.11}$$

where L_{st} is the stray inductance of the busbar connection between the inverter leg (IGBT) and the actual DC source (usually an electrolytic capacitor).

2.3.2.2.4 Avoiding Cross-Conduction

A fourth design constraint comes from the request to avoid cross-conduction due to the large (dv/dt). If we suddenly apply a large dv/dt across an IGBT device, currents may start to circulate within. The collector-gate current would become gate current and supply the IGBT turn-on through the actual gate resistor R_{Goff}.

It yields

$$i_{CG} = i_G + i_{GE} \tag{2.12}$$

$$C_{gc} \cdot \frac{dv_C}{dt} = i_G + C_{ge} \cdot R_g \cdot \frac{di_G}{dt} \tag{2.13}$$

$$C_{gc} \cdot \left(\frac{dv}{dt}\right)_{on} = \frac{V_G - V_{G\,min}}{R_{Goff}} + C_{ge} \cdot \frac{dv_G}{dt} \tag{2.14}$$

The variation of the gate voltage can be considered as being zero ($dv_G/dt = 0$) during the Miller plateau interval. This reduces at

$$C_{gc} \cdot \left(\frac{dv}{dt}\right)_{on} = \frac{V_G - V_{G\,min}}{R_{Goff}} \tag{2.15}$$

The peak of the induced gate voltage results as

$$v_{G\,max} = V_{G\,min} + R_{Goff} \cdot C_{ge} \cdot \left(\frac{dv}{dt}\right)_{on} \tag{2.16}$$

Equation 2.5 into 2.16 yields

$$R_{Goff} = R_{Gon} \cdot \frac{V_{GE}\left(I_L\right) - V_{Gmin}}{V_G^+ - V_{GE}\left(I_L\right)} \tag{2.17}$$

Appropriate value of the gate resistance can prevent this shoot-through conduction. When considering the "on" resistance defined by previous conditions, the "off" gate resistance can be calculated from this relationship for different gate voltage and load current values.

Automated design procedures can be implemented on computer software based on the above considerations as well as from considering parameter variations [18].

2.3.3 PROTECTION

A power converter equipment includes a set of protection circuits and features. Chapter 6 presents in detail the practical aspects of building a power converter. Before such a design can be accomplished, the datasheet information about the limits of operation should be understood.

The proper operation of an IGBT or MOSFET device is bounded by datasheet limitations. The collector current is limited to avoid latch-up. The maximum gate emitter voltage is set by the gate oxide breakdown considerations. The maximum current that can flow under short-circuit with a maximum gate-emitter voltage is four to ten times the nominal rated collector current. The maximum collector–emitter voltage of an IGBT device is set with the breakdown voltage of the internal pnp transistor. The maximum junction temperature is 150°C.

Special datasheet information refers to the safe operating area (SOA). Both IGBT and MOSFET devices have square SOAs for short switching times. If the conduction intervals are longer, thermal aspects modify the SOA, as shown in Figure 2.27. Modern IGBT devices can operate at the corner of the SOA for 10 µs (Figure 2.27).

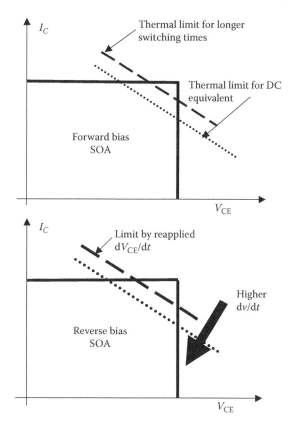

FIGURE 2.27   Ideal SOA and limitations due to special operation conditions.

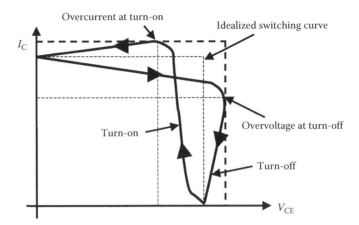

FIGURE 2.28   Switching characteristics.

This allows a protection circuit to trigger the gate signal and to protect the IGBT at high currents. In other words, the IGBT can operate in short-circuit for up to 10 μs [12].

The parasitic components of the collector–emitter circuit determine a real switching characteristic far from a square one. The designer should make sure that this real characteristic is always inside the SOA (Figure 2.28). These trajectories depend upon stray inductances, parasitic capacitances, and the MOSFET's switching performance as (d*i*/d*t*), (d*v*/d*t*). The IGBT package itself has a stray inductance of about 10–20 nH [12,19].

2.4   POWER LOSS ESTIMATION

As the switching characteristics are the results of nonlinear phenomena, there are different methods to estimate the power loss. Power loss and efficiency of the power stage are very important, given the use of MOSFET and IGBT devices in power-conversion circuits.

There are two major methods for loss estimation:

- Calculation of energy loss based on analytical investigation of the switching waveforms
- Estimation of the energy loss based on empirical models derived from bench measurements

Observing the collector current and voltage waveforms, switching loss can be derived by calculating the areas of VI regions that correspond to the switchings.

The switching-loss energy at IGBT turn-on (Figure 2.3) is [17,18]:

$$E_{\text{Ton}} = 0.5 \cdot \left[V_{\text{DC}} - L_{\text{st}} \cdot \left(\frac{\mathrm{d}i}{\mathrm{d}t} \right)_{\text{on}} \right] \cdot \frac{\left(I_L + I_{\text{RM}} \right)^2}{\left(\dfrac{\mathrm{d}i}{\mathrm{d}t} \right)_{\text{on}}} + 0.5 \cdot I_L \cdot \frac{V_{\text{DC}}^2}{\left(\dfrac{\mathrm{d}v}{\mathrm{d}t} \right)_{\text{on}}} \qquad (2.18)$$

where we assumed L_{st} as the stray inductance, V_{DC} as the bus voltage, I_{RM} as the peak recovery current of the adjacent diode, (di/dt) and (dv/dt) as the datasheet information about the selected semiconductor device.

The turn-off energy is calculated with (see Figure 2.5)

$$E_{Toff} = 0.5 \cdot \frac{V_{DC}^2}{\left(\dfrac{dv}{dt}\right)_{off}} \cdot I_L - 0.5 \cdot \left[V_{DC} - 2 \cdot L_{st} \cdot \left(\frac{di}{dt}\right)_{off}\right] \cdot \frac{\left(I_L\right)^2}{\left(\dfrac{di}{dt}\right)_{off}}$$
$$+ 0.5 \cdot k_t \cdot V_{DC} \cdot I_L \cdot t_{tail} \tag{2.19}$$

The last term corresponds to the tail current at the IGBT turn-off and should miss at the same calculation performed for a MOSFET device. It can be seen that the gate drive circuit (especially the gate resistance) influences the switching losses by di/dt, dv/dt, I_{RM}, and overvoltage.

Finally, the diode turn-off within an inverter leg is characterized by loss expressed by

$$E_{Doff} = 0.5 \cdot \left[V_{DC} + 2 \cdot L_{st} \cdot \left(\frac{di}{dt}\right)_{diode}\right] \cdot \frac{\left(I_L\right)^2}{\left(\dfrac{di}{dt}\right)_{diode}} \tag{2.20}$$

The conduction loss is calculated as

$$P_{COND} = \frac{1}{T} \cdot \int_0^T v_{on}(t) \cdot i_L(t)\, dt \tag{2.21}$$

Integral across the fundamental period T can be reduced to integrals across all conduction intervals during a period.

Total losses can be calculated by adding up the switching and conduction losses, by taking into account the inverter topology, the modulation function for each device, and the operation mode or load power factor.

The second approach to loss estimation is based on bench measurement of power losses under specified conditions, followed by interpolation of results from these measurements for the actual circuit operation.

An example for this procedure is provided in [20], where the power loss equations for a three-phase inverter power module are determined by experiment rather than analytical calculation. In this respect, it is first determined that the power loss depends on actual load current and steady-state measurements are made for each possible load current. Given the peculiar construction of the IPM module, an empirical calculation of the loss is preferred [20].

This empirical model considers each switch individually and calculates the power loss with the following set of equations (valid for IRAMS devices [20]):

$$E_{on} = (h_1 + h_2 \cdot I^x) \cdot I^k = [(7.69e - 4) + (2.99e - 2) \cdot I^{-1.159}] \cdot I^2 \tag{2.22}$$

$$E_{\text{off}} = (m_1 + m_2 \cdot I^y) \cdot I^n = [(1.76e - 2) + (4.34e - 2) \cdot I^{-0.492}] \cdot I^1 \quad (2.23)$$

$$V_{\text{CEon}} = V_T + a \cdot I^b = 0.51 + 0.46 \cdot I^{0.649} \quad (2.24)$$

The results of this method used for a conventional motor drive application built with a 20 A IPM module show 2.3 W power loss per switch, and 14.1 W per entire package, when operated in ambient temperature and trying to prevent a junction temperature close to 125°C.

2.5 ACTIVE GATE DRIVERS

The role of a passive gate-driver resistor has already been explained. It has also been shown that the value of the gate resistor influences different characteristics of the switching circuit (Figure 2.26). These performance aspects are not based on simultaneous phenomena and introduce the possibility of an idealistic control through a variable gate resistor. A variable value of the gate resistor can be achieved with an active gate control [21–30].

Historically, the use of active gate control was first mentioned in series connection of IGBT devices. This is required in medium-voltage applications, when the DC bus has values in thousands of voltage range. Series connection of IGBTs raises the problem of unequal voltage-sharing across these devices. The unequal voltage sharing across the IGBTs is due to

- Different delay times in gate driver and power semiconductor device
- Small parameter deviation among different devices
- Different reverse recovery behavior of the free-wheeling diodes
- Increased (dv/dt) with the number of series-connected devices

Voltage-balancing between series-connected devices is traditionally achieved with individual snubbering of each IGBT device. To reduce the passive component count and volume, modern active-snubbering methods have been reported to limit (dv/dt) and the overvoltage (Figure 2.29).

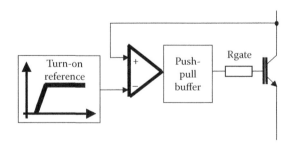

FIGURE 2.29   Active voltage balancing circuit used within series connected IGBTs.

Control of the collector voltage is achieved within an analog fast-feedback loop. Stability requirements imply design of a controller with poles at a frequency higher than gate circuitry, with a pole in the range of 1–10 MHz. Control bandwidth of 50–90 MHz is achieved with high-performance operational amplifiers. A significant loss reduction can be achieved by controlling the IGBT voltage in closed-loop operations only near the peak rating. Open-loop operation can be considered for the rest of the operation range. This obviously complicates the control circuit.

The downside of this solution can be seen at inductive loads. The IGBT voltage cannot respond to the gate voltage turn-on control until the free-wheeling diode has turned off. The closed-loop approach charges the gate quickly, producing a very high (di/dt).

Historically, the second step in active gate control was in short-circuit protection. If the collectors circuit experiences a short-circuit, the protection circuit tries to shut the gate down and cut the current. This produces a large (di/dt) and a large overshoot. The equivalent gate resistance increases when the protection acts to turn the IGBT off with a soft shutdown. This avoids the large (di/dt) and the large voltage overshoot.

Voltage-overshoot protection can be achieved by including an additional transistor stage in the gate driver (Figure 2.30) [31]. At turn-off, Q_{prot} is turned on and the current is discharged through it. When the collector voltage reaches the breakdown voltage of the Zener diode, a current will flow through the gate of Q_{prot} and will turn it off. The remaining current would flow through $R_{gate(off)}$ slowing down the (dv/dt) rate.

Modern gate drivers adjust (di/dt) and/or (dv/dt) independently according to criteria, such as electron magnetic interference (EMI) emission control or efficiency

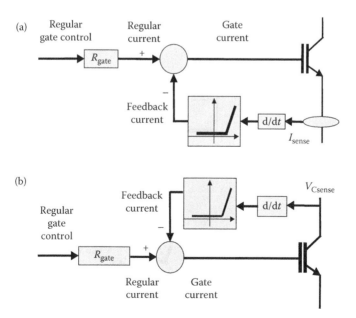

FIGURE 2.30 Principle circuits for active gate control: (a) based on (di/dt) optimization; (b) based on (dv/dt) optimization.

improvement through snubberless operation. For instance, the gate resistor can be optimized to reduce EMI emission through controlled (di/dt) and/or (dv/dt). Usually, this value increases loss. These two constraints require different values of the gate resistors during operation.

There are different methods for active control of (di/dt) and (dv/dt) (Figure 2.30). The simplest control is the feedback control of the gate current based on the device current or voltage slopes. Sensing the current or voltage slopes is carried out with a shunt resistor, on Kelvin emitter, on the information resulting from the Miller effect sensing.

The active gate control is not easy to implement. The circuit designer faces several constraints related to a fast event time scale that does not allow delays within the analog circuit and to a feedback dependence on IGBT parameters.

Consider the experiment shown in Figure 2.31 and the results in Figure 2.32 [21].

The gate voltage has an intermediate voltage level that decreases the gate current level on the first slope of turn-on. The voltage level ΔV_s and the length of the time interval Δt_{ts} within this voltage level are adjusted. The IGBT/MOSFET behavior is a result of variable inductance. Despite the clear demonstration of the principle of active gate control, this experiment is not easy to implement. Generally, the strong nonlinear character and the detection of the Miller plateau are a problem.

Different solutions were reported in the literature for active gate control even from late 1900s [32–42]. Because all of the active gate drivers need a design dependent on the actual IGBT device to be controlled, these technologies did not make it into too many IC solutions. Several recent IPM products are taking advantage of dynamically controlled gate drivers only.

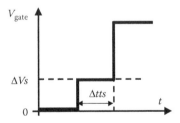

FIGURE 2.31 Principle of a simple experiment.

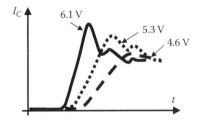

FIGURE 2.32 Results of the simple experiment.

2.6 GATE TURN-OFF THYRISTORS (GTOs)

As shown in the introduction to this chapter, gate turn-off thyristor (GTO) devices are used at high levels of current and voltage. They are a derivative of SCR devices, with a p–n–p–n structure, and can control both turn-on and turn-off processes. The operation is based on conventional recombination processes and the physics of junctions.

The major drawback of these devices is the large current required in the gate circuit for turn-off. Moreover, the GTO has a very low gain which means it requires a sophisticated and expensive gate drive. It is therefore impractical to use a charge extracting drive circuit, and so the GTO has a *"tail effect"* whereby the device still conducts while the minority carriers combine naturally.

2.7 ADVANCED POWER DEVICES

Despite the technology saturation in what concerns conversion circuits and converters, the power semiconductor sector is still dynamic. There is continuous development along existing devices like MOSFET and IGBT. New generations of IGBTs and MOSFETs are introduced each year to the market and their performance is continuously improving, especially through the design rule improvement (the pitch resolution in defining the shape of each semiconductor region).

However, the most exciting news about completely new devices is their ability to change performance patterns through disruptive innovation. These disruptive devices can be classified in three categories:

- New devices solving certain issues with conventional devices and hence dedicated to certain peculiar applications (IGCT, IGBT-RC, IGBT-RB, and so on).
- High-frequency, high-voltage devices aiming at increasing the operation frequency with a degree of magnitude, and therefore requiring a complete overhaul of the inverter assembly.
- Devices using emerging new substrate materials and aiming at efficiency improvements through a new class of devices. Their novelty may lead to changes in the design of the gate driver and/or protection circuitry and hence may not be useable as a drop-in substitute for the existing components.

2.7.1 SPECIALTY DEVICES

2.7.1.1 IGCT

A good example of device especially designed for medium and high-power applications is the *integrated gate commutated thyristor* (IGCT). Its architecture is combining the best features of an IGBT and a GTO. The new solid-state switch is for medium-voltage applications from 2 to 6.9 kV, with maximum ratings of 4000 A, which builds upon the drawbacks of IGBTs that have high conduction losses and GTOs that are slow and require additional circuitry.

Through this new architecture, the IGCT makes possible designs that have not been feasible in the past. Thus, engineers need not design around IGBT and GTO trade-offs, which often impose limits on starting torque and regeneration ability of motor drives.

2.7.1.2 IGBT-RC

The Reverse Conducting IGBT is a very new device proposed to the market by *Infineon* at the end of 2009 [43]. This new device addresses the motor drives market, and it is especially successful within the BLDC motor drives controlled with the 120° program. Other applications include the soft-switching converters used within the induction heating and induction cooking products. This device incorporates the diode monolithically along the IGBT, providing low conduction losses of both IGBT and the integrated diode. This helps reducing the overall losses.

2.7.1.3 IGBT-RB

The increasing use of power electronic converters in energy processing has brought into attention topologies like Current Source Inverter, Matrix Converter, or the Neutral Point Clamped Inverter that require AC switches. All of these topologies promise a better power density for a given application. The conventional solution for preventing reverse conduction consists in connecting a diode in series with the usual IGBT. This is increasing the voltage drop during conduction state.

An alternative solution was proposed in early 2000 s with the IGBT-RB device. This device has the capacity to block voltage of both polarities without adding any supplemental series diode. Figure 2.33 is based on data compiled from [44,45], and it shows the improvement of the trade-off between the turn-off energy and the voltage drop during the conduction state.

The dependency shown in Figure 2.33 is used for comparing different power semiconductor technologies. Each curve is typical for a certain technology, and

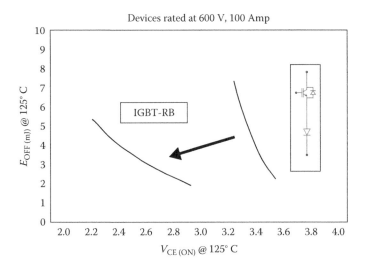

FIGURE 2.33 Performance improvement with the use of an IGBT-RB device instead of a conventional IGBT-Diode pair.

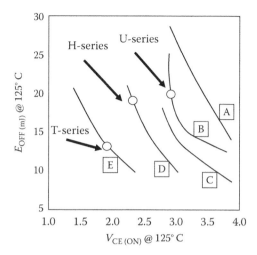

FIGURE 2.34 Evaluation of different Powerex IGBT technologies: A = conventional NPT, t_N = 250 mm, J_c = 100 A/cm²; B = PT planar gate uniform lifetime control, J_c = 100 A/cm²; C = thin drift region NPT, t_N = 150 mm, J_c = 100 A/cm²; D = PT planar gate, uniform lifetime control, J_c = 75 A/cm²; E = PT trench gate, local lifetime control, J_c = 140 A/cm².

different products are at certain points of the technology performance curve. For instance, an IGBT for motor drives can aim at a low voltage drop in the conduction state as it is switched at low switching frequencies and the efficiency can be improved with reducing the voltage drop. Such a device would sit in the top-left corner of the technology curve. By contrary a different IGBT product made with the same technology can be aimed at use within UPS and power supplies applications where switching at a high switching frequency would reduce the filter requirements. Such an IGBT device would sit in the bottom-right corner of the technology curve. Figure 2.34 illustrates further this figure of merit by showing *Powerex* devices and their technologies [46]. To avoid a direct comparison between products of different manufacturers (which is not the goal here), *Powerex* products and technologies of the year 1999 have been compared in Figure 2.34, and *Fuji* products and technologies for the year 2010 are shown in Figure 2.33. Both figures are given for illustration of this performance figure of merit.

2.7.2 HIGH-FREQUENCY, HIGH-VOLTAGE DEVICES

Another power semiconductor device that is picking up in the market is the CoolMOS, a MOSFET with a special structure, rated up to 600 V, and able to switch up to 50 A. These devices change the entire way we think about power converters.

Design of the power stage is limited by the parasitic of the implementation (printed circuit or busbar). The idea of switching, say, a 400 V bus at 250 kHz, pushes the designer to be very careful while designing circuit details.

Given the requirements of a general overhaul of the power electronics equipment, the penetration of these devices as a substitute for conventional IGBTs was not very spectacular. They remain attractive solutions for high-power density equipment like certain aviation systems.

2.7.3 USING NEW SUBSTRATE MATERIALS (SiC, GaN, AND SO ON)

A generic trend in emerging power semiconductor devices is the use of new substrate semiconductors [47].

The development of new device technology started from within academic laboratories and small-business companies, and hence their experimental character has recommended them onto lower power converter market. The first applications included industrial applications in the HEV automotive market, IT and consumer, grid connected low power converters, or certain appliances. According to [44], the most dynamic sector is represented by the power factor correction converters in sub-kW power range. The SiC device technology being a little more advanced was able to bring up devices for applications up to MW power range. The GaN technology started 10 years after the SiC technology, and is currently featuring mostly diodes.

The success of the new substrate material semiconductors is mostly seen in diodes replacing conventional recovery diodes. The first application was the power factor correction circuitry in early 2001. The use of SiC diodes in the PFC application leads to power conversion improvement of around 2.4% [47] and a more optimal packaging. The losses are reduced tremendously and there is virtually no cost of circuit adaptation. Moreover, different auxiliary components like small inductor snubbers can be removed from circuit. The remaining obstacle is yet the slight cost difference (to beat the $0.20/Amp barrier).

Another good example is the recent release of IGBT-Diode copack devices built of standard Si-based IGBT and emerging SiC diodes, and dedicated to motor drives as a drop-in replacement or for new designs.

Conversely, the application of the power semiconductor switch made on new substrate material is more limited. The new devices are also requiring special gate drivers for control and protection and this is complicating the replacement in existing designs. Another limitation is the lack of reliability information or qualifications that prohibit somehow the use of these devices in applications with a critical lifetime or ruggedness requirements.

The new substrate materials devices do not present a major improvement for medium voltage converters switched at low switching frequency. The operation of high voltage (multi kV-range) converters will benefit from these new devices. Especially here, the plasma science or physics instrumentation equipment with operation at 50–100 kV would see major improvement with the use of SiC diodes for a considerable voltage drop across the device.

What concerns the high switching frequency application, here the voltage level matters again. At very low voltage levels, the soft-switching operation can easily be achieved with resonant converters and the merits of the new substrate material devices fade. At medium voltage (100s V) the new devices have clear merits over the use of conventional semiconductors at high switching frequencies.

The major advantages of using a complete SiC based technology instead of conventional Si devices can be briefly stated as [48]

- Approximately half of the chip area is required for a SiC JFET in order to achieve the same efficiency as with a Si MOSFET.
- When operating SiC devices with a junction temperature limit above 165°C, the heat-sink temperature can be higher and the size of the cooling system decreases.
- Hence, the power density increases.
- The same power density could be achieved if the SiC devices have 30–40% of the Si MOSFET chip area.

2.8 DATASHEET INFORMATION

Successful design of power converters depends strongly on the information provided by the power semiconductor manufacturer within the datasheet. Each power device or class of devices is characterized with a peculiar technical document called *"datasheet."* Let us see here how the information from a device datasheet has been gathered and what does it mean.

The first section of a datasheet includes a brief description of the device and a drawing of the device. In most cases, the application field is also suggested herein.

Even if presented in other random order, a first important section contains information about the absolute maximum electrical ratings of the device. If the operation of the power semiconductor device exceeds these ratings even on transients—it voids the manufacturer's guarantee regardless of the duration or the conditions of the stress. Usually manufacturers use a test tolerance for these ratings when testing to establish the highest absolute-maximum rating. However, the circuit designer should never count on this tolerance band as it may vary from manufacturer to manufacturer. Alternatively, protective components can be used to completely alleviate operation beyond the absolute maximum ratings.

Any datasheet also provides three absolute maximum temperature ratings:

- Operating temperature is the maximum temperature over which you can operate the power device, even if there is no guarantee implied that electrical performance will be maintained over the entire operating temperature range.
- Junction temperature is the maximum temperature that the internal semiconductor die can reach under any environmental or operation condition.
- Storage temperature is the maximum temperature that the device may reach under a storage condition. This also voids the warranty offered by manufacturer for the device.

There is also another environmental warning regarding the ESD protection.

The most challenging information for a converter designer comes from the electrical characteristics tables. Each performance characteristics is reported along with test conditions for measurement, and it is provided with three fields: MIN (minimum), TYP (typical), and MAX (maximum). The manufacturer is trying to select the most significant test points. However, the circuit designer should exercise care in understanding the differences between the circuit under design and what the manufacturer is offering as test data. After initial assessment of their own power semiconductor devices, the manufacturer applies statistics to the data to obtain the mean value for each parameter. The following production batches are tested by statistical sampling for this datasheet information. The statistics yield the variance and sigma for the results distribution. The maximum and minimum values for each characteristics are selected at six times the value of sigma. These six sigma points become the minimum and maximum values for that parameter, and the mean is usually taken as the typical specification. Modern *Design for Reliability* concerns require the circuit designers to consider both the MAX and MIN values in order to cover all possible mishaps during operation rather than reducing the design to the use of TYP values. Alternatively, the designer may use its own statistical process of evaluation for the in-circuit tests of the new design.

A section of the datasheet contains graphs and table data for parameters that may vary with different operation points.

This applications related section often includes parameter measurement information, or unusual measurement circuits. The application section covers load-driving capability, layout and heat-sink suggestions, safe-operating area curves, special stabilization techniques for control circuits if included, or Spice models. All this information is provided by application engineer with experience in product and with a desire to present the product at its best. Such examples are eye-catching and not necessarily the best solutions for very large volume production. This is why the modern *Design for Reliability* concerns should be applied in all phases of design.

PROBLEMS

P2.1   Try to explain the variation of the gate-drain and gate-source capacitances?

P2.2   The on-resistance of a power MOSFET equals 120 mOhms at a junction temperature of 25°C and increases linearly with temperature up to 200 mOhms at 100°C. Calculate the conduction loss in function of the operation temperature if the load resistance is 10 Ohms and the supply voltage is 150 V for a chopper operation.

P2.3   Imagine a hybrid power switch made up of a bipolar transistor and a power MOSFET connected in parallel. What would be the benefits of such a device?

P2.4   Qualitatively sketch the collector current versus time during turn-off for a short lifetime IGBT and for a long lifetime IGBT and explain the differences.

P2.5   Qualitatively sketch the collector current change during the turn-on of an IGBT device controlled through different gate resistors.

P2.6   Qualitatively sketch the collector current change during the turn-off of an IGBT device controlled through different gate resistors.

P2.7   Write a computer program for power loss estimation based on the equations shown in this chapter and run this program for a simple case of a single IGBT switching a load resistor of 20 Ohms, at 20 kHz, from a

DC bus of 400 V_{dc}. Consider a real IGBT device along with the manufacturer datasheet and compare results with those given in datasheet.

P2.8   Consider a MOSFET and an IGBT with the same breakdown voltage and the same current rating. How would you compare the gate-drain and gate-source capacitances of these two devices?

REFERENCES

1. Anon. 2012. About the World Power Semiconductor Discretes and Modules Report, 2012 edition, IMS Research, July.
2. Anon. 2008. Power Semiconductors Market: Latest Trends 2008, Yano research.
3. Brown, J. and Moxey, G. 2003. Power MOSFET Basics: Understanding MOSFET Characteristics Associated With The Figure of Merit, Vishay-Siliconix Application Note AN-605, September.
4. Seki, Y. 2001. Present status and trends of power semiconductor devices. *Fuji Electr. Rev.* 47(2), 34–36.
5. Seki, Y., Hosen, T., and Yamazoe, M. 2010. The current status and future outlook for power semiconductors. *Fuji Electr. Rev.* 56(2), 47–50.
6. Fujihira, T., Kaneda, H., and Kuneta, S. 2006. Fuji' electric semiconductor: Current status and future outlook. Fuji Electr. Rev. 52(2), 42–47.
7. Yamamoto, T., Yoshiwatari, S., and Ichikawa, H. 2012. Expanded lineup of high-power 6th generation IGBT module families. *Fuji Electr. Rev.* 58(2), 60–64.
8. Gutsmann, B., Kanschat, P., Münzer, M., Pfaffenlehner M., and Laska, T. 2003. Repetitive Short-Circuit Behavior of Trench/FieldStop IGBTs, PCIM Europe 2003.
9. Schreiber, D. 2005. New power semiconductor technology for renewable energy sources application, Semikron Seminar in Sevilla, Spain, Internet documentation.
10. Kang, X., Caiafa, A., Santi, E., Hudgins, J.L., and Palmer, P.R. 2003. Characterization and modeling of high-voltage field-stop IGBTs. *IEEE Trans. Indust. Appl.* 39(4), 922–929.
11. Dodge, J. 2004. IGBT Technical Overview, Application Note APT0408, Advanced Power Technology Corporation, November.
12. Anon. 1998. Using IGBT Modules, Mitsubishi Electric, September.
13. Motto, E. 1996. Gate drive techniques for large IGBT modules, *PCIM Magazine*.
14. Zverev, I., Konrad, S., Voelker, H., Petzoldt, J., and Klotz, F. 1997. Influence of the gate drive technique on the conducted EMI behaviors of a power converter. *IEEE PESC* 2, 1522–1528.
15. Hefner, A.R. 1991. An investigation of the drive circuit requirement for the power insulated gate bipolar transistor. *IEEE Trans. Power Electron.* 4, 208–218.
16. Neacsu, D., Takahashi, T., and Nguyen, H.H. 2000. Using IR 2137, International Rectifier Design Tip 00-1.
17. Blaabjerg, F. and Pedersen, K. 1997. Optimized design of a complete three-phase PWM-VS inverter. *IEEE Trans. Power Electron.* 12(3), 567–577.
18. Neacsu, D.O. and Takahashi, T. 2000. Computer-aided design of a low-cost low-power snubberless three-phase inverter, *IEEE Workshop on Computers in Power Electronics, Blacksburg*, VA, June, pp. 204–210.
19. Neacsu, D., Takahashi, T., and Nguyen, H.H. 2000. Using IR 2137, IR Design Tip 00-1.
20. Wood, P., Battello, M., Keskar, N., and Guerra, A. IPM Application Overview—Integrated Power Module for Appliance Motor Drives, International Rectifier AN-1044.
21. McNeil, N., Sheng, K., Williams, B.W., and Finley, S.J. 1998. Assessment of OFF-state negative gate voltage for IGBTs. *IEEE Trans. Power Electron.* 13, 436–440.
22. Ferrieux, J.P., Forest, F., and Lienart, P. 1989. The insulated gate bipolar transistor: switching mode, *EPE Conference*, Aachen, pp. 171–175.

23. Gerster, C.H. and Hofer-Noser, P. 1996. Gate controlled dv/dt and di/dt—Limitation in high power IGBT converters. *EPE J.* 5(3/4), 11–16.

24. Musumeci, S., Raciti, A., Testa, A., Galluzo, A., and Melito, M. 1997. Switching behavior improvement of insulated gate-controlled devices. *IEEE Trans. Power Electron.* 12(4), 645–653.

25. Takizawa, S., Igarashi, S., and Kuroki, K. 1998. A New di/dt control gate drive circuit for IGBTs to reduce EMI noise and switching losses. *IEEE Power Electronics Specialists Conference*, Fukuoka, Japan, vol. 2, pp. 1443–1449, July.

26. Lee, H.G., Lee. Y.H., Suh, B.S., and Hyun, D.S. 1997. An improved gate control scheme for snubberless operation of high power IGBTs, *IEEE Industry Applications, Annual Meeting*, New Orleans, Louisiana, USA, vol. 2, pp. 975–982, 5–9 October.

27. Hong, S. and Lee, Y.G. 1999. Active gate-control strategy of series connected IGBTs for high power PWM inverter, *IEEE PEDS*, Hong Kong, vol. 2, pp. 646–452, 27–29, July.

28. Gerster, C., Hofer, P., and Karrer, N. 1996. Gate-control strategies for snubberless operation of series connected IGBTs. *IEEE PESC Conference*, Baveno, Italy, vol. 2, pp. 1739–1742, 23–27 June.

29. Idir, N., Franchaud, J.J., and Bausiere, R. 2000. New control technique achieves low di/dt and dv/dt. *PCIM Magazine*, February.

30. John, V., Suh, B.S., and Lipo, T.A. 1999. High performance active gate drive for high power IGBTs. *IEEE Trans. Indust. Appl.* 35(5), 1108–1117.

31. Heath, D. and Wood, P. 1999. Overshoot voltage reduction using IGBT modules with special drivers. International Rectifier Design Tip no. 99-1.

32. McNeil, N., Sheng, K., Williams, B.W., and Finley, S.J. 1998. Assessment of OFF-state negative gate voltage for IGBTs. *IEEE Trans. Power Electron.* 13, 436–440.

33. Ferrieux, J.P., Forest, F., and Lienart, P. 1989. The Insulated Gate Bipolar Transistor: Switching Mode, *EPE Conference*, Aachen, pp. 171–175.

34. Gerster, C.H. and Hofer-Noser, P. 1996. Gate controlled dv/dt and di/dt—Limitation in high power IGBT converters. *EPE J.*, 5(3/4), 11–16.

35. Musumeci, S., Raciti, A., Testa, A., Galluzo, A., and Melito, M. 1997. Switching behavior improvement of insulated gate-controlled devices. *IEEE Trans. Power Electron.* 12(4), 645–653.

36. Takizawa, S., Igarashi, S., and Kuroki, K. 1998. A New di/dt control gate drive circuit for IGBTs to reduce EMI noise and switching losses. *IEEE Power Electronics Specialists Conference*, Fukuoka, Japan, vol. 2, pp. 1443–1449, July.

37. Lee, H.G., Lee, Y.H., Suh, B.S., and Hyun, D.S. 1997. An improved gate control scheme for snubberless operation of high power IGBTs, *IEEE Industry Applications, Annual Meeting*, New Orleans, Louisiana, USA, vol. 2, pp. 975–982, 5–9 October.

38. Hong, S. and Lee, Y.G. 1999. Active gate-control strategy of series connected IGBTs for high power PWM inverter, *IEEE PEDS*, Hong Kong, vol. 2, pp. 646–452, 27–29, July.

39. Gerster, C., Hofer, P., and Karrer, N. 1996. Gate-control strategies for snubberless operation of series connected IGBTs. *IEEE PESC Conference*, Baveno, Italy, vol. 2, pp. 1739–1742, 23–27 June.

40. Idir, N., Franchaud J.J., and Bausiere, R. 2000. New control technique achieves low di/dt and dv/dt. *PCIM Magazine*, February.

41. John, V., Suh, B.S., and Lipo, T.A. 1999. High performance active gate drive for high power IGBTs. *IEEE Trans. Indust. Appl.* 35(5), 108–1117.

42. Luniewski, P., Jansen, U., and Hornkamp, M. 2005. Dynamic voltage rise control, the most efficient way to control turn-off switching behaviour of IGBT transistors, *pelincec Conference*, October 16–19, 2005, Warsaw, Poland, pp. 80.1–80.6.

43. Rebec, M. 2012. "RC-D Fast": RC-Drives IGBT optimized for high switching frequency, Infineon Application Note, July.

44. Hosen, T. and Yanagisawa, K. 2011. Fuji electric's semiconductors: Current status and future outlook. *Fuji Electr. Rev.* 57(3), 68–71.
45. Komatsu, K., Harada, T., and Nakazawa, H. 2011. IGBT module for advanced NPC topology. *Fuji Electr. Rev.* 57(3), 72–76.
46. Motto, E.R., Donlon, J.F., Takahashi, H., Tabata, M., and Iwamoto, H. 1999. Characteristics of a 1200 V PT IGBT with trench gate and local life time control, *IEEE IAS Annual Meeting*, pp. 811–816 vol. 2.
47. Anon. 2009. SiC and GaNi Power Electronics: Focus on PFC Market, Yole Development Report.
48. Schweizer, M., Waffler, S., and Kolar, J.W. 2011. SiC versus Si—Evaluation of potentials for performance improvement of inverter and DC–DC converter systems by SiC power semiconductors. *IEEE Trans. Indust. Electron.* 58(7), 2872–2882.

3 Basic Three-Phase Inverters

3.1 HIGH-POWER DEVICES OPERATED AS SIMPLE SWITCHES

Using semiconductor devices such as bipolar transistors and MOSFETs at somewhat higher power levels in the active region yields low efficiency. A voltage regulator built with a bipolar transistor working in the active region can barely provide an efficiency of 30–40% because of the large collector–emitter voltage drop. In contrast, operation of the same device in switch mode allows efficiencies of 80–99%, depending on the converter topology and the number of passive elements. The whole idea of using switches is to operate at high frequency with a constant or variable duty cycle followed by low-pass filtering with a passive filter. The DC or low-frequency component of the converter waveform is therefore derived and used as a result of the power transfer.

Take the example of a switch controlled to produce a voltage on the load (Figure 3.1). This is the simple case of a step-down DC/DC conversion. If a low-frequency AC component is injected in the duty cycle of the switch, a low-frequency variation is seen on the load (Figure 3.1). In other words, the low-frequency component of the load voltage reflects the control reference. Figure 3.2 illustrates this principle of modulating the pulse width (or duty cycle) of a buck (step-down) converter according to a variable control function.

Figure 3.2 also helps us to understand the difference in terminology between duty cycle and modulation index. The former characterizes the variation of the whole control reference, whereas the latter concerns the control of each individual switch. Therefore, we can modify the load voltage root mean square (RMS) on the fundamental frequency through the modulation index.

Many power loads are actually supplied in current, and the goal of power conversion is to provide a smooth variation in the load current. In such a case, the filtering task is partly laid on the inductive component in the load circuit, which is able to transform pulses of voltage into smooth current. If the load already has enough inductive components, it is not necessary to add an inductive filter in the system; the load itself will take care of the filtering task.

Conversion at higher power levels requires drawing power from a three-phase grid system and delivering it to a three-phase load. Special topologies have been developed for three-phase power conversion and their operation is more complex than a single switch-based DC/DC converter. Particularities within the operation of the three-phase power converter are presented in this chapter. The main power devices used today in building power converters have already been introduced in the previous chapter.

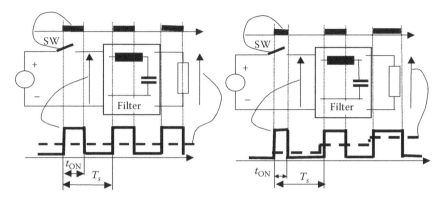

FIGURE 3.1 Basic switch operation: constant and variable duty cycle.

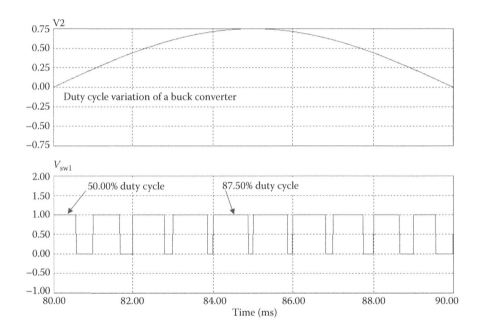

FIGURE 3.2 Duty cycle with an AC variation.

3.2 INVERTER LEG WITH INDUCTIVE LOAD OPERATION

The simplest circuit to produce an AC waveform on the output is shown in Figure 3.3.

To simplify our explanation, a pure inductive load is first considered. The high-side switch S1 and the low-side diode D1 constitute a buck converter during the positive half-wave of the inductor current. Analogously, the low-side switch S2 and the high-side diode D2 form a buck converter during the negative half-wave of the inductor current. During the ON-time of any switch (S1 or S2), energy is charged on the

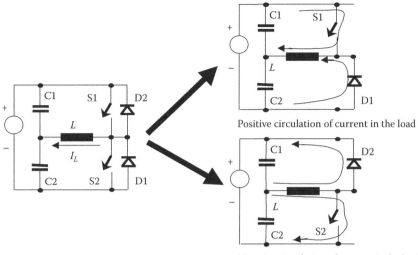

FIGURE 3.3 Current circulation for control of a converter leg.

inductor depending on the width of the ON-time interval. Diodes (D1 or D2) ensure a circulation path for the currents during the OFF state of the switches. Overall, the current through the load inductor follows the control reference, with the delay given by the load power factor. Because of this phase-shift between the control reference and the actual load current, it becomes mandatory to control both switches in a complementary manner. If each switch is controlled on its appropriate buck converter half-wave, an interval without any switching of the load voltage will be necessary for inductor current to discharge through the diodes (Figure 3.4). A complementary control is shown in Figure 3.5. The load current is always under control and it follows a quasi-sinusoidal waveform with a constant phase-shift from the control reference.

3.3   WHAT IS A PWM ALGORITHM?

Both cases presented in Figures 3.4 and 3.5 show a continuous variation of the width of voltage pulses applied to the load. This is called pulse width modulation (PWM). The modulating waveform coincides with the control waveform and pulses are produced at a constant frequency. The carrier pulses are ensured by a waveform called carrier signal. The way the modulating and carrier signals are produced and compared differentiates between PWM algorithms [10–14].

Figure 3.2 presents a very generic power converter. In practical applications, other topologies may be employed and the load might not be a simple inductance. The principle of transferring energy from a source in a switched operation mode ensuring high efficiency of the conversion always remains the same. It has already been shown that the controls for such operation are called PWM techniques.

In the AC/DC conversion case, it is obvious that the line or boost inductance acts as a low-pass filter (LPF) for the applied voltage. Considering an AC drive, the

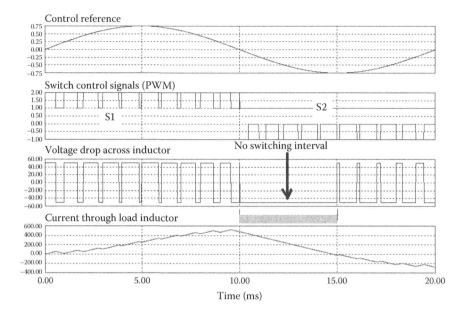

FIGURE 3.4 Control of each switch only on its appropriate half-wave (1 kHz switching frequency, 100 VDC, *m* = 0.75 and 0.5 mH inductor).

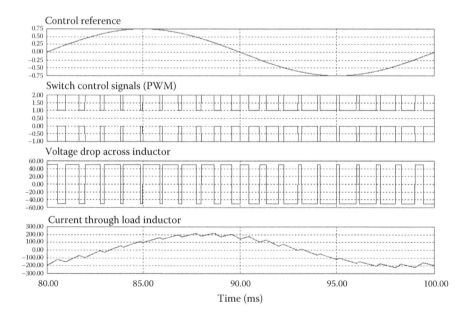

FIGURE 3.5 Control of each switch on the whole waveform (1 kHz switching frequency, 100 VDC, *m* = 0.75 and 0.5 mH inductor).

machine inductance filters the harmonics provided by the discontinuous power flow at switching. The flux linkage in the machine's windings is approximately equal to the time integral of the impressed voltage, if a drop in voltage across resistance and leakage inductance of the stator windings are neglected. No matter how the inductance appears on the load circuit, the power transfer is ensured through PWM.

The great advantage of the PWM algorithm is its ability to control the content in fundamental voltage across the load. This is ensured by the modulation index that is related to the magnitude of the control reference waveform. For instance, the examples shown in Figures 3.4 and 3.5 are based on a modulation index of 0.75. The pulse width variations are based not only on the shape of the control reference but also on this modulation index. The pulse widths modify the charge and discharge intervals of the energy within the load inductor. Accordingly, the ripple of the current through load will depend on the variation of the pulse width. It is important to note that the duty cycle of the switch control coincides with the modulation index in the converter shown in Figure 3.2. Figure 3.6 shows how all waveforms depend on the modulation index. The load current ripple depends on the modulation index value.

Fourier or frequency analysis provides an instrument with which to compare the amount of ripple or harmonics that results from operating at one modulation index or another with one converter topology or another. Fast Fourier transform (FFT) results for the voltage applied to the load at different modulation indices and different switching frequencies are shown in Figures 3.7 and 3.8 for the converter shown in Figure 3.2.

If no modulation is involved, the harmonic spectrum of the output voltage shows a strong harmonic component at the switching frequency. When the modulation

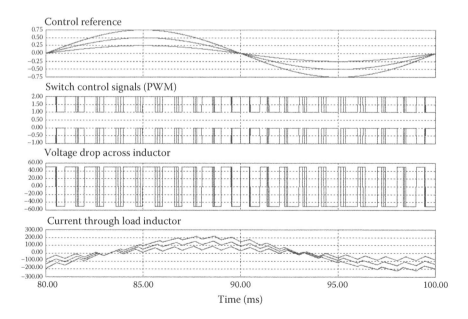

FIGURE 3.6 Waveforms characterizing the converter operation with different modulation indices.

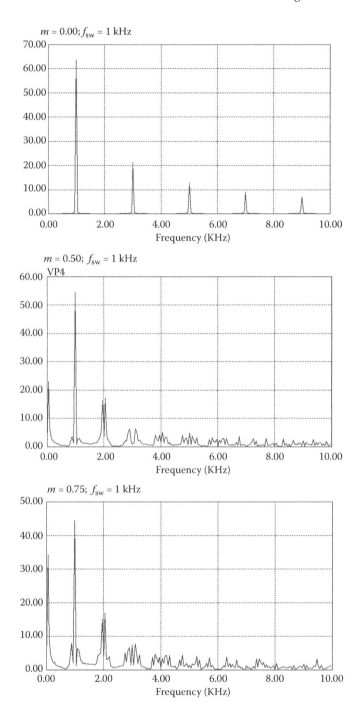

FIGURE 3.7 Output voltage spectra of different operation modes of the converter from Figure 3.2 at 1 kHz switching frequency.

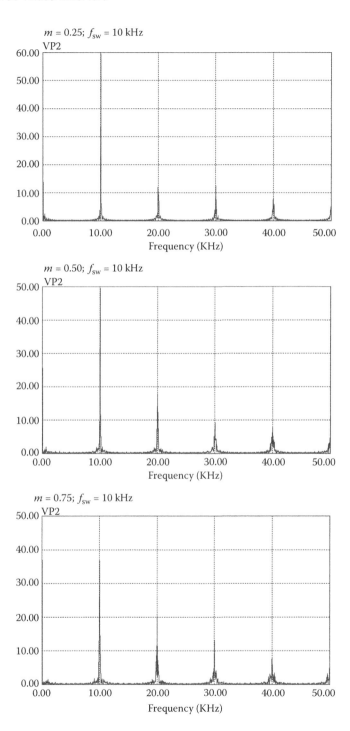

FIGURE 3.8 Spectra of different operation modes of the converter from Figure 3.2 (10 kHz).

index increases, the component at the switching frequency decreases, and the low-frequency component at the frequency of the modulating signal increases. A part of the power transfers from the switching frequency component to the fundamental. The best case, with the most pure low-frequency component synthesis, is achieved at higher modulation indices. However, the modulation index can increase only until the wider pulse in the pattern fits the whole sampling (or switching) period interval. After this point, any increase of the modulation index will produce distortion of the synthesized fundamental frequency waveform.

Conclusions of these harmonic results can be grouped as follows:

* The main effect of modulation is to move the harmonics toward higher frequencies where they can be filtered easier.
* The load voltage spectra contain no DC component, but only components of fundamental frequency and multiplies of the switching frequency.
* The magnitude of the components at multiples of the switching frequency is smaller when frequency increases.
* More harmonics are observed at lower modulation indices.

Global harmonic performance coefficients provide a better comparative image. Section 3.8 makes a complete discussion of these coefficients.

3.4   BASIC THREE-PHASE VOLTAGE SOURCE INVERTER: OPERATION AND FUNCTIONS

Figure 3.2 has shown a possible solution for energy conversion from a DC power supply to a single-phase AC load. AC has been achieved on the load by PWM. At higher power levels, it is common to receive or deliver power on a three-phase system. Power is, therefore, distributed equally on the three phases, and the current in each phase is smaller.

The simplest solution to define a three-phase converter is to multiply the structure of Figure 3.2 on three similar circuits. Figure 3.9 shows such a converter. The operation of each inverter leg is identical to the operation of the converter from Figure 3.3 with reference voltages shifted 120° from one leg to another.

A star-connected three-phase load is represented in Figure 3.9. The load neutral point is connected to the mid-point of the capacitor bank. As the load voltages follow the three-phase reference system, the fundamental voltages applied to the load also form a three-phase system. If the load is symmetrical, the phase currents form a system with no zero sequence, which means that there is no current circulation from the load to the capacitor bank mid-point. The load can be disconnected from the capacitor bank mid-point and an alternative delta-connected load can also be considered (Figure 3.10). Because the phase currents must add up to zero, there is no zero-sequence in the currents for the star-connected load without connection to the DC capacitor bank mid-point.

It has been shown that switches have been controlled on each phase with sinusoidal references modulating the duty cycle of the pulses. The phase voltages can be expressed with dependence on the switch voltage drop.

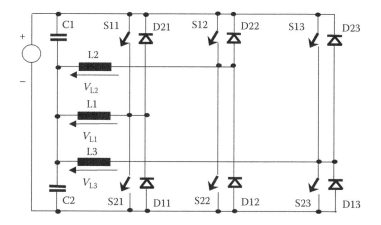

FIGURE 3.9   Three-phase inverter derived from the single-phase topology.

The main constraint for PWM generation is the need to produce a symmetrical set of voltages on the load (Figure 3.11). This can be expressed as

$$V_{\text{ph A}} + V_{\text{ph B}} + V_{\text{ph C}} = 0 \tag{3.1}$$

The circuit equations are

$$\begin{cases} V_{\text{ph A}} = V_{\text{pole A}} - V_{\text{NO}} \\ V_{\text{ph B}} = V_{\text{pole B}} - V_{\text{NO}} \\ \quad \Rightarrow V_{\text{pole A}} + V_{\text{pole B}} + V_{\text{pole C}} = 3V_{\text{NO}} \\ \quad \Rightarrow V_{\text{NO}} = \dfrac{1}{3}[V_{\text{pole A}} + V_{\text{pole B}} + V_{\text{pole C}}] \\ V_{\text{ph C}} = V_{\text{pole C}} - V_{\text{NO}} \end{cases} \tag{3.2}$$

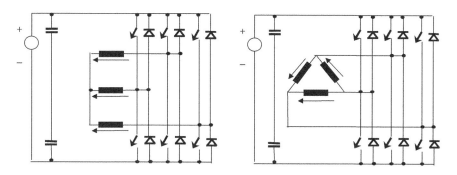

FIGURE 3.10   Different load connections.

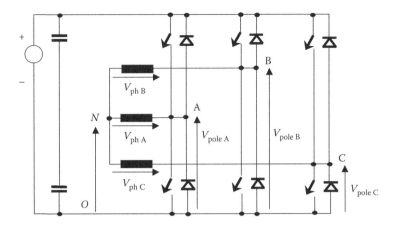

FIGURE 3.11 Voltage construction on the load.

These equations can be seen both in average and instantaneous values. The average analysis neglects all switching processes. If the pole voltages follow sinusoidal references superimposed to half of the DC voltage, the V_{NO} voltage always equals half the DC voltage. If we consider a third or a multiple of the third harmonic injected identically within the pole voltages $V_{pole A}$, $V_{pole B}$, $V_{pole C}$, this harmonic will be found on the V_{NO} voltage. Any shape of a repetitive signal on the third harmonic frequency will satisfy the same remark. Furthermore, as the same signal is a part of the pole voltages and the neutral voltage, it is not seen on the output phase voltage. This leads to a very important conclusion: third and multiple of three harmonics in the modulator reference voltages are not seen on the output phase voltages. It will be shown later that this conclusion helps increase the maximum modulation index.

The previous equations in instantaneous values help demonstrate that the phase voltages equal $1/3 \, V_{DC}$, $2/3 \, V_{DC}$, $-1/3 \, V_{DC}$, $-2/3 \, V_{DC}$, or 0. This also implies that the VNO voltage does not maintain a stiff constant DC voltage, but changes between different levels of voltage at each switching.

The operation of the ideal converter with a six-step modulation or PWM outlines the following conclusions for the output phase voltages:

- There is no even order harmonics
- There is no 3rd harmonic or harmonic multiple of three
- There is no DC component

Accordingly, a typical spectrum of the load voltage is characterized by fundamental, pairs of $6 \, k \pm 1$ order harmonics (5,7,11,13,17,19,...), up to the switching frequency and its multiplies. Examples of spectra of the load voltage are included in Figures 3.12 through 3.14. A low ratio between switching frequency and the frequency of the sinusoidal reference has been assumed in order to outline the lower-order harmonics.

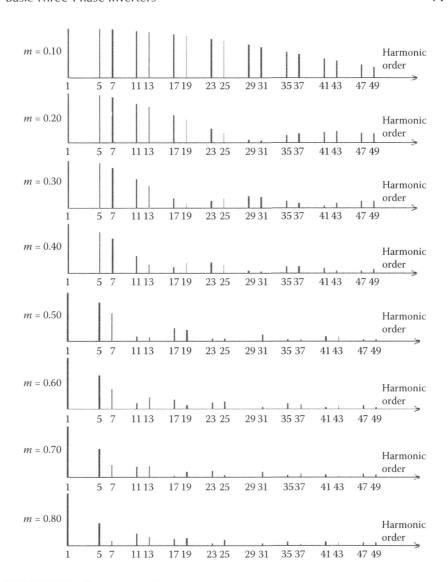

FIGURE 3.12 Six-step operation.

The six-step operation (without modulation) of the three-phase inverter produces large harmonics of the load voltages. Different methods for harmonic improvement have been introduced:

- Connection of several identical power stages through transformers and control with phase shift in order to add up voltage or current waveforms on the load (Figure 3.15)

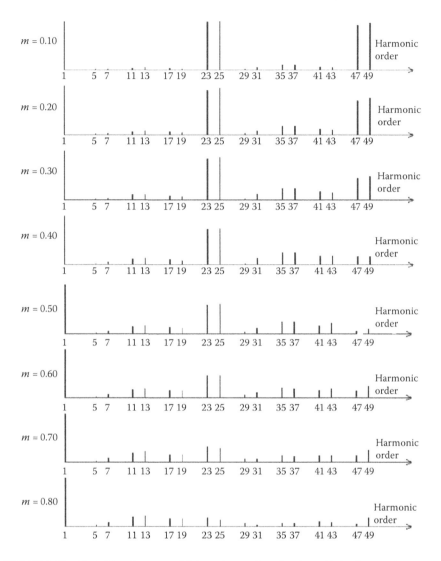

FIGURE 3.13 Switching frequency 24 times larger than the sinusoidal reference frequency.

- Control with PWM algorithms such as
  - Programmed pattern calculated to optimize a harmonic coefficient for low-frequency range or for reduction of certain low-frequency harmonics
  - Triangle-intersection or direct digital pulse-programming techniques to achieve carrier-based PWM methods (carrier-based PWM algorithms)
  - Vectorial PWM methods

Each of these methods will be detailed later with specific examples of applications. Before pursuing such analysis, let us understand what the requirements are for performance indices in three-phase inverters.

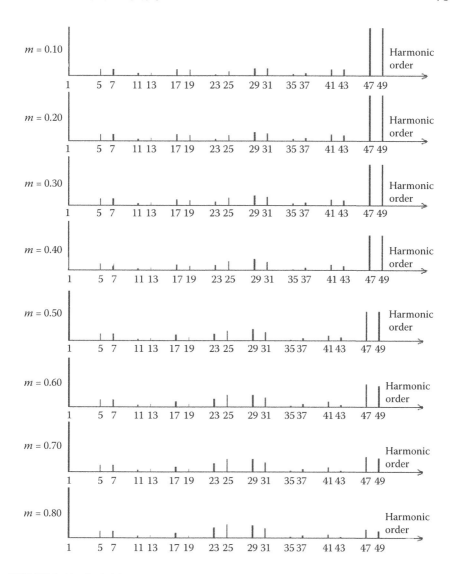

FIGURE 3.14   Switching frequency 48 times larger than the sinusoidal reference frequency.

3.5   PERFORMANCE INDICES: DEFINITIONS AND TERMS USED IN DIFFERENT COUNTRIES

In order to compare the results from different PWM methods and different power converters, several performance indices are defined based on frequency analysis. These analyses of the voltages and currents at the input or output of a power converter can be performed with coefficients of the Fourier series or with Fourier transform.

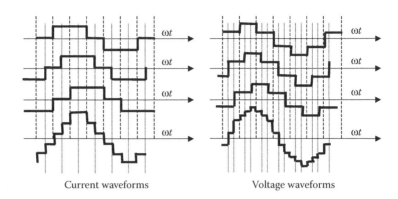

Current waveforms   Voltage waveforms

FIGURE 3.15 Harmonic improvement by adding-up voltage or current waveforms from individual power converters. Off-line optimization of the phase-shift is required.

3.5.1 Frequency Analysis

Any periodic function can be developed in a constant value and an infinite series of sin and cos functions on even and odd multiples of the fundamental frequency.

This is called Fourier series and it is easy to apply for signals defined analytically.

$$u(\omega t) = \frac{A_0}{2} + A_1 \sin(\omega t) + A_2 \sin(2\omega t) + \cdots + A_n \sin(n\omega t) + \cdots$$
$$+ B_1 \cos(\omega t) + B_2 \cos(2\omega t) + \cdots + B_n \cos(n\omega t) + \cdots \tag{3.3}$$

where

$$A_0 = \frac{1}{\pi} \int_0^{2\pi} u(\omega t)\, d\omega t \tag{3.4}$$

$$A_n = \frac{1}{\pi} \int_0^{2\pi} u(\omega t) \sin(n\omega t)\, d\omega t \tag{3.5}$$

$$B_n = \frac{1}{\pi} \int_0^{2\pi} u(\omega t) \cos(n\omega t)\, d\omega t \tag{3.6}$$

Components on the same frequency determine the magnitude of the respective harmonic.

$$V_n = \sqrt{A_n^2 + B_n^2} \tag{3.7}$$

with the RMS value of

$$V_n^{\text{RMS}} = \frac{V_n}{\sqrt{2}} \tag{3.8}$$

Previous results shown in Figures 3.12 through 3.14 have used the Fourier series.

If the waveform is not defined analytically, but as a set of measurements or simulation results, then it is worthwhile to calculate the Fourier transform. This converts a time domain periodic function into a frequency domain function called spectral function.

$$S(n\omega) = \frac{1}{T} \int_0^T u(t)e^{-j\omega nt}\, dt \tag{3.9}$$

where $n\varepsilon\ (-\infty,\infty)$.

The reverse transform is given by

$$u(t) = \sum_{n=-\infty}^{\infty} S(n\omega)e^{-j\omega nt} \tag{3.10}$$

Numeric calculus is achieved for an approximation of the Fourier integral when samples of the measured signal are known as $u(kT) = u_k$. In this respect, let us consider $T = NT_s$ and $dt = T_s$.

$$S(n\omega) = \frac{1}{NT_s} \sum_{k=0}^{N-1} u_k e^{(-j\omega kn/N)T_s} \qquad T_s = \frac{1}{N} \sum_{k=0}^{N-1} u_k e^{(-j\omega kn/N)T_s} \tag{3.11}$$

The sampling theorem states that N samples of a signal can define $(N/2) - 1$ positive spectral components and $(N/2) - 1$ negative spectral components. Function $u(t)$ is periodic and this implies $S(-n\omega) = S((N - n)\omega)$. This further allows conversion of the negative spectral components into the upper range $(N/2, N - 1)$ and calculation of N spectral components from the N samples of the waveform in time domain.

Computer calculation of the spectral function $S(n\omega)$ can be done after evaluation of the expression:

$$w = e^{-j2\pi/N}$$

It yields:

$$S(n\omega) = \frac{1}{N} \sum_{k=0}^{N-1} u_k w^{kn} \tag{3.12}$$

The sequence of calculus can be reduced when considering the number of samples as a power of two. The outcome is also named FFT. For instance, choosing 1024 samples and the advantages of the FFT algorithm reduces the running time to only 1% of the time required for conventional Fourier transform calculation.

Because of the limited resolution of the Fourier transform methods, each component is shown for an interval adjacent with a triangular shape. The base of this triangle will be smaller for a finer sampling of the original signal and its magnitude will better approximate the magnitude of the frequency component. Previous results shown in Figures 3.7 and 3.8 have been determined with FFT.

3.5.2 MODULATION INDEX FOR THREE-PHASE CONVERTERS

For a three-phase inverter, performance indices are defined with respect to the modulation index

$$m = \frac{V_s}{(2/3)V_{DC}} \tag{3.13}$$

3.5.3 PERFORMANCE INDICES

Commonly used performance indices are introduced next. Calculated results are introduced as examples, but details on how these results have been achieved are overlooked in order to simplify the presentation. Precise differences in results from different PWM methods will be shown in Chapters 4 and 5.

3.5.3.1 Content in Fundamental (z)

It represents the ratio between the RMS value of the fundamental of the output phase voltage (V_{L1}) and the RMS value of the output phase voltage (V_L). It is used mostly in Europe.

3.5.3.2 Total Harmonic Distortion (THD) Coefficient

$$\text{THD}(\%) = \frac{100}{V_{(1)}} \sqrt{\sum_{n=2}^{\infty} [V_{(n)}]^2} \tag{3.14}$$

Results of this coefficient depend on the number of harmonics considered in calculus. It is a good practice to consider a number of harmonics several times larger than the switching frequency.

Figure 3.16 shows THD for different switching frequencies when the modulation index varies between 0.1 and 0.8 and when calculus is performed for an extremely large number of samples. It is obvious that for the same modulation index, the results are approximately identical, no matter what the switching frequency when the frequency ratio is a multiple of six. This certifies that a PWM algorithm is just moving harmonics from lower frequency to higher frequency without altering the power delivered on the load.

3.5.3.3 Harmonic Current Factor (HCF)

As the inductive load is basically a low pass filter (LPF), the higher order current harmonics will be attenuated. The remaining spectrum of the current will be different from one PWM method to another and from one switching frequency to another.

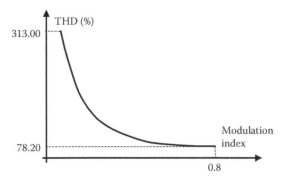

FIGURE 3.16 THD variation with the modulation index for 1.2; 2.4 and 3.6 kHz switching frequency when harmonics up to 150 kHz are considered.

A coefficient regarding current harmonics would better define the performance of a PWM method. Such a coefficient is called HCF and it can be expressed also based on the voltage harmonics at the converter output:

$$HCF(\%) = \frac{100}{I_{(1)}} \sqrt{\sum_{n=5}^{\infty} [I_{(n)}]^2} = \frac{100}{(V_{(1)}/\omega L)} \sqrt{\sum_{n=5}^{\infty} \left[\frac{V_{(n)}}{n\omega L} \right]^2}$$

$$= \frac{100}{V_{(1)}} \sqrt{\sum_{n=5}^{\infty} \left[\frac{V_{(n)}}{n} \right]^2}$$

(3.15)

After some calculation, Figure 3.17 presents the HCF coefficient dependence on the modulation index for the three-phase converter shown in Figure 3.10. The larger the switching frequency, the lower the HCF coefficient.

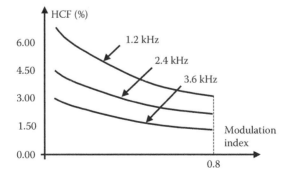

FIGURE 3.17 HCF variation with the modulation index for 1.2; 2.4 and 3.6 kHz switching frequency when harmonics up to 150 kHz are considered.

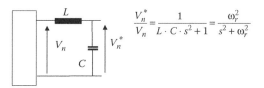

FIGURE 3.18 Output with filter of a power supply.

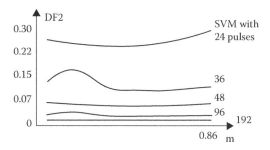

FIGURE 3.19 DF2 for regular SVM with different number of pulses on the fundamental period. (Adapted from Lucanu, M., Neacsu, D., and Donescu, V. 1995. Optimal Power Control Strategies for Space Vector PWM Inverters, Technical Bulletin of IPlasi, Romania, Tomme XLI (XLV), *Fasc.* 1–2, pp. 97–102.)

3.5.3.4 Current Distortion Factor

$$\text{DF} = \frac{I_{\text{harm,rms}}}{I_{\text{harm,6-step}}} \tag{3.16}$$

This performance index is equivalent with HCF.

The requirements for AC power supplies consist of low output impedance and less than 5% voltage THD at load terminals. An output LC filter is necessary to decrease the THD content of the output voltage (Figure 3.18).

Let us note the harmonics of the filter output voltage V_n^*. Taking into account the effect of the filter and the transfer function between the filter output voltage and the inverter voltage, DF yields (Figure 3.19):

$$\text{DF}^2 = \frac{100}{V_{(1)}^*} \sqrt{\sum_{n=5}^{\infty} \left[\frac{V_{(n)}^*}{n}\right]^2} = \frac{100}{V_1} \sqrt{\sum_{n=5}^{\infty} \left[\frac{V_n}{n^2}\right]^2} \tag{3.17}$$

3.6 DIRECT CALCULATION OF HARMONIC SPECTRUM FROM INVERTER WAVEFORMS

Any version of FFT can be calculated based on the samples of the waveform, but it requires extensive calculation. Calculation of the harmonic coefficients based on

Fourier definitions is also complicated. This section introduces two methods for a quick estimation of the harmonic spectrum without integral calculation.

3.6.1 DECOMPOSITION IN QUASI-RECTANGULAR WAVEFORMS

Let us start with the quasi-rectangular waveform shown in Figure 3.20.

Applying Equation 3.5, the voltage harmonics can be expressed as

$$V_n = \frac{4}{\pi} \int_0^\alpha V_{DC} \cos(n\omega t)\,d\omega t = \frac{4}{\pi} V_{DC} \frac{1}{n}[\sin(n\alpha)] \tag{3.18}$$

All waveforms in power converters can be characterized with rectangular shapes. Moreover, such waveforms can be decomposed in periodic elementary quasi-rectangular waveforms, as shown in Figure 3.20. This approach can be applied to all staircase, two-level, and three-level waveforms.

Figure 3.21 shows the decomposition of a staircase waveform in quasi-rectangular waveforms.

Each harmonic component can be calculated by simple addition of the harmonics of the same order from the individual quasi-rectangular waveforms. The previous Fourier series development helps in the calculation of harmonics through simple addition.

$$V_n = V_{DC} \frac{1}{n}[\sin(n\alpha_1)] + V_{DC} \frac{1}{n}[\sin(n\alpha_2)] + V_{DC} \frac{1}{n}[\sin(n\alpha_3)] \tag{3.19}$$

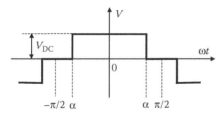

FIGURE 3.20 Waveform with parameter α.

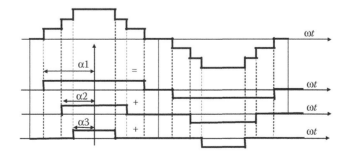

FIGURE 3.21 Decomposition in quasi-rectangular waveforms.

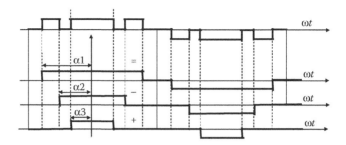

FIGURE 3.22   Decomposition of a PWM waveform.

Similarly, a PWM waveform can be decomposed into an algebraic sum of components (Figure 3.22).

The mathematical form of this decomposition is:

$$V_n = V_{DC}\frac{1}{n}[\sin(n\alpha_1)] - V_{DC}\frac{1}{n}[\sin(n\alpha_2)] + V_{DC}\frac{1}{n}[\sin(n\alpha_3)] \qquad (3.20)$$

3.6.2   VECTORIAL METHOD

Any periodic waveform can be decomposed into simple periodic rectangular waveforms with the same shape but phase shifted. The Fourier series of each such simple waveform is well known. Moreover, each harmonic component can be represented with a vector. Figure 3.23 shows an example of the waveform composed of simple square-waves. Adding up the appropriate waveforms is equivalent to adding up their corresponding vectors for fundamental frequency and generic harmonics of order n.

Simple mathematical relationships can be written for this vectorial composition. The magnitude and phase of a vector that results from composing two other vectors is well defined in mathematics textbooks. Definitely, this method is appealing for a reduced number of square-waves in decomposition.

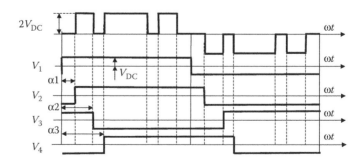

FIGURE 3.23   Decomposition in simple square-waves: (a) fundamental frequency; (b) n-th harmonics.

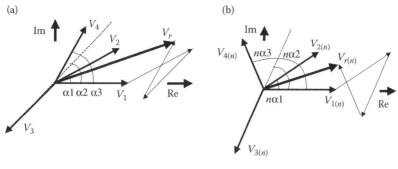

FIGURE 3.24 Vectorial composition.

The magnitude of the nth harmonics in the development of Fourier series for each individual square wave is provided by

$$V_n = \begin{cases} \dfrac{4V_{DC}}{n\pi}, \text{ for } n = 1,5,9,\ldots \\[3mm] -\dfrac{4V_{DC}}{n\pi}, \text{ for } n = 3,7,11,\ldots \end{cases} = (-1)^k \dfrac{4V_{DC}}{(2k+1)\pi}, \text{ for } k = 0,\ldots,\infty \quad (3.21)$$

The vectorial result can be computed easily by the decomposition of each particular vector on the real and imaginary axes, followed by algebraic operations on each of these two axes (Figure 3.24). It yields:

$$\text{Re} : V_{2k+1}^{\text{All,Re}} = V_{2k+1}^1 \cos\alpha_1 + V_{2k+1}^2 \cos\alpha_2 + V_{2k+1}^3 \cos\alpha_3 + \cdots$$
$$\text{Im} : V_{2k+1}^{\text{All,Im}} = V_{2k+1}^1 \sin\alpha_1 + V_{2k+1}^2 \sin\alpha_2 + V_{2k+1}^3 \sin\alpha_3 + \cdots + V_{2k+1}^n \sin\alpha_n + \cdots \quad (3.22)$$

3.7 PREPROGRAMMED PWM FOR THREE-PHASE INVERTERS

The application of PWM methods in different industrial systems aims to improve global harmonic factors, reduce losses in the power converter or load, reduce torque pulsations in the motor drive applications, and reduce noise and vibrations.

It is easy to imagine a direct method of achieving this by optimal off-line definition of the switching instants. Results from all possible optimization criteria have reduction of low harmonics in common. This PWM can therefore operate without a fixed frequency but according to a preprogrammed pattern. The drawback of this approach is extensive computing. In comparison with carrier-based PWM or vectorial PWM, preprogrammed PWM can offer:

- Reduction of the inverter switching frequency by about 50%
- Direct operation into overmodulation providing more output voltage

- Reduced ripple of the DC current and elimination of the possibility of oscillations within the output filter
- Simpler implementation from a memory look-up table

3.7.1 Preprogrammed PWM for Single-Phase Inverter

Different topologies for single-phase voltage generation can be operated with bipolar PWM (two-level) or unipolar PWM (three-level) (Figure 3.25) [7,8,14,15]. The bipolar waveform can also be mathematically derived as a difference between a unipolar PWM and a square wave of half the amplitude. The following harmonic analysis supposes that waveforms are synchronized with a cos function and the B_n term equals zero.

The Fourier series for the three-level (unipolar) PWM can be expressed as

$$A_n = \frac{1}{n}\frac{4V_{DC}}{\pi}\left[\sum_{k=1}^{N}(-1)^{k-1}\cos(k\alpha_k)\right] = \frac{1}{n}\left[\sum_{k=1}^{N}(-1)^{k-1}\cos(k\alpha_k)\right] \quad (3.23)$$

A DC bus voltage of $\pi/4$ has been considered for normalization in order to simplify calculation. The fundamental component can therefore be expressed as

$$A_1 = \left[\sum_{k=1}^{N}(-1)^{k-1}\cos(\alpha_k)\right] \quad (3.24)$$

The first optimization constraint consists in setting a desired level of the fundamental. Canceling the 3rd, 5th, 7th, 9th, 11th, … harmonics require the appropriate Fourier coefficients to be zero. The number of degrees of freedom is provided by the number of angular coordinates α_k. For instance, controlling the fundamental and cancelation of the first five odd harmonics is achieved when the output voltage has six level changes within a 90° interval.

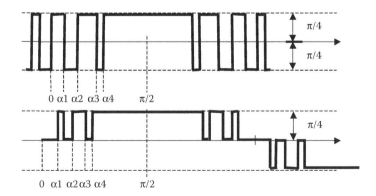

FIGURE 3.25 Bipolar and unipolar PWM waveforms.

If only cancelation of the first five harmonics is our goal, the following can be written:

$$A_3 = \frac{1}{3}\left[\sum_{k=1}^{5}(-1)^{k-1}\cos(3\alpha_k)\right] = 0$$

$$A_5 = \frac{1}{5}\left[\sum_{k=1}^{5}(-1)^{k-1}\cos(5\alpha_k)\right] = 0$$

$$A_7 = \frac{1}{7}\left[\sum_{k=1}^{5}(-1)^{k-1}\cos(7\alpha_k)\right] = 0 \quad\quad (3.25)$$

$$A_9 = \frac{1}{9}\left[\sum_{k=1}^{5}(-1)^{k-1}\cos(9\alpha_k)\right] = 0$$

$$A_{11} = \frac{1}{11}\left[\sum_{k=1}^{5}(-1)^{k-1}\cos(11\alpha_k)\right] = 0$$

Solving this system yields the following values:

$$\alpha_1 = 18.17°$$
$$\alpha_2 = 26.64°$$
$$\alpha_3 = 36.87° \quad\quad (3.26)$$
$$\alpha_4 = 52.90°$$
$$\alpha_5 = 56.69°$$

Replacing these values for the fundamental component yields:

$$A_1 = \cos(18.17) - \cos(26.64) + \cos(36.87) - \cos(52.90) + \cos(56.69) = 0.74 \quad (3.27)$$

A proper adjustment of the V_{DC} voltage can modify the content in fundamental A1. Similar calculus can be performed for any other harmonic condition (Table 3.1).

The bipolar (two-level) PWM wave has the following development in the Fourier series:

$$A_n = \frac{1}{n}\left[-1 + 2\sum_{k=1}^{N}(-1)^{k-1}\cos(k\alpha_k)\right] \quad\quad (3.28)$$

A similar system of equations can be written for specific harmonic elimination or fundamental component control. For instance, elimination of the 5th and 7th harmonics along with the control of fundamental needs three angular variables.

TABLE 3.1

Eliminated Harmonics without Restriction on the Fundamental Content and the Appropriate Switching Angles

Eliminated Harmonics	5th	7th	11th	13th		
Switching angles	18.00°	21.43°	24.54°	25.39°		
	30.00°	30.00°	30.00°	30.00°		
	42.00°	38.57°	35.45°	34.61°		
Eliminated Harmonics	5th and 7th	5th and 11th	5th and 13th	7th and 11th	7th and 13th	11th and 13th
Switching angles	7.93°	7.93°	12.96°	15.24°	16.59°	19.03°
	13.75°	13.75°	19.14°	19.37°	20.80°	21.76°
	30.00°	30.00°	30.00°	30.00°	30.00°	30.00°
	46.25°	46.25°	38.88°	40.63°	39.20°	38.24°
	52.07°	52.07°	45.52°	44.76°	43.41°	40.97°
Eliminated Harmonics	5th 7th 11th	5th 13th 11th	7th 13th 11th			
Switching angles	2.25°	7.82°	9.48°			
	5.61°	11.04°	11.61°			
	21.26°	22.13°	23.26°			
	30.00°	30.00°	30.00°			
	38.74°	37.87°	36.74°			
	54.39°	48.96°	48.39°			
	57.75°	52.18°	50.52°			

The equations yield:

$$A_1 = [-1 + 2(\cos \alpha_1 - \cos \alpha_2 + \cos \alpha_3)] = V$$

$$A_5 = [-1 + 2(\cos(5\alpha_1) - \cos(5\alpha_2) + \cos(5\alpha_3))] = 0 \qquad (3.29)$$

$$A7 = [-1 + 2(\cos(7\alpha_1) - \cos(7\alpha_2) + \cos(7\alpha_3))] = 0$$

As these equations are similar to those from the three-phase inverter analysis, a practical result will be shown later for the more popular case of a three-phase inverter.

3.7.2 PREPROGRAMMED PWM FOR THREE-PHASE INVERTER

For a three-phase voltage source inverter, elimination of low harmonics in the pole voltage (switching function) implies elimination of low harmonics in the phase voltage.

The harmonics in the line-to-line voltage (V_{LL}) are related to the harmonics in the phase voltage through the following relationship:

$$V_n^{L-L} = V_n^{\text{ph A}} - V_n^{\text{ph B}} = \frac{1}{n} \sum_n \left[\cos(n\alpha_k) - \cos n\left(\alpha_k - \frac{2\pi}{3} \right) \right] \qquad (3.30)$$

$$= \frac{1}{n} \sum_n \left[\cos(n\alpha_k) - \cos n\left(\alpha_k + \frac{\pi}{3} - \pi \right) \right]$$

$$= \frac{1}{n} \sum_n \left[\cos(n\alpha_k) + \cos n\left(\alpha_k + \frac{\pi}{3} \right) \right] \qquad (3.31)$$

$$= \frac{1}{n} \sum_n \left[2 \cos n\left(\alpha_k + \frac{\pi}{6} \right) \cos n\left(\frac{\pi}{6} \right) \right]$$

The Fourier coefficients of the line-to-line voltage can be expressed by

$$A_n = \frac{4}{n\pi} \left[-1 - 2 \sum_{\substack{k=1 \\ B_n = 0}}^{N} (-1)^k \cos(k\alpha_k) \right] \qquad (3.32)$$

that is identical with the Fourier series for the unipolar PWM in the single-phase case.

However, symmetries for a three-phase system should be taken into account. Because of these symmetries, the switching-pattern calculation can be reduced to 30°. The three-phase switching pattern optimally defined for a 30° interval can then be used in the definition of the whole switching pattern. Switching instants within the first 30° interval are defined by an angular coordinate α measured from the beginning of the interval. Optimization can be set up based on appropriate waveforms for the phase voltages, V_{LL}, or pole voltages in a three-phase voltage source inverter and for the phase currents within a current source three-phase inverter.

Similarly, operation of a single-phase preprogrammed PWM needs a pattern definition for 90°.

Therefore, it can be demonstrated that voltage reversals within the first 60° need mirror symmetries around the middle point situated at 30° from the beginning. For instance, a voltage transition from 0 to V_{DC} at α_1 implies a voltage transition from V_{DC} to 0 at $60 - \alpha_1$. Next, there should be no switching at the top of the waveform for a 60° interval (between 60° and 120° from the beginning of the waveform). The symmetry on the second harmonic imposes transitions on the following 60° with the same angular delays as on the first 60° interval. If all these conditions are respected, one can analyze only a single-phase waveform and account automatically for the cancelation of the second and third harmonics.

If the considered pattern has N switching instants within a 30° sector, N variables can be defined. For a given content in fundamental (A_1), there are $N - 1$ degrees of

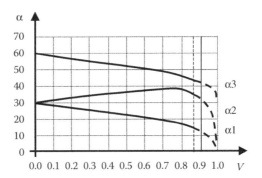

FIGURE 3.26 Solution for 5th and 7th harmonics eliminated with control of fundamental.

freedom for harmonic elimination. The nonlinear form of these constraints provides
the complexity of the optimization calculus.

$$\begin{cases} A_1 = V \\ A_5 = 0, \quad A_7 = 0 \\ A_{11} = 0, \quad A_{13} = 0 \\ A_{6M-1} = 0, \quad A_{6M+1} = 0 \end{cases} \tag{3.33}$$

where $2M$ is the number of switching instants over a $30°$ interval and V is the desired
fundamental voltage (current). Solving this system provides a set of values for α_k at
each V. Accordingly, the solution of Equation 3.27 or 3.33 can be presented graphically, as in Figure 3.26. Possible shapes of waveforms within a three-phase system
are displayed in Figure 3.27.

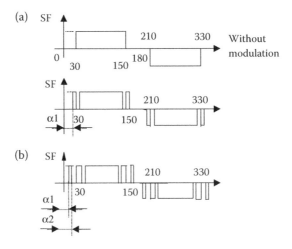

FIGURE 3.27 Examples of line-to-line voltage with eliminated harmonics.

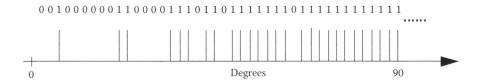

0 0 1 0 0 0 0 0 0 1 1 0 0 0 0 1 1 1 0 1 1 0 1 1 1 1 1 1 1 0 1 1 1 1 1 1 1 1 1 1 1 ......

0   Degrees   90

FIGURE 3.28   Binary programmed PWM.

This section presented methods for cancelation of specific harmonics. Other optimization criteria can be considered for minimization of THD, HCF, or torque harmonics [18]. They will lead to more complex calculus and require extensive use of computer programs, as MATHCAD or MATHEMATICA.

3.7.3   BINARY-PROGRAMMED PWM [1]

A version of the harmonic elimination principle can be achieved for a three-phase system with a division of the 30° interval in a fixed number of equal intervals. A variable is inserted at each of these sampling instants and optimization calculus is performed to define positive or zero values for these variables. Using symmetry, the whole waveform is finally built-up in the microcontroller memory.

For instance, Figure 3.28 applies this principle to a single-phase system required to cancel all harmonics up to the 13th and to maximize the content in fundamental [1]. The switching waveform results for 45 samples over an interval of 90° and the remaining higher harmonics are below 0.45% of fundamental. If applied to an induction machine, the torque harmonics result is below 3% of torque fundamental component.

3.8   MODELING A THREE-PHASE INVERTER WITH SWITCHING FUNCTIONS

Understanding the operation of each single-phase circuitry (Figure 3.3) helps define the controller for the three-phase converter. A very good tool for mathematical modeling of the operation of a three-phase converter is based on the switching functions concept. Given the repetitive manner of switching power devices within the three-phase power converter, one can define switching functions as periodical functions built up of rectangular pulses. The switching functions can commute within a limited number of states. This mathematical representation is possible only when switching of the power devices is not dictated by circuit operation (as in the SCR).

A conventional analysis of the converter presented in Figure 3.10 would need 36 circuit equations to be written for all currents and voltages. This system of equations can be further reduced to three if symmetries of the three-phase circuitry are considered. To simplify the mathematical model, switching functions are introduced.

The definitions of the switching functions are not unique [2–4]. Let us consider several examples:

$$f_1 = \begin{cases} 1, & S11 = \text{on and } S21 = \text{off} \\ 0, & S11 = \text{off and } S21 = \text{on} \end{cases}$$

$$f_2 = \begin{cases} 1, & S12 = \text{on and } S22 = \text{off} \\ 0, & S12 = \text{off and } S22 = \text{on} \end{cases} \tag{3.34}$$

$$f_3 = \begin{cases} 1, & S13 = \text{on and } S23 = \text{off} \\ 0, & S13 = \text{off and } S23 = \text{on} \end{cases}$$

Load voltages can be expressed with dependency on these switching functions:

$$\begin{bmatrix} v_{L1} \\ v_{L2} \\ v_{L3} \end{bmatrix} = \frac{2}{3} V_{DC} \begin{bmatrix} f_1 - 0.5(f_2 + f_3) \\ f_2 - 0.5(f_1 + f_3) \\ f_3 - 0.5(f_1 + f_2) \end{bmatrix} \tag{3.35}$$

The DC current can also be calculated based on these functions:

$$i_{DC} = i_{ph\,A} f_1 + i_{ph\,B} f_2 + i_{ph\,C} f_3 \tag{3.36}$$

Another possibility:

$$f_1 = \begin{cases} 1, & \text{if } S11 = \text{on and } S21 = \text{off and } S31 = \text{off} \\ -1, & \text{if } S12 = \text{on and } S22 = \text{off and } S32 = \text{off} \\ 0, & \text{any other situation} \end{cases}$$

$$f_2 = \begin{cases} 1, & \text{if } S12 = \text{on and } S11 = \text{off and } S31 = \text{off} \\ -1, & \text{if } S22 = \text{on and } S12 = \text{off and } S32 = \text{off} \\ 0, & \text{any other situation} \end{cases} \tag{3.37}$$

$$f_3 = \begin{cases} 1, & \text{if } S31 = \text{on and } S11 = \text{off and } S21 = \text{off} \\ -1, & \text{if } S32 = \text{on and } S12 = \text{off and } S22 = \text{off} \\ 0, & \text{any other situation} \end{cases}$$

Load voltages can now be expressed as

$$\begin{bmatrix} v_{L1} \\ v_{L2} \\ v_{L3} \end{bmatrix} = \frac{2}{3} V_{DC} \begin{bmatrix} f_1 - 0.5(f_2 + f_3) \\ f_2 - 0.5(f_1 + f_3) \\ f_3 - 0.5(f_1 + f_2) \end{bmatrix} \tag{3.38}$$

The DC current can also be calculated on the basis of these functions:

$$i_{DC} = i_{ph\,A}f_1 + i_{ph\,B}f_2 + i_{ph\,C}f_3 \qquad (3.39)$$

This method provides a mathematical relationship between the PWM control algorithm and the load voltages and DC current in a three-phase converter. Simulation tools can be built-up based on this concept, and they can provide a quick simulation without taking into account all the transient details pertaining to any peculiar power device. For instance, a direct implementation of these equations can be made in MATLAB-SIMULINK [2] environment, whereas an implementation with current and voltage sources can be accomplished in PSPICE [3].

3.9   BRAKING LEG IN POWER CONVERTERS FOR MOTOR DRIVES

Motor drives represent the greatest application for three-phase inverters. Power converters are manufactured especially for this application in the topology with six switches, presented in Figure 3.10. The braking deceleration of these motors transfers power to the intermediary circuit of these power converters. The basic requirements for a braking module have been analyzed in the introductory chapter. The DC voltage rises until the frequency converter trips for protection and it requires, sometimes, a special brake module to absorb this braking power. For power levels above 6.8 kW and less than 20 kW, the power converter itself includes an internal brake circuit and can accept an external brake resistor [4,7]. This resistor should be mounted on a heatsink and covered. For higher power levels, such braking modules can be attached outside [19].

This circuit is useful for dynamic regeneration during power dissipation, avoiding overcharging of the DC capacitor. This circuit is not rated for a continuously overhauling load, but it needs to absorb the peak brake power during large dynamics. It is rated for the average power calculated over a complete cycle.

$$P_{pk} = \frac{0.0055J(n_1^2 - n_2^2)}{t_b}[W] \qquad (3.40)$$

$$P_{av} = P_{pk}\frac{t_b}{t_c} \qquad (3.41)$$

where J is total inertia (kg m^2); n_1 the initial speed (RPM); n_2 the final speed (RPM); t_b the brake time (s); t_c the cycle time (s).

A minimum value of the brake resistor must be defined to limit its current and power.

Then, we derate this calculus based on the ambient temperature at the moment of braking (Figure 3.29).

Another way to brake a motor is to use the DC brake. A DC voltage is applied across two motor phases to produce a stationary magnetic field in the stator. The braking power remains in the motor, which may overheat. For this reason, this method is used only in low-speed ranges.

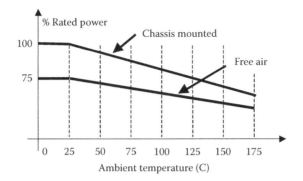

FIGURE 3.29 Derating based on temperature.

3.10 DC BUS CAPACITOR WITHIN AN AC/DC/AC POWER CONVERTER

The introductory chapter showed the role and features of the DC capacitor bank. The selection of the bus capacitors and the associated ripple aspects are next discussed. One of the functions of the DC capacitive bank is to reduce ripple. The input current to a three-phase inverter is composed of current pulses according to the switching sequence. An example is shown in Figure 3.30. The average value of these pulses represents the active power delivered to the three-phase inverter. The other harmonics compose the ripple to be filtered by the capacitor bank [5,6,16,17].

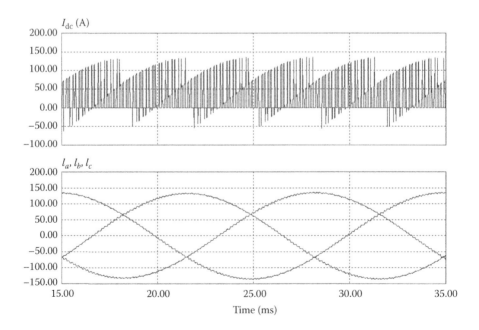

FIGURE 3.30 DC current and the phase currents of a three-phase inverter.

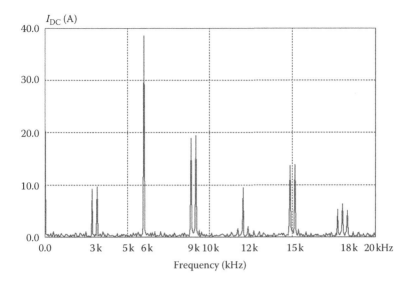

FIGURE 3.31   Spectrum of the DC current for carrier sinusoidal modulation at 3 kHz.

Aluminum electrolytic capacitors are usually selected in order to filter the front-end rectification waveform and to store the energy necessary for the dynamics of the load (Figure 3.31). This type of capacitor has an anode foil with an aluminum oxide layer acting as the dielectric, a cathode foil with no oxidation process, and a separator paper. All of them are wound together and impregnated with an electrolyte.

The equivalent circuit of an aluminum electrolytic capacitor is shown in Figure 3.32. The most important parameter is definitely the capacitance, and it is expressed by [5]

$$C = 8.855 \times 10^{-8} \frac{\varepsilon S}{d} \tag{3.42}$$

where ε, is the dielectric constant, S the surface area of dielectric (cm^2), d the thickness of the dielectric (cm). The dielectric constant is [8–10] within any aluminum electrolytic capacitor, whereas the dielectric layer is very thin, in the range of about 15 A per volt. The surface area is increased by electrochemically etching the aluminum foil up to 100 times in low-voltage foil and 25 times in high-voltage foil. This

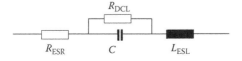

FIGURE 3.32   Equivalent circuit for an electrolytic capacitor.

is the major advantage of aluminum capacitors, as they provide a larger capacitance when compared with other types of electrolytic capacitors.

The equivalent series resistance (ESR) represents the resistance that produces heat in the capacitor when the AC ripple current is applied. It is a combination of the resistive losses because of aluminum oxide thickness, electrolytic spacer combination, and resistance due to materials and material characteristics, such as foil length, tabbing, lead wires, and contact resistance.

The leakage current (DCL) corresponds to the DC current leaking through the capacitor. Ideally, it is well known that the capacitor is supposed not to allow any circulation of a DC current. However, small leakage of current occurs; it is proportional with capacitance and decreases when the applied voltage reduces.

The inductance of a capacitor (equivalent series inductance—ESL) is a constant and it depends on the mechanical mounting of terminals. The ESL is in the range of 3–40 nF and it influences the capacitor operation only at very high frequencies.

All these components contribute to the capacitor impedance. The frequency characteristic of this impedance has usually a notch in tens of kilohertz range, whereas the magnitude at lower and higher frequencies is higher. The lowest impedance value corresponds to the resonant frequency that turns the electrolytic capacitor into an inductor [6]. The frequency components within this frequency range will not be filtered by the electrolytic capacitor. Therefore, small low-inductance film capacitors (polyester or polypropylene) are generally used in differential or common mode to compensate the frequency characteristic of the electrolytic capacitor. A simple solution consists of using parallel connections with electrolytic capacitors in order to filter the harmonic components from current ripple and electromagnetic inference. The use of high-frequency film capacitors as DC bus snubber is suggested at the electrolytic capacitor terminals to minimize the connection inductance.

Finally, let us note two other important parameters in the selection of the DC bus electrolytic capacitor: the rated voltage and the ripple current. The rated voltage is calculated as the sum of the DC and AC voltages applied to the capacitor. If the ripple current is larger than a specified value, the life of the capacitor becomes shorter because of the heat generated by the excessive ripple current. Accordingly, there is an inverse relationship between the current ripple and the capacitor's ESR.

3.11   CONCLUSION

This chapter introduces the single-phase and three-phase inverters and explains the challenges in meeting harmonic performance requirements. Details of preprogrammed PWMs are provided along with mathematical tools to calculate harmonics. Among all possible algorithms, the most used are the carrier-based PWM and the vectorial PWM that are explained in later chapters.

Finally, the brake leg and the DC capacitor bank are shown as possible auxiliary components of a three-phase inverter. Chapter 6 will provide the details on protection and building a three-phase inverter for different power levels.

PROBLEMS

P3.1 Figure 3.16 shows no difference between the THD of the output voltage for different switching frequencies at any modulation index. How can this be explained?

P3.2 Consider Figure 3.21 with a DC voltage of 100 V. Write the mathematical constraints for achieving a fundamental voltage of 48 V and cancelation of the third and fifth harmonics (use Equation 3.19). Solve this system of equations and look for a solution with $\alpha 1 < \alpha 2 < \alpha 3$.

P3.3 Consider Figure 3.22 with a DC voltage of 100 V. Write the mathematical constraints for achieving a fundamental voltage of 48 V and cancelation of the fifth and seventh harmonics (use Equation 3.20). Solve this system of equations and look for a solution with $\alpha 1 < \alpha 2 < \alpha 3 < 30°$. This condition also ensures that third harmonic vanishes.

P3.4 Write Equation 3.22 for the case of Figure 3.23. Consider $\alpha 1 = 12$, $\alpha 2 = 18$, $\alpha 3 = 23$, and calculate the first seven harmonics.

P3.5 Consider $V = 0.4$ and read $\alpha 1$, $\alpha 2$, $\alpha 3$ from Figure 3.26. Calculate the first 10 harmonics using these values in Equation 3.27.

P3.6 Imagine a new definition of the switching functions for a three-phase inverter and write the appropriate dependency of the phase voltages and DC current on these switching functions.

REFERENCES

1. Said, W. 1989. Torque pulsations harmonics in PWM inverter induction motor drives. *etzArchiv* 11, 267–269.
2. Neacsu, D., Yao, Z., and Rajagopalan, V. 1995. Switching Function Analysis of Power Converters in MATLAB. Research Report, UQTR, Canada.
3. Salazar, L. and Joos, G. 1995. PSPICE simulation of three-phase inverters by means of switching functions. *IEEE Trans. Power Electron.* 9(1), 35–42.
4. Mohan, N., Undeland, T., and Robbins, W. 2002. *Power Electronics*, 3rd edition. John Wiley and Sons, New York.
5. Anon. 2000. United Chemicon Catalog H9.
6. Lai, J.S., Kouns, J.S., and Bond, J. 2002. A low-inductance DC bus capacitor for high-power traction motor drive inverter. *IEEE Industry Applications Annual Meeting,* 2002, Vol. 2 , pp. 826–831.
7. Thornborg, K. 1988. *Power Electronics.* Prentice Hall, London, UK.
8. Brichant, F. 1984. *Force-Commutated Inverters—Design and Industrial Applications.* MacMillan Publishing Company, New York, USA.
9. Alex, D., Turic, L., and Stiurca, D. 1985. Umrichtersystem mit hoherem grundschwingungsge-halt fur die Drehstromtraktion. Bulletin SEV/VSE, Zurich, Switzerland, pp. 490–492.
10. Holtz, J. 1992. Pulsewidth modulation—A survey. *IEEE Trans. IE* 39, 410–420.
11. Holtz, J. 1994. Pulsewidth modulation for electronic power conversion. *Proc. IEEE* 82, 1194–1212.
12. Buja, G.S. and Indri, G.B. 1977. Optimal PWM for feeding AC motors. *IEEE Trans. IA* 13, 38–44.
13. Patel, H.S. and Hoft, R.G. 1973. Generalized technique of harmonic elimination and voltage control in thyristor inverters. *IEEE Trans. Ind. Appl.* 9, 310–317.

14. Neacsu, D. 2001. Space Vector Modulation. IEEE IECON Tutorial.
15. Enjeti, P., Ziogas, P., and Lindsay, J. 1990. Programmed PWM techniques to eliminate harmonics: A critical evaluation. *IEEE Trans. IA* 26, 302–316.
16. Enjeti, P. and Shireen, W. 1990. A new technique to reject DC-link voltage ripple for inverters operating on programmed PWM Waveforms. *IEEE PESC* 705–711.
17. Dahono, P.A., Sato, Y., and Kataoka, T. 1995. Analysis and minimization of ripple components of input current and Voltage of PWM Inverter. *IEEE IAS Annual Meeting* 3, 2444–2450.
18. Sun, J., Beineke, S., and Grotstollen, H. 1994. DSP-based real-time harmonic elimination of PWM inverters. *IEEE PESC* 679–685.
19. Anon. 2001. Drives 101—Lessons. and Danfoss Internet documentation.
20. Lucanu, M., Neacsu, D., and Donescu, V. 1995. Optimal Power Control Strategies for Space Vector PWM Inverters, Technical Bulletin of IPlasi, Romania, Tomme XLI (XLV), *Fasc.* 1–2, pp. 97–102.

4 Carrier-Based Pulse Width Modulation and Operation Limits

4.1 CARRIER-BASED PULSE WIDTH MODULATION ALGORITHMS: HISTORICAL IMPORTANCE

Figure 3.2 shows the principle of pulse width modulation (PWM) control, but it does not provide any information on how we can produce the modulation within a real hardware circuitry. In order to keep the pulse frequency constant, a carrier signal is necessary. Switches change their conduction states at moments of time that are determined by the intersections between our reference voltage and a triangular high-frequency signal with a fixed magnitude equal to unity. This operation is shown in Figure 4.1.

The frequency of the pulses is kept constant while their duty cycle is modulated. This is known as *natural sampling, suboscillation* or the *subcycle* method. The method and the names are derived from the initial hardware implementation. In the 1970s, when engineers were already using power semiconductor switches fast enough to support modulation, the only control hardware available was analog circuitry. It was therefore easy to generate the carrier signal as a triangular waveform and to achieve modulation by comparison with a variable reference.

The method described in Figure 4.1 allows several possible shapes for the triangular waveform (Figure 4.2):

- Center-aligned (*a*)
- Left-aligned (*c*)
- Right-aligned (*b*)

Secondly, the ratio (*q*) between the frequency of the carrier and the frequency of the reference signal can have different values. If *q* is small, harmonics can be improved if the following are taken into account:

- *q* should be an integer in order to have synchronous waveforms leading to a periodical train of pulses.
- *q* should be odd in order to produce the same number of pulses on both positive and negative half-waves. The output voltage, therefore, contains no even harmonics.
- The harmonic of order *q* is the dominant one.
- In a three-phase system, the third and multiple-of-three harmonics vanish when *q* is selected as a multiple of three.

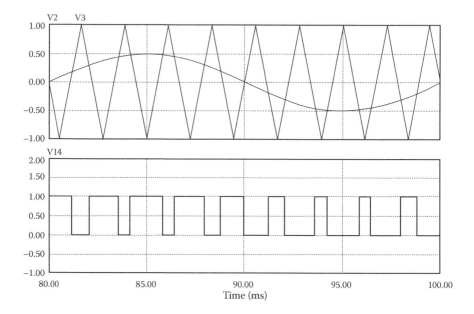

FIGURE 4.1   Sinusoidal PWM based on intersection between a triangular carrier signal and the reference.

The modulation signals could also have other shapes, such as trapezoidal and staircase (Figure 4.3), which provide advantages in the hardware implementation of the controller and in optimal switching performance at the power stage [1–7].

Finally, let us note here the advantage of a uniform-sampled PWM in digital implementation. This method samples and holds the sinusoidal reference at the same frequency as the carrier. The resulting reference looks like a staircase wave-form with elementary steps equaling the width of the carrier period. Figure 4.4 illustrates this method for a low-frequency carrier. Control pulses are obtained at the intersection of the *S/H* signal and the triangular waveform. A proper synchronization of these signals produces symmetrical signals.

A mathematical description of the harmonics of these carrier-based methods is difficult because the intersection moments are not linearly or equally spread over the reference cycle. Harmonics can be observed by FFT applied to simulation or measured data (Figure 4.2, right). The best harmonic content is achieved for the center-aligned pulses.

The case of a three-phase system can be treated as three individual modulators (Figure 4.5) with a star connected load. This load connection modifies the shape of the load voltage in each phase. Since the ratio of the switching and fundamental frequencies is chosen as a multiple of three (denoted with 6 *k*), high-frequency harmonics are seen in pairs at orders of 6 *k* ± 1. Frequency components at multiples of the switching frequency vanish.

Center-aligned PWM is also known as symmetrical PWM, whereas the left- and right-aligned methods are called asymmetrical PWM. Figure 4.2 showed the single-phase generation for each of these methods. The three-phase converter uses the same

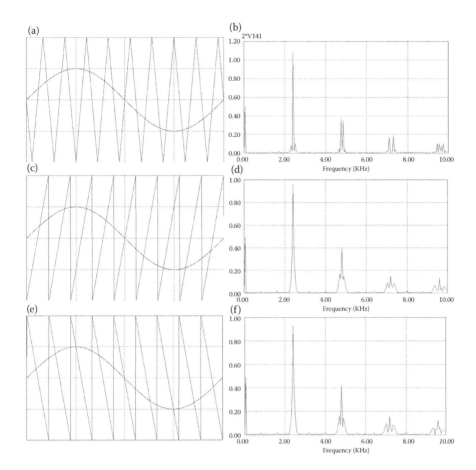

FIGURE 4.2 Different shapes of the triangular signal for $q = 9$ and modulation index of 0.5, for a single-phase inverter switched at 2.4 kHz: (a) time-domain waveforms for center-aligned triangular carrier; (b) spectrum of output voltage with modulation defined in (a); (c) time-domain waveforms for right-aligned triangular carrier; (d) spectrum of output voltage with modulation defined at (c); (e) time-domain waveforms for left-aligned triangular carrier; (f) spectrum of output voltage with modulation defined at (e).

modulators, but the special load-connection modifies the shape of the load voltage and its spectrum.

4.2 CARRIER-BASED PWM ALGORITHMS WITH IMPROVED REFERENCE

The magnitude of the sinusoidal reference can be extended up to the magnitude of the carrier waveform. This situation corresponds to a maximum modulation index:

$$m_{\max} = \frac{u_{(1)}}{u_{\text{six-step}}} = \frac{\pi}{4} = 0.785 \tag{4.1}$$

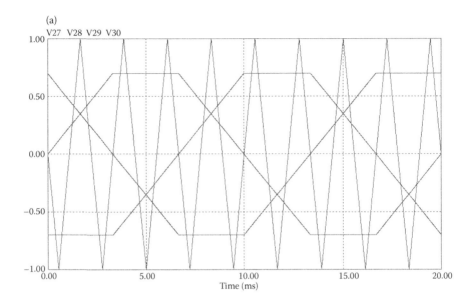

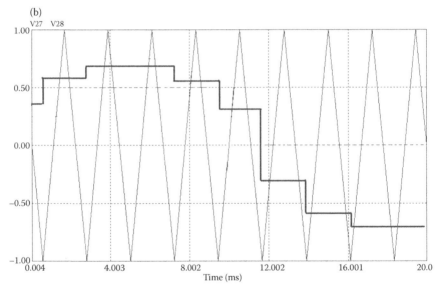

FIGURE 4.3 Trapezoidal and staircase references.

This is less than what is obviously possible when the pole (leg) voltage bounces on the full DC bus. This deficiency—of limited modulation index—is corrected by modified reference waveforms.

In Section 3.4, we have seen that a three-phase inverter can have third or higher order harmonics on the pole voltage without seeing them on the load voltage. Generally, any zero-sequence waveform can be added to the reference without being

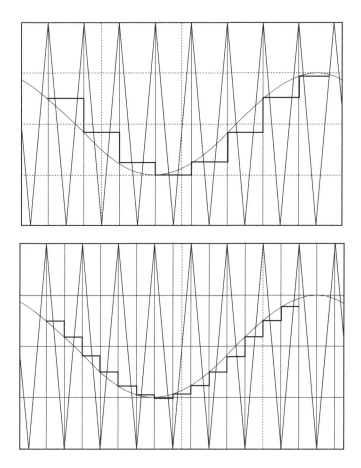

FIGURE 4.4   Uniform sampling of the reference signals.

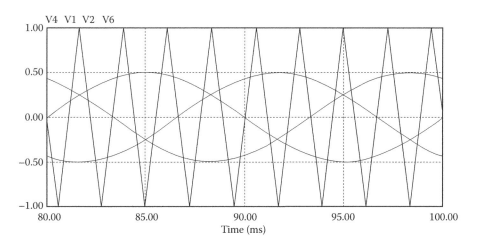

FIGURE 4.5   PWM generation in a three-phase system.

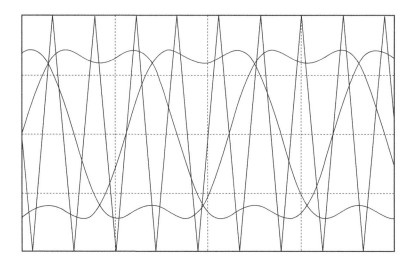

FIGURE 4.6   Injection of the third harmonic in the reference signal.

noticed on the load. This feature is used in applications that limit the peak-to-peak voltage applied to the load while preserving a high content in fundamental for this voltage. There is an infinite number of possible additions to the reference waveform that satisfy this condition.

First, let us consider a simple third-harmonic injection with the phase selected to suppress the peak of the sinusoidal reference (Figure 4.6).

$$\begin{cases} v_{\text{ref}}, A = V\sin(\omega t) + kV\sin(3\omega t) \\ v_{\text{ref}}, B = V\sin(\omega t + 120) + kV\sin(3\omega t) \\ v_{\text{ref}}, C = V\sin(\omega t + 240) + kV\sin(3\omega t) \end{cases} \qquad (4.2)$$

Considering k as a parameter, we can define the amount of the third harmonic to be injected in the reference signals in order to optimize performance indices such as maximum content in fundamental or minimum harmonic current factor (HCF) coefficient.

In [1], a minimization of the total harmonic distortion (THD) is considered and $k = 1/4$ is found as the optimal solution. Separate works [2,3] demonstrate that a third harmonic with $k = 1/3$ maximizes the fundamental voltage.

As the shape of the pole voltage is different from the shape of the phase voltage (Section 3.4), it is possible to keep one inverter leg unswitched and to produce the load three-phase system of sinusoidal waveforms out of the other two phases. Due to symmetries within a three-phase system, each interval of no switching can last for 60°. Accordingly, the reference function needs to be defined by six sets of functions, each one valid for 60°. A 50% theoretical switching loss reduction is therefore possible by not switching each switch for 60°. This opportunity was first explored in [4,5] and it was shown that discontinuous references produced by discontinuous zero sequence components can extend linearity above 0.785 up to the six-step operation fundamental and reduce switching losses. Later, other researchers proposed other shapes (or functions) for the zero sequence component. Figure 4.7 provides several examples [8–15].

The difference between these methods is the way in which we have selected the 60° interval when an inverter leg is not switched. It is advantageous to select between these methods based on each application and taking into account that switching losses strongly depend on the current through the insulated gate bipolar transistors (IGBTs). Power converters working on the grid application would benefit by using the first method in which the no-switching interval is produced around the peak of the voltage reference and phase current. Motor drive applications are typically characterized by a lagging current and the second method would be better, as it has the no-switching interval after the peak of the voltage reference. If the high-side IGBTs and the low-side IGBTs are identical in the inverter building, using the last two methods would produce different power loss and heat on the high side and the low side. The last two methods are not used, since they do not share equally losses between power switches.

Closely observing the mathematics of these methods enables us to define a generalized modulator configurable in specific applications. Figure 4.8 illustrates generalized PWM generation. The injected zero sequence is calculated as the difference between the sinusoidal references and the peak of the carrier waveforms considered here as unity. This difference represents how much we should add to the top of the existing reference system in order to saturate the modulator and to not have any switching during a 60° interval. It becomes obvious that the range of φ lies within 0° and 60°. Due to symmetries in a three-phase system and the condition of equally sharing losses, alternative use of saturation at maximum and minimum of the reference is considered.

Despite the obvious theoretical advantages of these methods, they are not often used in practical systems. Carrier-based PWM algorithms have emerged in analog hardware support. Using discontinuous zero-sequence injection highly increases the complexity and cost of the analog control circuits. Moreover, the performance at low-modulation index is very poor due to the narrow pulse limitations and transition instabilities during the change of the modulation function. As an alternative, engineers have used this method in high-modulation indexes only, keeping the conventional continuous modulation for low-modulation indexes.

Understanding the principles of carrier-based PWM generation with discontinuous reference functions later helped define reduced loss space vector PWM methods. These are analyzed extensively in Chapter 5 and are based exclusively on digital implementation. Because of their usefulness in extended linear ranges and reduced switching losses, combined with the advantages of digital implementation, they are used nowadays by industry.

4.3 PWM USED WITHIN VOLT/HERTZ DRIVES: CHOICE OF NUMBER OF PULSES BASED ON THE DESIRED CURRENT HARMONIC FACTOR

One of the major applications for three-phase inverters operated with PWM is within motor drives. The simplest and still most used control of an induction motor considers the constant flux within the machine on the basis of a constant Volt/Hertz (V/Hz)

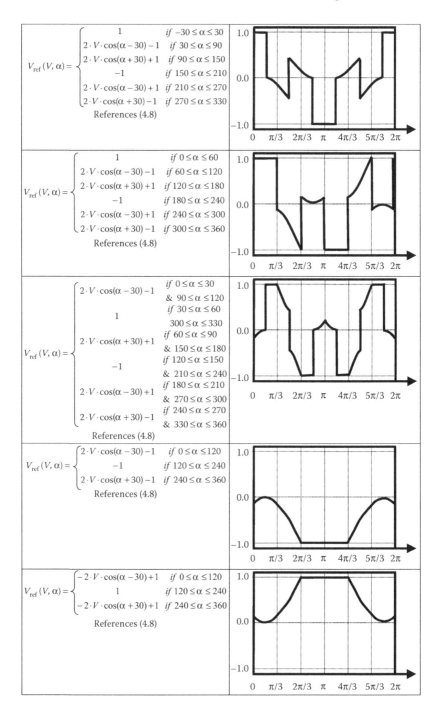

$$V_{\text{ref}}(V,\alpha)=\begin{cases}1 & \text{if } -30\leq\alpha\leq30\\ 2\cdot V\cdot\cos(\alpha-30)-1 & \text{if } 30\leq\alpha\leq90\\ 2\cdot V\cdot\cos(\alpha+30)+1 & \text{if } 90\leq\alpha\leq150\\ -1 & \text{if } 150\leq\alpha\leq210\\ 2\cdot V\cdot\cos(\alpha-30)+1 & \text{if } 210\leq\alpha\leq270\\ 2\cdot V\cdot\cos(\alpha+30)-1 & \text{if } 270\leq\alpha\leq330\end{cases}$$

References (4.8)

$$V_{\text{ref}}(V,\alpha)=\begin{cases}1 & \text{if } 0\leq\alpha\leq60\\ 2\cdot V\cdot\cos(\alpha-30)-1 & \text{if } 60\leq\alpha\leq120\\ 2\cdot V\cdot\cos(\alpha+30)+1 & \text{if } 120\leq\alpha\leq180\\ -1 & \text{if } 180\leq\alpha\leq240\\ 2\cdot V\cdot\cos(\alpha-30)+1 & \text{if } 240\leq\alpha\leq300\\ 2\cdot V\cdot\cos(\alpha+30)-1 & \text{if } 300\leq\alpha\leq360\end{cases}$$

References (4.8)

$$V_{\text{ref}}(V,\alpha)=\begin{cases}2\cdot V\cdot\cos(\alpha-30)-1 & \text{if } 0\leq\alpha\leq30\\ & \&\ 90\leq\alpha\leq120\\ 1 & \text{if } 30\leq\alpha\leq60\\ & 300\leq\alpha\leq330\\ 2\cdot V\cdot\cos(\alpha+30)+1 & \text{if } 60\leq\alpha\leq90\\ & \&\ 150\leq\alpha\leq180\\ & \text{if } 120\leq\alpha\leq150\\ -1 & \&\ 210\leq\alpha\leq240\\ & \text{if } 180\leq\alpha\leq210\\ 2\cdot V\cdot\cos(\alpha-30)+1 & \&\ 270\leq\alpha\leq300\\ & \text{if } 240\leq\alpha\leq270\\ 2\cdot V\cdot\cos(\alpha+30)-1 & \&\ 330\leq\alpha\leq360\end{cases}$$

References (4.8)

$$V_{\text{ref}}(V,\alpha)=\begin{cases}2\cdot V\cdot\cos(\alpha-30)-1 & \text{if } 0\leq\alpha\leq120\\ -1 & \text{if } 120\leq\alpha\leq240\\ 2\cdot V\cdot\cos(\alpha+30)-1 & \text{if } 240\leq\alpha\leq360\end{cases}$$

References (4.8)

$$V_{\text{ref}}(V,\alpha)=\begin{cases}-2\cdot V\cdot\cos(\alpha-30)+1 & \text{if } 0\leq\alpha\leq120\\ 1 & \text{if } 120\leq\alpha\leq240\\ -2\cdot V\cdot\cos(\alpha+30)+1 & \text{if } 240\leq\alpha\leq360\end{cases}$$

References (4.8)

FIGURE 4.7 Possible modulator references for discontinuous PWM for a modulation index of 0.5.

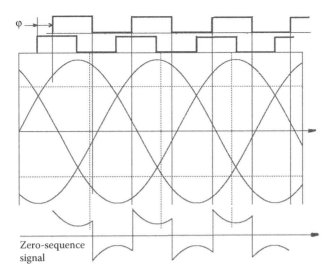

FIGURE 4.8   Generalized PWM generator.

ratio. Without getting into machine drive details, let us see what are the consequences of applying PWM inverters for this class of applications.

First, note that the V/Hz characteristic used in control is not linear over the entire frequency range (Figure 4.9). At low frequencies, the magnitude of the reference voltage is small and the voltage drop on the resistive component of the stator is larger than the inductive voltage component. The reference magnitude, therefore, actually increases within the controller, accounting for the resistive voltage drop while the machine flux remains constant. Moreover, at high frequencies, the characteristic is limited in order to keep a constant voltage. This ensures control with a field weakening.

From the control and implementation of control perspectives, V/Hz control considers phase voltage generation in phase measures, with variations on a time scale referring more to the shape over the fundamental period. In contrast, modern vector control or field-orientation control assumes a high-frequency sample-and-hold of the drive system and a control independent of phase measures. Within a vector control

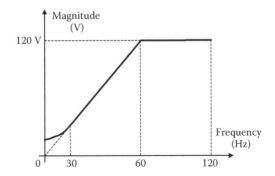

FIGURE 4.9   Real V/Hz characteristics.

system, the digital PWM generation does not use the desired shape of the phase voltage over a cycle, but the calculation of the instantaneous changes of these voltages (or currents) only. Volt per hertz control changes magnitude and phase while the vector-control method changes current references at a fixed sampling interval. Understanding this big difference emphasizes the advantage of using carrier based PWM in V/Hz control methods [4–6,12,16].

It has been shown that for very large ratios between the carrier frequency and the frequency of the reference waveform, these two waveforms do not need to be synchronized to each other. However, for low-frequency ratios, waveform synchronization becomes mandatory and the frequency ratios may take particular values only (multiple of six).

4.3.1 Operation in the Low-Frequencies Range (Below Nominal Frequency)

In a motor drive application, the fundamental frequency varies over a wide range. On the other hand, the switching frequency must be contained due to power loss and thermal considerations. It is very difficult to satisfy both constraints using a single pattern of the PWM generator over the whole frequency range. The frequency ratio is generally selected to take different values on different fundamental frequency intervals, on the basis of the *limits of the switching frequency* or *optimization of the HCF factor* [4–6,12,16]. If we extend this method for power supplies that generate voltages with variable frequency and magnitude, we can optimize the number of pulses on the basis of *filter requirements optimization*. Let us analyze each method separately.

- Limit of the switching frequency: A possible solution is shown in Figure 4.10. For each frequency interval there is a linear relationship between the fundamental frequency of the reference and the carrier frequency (*fsw*). The slope of the characteristics equals the frequency ratio. To avoid system oscillations, transition between operation modes is ensured with a hysteresis. The solution shown in Figure 4.10 changes operation modes when frequency doubles, but similar controls can be defined for ratio values in a series like $24 \to 36 \to 48 \to 60 \to 72$ or $18 \to 36 \to 72 \to 144 \to 288$. Many industrial solutions go up to a frequency ratio of six for the last interval, ending up with a six-step not-modulated operation. The solution

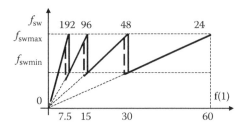

FIGURE 4.10   Use of different number of pulses on different intervals.

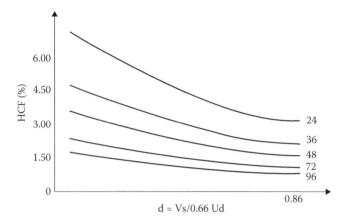

FIGURE 4.11   HCF for PWM with different number of pulses.

in which intervals are selected at double frequency is generally preferred because that reduces the number of transitions between operation modes and is simple to implement (see Chapter 8).

- Optimization of the HCF factor: HCF has been calculated for different frequency ratio methods and results are shown in Figure 4.11. If the application requires limiting this coefficient to below a given level, we can optimize the number of switching processes while still satisfying the harmonic requirement by selecting different frequency ratios for different frequency variation intervals (Figure 4.11) (see Chapter 3, Ref. 20). For instance, when a minimum value of 4% is considered, the maximum switching frequency is about 1.5 kHz. A limit to the frequency ratio must be assumed in the very low-speed range in order to simplify the digital implementation. Therefore, the HCF constraints will be relaxed for frequencies less than a few hertz. The proposed dependence of the sampling frequency f_{sw} on the fundamental frequency f_1 is presented in Figure 4.12 (see Chapter 3, Ref. 20).

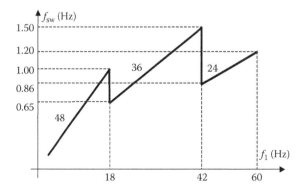

FIGURE 4.12   Sampling frequency versus output frequency.

- Filter requirements optimization for power supply applications: High-power power supplies operate at a constant frequency and with variable-voltage and are required to provide a low output impedance and less than 5% THD voltage content at load terminals. Their harmonic performance analyses have been considered in Section 3.5 and computer analysis results for DF in frequency ratios that equal 24, 36, 48, 72, and 96 have been shown in Figure 3.19. The proper frequency ratio can be selected from these results in order to keep DF below a certain value for any modulation index.

4.3.2 HIGH FREQUENCIES (>60 Hz)

In high frequencies that range from (60–120 Hz), a unique PWM pattern with a low number of pulses must be used independent of frequency and with a constant modulation index. This is generally optimized to reduce harmonics with lowest orders:

- Elimination of selective harmonics, such as the 5th or the 5th and 7th and so on, in the output phase voltage (Section 3.6).
- Global reduction of several low harmonics; for instance, minimization of the $V_5^2 + V_7^2$.

4.4 IMPLEMENTATION OF HARMONIC REDUCTION WITH CARRIER PWM

Chapter 3 has provided solutions for specific harmonic elimination within three-phase inverters. Their implementation requires extensive off-line calculation and storage in memory look-up tables. This solution is not very advantageous for the control engineer. In contrast, this chapter has shown that carrier-based PWM is easily understood and implemented in modern microcontrollers without off-line calculation. Chapter 7 will further detail different implementation strategies within modern controllers. Some of these devices already have special peripherals for natural-sampled or regular-sampled carrier-based PWM. In the early 1990s, researchers considered these digital hardware platforms for implementation of harmonic elimination strategies [17].

Let us reconsider Figure 3.26 in Figure 4.13. Solutions for harmonic elimination for the line-to-line voltage (V_{LL}) can be achieved below a modulation index of 0.866. The operation outside 0.866 can be artificially achieved by extending the angular solutions, as shown in Figure 4.13. As a coincidence, this also represents the maximum modulation index for a space vector modulation algorithm. Moreover, the dependency of the switching angles on the modulation index is linear up to approximately 0.69 and nonlinear between 0.69 and 0.866. Figure 4.13 also shows that odd switching angles have a negative slope whereas the even switching angles have positive slopes. All these characteristics start and end at points with a separation of

$$\omega Ts = \frac{2\pi}{3(N + 1)} \tag{4.3}$$

where N is the number of switchings within a 60° interval (3 in our example).

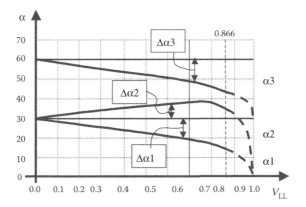

FIGURE 4.13 Switching angles for harmonic cancelation in V_{LL} of a three-phase inverter.

Let us consider a regular-sampled PWM with this sampling interval (T_s). During each sampling period, control of the trailing and leading edges are supposed to be achieved separately according to the regular-sampled PWM approach. For each of these controls, sinusoidal reference waveforms are considered, as shown in Figure 4.14 and the switching angles are given by Equation 4.4 in absolute values from the beginning of the whole cycle.

$$k = \text{odd} \quad \alpha_k = (k+1)\frac{T_s}{2} - m\frac{T_s}{2}\sin\left[(k+1)\frac{T_s}{2} + \phi_1\right]$$

$$k = \text{even} \quad \alpha_k = (k)\frac{T_s}{2} + m\frac{T_s}{2}\sin\left[k\frac{T_s}{2} + \phi_2\right] \tag{4.4}$$

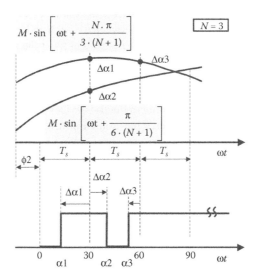

FIGURE 4.14 Equivalence between harmonic elimination and regular-sampled PWM.

The sinusoidal term is required for a microcontroller implementation of Equation 4.4, since the first term is already incrementally generated within the natural PWM algorithm. The magnitude of the reference sinusoidal waveform represents the modulation index (m) and it follows closely the desired amount of voltage on the load with an error less than 3.5%. The phase of these sinusoidal references can be determined with curve fitting at each sampling moment or with an approximate solution. The approximate solution uses:

$$k = \text{odd} \quad \phi_1 = N\frac{T}{2}$$
$$k = \text{even} \quad \phi_2 = \frac{T}{4} \tag{4.5}$$

The resulting errors of this approximation are less than $0.25°$ at any operation point, leading to "cancelled" harmonics less than 2% [17].

For modulation indices between 0.69 and 0.866, the switching angle dependency on the modulation index is nonlinear, and the implementation on a regular-sampled PWM can be achieved by a nonlinear variation of the position of the sampling instants. The sampling period becomes smaller and smaller when the modulation index increases and all the sampling intervals are crowded toward $0°$. Finally, all sampling moments coincide at $0°$ for the maximum modulation index, and square-wave operation is achieved.

4.5 LIMITS OF OPERATION: MINIMUM PULSE WIDTH

The circuit presented in Figure 3.2 is ideal and differences in operation will occur when switches are implemented with real semiconductor devices. Figure 3.2 also showed how current always commutes between a switch and a diode on the same leg. The PWM method can sometimes lead to short widths of the pulses to be applied on the load [18,19].

Transients in a real semiconductor device delay and narrow the shape of the voltage pulses achieved on the load. In extreme cases, the load does not see any clear voltage pulses while the power semiconductor devices are still switching. In other words, losses remain but the expected harmonic improvement on the load is not realized. Waveforms pertaining to such a case are shown in Figure 4.15.

In order to avoid the commutation at the power stage of short pulses, the most common methods employed are

- Pulse deletion: pulses are deleted from the PWM waveform if narrower than a certain amount and intervals with no switchings on the power converter occur. Waveforms resulting from this method are shown in Figures 4.16 and 4.17. One can see that the resulting load current has an increased content in fundamental and larger harmonics.

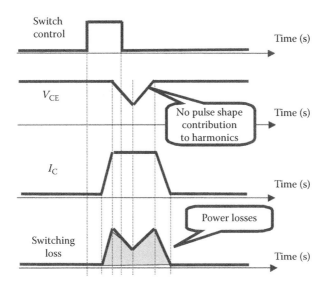

FIGURE 4.15   Switching waveforms for a short pulse.

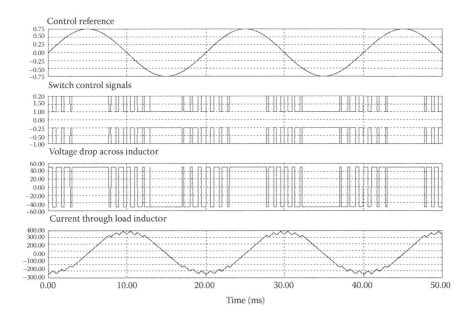

FIGURE 4.16   Effect of pulse dropping in the converter waveforms (converter from Figure 3.3).

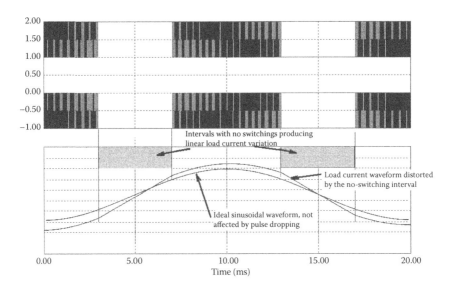

FIGURE 4.17   Effect of a very large pulse dropping interval (same converter at 10 kHz).

- Pulse limitation: pulses are limited to a predefined value if narrower than a certain amount and switches are transitioning at fixed pulse widths for some time. Waveforms resulting from this method are shown in Figures 4.18 and 4.19. The loss in fundamental is noticeable; the worsened content in harmonics is also noticeable.

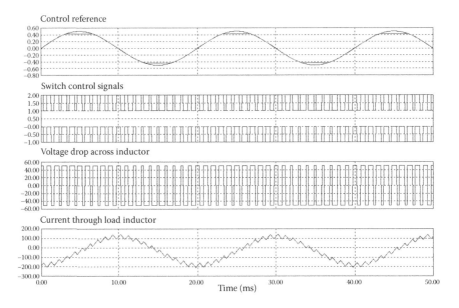

FIGURE 4.18   Effect of pulse limitation in the converter waveforms (converter from Figure 3.3).

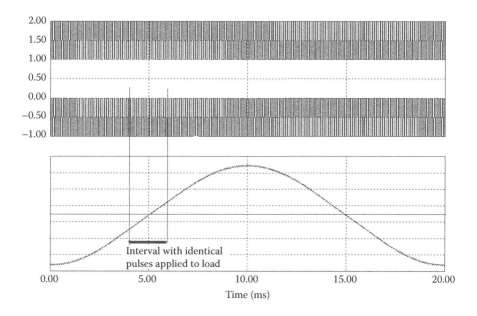

FIGURE 4.19 Effect of a very large pulse limitation interval (same converter at 10 kHz).

The pulse dropping region occurs at an especially high modulation index and the dependence of the fundamental component of the phase voltage on the modulation index is shown in Figure 4.20. Loss or gain of characteristics is more obvious at the high modulation index where generally high currents are also present.

The pulse-dropping region poses important problems for the control engineer. The system becomes nonlinear and controllability is lost during the intervals when pulses are limited or deleted. This ultimately leads to instabilities.

Many researchers have tried to find solutions to compensate for the *pulse dropping effect* or to avoid operation with narrow pulses. Compensation of the *pulse dropping effect* can extend linearity of the inverter transfer characteristics in high-modulation indices range.

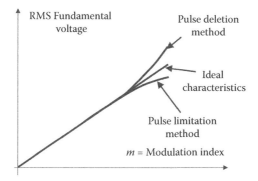

FIGURE 4.20 Fundamental voltage dependency on the modulation index.

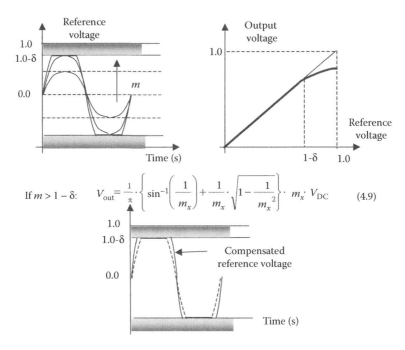

FIGURE 4.21 Compensation of the control (reference) voltage in order to achieve the desired linear dependence on the fundamental component.

Figure 4.21 shows how the real reference voltage signal is saturated at a level equal to $1 - \delta$, where δ is the accepted minimum pulse normalized. This normalization is made to the maximum available modulation index, which is 0.785 (or $\pi/4$).

To calculate its component on the fundamental frequency, we have to use the Fourier coefficient relationship:

$$A_{(1)} = \frac{1}{\omega T} \int_0^{\omega T} v(\omega t)\cos(\omega t)\mathrm{d}\omega t$$

$$\Rightarrow v_{\mathrm{ref}} = \sqrt{A_{(1)}^2 + B_{(1)}^2} \qquad (4.6)$$

$$B_{(1)} = \frac{1}{\omega T} \int_0^{\omega T} v(\omega t)\sin(\omega t)\mathrm{d}\omega t$$

Let us denote m_x as the magnitude of the voltage reference before saturation (larger than 1 for the considered normalization). The relationship between the modulation index and m_x is given by

$$m_x = \frac{m}{0.785} = \frac{4m}{\pi} \qquad (4.7)$$

The component on the fundamental frequency after saturation is calculated as

$$
v_{out}^{(1)} = \frac{1}{\pi} \left\{ \sin^{-1}\left(\frac{\pi}{4m}\right) + \left[\frac{\pi}{4m}\right]\sqrt{1 - \left[\frac{\pi}{4m}\right]^2} \right\} \frac{4m}{\pi} V_{DC}
$$

$$
= \left\{ \left[\frac{4m}{\pi}\right]\sin^{-1}\left(\frac{\pi}{4m}\right) + \sqrt{1 - \left[\frac{\pi}{4m}\right]^2} \right\} \frac{1}{2}\left[\frac{2V_{DC}}{\pi}\right]
$$

$$
\Rightarrow \frac{v_{out}^{(1)}}{[(2V_{DC}/\pi]} = \frac{1}{2}\left\{ \left[\frac{4m}{\pi}\right]\sin^{-1}\left(\frac{\pi}{4m}\right) + \sqrt{1 - \left[\frac{\pi}{4m}\right]^2} \right\} \qquad (4.8)
$$

Several methods have proven useful:

* Increase of the voltage reference in inverse proportion with the falling transfer characteristic [11] with drawbacks on the dynamic range (Figure 4.21). Observing the limitation of the real reference voltage due to limitation of the pulse width provides a mathematical relationship between the modulation index and the real output voltage. A memory look-up table of the inverse relationship helps to compensate for the truncation of the transfer characteristics in order to get the desired fundamental component on the load.
* Adding a square-wave to the modulating voltage command [11] with drawbacks on the harmonic content of the phase currents, inverter, and machine losses (Figure 4.22). The magnitude of the square-wave (*x*) is calculated based on the magnitude of desired waveform (*V*) so that, after saturation at "1," the same level of the component at the fundamental frequency can be kept.
* Other reshaping of the modulating commands with increased computational effort (Figure 4.23).

All these methods based on reshaping the reference voltage are not very suitable for motor drive applications due to the intensive computation required in real time. At each operating point, we would need to correct the reference waveform for linearity.

A completely different method consists of using *staircase PWM* or another PWM method that does not produce narrow pulses. The idea is to use a reference voltage with changes in the low frequency and sampled by the PWM generator at high frequency (Figure 4.24).

Using this method results in loss of resolution in defining the pulse width based on the sinusoidal reference waveform. This has effects on the harmonic content of the load voltage. Figure 4.25 shows what we gain by using this method while limiting

(a) Waveforms

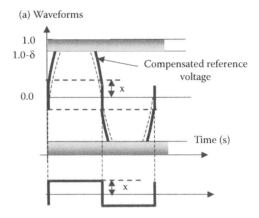

(b) Controller

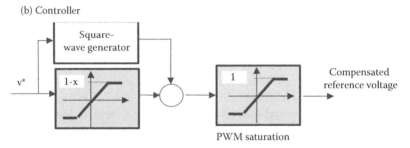

FIGURE 4.22   Reshaping the control (reference) voltage by adding square-wave functions: (a) waveforms; (b) controller.

pulses to a minimum pulse width and what we lose by using this staircase reference instead of a sinusoidal reference.

4.5.1   Avoiding Pulse Dropping by Harmonic Injection [18]

The last solution reviewed in this chapter refers to a harmonic injection able to avoid small pulse widths. It has already been shown that the presence of the third harmonic in the switching function or pole voltage does not show up in the phase voltage.

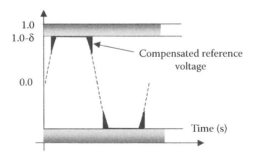

FIGURE 4.23   Reshaping the control (reference) voltage by adding square-wave functions.

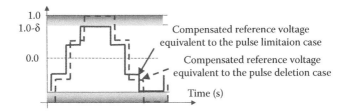

FIGURE 4.24   Principle of generating staircase PWM.

Since many years, this has been used to increase the linear transfer range of a three-phase inverter. It was originally required because only a low switching frequency was available from power semiconductor devices.

The same idea can be used along with high-frequency power semiconductor devices to avoid pulse dropping. This will modify the modulation waveform so that the pulse width is kept inside a desired bandwidth, as shown in Figure 4.26.

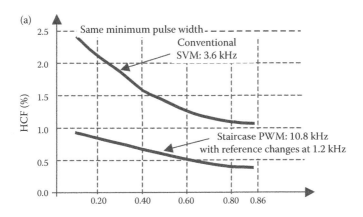

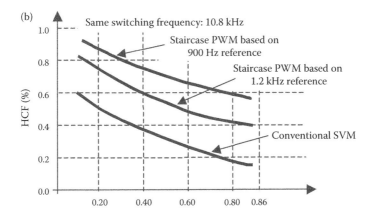

FIGURE 4.25   Different harmonic results for the load current. (Printed with permission of IEEE.)

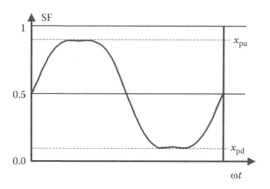

FIGURE 4.26   Limit the pulse width by third harmonic injection.

The pulse widths in the presence of the third harmonic are calculated with

$$\text{PWMA}^* = \frac{m[\sin\theta + k\sin(3\theta)] + 1}{2} \tag{4.9}$$

$$\text{PWMB}^* = \frac{m\left[\sin\left(\theta - \dfrac{2\pi}{3}\right) + k\sin(3\theta)\right] + 1}{2} \tag{4.10}$$

$$\text{PWMC}^* = \frac{m\left[\sin\left(\theta - \dfrac{4\pi}{3}\right) + k\sin(3\theta)\right] + 1}{2} \tag{4.11}$$

where θ is the angular coordinate. This is known from the methods using the third harmonic to increase the modulation range. These equations are easily implemented in a digital controller with center-aligned PWM generator. Examples of implementation will be shown in Chapter 9.

Considering that the only goal of this third harmonic injection consists of limiting the pulse width to avoid pulse deletion, the amount of the third harmonic can be calculated from the constraint of the minimum pulse.

The extreme points of any of PWMA, PWMB, or PWMC are given by

$$\frac{\partial \text{PWMA}}{\partial \theta} = 0 \tag{4.12}$$

The following solutions yield $\cos\theta = \sqrt{(9k-1)/12k}$ and $\sin\theta = \sqrt{(3k+1)/12k}$ [15] for $\theta < 90°$, and PWMA hits its maximum at this point.

Generally, we have min PW < PWMA < max PW (Figure 4.26). For specific operating switching frequency (pulse period) and minimum pulse width, the maximum accepted pulse width will result automatically. Let us denote:

$$\text{PWMA(max)} = \frac{T_s - T_{min}}{T_s} = x_{pu} \tag{4.13}$$

Replacing the solutions sin θ and cos θ in the definition of PWMA (Equation 4.9) yields the following [15]:

$$x_{pu} = \frac{m[\sin\theta + k[3\sin\theta - 4\sin^3\theta]] + 1}{2} \Leftrightarrow \frac{(2x_{pu} - 1)}{m} \tag{4.14}$$

$$= [\sin\theta + k[3\sin\theta - 4\sin^3\theta]]$$

$$\frac{(2x_{pu} - 1)}{m} = \sqrt{\frac{3k+1}{12k}}\left[1 + k\left[3 - 4\frac{3k+1}{12k}\right]\right] \Leftrightarrow \frac{(2x_{pu} - 1)}{m} \tag{4.15}$$

$$= \sqrt{\frac{3k+1}{12k}}\frac{2}{3}(3k+1)$$

$$\left[\frac{(2x_{pu} - 1)}{m}\right]^2 = \frac{3k+1}{12k}\frac{4}{9}(3k+1)^2 \Leftrightarrow \left[\frac{(2x_{pu} - 1)}{m}\right]^2 \tag{4.16}$$

$$= \frac{3k+1}{12k}\frac{4}{9}(9k^2 + 6k + 1)$$

$$\frac{12k \cdot 9}{4}\left[\frac{(2x_{pu} - 1)}{m}\right]^2 = (27k^3 + 27k^2 + 9k + 1) \tag{4.17}$$

$$\Leftrightarrow 27k^3 + 27k^2 + \left[9 - 27\left[\frac{(2x_{pu} - 1)}{m}\right]^2\right]k + 1 = 0 \tag{4.18}$$

This polynomial equation has three solutions for each set of numerical values for the period of the PWM cycle, the desired minimum pulse width, and the modulation index. The smaller solution in absolute value is preferred in order not to increase the inverter ratings.

Let us take an example for a switching frequency of 12 kHz (83.3 μs), a modulation index of 1.0, and different constraints for the minimum pulse width. Table 4.1 [18] presents solutions for Equation 4.18 in each case. It can be verified that the sum of all solutions equals –1.

For modulation indices less than unity and identical constraints, the amount of the injected third harmonic is smaller. At low-modulation index, there is no need for harmonic injection. For instance, the dependence of k on the modulation index for a desired minimum pulse of 5 μs is presented in Figure 4.27.

This kind of dependency can be stored in a look-up table. The third harmonic needs to be injected in all three reference voltages and this can seem difficult in

TABLE 4.1
Solutions of Polynomial Equation

Min PW [µs]	x_{pu}	Solution K3	Solution K1	Solution K2
0	1.0000	−1.4705	0.4089	0.0616
0.5	0.9939	−1.4580	0.3935	0.0646
1.0	0.9879	−1.4458	0.3780	0.0678
1.5	0.9819	−1.4335	0.3622	0.0713
2.0	0.9759	−1.4212	0.3459	0.0753
2.5	0.9699	−1.4089	0.3291	0.0799
3.0	0.9639	−1.3967	0.3115	0.0851
3.5	0.9579	−1.3844	0.2931	0.0913
4.0	0.9519	−1.3721	0.2733	0.0988
4.5	0.9459	−1.3598	0.2215	0.1083
5.0	0.9399	−1.3475	0.2257	0.1218
5.5	0.9339	−1.3352	0.1861	0.1491

Source: From Neacsu, D.O., Rajashekara, K., and Gunawan, F. Linear Control of PWM Inverter by Avoiding the Pulse Dropping, *IEEE Workshop on Power Electronics in Transportation*, Detroit, MI, USA, October 24–25, pp. 31–38. © (2002) With permission of IEEE.

a vector-controlled converter where the *(x,y)* coordinates are available, but not the phase references. Moreover, the phase of the injected third harmonic can be difficult to estimate even from the phase references. However, the third harmonic can be calculated from the *(x,y)* coordinates with

$$vt = mk\sin(3\theta) = mk\sin\theta(3 - 4\sin^2\theta)$$

$$= k\sqrt{v_{sx}^2 + v_{sy}^2}\ \frac{v_{sx}}{\sqrt{v_{sx}^2 + v_{sy}^2}}\left[3 - 4\frac{v_{sx}^2}{v_{sx}^2 + v_{sy}^2}\right] \tag{4.19}$$

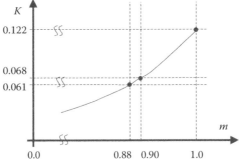

FIGURE 4.27 Optimal *k*-value for the considered example.

$$vt = kv_{sx} \frac{3v_{sy}^2 - v_{sx}^2}{v_{sx}^2 + v_{sy}^2} \tag{4.20}$$

This term should be added to each phase reference before using the center-aligned PWM generator hardware.

Using this approach within a motor drive helps to improve the phase current waveforms and removes the undesired distortion produced by the pulse width limitation algorithm. Comparative results are shown in Figure 4.28.

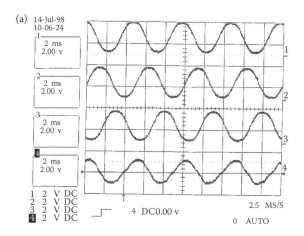

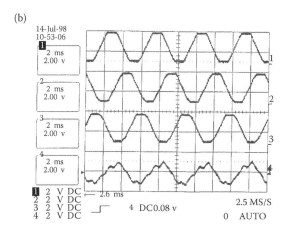

FIGURE 4.28 Vector control operation waveforms outlining the improved current waveform with the new algorithm. (a) The new algorithm for $ids = 25$ A, $iqs = 80$ A, $m = 0.84$; (b) pulse limitation method for sinusoidal PWM with $ids = 25$ A, $iqs = 80$ A, $m = 0.84$. (From Neacsu, D.O., Rajashekara, K., Gunawan, F. Linear Control of PWM Inverter by Avoiding the Pulse Dropping, *IEEE Workshop on Power Electronics in Transportation*, Detroit, MI, USA, October 24–25, pp. 31–38. © (2002) With permission of IEEE.)

4.6 LIMITS OF OPERATION

4.6.1 DEADTIME

The previous analysis of inverter operation considered ideal switchings within the inverter legs. However, Chapter 2 showed the ON/OFF transients of different semi-conductor power devices. Any change of state requires a finite interval of time that should be considered in the design of the control circuitry.

For instance, if the command for turning on the low-side IGBT comes quickly after the command for turn-off of the high-side IGBT, a short circuit occurs through both devices. To avoid this, a delay is introduced in the control of the turning-on device after the other device is turned-off. This provides enough time for the turn-off process to finish. During this deadtime interval, both switches are assumed OFF [20–27] (Figure 4.29).

When IGBTs are used and not MOSFETs, this delay needs to be longer due to the tail of the collector current. This tail is produced by the charges stored in the p–n junction of the bipolar transistor. As the MOSFET channel stops conducting, electron current ceases, and the IGBT current drops rapidly to the level of the recombination current at the inception of the tail. Different modern IGBTs are optimized to reduce this tail current by speeding-up the recombination time with different lifetime-killing techniques. As they reduce the gain of the bipolar transistor, these techniques also increase the voltage drop and turn-on losses. Accordingly, there are some limits or constraints to speeding up the turn-off process. The tail current interval cannot also be improved through the gate control. Finally, transient characteristics are worse at higher temperature.

Introducing this delay in the switching sequence modifies the width of the pulses applied on the load and their average value. Accordingly, the waveform of the load current and its harmonic spectrum are also altered.

Keeping these in mind, a proper definition of the deadtime interval is mandatory. If the deadtime interval is too short, the possible short circuit can absorb a large current and the heat produced may damage the power semiconductor. If the deadtime

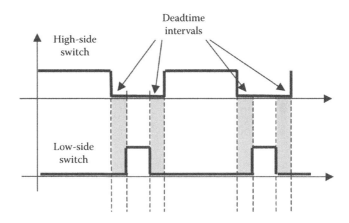

FIGURE 4.29 Deadtime.

is too long, the pulse shapes are more compromised and the current waveform more altered.

A good practical method of estimating the length of the deadtime interval is to measure the DC current at the inverter input when no load is connected. Then, the only possible current would result from short circuits due to cross-conduction. This experiment can be carried out in several steps:

- Use a very large deadtime interval and start switching on one inverter leg. A small pulse of current will be seen on the DC-side, due to the dv/dt of the pole voltage through the Miller capacitance when switching occurs. This can account for about 5% of the power semiconductor rated current.
- Use a small deadtime interval. A larger current will be noticed on the DC side. It will depend on the length of the tail current remaining active when the other switch tries to turn on.
- Increase deadtime interval from this small value and observe the different shape and value of the pulse current through the DC side wires. The value of the deadtime interval is best selected when this current approaches the initial value.

These tests are more useful when done at high temperature. Usually, deadtime is generated with a delayed turn-on event, as shown in Figure 4.30.

After the deadtime interval value has been properly selected, it becomes important to understand how much performance is lost due to the delay in the switching intervals. As both switches on the same leg are supposed to be turned-off during the deadtime interval, the load current temporarily turns-on the antiparallel diode that maintains the current circulation. If the load current is positive (it circulates from power converter to the load), the low-side diode will turn on after the high-side switch is turned off (Figure 4.31). This determines the loss in the load voltage as compared to the ideal switching pattern. The pole voltage error during the pulse depends on the width of the deadtime and DC bus voltage.

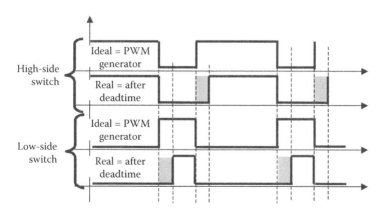

FIGURE 4.30 Practical generation of deadtime by delaying the turn-on of each power device.

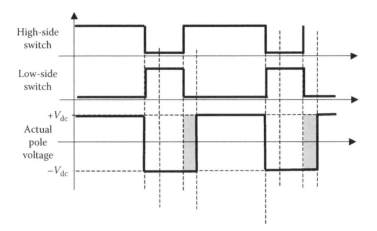

FIGURE 4.31   Effect of positive load current during deadtime.

In contrast, if the load current is negative (it circulates from load to the power converter), the load voltage gains something on its positive side (Figure 4.32). The amount of voltage error again depends on the width of the deadtime and the DC bus voltage.

On both positive and negative load currents, the amount of voltage error is constant throughout the modulation cycle and does not depend on the desired width of the pulses. The error voltage is shown in Figure 4.33. It is important to note that the amount of error voltage does not depend on the magnitude of the reference voltage (*modulation index*). The voltage error is larger at low modulation indices, such as in the case of V/Hz drives operated at low speeds.

This analysis assumes identical transients of the power switches at all moments during the cycle. This approximation is not exact as it has been proven that the transient slopes and delays of the power devices depend on the level of the current to be switched and the voltage on the DC bus. Voltage error is best analyzed by direct measurement of the load voltage.

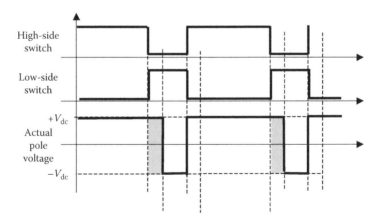

FIGURE 4.32   Effect of negative load current during deadtime.

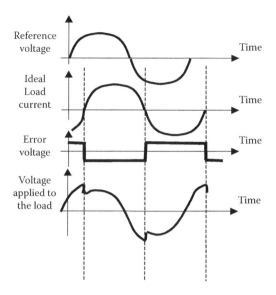

FIGURE 4.33 Voltage error is based on load current sign and load power factor (phase shift).

Applying voltage waveforms distorted by deadtime to a three-phase load would produce distortion of the current at each 60°. Each phase current will tend to be distorted twice during the modulation cycle and the effect of this distortion on the load current will obviously pass through the other two phases when the power converter from Figure 3.11a is considered.

This analysis outlines the following major drawbacks produced by deadtime:

- There is a clear difference between the voltage reference and the actual voltage applied on the load.
- Inverter output current has a 6th harmonic component that can be seen on the phase, the (d, q) components of the current, or on the torque ripple.

Several methods have been proposed for deadtime compensation:

- Voltage compensator based on hardware circuitry: A circuit is built to measure the actual load voltage and to compare it with the desired reference based on the current sign. The difference is always applied to the next pulse. There is an intrinsic phase shift in the applied voltage and this is the main problem of this method.
- Voltage feedback through a proportional-integral (PI) compensator: A real PI controller is built up from the error between the measured and reference voltages. The result is used as a compensation voltage added to the actual reference signal. This method is not suitable for fast switching converters, but for converters based on slow devices such as gate turn-off thyristors (Figure 4.34).

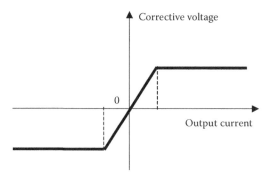

FIGURE 4.34 Modified corrective voltage for deadtime compensation.

- Pulse-based deadtime compensator: This method adjusts each pulse width on the basis of the previous pulse voltage by using symmetry of the output waveform. It is not very sensitive to fast load current changes.
- Current feedback compensator: This method adds or subtracts an amount from the desired pulse width based on the sign of the current. It works on open-loop and the goal is to approximate the deadtime effect for quasi-identical transient events. An improvement of this method is to modify the corrective amount added to the reference voltage depending on the current level. The same amount of pulse width is added for larger currents, but small currents imply a reduction of the corrective voltage.

The current feedback compensator method is the simplest to be implemented in a PWM converter working with a switching frequency of 8–20 kHz and controlled from a digital signal processor or microcontroller device. Compensation is achieved at each sampling interval with sign detection for each phase current and algebraic addition of a constant in the voltage reference waveform. This three-phase power converter is used for a motor drive application and results can be noticed at any motor speeds.

4.6.2 ZERO CURRENT CLAMPING

Another distortion of the output current in a real three-phase power converter refers to the *zero current clamping* (Figure 4.35). It is already known that a power semiconductor

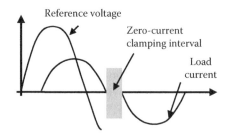

FIGURE 4.35 Zero current clamping.

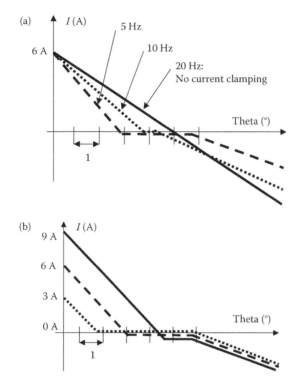

FIGURE 4.36 Zero-current clamping dependence on current frequency and magnitude.

device cannot conduct very small load currents. When current is near zero, the power device remains on the OFF state and the load current is "clamped" at zero. This state lasts for a time interval that depends on the power device used within the three-phase converter, the amount of inductive components on the circuit, the operating frequency, and the magnitude of the current. For a given power semiconductor device and circuit, the slope of the current at zero crossing defines the length of the zero current clamping interval. Assuming a quasi-sinusoidal waveform for the current, this slope (di/dt) depends on the current magnitude and the frequency ($di/dt = 2 * \pi * f * I$) (Figure 4.36). This figure does not include the effect of dead-time. The presence of a large deadtime increases the width of the zero-current clamping interval.

4.6.3 OVERMODULATION

Carrier-based modulation provides a linear characteristic up to the modulation index of 0.785. Deadtime and minimum pulse width constraints reduce further the linear region, as has been shown earlier. The interval between the maximum obtainable modulation index and the six-step operation is called *overmodulation* [28–32]. Inverter operation during this interval is characterized by nonlinearity and instabilities of the feedback controllers. This and other problems led engineers to design special PWM algorithms able to provide full inverter voltage utilization and linearity up to the six-step operation.

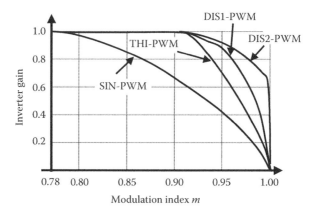

FIGURE 4.37   Inverter voltage gain.

Analysis of the saturation of any of the voltage references presented earlier is made with the same mathematics. An *inverter voltage gain* is defined and its dependence on the desired modulation index is shown in Figure 4.37. Another form for the same results is presented in Figure 4.38 and it can be considered as a zoom on the final part of the transfer characteristic between the control (desired) reference and the real modulation index calculated with respect to a six-step operation.

These PWM algorithms need very large reference signals in order to operate in the overmodulation range. For instance, conventional sinusoidal PWM requires a magnitude more than four times that of the sinusoidal reference to operate at six-step. The quickest variation is achieved for the discontinuous modulation: six-step operation is achieved for a modulator magnitude of 1.81.

4.6.3.1   Voltage Gain Linearization

The simplest solution to achieving six-step operation is to increase the magnitude of the reference voltage [33]. Another solution can be defined analogously with the

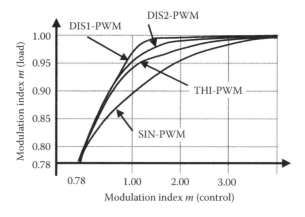

FIGURE 4.38   Inverter voltage gain.

minimum pulse compensation methods by adding square-waves to the reference. Both solutions require extensive off-line calculation and storage in a look-up table.

Compared with the uniform-sampled PWM implementation, the sine-triangle intersection method provides simpler overmodulation algorithms. Digital implementation of the uniform-sampled PWM method requires software preparation of the overmodulation operation based on calculation of the function to be added to the reference.

Even if overmodulation can be carried out with one of these methods, the performance is strongly affected during overmodulation operation and it is recommended only for dynamic or transient operation. Among the methods studied, discontinuous PWM provides better performance and less influence on the deadtime and minimum pulse. Finally, vectorial analysis of PWM algorithms and definition of space vector PWM allows easier definition of the overmodulation algorithm.

4.7   CONCLUSION

Chapter 3 has presented a simple method for generation of PWM signals based on the intersection between a sinusoidal and a triangular waveform. Problems associated with implementation and use of this principle in a three-phase inverter are detailed in this chapter. Inverter operation is unfortunately affected by minimum pulse width, deadtime selection, and maximum available voltage. Solutions for compensation for each of these effects have been shown and improvements also presented.

PROBLEMS

P4.1   The conventional PWM method produces switchings of the inverter's IGBTs at intersection of a sinusoidal waveform and a high-frequency triangular signal. The intersection moments are not easy to be expressed mathematically. How are they influencing the harmonic content of the output phase voltage?

P4.2   Use a computer program with graphical features and draw the dependence of Equation 4.8 or 4.9 for large modulation indices. Consider $\delta = 0.10$.

P4.3   For the same numerical example, calculate the magnitude of the compensation square-wave of Figure 4.22.

P4.4   How should the Staircase PWM be synchronized with the fundamental in order to obtain the most favorable value of the minimum pulse width?

P4.5   Remake all calculus shown by Equations 4.15 through 4.19 for a PWM with 20 kHz and a minimum pulse of 3 μs. How much is the resulting third harmonics at a modulation index of 1.00? Use a computer program and draw the appropriate dependence on modulation index.

P4.6   Equation 4.8 has been determined for sinusoidal modulation when the magnitude of the sinusoidal reference is exceeding 1.00 (or a modulation index of 0.785). Determine a similar relationship for the component on the fundamental frequency after saturation when a third harmonic injection is considered with a magnitude of 0.25 of fundamental.

P4.7   Consider a high power three-phase IGBT inverter operated at 10 kHz
        with a fixed deadtime of 4 μs and a fixed modulation index of 0.5 (modu-
        lation index is defined in respect with the six-step operation). Neglect
        the actual transients of voltage and currents at IGBT switchings and cal-
        culate the RMS value of the error in the phase voltages. Represent this
        voltage as percentage of the actual load phase voltage.

P4.8   Consider a sinusoidal function with a magnitude of 2.00 but limited on
        both positive and negative segments at 1.00. Calculate the RMS value of
        the fundamental of the waveform thus obtained. If the sinusoidal wave-
        form magnitude equal to 1.00 corresponds to a modulation index of 0.78,
        what value of the modulation index corresponds to the new waveform?

REFERENCES

1. Buja, G.S. and Fiorini, P. 1980. A microprocessor based quasi-continuous output controller for PWM inverters. *IEEE International Conference on Industrial Electronics*,
   Philadelphia, PA, pp. 107–111, March.
2. Buja, G.S. and Indri, G.B. 1975. Improvement of pulse width modulation techniques.
   Archiv fur Elektrotechnik 57, 281–289.
3. Houndsworth, J.A. and Grant, D.A. 1984. The use of harmonic distortion to increase
   voltage of three-phase PWM inverter. *Trans. IA* 20(5), 1224–1228.
4. Scorner, J. 1975. Bezugsspanung zur umrichterssteuerung. *ETZ-b* 27, 151–152.
5. Depenbrock, M. 1977. Pulse width control of a three-phase inverter with nonsinu-
   soidal phase voltages. Conference record, *IEEE International Semiconductor Power
   Conversion Conference*, pp. 399–403.
6. Rajashekara, K.S. and Vithayathil, J. 1982. Microprocessor based sinusoidal PWM
   inverter by DMA transfer. *IEEE Trans. IE* 29, 46–58.
7. Lucanu, M. and Neacsu, D. 1995. Optimal voltage/frequency control for space vector
   PWM three-phase inverters. *ETEP* 5(2), 115–120.
8. Legowski, S. and Trzynadlowski, A. 1993. Minimum loss vector PWM strategy for
   three-phase inverters. IEEE IECON, pp. 785–792.
9. Chung, D.W. and Sul, S.K. 1997. Minimum loss PWM strategy for three-phase PWM
   rectifier. *IEEE PESC*, pp. 1020–1026.
10. Trzynadlowski, A. and Legowski, S. 1993. Minimum loss vector PWM strategy for
    three-phase inverters. *Applied Power Electronics Conference and Exposition, APEC'93*,
    San Diego, CA, USA, pp. 785–792, 7–11 March.
11. Hava, A., Kerkman, R., and Lipo, T.A. 1998. A high-performance generalized discon-
    tinuous PWM algorithm. *IEEE Trans. IA* 34(5), 1059–1071.
12. Narayanan, G. and Ranganathan, T. 2002. Two novel synchronized bus-clamping PWM
    strategies based on space vector approach for high power devices. *IEEE Trans. PE* 17,
    84–93.
13. Lai, R.S. and Ngo, K.D.T. 1995. A PWM method for reducing of switching loss in a
    full-bridge inverter. *IEEE Trans. PE* 10, 326–332.
14. Trzynadlowski, A., Kirlin, R., and Legowski, S. 1997. Space vector PWM technique
    with minimum switching losses and a variable pulse rate. *IEEE Trans. IE* 44, 173–181.
15. Faldella, E. and Rossi, C. 1994. High efficiency PWM technique for digital control of
    DC/AC converters. *IEEE APEC*, pp. 115–121.
16. Bose, B.K. and Sunderland, A. 1983. A high performance PWM for an inverter fed drive
    system using a microcomputer. *IEEE Trans. IA* 19, 235–243.

17. Bowes, S. and Clark, P. 1995. Regular-sampled harmonic elimination PWM control of inverter drives. *IEEE Trans. PE* 10(5), 521–531.
18. Neacsu, D., Rajashekara, K., and Gunawan, F. 2002. Linear control of a PWM inverter in the pulse dropping region. *IEEE WPET*, pp. 37–47.
19. Kerkman, R., Seibel, B., and Legate, D. 1992. PWM control in the pulse dropping region. U.S. Patent 5,121,043.
20. Kimball, J. and Krein, P. 1997. Real-time optimization of deadtime for motor control inverters. *IEEE PESC*, vol. 1, pp. 597–600.
21. Murai, Y., Riyanto, A., Nakamura, H., and Matsui, K. 1992. PWM strategy for high-frequency carrier inverters eliminating current-clamps during switching dead-time. *Industry Applications Society Annual Meeting*, Houston, TX, vol. 1, pp. 317–322, 4–9 October.
22. Choi, J.W. and Yong, S.I. 1994. Inverter output voltage synthesis using novel dead-time compensation. *IEEE PESC*, pp. 100–106.
23. Choi, J.W. and Sul, S.K. 1994. New dead time compensation eliminating zero current clamping in voltage-fed PWM inverter. *IEEE IECON*, pp. 977–984.
24. Ben-Brahim, L. 1998. Analysis and compensation of dead-time effects in three-phase PWM inverters. IEEE Industrial Electronics Society, Aachen, Germany, vol. 2, no. 31, pp. 792–797, 31 August–4 September.
25. Legate, D. and Kerkman, R. 1997. Pulse-based dead-time compensator for PWM voltage inverters. *IEEE Trans. IE* 44(2), 191–197.
26. Munoz, A. and Lipo, T.A. 1999. On-line dead-time compensation technique for open-loop PWM-VSI drives. *IEEE Trans. PE* 14(4), 683–689.
27. Kameyama, T. 1999. Inverter control device. U.S. Patent 5,872,710.
28. Hava, A., Kerkman, R., and Lipo, T.A. 1998. Carrier-based PWM-VSI overmodulation strategies: Analysis, comparison and design. *IEEE Trans. PE* 13(4), 674–689.
29. Kerkman, R., Leggate, D., Seibel, B., and Rowan, T. 1993. An overmodulation strategy for PWM voltage inverters. *IEEE IECON*, pp. 1215–1221.
30. Floricau, D., Fodor, D., Ionescu, F., and Hapiot, J. 1996. Extension of the two-level SVM control strategy for the overmodulation area including the six-step mode, *Int'l Conference OPTIM*, Brasov, Romania, pp. 1471–1478.
31. Khambadkone, A. and Holtz, J. 2002. Compensated synchronous PI current controller in overmodulation range and six-step operation of space vector modulation based vector controlled drives. *IEEE Trans. IE* 49(1), 574–581.
32. Kaura, V. and Blasko, V. 1996. A method to improve linearity of a sinusoidal PWM in the overmodulation region. *IEEE PEDES*, pp. 325–331.
33. Kerkman, R., Leggate, D., Seibel, B., and Rowan, T. 1993. Inverter gain compensation for open-loop and current regulated PWM controllers. *IMACS-TCI*, pp. 7–12.

5 Vectorial PWM for Basic Three-Phase Inverters

Previous chapters have explained the need for pulse width modulation (PWM), control of three-phase converters and presented some conventional methods that generate PWM on each phase independently. Some limitations in their practical implementation have stimulated efforts to find other principles to generate PWM controls. The most remarkable is the analysis of the three-phase inverter in the complex plane by the Vector Space theory. Understanding the mathematical representation of the inverter operation in the complex plane provided a tool for the generation of a new PWM algorithm called space vector modulation (SVM). (Note the difference between a vector space, which means a space of vectors, and a space vector, which means a vector with a spatial displacement.)

SVM has become a standard for medium- and high-power switching converters in both industry and university. The last 20 years have provided a large volume of publications that fully define SVM theory. Implementation on different digital platforms has been considered and some dedicated integrated circuits have already been developed on the basis of this principle. The SVM theory initially developed for three-phase voltage-source inverters has been extended with new applications to other three-phase topologies. Such methods will be presented in later chapters.

5.1 REVIEW OF SPACE VECTOR THEORY

5.1.1 History and Evolution of the Concept

The first vectorial representation of three-phase systems was introduced by Park [1], Kron [2], and Stanley [3]. They showed the separation of effects on two axes at the operation of a three-phase electrical drive. First, Park [4] replaced the variables associated to the stator windings with variables of fictitious windings rotating with the rotor. This work can be considered the base for the well-known theory of vector control or field-orientation control for induction and synchronous drives. In the late 1930s, Stanley [3] used the same idea for induction machine drives but he replaced the rotor variables to a frame fixed on the stator. During the 1960s, the advent of thyristors led to the systematic use of the space vector theory to analyze and control three-phase electrical drives. Kovacs and Racz [4] provided mathematical treatment along with a physical description and understanding of machine drive transients even in the cases in which machines were fed through electronic converters.

Space vector-derived methods were widely used by the industry in the early 1970s and numerous books have presented this theory. Stepina [5] and Serrano-Iribarnegaray [6] suggested the use of the space phasor to analyze electrical machines.

The space phasor concept is now used mainly for current and flux measures when analyzing electrical machines.

It was again the semiconductor technology that pushed for more consideration of the vectorial analysis theory in the early 1980s. Development and intensive use of the first microprocessors in industry opened new research areas to find the most appropriate implementation algorithms for conventional issues of PWM generation, current control, or field-orientation-based control of electrical drives. Murai and Tsunehiro [7] reported in 1983 an improved PWM method derived from vectorial analysis of the operation of a three-phase inverter. A few years later, different researchers [8–11], considering all three phases in a unique vector, developed the SVM theory further to control the inverter in open loop and, later, in closed loop. The new method provided great advantages in digital implementation with the newly arrived microcontrollers.

5.1.2 Theory: Vectorial Transforms and Advantages

A three-phase system defined by $u_x(t)$, $u_y(t)$, and $u_z(t)$ can be represented uniquely by a rotating vector u_S in the complex plane:

$$\underline{u_S} = \frac{2}{3} \cdot \left[u_X(t) + \underline{a} \cdot u_Y(t) + \underline{a}^2 \cdot u_Z(t) \right] \tag{5.1}$$

where $\underline{a} = e^{j \cdot \frac{2\pi}{3}}$ and $\underline{a}^2 = e^{j \cdot \frac{4\pi}{3}}$.

All vectors in the complex plane form a space of vectors [12]. The mathematical theory of vector spaces is next employed to provide an instrument for analysis.

A base within a vector space consists of a system of vectors b (b_1, b_2, b_3) that is a unique representation of any member V of that vector space as a linear combination of vectors from b. For instance,

$$\vec{V} = b_1(\omega_j) \cdot v_d + b_2(\omega_j) \cdot v_q + b_3(\omega_j) \cdot v_0 \tag{5.2}$$

or

$$\vec{V} = b_1(\omega_j) \cdot v_d + b_2(\omega_j) \cdot v_q \tag{5.3}$$

where v_d, v_q, and v_0 are also called coordinates. If the vector space has a finite dimension, then all possible bases have the same number of elements. The dimension of a vector space refers to the number of vectors within any base. When applied to three-phase power systems or power converters, the dimension of the vector space is three, which means that any base has three elements. A base is ortho-normalized if all its vectors are unitary and any two different vectors are orthogonal. The mathematical theory of vector spaces also provides tools for making transformations between different bases of the same vector space. These transformations are unique and reversible and can have linear or rotational effects on the vectors.

All the previously reported papers in the three-phase power electronics indirectly consider the orthogonal base vectors as

$$\left[b_1(\omega_i)\right]^T = \frac{3}{2} \cdot \left[\cos\omega_i t \quad \cos\left[\omega_i t - \frac{2\pi}{3}\right] \quad \cos\left[\omega_i t - \frac{4\pi}{3}\right]\right] \tag{5.4}$$

$$\left[b_2(\omega_i)\right]^T = \frac{3}{2} \left[\sin\omega_i t \quad \sin\left[\omega_i t - \frac{2\pi}{3}\right] \quad \sin\left[\omega_i t - \frac{4\pi}{3}\right]\right] \tag{5.5}$$

$$\left[b_3(\omega_i)\right]^T = \frac{3}{2} \cdot \left[1 \quad 1 \quad 1\right] \tag{5.6}$$

The selection of this set of vectors is not unique. Coefficients and functions within each term may have other forms depending on where they are to be applied.

In our case, in a three-phase system, we would like to take advantage of sin and cos functions because we know that our phase voltages have this type of variation and we hope to separate DC quantities (constant numbers) as coordinates of Equation 5.2. The discontinuous PWM algorithms presented in Chapter 3 can enlarge the field of application of this theory. The ON-time variation is represented by the discontinuous function, depending on the phase coordinates and the conventional outputs of the vector control algorithm. Selecting base vectors characterized by that type of variation would provide a mathematical instrument for directly transforming the quasi-DC quantities of the vector control algorithm into an ON-time variation function that can be used to control the PWM generator.

Another example can refer to the selection of the coefficients in [2]. Engineers analyzing power systems take advantage of this property by defining base vectors and transform coefficients from either power conservation or magnitude conservation constraints. Thus, the coefficient (3/2) from Equations 5.4 and 5.5 is calculated to preserve the magnitude of the vector in the complex plane, but some engineers use a coefficient of *sqrt*(3/2) to preserve power through the transform.

Basically, this theory says that we can decompose any vector in the complex plane in a form shown by Equation 5.2 where the base vectors may have a form as in Equations 5.4 through 5.6. The coordinate v_d of Equation 5.2 is the result of that part of the first phase voltage that follows $\cos(\omega_i t)$ added to that part of the second phase that follows $\cos(\omega_i t - 2\pi/3)$ and to that part of the third phase that follows $\cos(\omega_i t - 4\pi/3)$.

Similarly, the coordinate v_q of Equation 5.2 is the result of that part of the first phase voltage that follows $\sin(\omega_i t)$ added to that part of the second phase that follows $\sin(\omega_i t - 2\pi/3)$ and to that part of the third phase that follows $\sin(\omega_i t - 4\pi/3)$.

Finally, the quasi-DC components in all the phases are added in v_0. Replacing Equations 5.4 through 5.6 in 5.2 yields a relationship between coordinates in different bases:

$$
\vec{V} = \begin{bmatrix} u_X \\ u_Y \\ u_Z \end{bmatrix} = \frac{3}{2} \cdot \begin{bmatrix} \cos\omega_i t \\ \cos\left(\omega_i t - \dfrac{2 \cdot \pi}{3}\right) \\ \cos\left(\omega_i t - \dfrac{4 \cdot \pi}{3}\right) \end{bmatrix} \cdot v_d + \frac{3}{2} \cdot \begin{bmatrix} \sin\omega_i t \\ \sin\left(\omega_i t - \dfrac{2 \cdot \pi}{3}\right) \\ \sin\left(\omega_i t - \dfrac{4 \cdot \pi}{3}\right) \end{bmatrix} \cdot v_q
$$

$$
+ \frac{1}{2} \cdot \begin{bmatrix} 1 \\ 1 \\ 1 \end{bmatrix} \cdot v_0 \Rightarrow \begin{bmatrix} u_X \\ u_Y \\ u_Z \end{bmatrix} = \frac{3}{2} \cdot \begin{bmatrix} \cos\omega_i t & \sin\omega_i t & \dfrac{1}{3} \\ \cos\left(\omega_i t - \dfrac{2 \cdot \pi}{3}\right) & \sin\left(\omega_i t - \dfrac{2 \cdot \pi}{3}\right) & \dfrac{1}{3} \\ \cos\left(\omega_i t - \dfrac{4 \cdot \pi}{3}\right) & \sin\left(\omega_i t - \dfrac{4 \cdot \pi}{3}\right) & \dfrac{1}{3} \end{bmatrix} \cdot \begin{bmatrix} v_d \\ v_q \\ v_0 \end{bmatrix}
$$

$$(5.7)$$

This is the core idea of a vector transformation from coordinates $u_x(t)$, $u_y(t)$, $u_z(t)$ to coordinates $u_d(t)$, $u_q(t)$, $u_0(t)$ or vice versa. This operation of transforming a three-phase system in a unique vector followed up by the transformation of the orthogonal vector coordinates in quasi-DC coordinates $(d,q,0)$ is known as Park/Clarke *transforms* for three phase systems.

5.1.2.1 Clarke Transform

First let us transform the phase measures into orthogonal coordinates with a third homopolar (or zero sequence) coordinate. Any vector in the complex plane can be decomposed in two orthogonal coordinates (U_α, U_β, U_0) and a homopolar coordinate U_0.

$$
\begin{bmatrix} U_\alpha \\ U_\beta \\ U_0 \end{bmatrix} = \frac{2}{3} \cdot \begin{bmatrix} 1 & -\dfrac{1}{2} & -\dfrac{1}{2} \\ 0 & \dfrac{\sqrt{3}}{2} & -\dfrac{\sqrt{3}}{2} \\ \dfrac{1}{2} & \dfrac{1}{2} & \dfrac{1}{2} \end{bmatrix} \cdot \begin{bmatrix} u_X \\ u_Y \\ u_Z \end{bmatrix}
$$

$$(5.8)$$

where (U_α, U_β) are forming an orthogonal two-phase system and $\underline{u_S} = U_\alpha + j \cdot U_\beta$. If the system of phase voltages is symmetrical, the homopolar term equals zero and it can miss within the previous transform. Moreover, this transform is unique as shown in Figure 5.1.

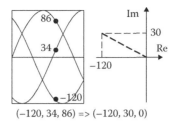

(−120, 34, 86) => (−120, 30, 0)

FIGURE 5.1   Derivation of a vector equivalent to a three-phase system.

5.1.2.2   Park Transform

This transform is not directly necessary for the presentation of space vector modulation algorithm, but is very used by electrical engineers in vector control of power converters. The two coordinates (α, β) are next transformed through a vector rotation with the rotational frequency identical to the electrical frequency.

$$
\begin{bmatrix} U_d \\ U_q \\ U_0 \end{bmatrix} = \begin{bmatrix} \cos\theta & \sin\theta & 0 \\ -\sin\theta & \cos\theta & 0 \\ 0 & 0 & 1 \end{bmatrix} \cdot \begin{bmatrix} U_\alpha \\ U_\beta \\ U_0 \end{bmatrix} \tag{5.9}
$$

Frequently, these two transforms are used in a single transform stage that coincides with the vector space theory introduced in the beginning (5.7):

$$
\begin{bmatrix} U_d \\ U_q \\ U_0 \end{bmatrix} = \frac{2}{3} \cdot \begin{bmatrix} \cos\theta & \cos\left[\theta - \dfrac{2 \cdot \pi}{3}\right] & \cos\left[\theta - \dfrac{4 \cdot \pi}{3}\right] \\ \sin\theta & \sin\left[\theta - \dfrac{2 \cdot \pi}{3}\right] & \sin\left[\theta - \dfrac{4 \cdot \pi}{3}\right] \\ \dfrac{1}{2} & \dfrac{1}{2} & \dfrac{1}{2} \end{bmatrix} \cdot \begin{bmatrix} u_X \\ u_Y \\ u_Z \end{bmatrix} \tag{5.10}
$$

Each of these transforms has an inverse that allows transformation to the phase measures from the orthogonal coordinates. The general inverse transform is given by

$$
\begin{cases} u_X(t) = \mathrm{Re}\left[\underline{u_S}\right] + u_0(t) \\ u_Y(t) = \mathrm{Re}\left[\underline{a^2} \cdot \underline{u_S}\right] + u_0(t) \\ u_Z(t) = \mathrm{Re}\left[\underline{a} \cdot \underline{u_S}\right] + u_0(t) \end{cases} \tag{5.11}
$$

where $u_0(t) = (1/3) \cdot [u_X(t) + u_Y(t) + u_Z(t)]$ represents the homopolar component. Particular forms are also available for Park and Clarke inverse transforms.

5.1.3 Application to Three-Phase Control Systems

This mathematical representation of three-phase measures treats the analysis of the three-phase system as a whole, instead of considering equations for each phase individually. Many control methods for three-phase systems have been derived from this mathematical approach.

Electrical drives (induction machine or synchronous machine drives) are controlled by the so-called field-orientation principle (Figure 5.2a). The three-phase grid interfaces or AC/DC converters (Figure 5.2b) are nowadays seen as active filtering systems controlled by the instantaneous power components (p–q) theory. All these systems use PWM algorithms in the final control stage.

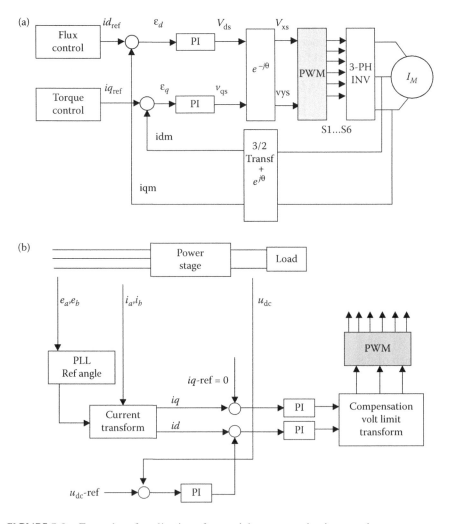

FIGURE 5.2 Examples of application of vectorial representation in control.

5.2 VECTORIAL ANALYSIS OF THE THREE-PHASE INVERTER

5.2.1 MATHEMATICAL DERIVATION OF CURRENT SPACE VECTOR TRAJECTORY IN COMPLEX PLANES FOR SIX-STEP OPERATION (WITH RESISTIVE AND RESISTIVE-INDUCTIVE LOADS)

The three-phase inverter presented in Figure 5.3 is built of six insulated gate bipolar transistors (IGBTs), but it can be made with any other power-switching device, depending on the voltage and current ratings.

Figure 5.4 presents the appropriate output voltages without PWM. This is the so-called six-step operation and it is also the simplest and oldest control method for this type of power converter. Different switching states are given in the figure with a digital code (for example: 1 0 1), showing whether the high-side IGBT is ON (for 1) or the low-side device is ON (for 0). Another possible notation uses a sign to show where the pole terminals (A, B, C) are connected.

Each state of the power converter leads to a switching vector in the complex plane. There are thus six active switching vectors $V_1,...,V_6$ equally sharing six sectors within the complex plane (Figure 5.5).

The vectorial analysis of the operation of this system is first developed for an inductive three-phase load. Each phase current waveform can be derived by the integral of each phase voltage equation.

$$\begin{cases} v_a = L_s \cdot \dfrac{di_a}{dt} \\[2mm] v_b = L_s \cdot \dfrac{di_b}{dt} \\[2mm] v_c = L_s \cdot \dfrac{di_c}{dt} \end{cases} \tag{5.12}$$

Applying transform (5.2) to these allows writing the same equations for the vector space variables.

$$\underline{v}_s = L \cdot \frac{d\underline{i}_s}{dt} \tag{5.13}$$

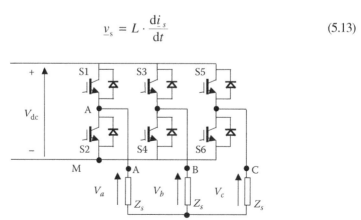

FIGURE 5.3 Basic topology for the three-phase voltage-source inverter.

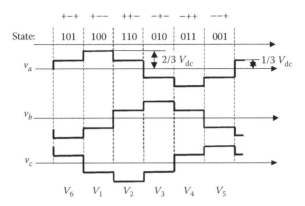

FIGURE 5.4 Output voltage waveforms and state coding for the six-step operation.

The variation slope of the phase current is doubled when the phase voltage gets doubled. The maximum value of the phase current is denoted here by I_M. During the time interval $[t_1, t_2]$, the output voltage vector is V_1 and the phase voltages are $2/3 \, V_{dc}$, $-1/3 \, V_{dc}$, and $-1/3 \, V_{dc}$.

Choosing the time origin in $t_1(t_1 = 0)$, the load current and voltage expressions can be mathematically expressed as linear variations during the interval (t_1, t_2). From Figure 5.6, it yields:

$$\begin{cases} i_a(t) = -0.5 \cdot I_M + \dfrac{2 \cdot V_{dc}}{3 \cdot L} \cdot t \\[2mm] i_b(t) = -0.5 \cdot I_M - \dfrac{V_{dc}}{3 \cdot L} \cdot t \\[2mm] i_c(t) = I_M - \dfrac{V_{dc}}{3 \cdot L} \cdot t \end{cases} \quad (5.14)$$

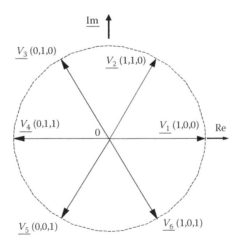

FIGURE 5.5 Switching vectors corresponding to the six-step operation of the inverter.

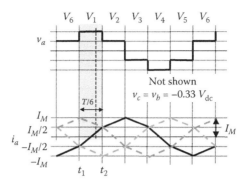

FIGURE 5.6 Output phase current and voltage waveforms: (a) calculation; (b) full-cycle trajectory.

Applying definition (5.1) to the space vectors associated to the current and voltage waveforms yields:

$$\underline{v}(t) = \frac{2}{3} \cdot \left[v_a(t) + \underline{a} \cdot v_b(t) + \underline{a}^2 \cdot v_c(t) \right] \tag{5.15}$$

$$\underline{i}(t) = \frac{2}{3} \cdot \left[i_a(t) + \underline{a} \cdot i_b(t) + \underline{a}^2 \cdot i_c(t) \right] \tag{5.16}$$

where

$$\underline{a} = -\frac{1}{2} + j \cdot \frac{\sqrt{3}}{2} \quad \underline{a}^2 = -\frac{1}{2} - j \cdot \frac{\sqrt{3}}{2} \tag{5.17}$$

leading to

$$\underline{v}(t) = \frac{2}{3} \cdot V_d + j \cdot 0 \tag{5.18}$$

and

$$\underline{i}(t) = \left[-\frac{I_M}{2} + \frac{2}{3} \cdot \frac{V_d}{L} \cdot t \right] + j \cdot \left[-\frac{\sqrt{3}}{2} \cdot I_M \right] \tag{5.19}$$

It can be seen from Equation 5.18 that the magnitude of each voltage switching vector is $(2/3)V_{dc}$. The tip of the current space vector has a linear variation in the complex plane, with the *Real* part varying linearly from $-0.5I_M$ to $0.5I_M$ during the time interval $[t_1, t_2]$ of duration $T/6$ (Figure 5.7).

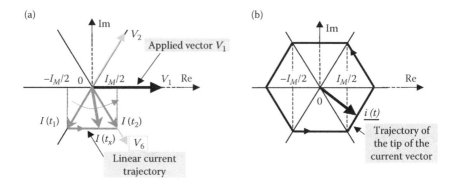

FIGURE 5.7   Current vector trajectory in the complex plane for a pure L load.

This trajectory is oriented so that the current space vector is quasi-perpendicular on the voltage space vector and this was expected from the integral form of the inductive load equation. The vector projection on the *Real* axis represents the value of the first phase current. Considering the variation of the phase current during the interval $[t_1, t_2]$ allows determination of the maximum value of the current (I_M). It yields:

$$\frac{2}{3} \cdot V_{dc} = L \cdot \frac{0.5 \cdot I_M - (-0.5) \cdot I_M}{\dfrac{T}{6}} \tag{5.20}$$

$$I_M = \frac{2}{3L} \cdot \frac{T}{6} \cdot V_{dc} \tag{5.21}$$

The voltage space vector is identical with the switching vector V_2 at the next interval. The current space vector moves between the position along the vectors V_6 (at t_2) and V_1 (at next interval, t_3). Similar calculation proves the linearity of this trajectory. Extending the same reasoning for all six possible voltage-switching vectors defines the trajectory of the tip of the current vector in the complex plane.

The resulting trajectory is a hexagon oriented along the voltage-switching vectors.

The projection on the real axis is at any time equal to the current on the first phase. Currents or voltages on the other two phases can be graphically derived by the projection on two fictitious axes at 120° and 240°, respectively. This vectorial analysis provides information about all the currents and voltages in the system. Moreover, because of the 60° symmetry of the operation, it is enough to limit the vectorial analysis to a 60° sector.

Extending this analysis to the general case of an R–L load (Figure 5.8), consider the vectorial equation for the load circuit

$$\underline{V}_s = \underline{i}_s \cdot R_s + L_s \cdot \frac{d}{dt} \underline{i}_s \tag{5.22}$$

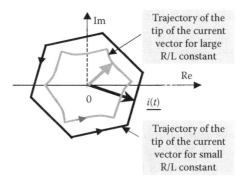

FIGURE 5.8 Current vector trajectory in the complex plane for an *R–L* load.

with the generic solution

$$\underline{i}_s(t) = \frac{1}{R} \cdot \underline{V}_s + \underline{C} \cdot e^{\frac{-t}{\tau}} \qquad (5.23)$$

where $\underline{C}$ is a complex constant and $\tau = (R/L)$ represents the time constant of the load. In other words, for each switching vector applied to the load, the current trajectory is a portion of exponential. To better define such trajectory, let us suppose the initial value as being equal to $i(0) = I$ and the final value after $T/6$ is

$$\underline{i}\left[\frac{2 \cdot \pi}{6 \cdot \omega}\right] = I \cdot e^{\frac{-j\pi}{3}}$$

These conditions help defining the constant C and initial value I:

$$\left.\begin{array}{l} t = 0 \Rightarrow \underline{I} = \dfrac{1}{R} \cdot \underline{V}_s + \underline{C} \\[2mm] t = \dfrac{\pi}{3} \Rightarrow \underline{I} \cdot e^{j\frac{\pi}{3}} = \dfrac{1}{R} \cdot \underline{V}_s + \underline{C} \cdot e^{\frac{-\pi}{3 \cdot \omega \cdot \tau}} \end{array}\right\} \Rightarrow$$

$$\Rightarrow \left\{\begin{array}{l} \underline{C} = \underline{I} - \dfrac{1}{R} \cdot \underline{V}_s \\[2mm] \underline{I} \cdot e^{j\frac{\pi}{3}} = \dfrac{1}{R} \cdot \underline{V}_s + \underline{C} \cdot e^{\frac{-\pi}{3 \cdot \omega \cdot \tau}} \end{array}\right. \Rightarrow \left\{\begin{array}{l} \underline{C} = \underline{I} - \dfrac{1}{R} \cdot \underline{V}_s \\[2mm] \underline{I} \cdot e^{j\frac{\pi}{3}} = \dfrac{1}{R} \cdot \underline{V}_s + \left(\underline{I} - \dfrac{1}{R} \cdot \underline{V}_s\right) \cdot e^{\frac{-\pi}{3 \cdot \omega \cdot \tau}} \end{array}\right. \Rightarrow$$

$$\Rightarrow \left\{\begin{array}{l} \underline{C} = \underline{I} - \dfrac{1}{R} \cdot \underline{V}_s \\[2mm] \underline{I} \cdot \left(e^{j\frac{\pi}{3}} - e^{\frac{-\pi}{3 \cdot \omega \cdot \tau}}\right) = \dfrac{1}{R} \cdot \underline{V}_s \cdot \left(1 - e^{\frac{-\pi}{3 \cdot \omega \cdot \tau}}\right) \end{array}\right. \Rightarrow \left\{\begin{array}{l} \underline{C} = \dfrac{1}{R} \cdot \underline{V}_s \cdot \left(\dfrac{1 - e^{\frac{-\pi}{3 \cdot \omega \cdot \tau}}}{e^{j\frac{\pi}{3}} - e^{\frac{-\pi}{3 \cdot \omega \cdot \tau}}} - 1\right) \\[4mm] \underline{I} = \dfrac{1}{R} \cdot \underline{V}_s \cdot \dfrac{1 - e^{\frac{-\pi}{3 \cdot \omega \cdot \tau}}}{e^{j\frac{\pi}{3}} - e^{\frac{-\pi}{3 \cdot \omega \cdot \tau}}} \end{array}\right.$$

$$(5.24)$$

Replacing complex constant C in Equation 5.23 yields:

$$\underline{i}(t) = \frac{1}{R} \cdot \underline{V}_s \cdot \left[1 - e^{\frac{-t}{\tau}} \right] + \underline{I} \cdot e^{\frac{-t}{\tau}} \qquad (5.25)$$

5.2.2 DEFINITION OF FLUX OF A (VOLTAGE) VECTOR AND IDEAL FLUX TRAJECTORY

Three-phase power-switching inverters are often used to supply a machine drive or a strongly inductive load. The leakage inductance of the machine and the inertia of the mechanical system account for low-pass filtering of the harmonics provided by the discontinuous power flow at switching. If a voltage drop across the resistance and leakage inductance of the stator windings is neglected, the flux linkage in the machine is approximately equal to the time integral of the impressed voltage. The flux vector yields

$$\underline{\lambda} = \int \underline{V}_s dt \qquad (5.26)$$

A very similar definition can be made for AC/DC applications, where the boost input inductance accounts for low-pass filtering of the voltage pulses resulting from the PWM operation of the power stage.

All these applications will work properly if the magnitude of the flux linkage is kept constant, which denotes a circular trajectory of the flux. As V_s occupies different discrete positions in the complex plane, its time integral leads to a polygon close to a circle, as anticipated by Figure 5.9. What was not explained previously is the existence of zero vectors in the flux trajectory at control with PWM. A zero vector, also called homopolar vector, is achieved when all power devices are connected to the same DC bus terminal, positive or negative. Any of the PWM methods explained previously in Chapter 3 uses zero-vector states during operation. Voltages applied on each phase of the load equal zero in this case and the integral of these voltages show no displacement of the flux trajectory. A direct consequence of this is the possibility of using zero vectors to regulate the speed of the flux trajectory (Figure 5.10).

A real trajectory of the flux achieved by PWM operation presents both radial and angular errors. The radial errors are variations along the radius of the circular locus, whereas the angular errors are variations from a constant rotational speed. In a motor

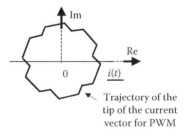

FIGURE 5.9 Current space vector trajectory for a simplified PWM case.

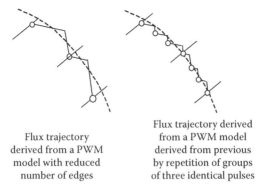

Flux trajectory
derived from a PWM
model with reduced
number of edges

Flux trajectory derived
from a PWM model
derived from previous
by repetition of groups
of three identical pulses

FIGURE 5.10 Flux trajectory improved by increasing the switching frequency even if the same shape of pulses is used.

drive, any error in the flux trajectory has a direct influence on the torque ripple. The difference between the reference λ_0 having a circular locus in the complex plane and the actual trajectory λ produced by a PWM inverter causes the torque oscillations.

The dependence of the torque pulsation on radial and angular components of the flux error has been analyzed and presented in [13–17], and it has been stated that the torque oscillation is more sensitive to the angular error than the radial error. The angular error can be reduced by operating with a smoother rotational speed that can be achieved when employing more zero states on the flux locus. The radial errors can be reduced with an optimal polygonal flux locus having all edges staying as close to the desired circular locus.

As the angular error is more important than the radial error, the torque ripple is lower when a higher carrier frequency (more zero vector states) is employed, even though this splits a polygonal flux locus to a reduced number of edges. A limited switching frequency is desired for completely different reasons, such as reduction of the switching loss.

5.3 SVM THEORY: DERIVATION OF TIME INTERVALS ASSOCIATED TO ACTIVE AND ZERO STATES BY AVERAGING

SVM was developed in [7–11] and the importance of this method has been outlined in many publications [18–21]. The three-phase inverter presented in Figure 5.3 produces a symmetrical three-phase system of voltages on the load. If the magnitude of the phase voltages is V_s, this is equivalent to generating a circular locus with a constant radius equaling the same magnitude. Such an ideal locus cannot be achieved with a switching power converter that leads to six discrete positions of only the voltage space vector. Each desired position on the circular locus can be synthesized only through an average relationship between two neighboring active vectors and zero vectors

$$\underline{V_S} \cdot T_S = \underline{V_a} \cdot t_a + \underline{V_b} \cdot t_b + \underline{V_0} \cdot t_0 \tag{5.27}$$

where T_s is the sampling period of the given circular locus (as, usually, the switching frequency is not equal to the carrier frequency, it is preferable to call T_s the sampling period) and t_a, t_b are the time intervals allocated to the neighboring vectors V_a, V_b. The averaging process is a result of the low pass filtering action of the load on the voltage pulses. Zero vectors are necessary to keep the sampling interval constant and they are calculated by

$$t_0 = T_s - t_b - t_a \tag{5.28}$$

In order to calculate the time intervals associated with a desired position of the voltage vector in the complex plane, the symmetry after a $60°$ interval is first noticed. This opens up the possibility that the analysis can be limited to a generalized sector of $60°$, repeated six times. To simplify the calculation, such a sector is considered superimposed with the first sector of the complex plane.

Decomposition of Equation 5.27 on both *Real* (*Re* in all figures) and Imaginary (*Im* in all figures) axes yields:

$$\begin{cases} \text{Re} : V_s \cdot \cos\alpha \cdot T_S = \left(\frac{2}{3} \cdot V_{dc} \right) \cdot t_a + \left(\frac{2}{3} \cdot V_{dc} \right) \cdot \frac{1}{2} \cdot t_b \\[3mm] \text{Im} : V_s \cdot \sin\alpha \cdot T_S = \left(\frac{2}{3} \cdot V_{dc} \right) \cdot \frac{\sqrt{3}}{2} \cdot t_b \end{cases} \tag{5.29}$$

Time intervals involved in the PWM generation are calculated by (Figure 5.14):

$$\begin{cases} t_a = \frac{3V_s}{2V_{dc}} \cdot T_S \cdot \left(\cos\alpha - \frac{1}{\sqrt{3}} \sin\alpha \right) = \frac{\sqrt{3} \cdot V_s}{V_{dc}} \cdot T_S \cdot \sin\left[\frac{\pi}{3} - \alpha \right] \\[3mm] t_b = \frac{\sqrt{3} \cdot V_s}{V_{dc}} \cdot T_S \cdot \sin\alpha \\[3mm] t_0 = T_S - t_a - t_b \end{cases} \tag{5.30}$$

Calculation of the time intervals associated with each active state is only the first part of the generation of a PWM algorithm. Determination of the switching sequence is the second stage. Modern microcontrollers achieve both functions and take advantage of special expressions for the time intervals associated with the conduction of each switch. These allow the implementation of the SVM method within the center aligned PWM hardware. Calculation of the pulse widths is based on a memory lookup table for the sine function within a $60°$ interval or of the $\sin\alpha$ and $\sin(60 - \alpha)$ functions within a $30°$ interval. Alternative solutions are based on real-time interpolation of a minimized look-up table. This interpolation can be carried out by fuzzy logic as well [22,23]. Chapter 8 will present hardware solutions available in the market to implement SVM.

Observing Equation 5.30 and Figure 5.11, one can derive the maximum modulation. It will correspond to the circular locus with the maximum radius and it equals:

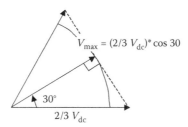

$V_{max} = (2/3\ V_{dc})^* \cos 30$

$30°$

$2/3\ V_{dc}$

FIGURE 5.11 Definition of the maximum modulation index. (From Neacsu, D.O. Tutorial presented at IECON'01: The 27th Annual Conference of the IEEE Industrial Electronics Society 2001, IEEE Paper 0-7803-7108-9/01. © (2001) With permission of IEEE.)

$$V_{max} = \frac{2}{3} \cdot V_{dc} \cdot \cos\frac{\pi}{6} = \frac{1}{\sqrt{3}} \cdot V_{dc} = 0.577 \cdot V_{dc} = 0.866 \cdot \left[\frac{2 \cdot V_{dc}}{3}\right]$$

$$m_{max} \xrightarrow{\text{definition}} \frac{V_{max}}{V_{magn\_six\_step}} \Rightarrow m_{max} = 0.866$$

(5.31)

This value is 15% higher than the maximum modulation index from the sinusoidal modulation.

The sampling period T_s is shown in Equation 5.30. Note the nomenclature difference between sampling interval/sampling frequency and switching interval/switching frequency. A new PWM method can be derived by changing the sampling frequency while using the same equations. The frequency modulation is the result of adjusting T_s during the fundamental or cycle period [10,17]. Several methods have been developed by using the sampling frequency as a degree of freedom in optimization after one or more criteria: low harmonics reduction, harmonic current factor (HCF) reduction, and so on, but all redefine T_s as

$$T_s = \frac{1}{N \cdot f} \cdot (1 + \delta \cdot \cos(6 \cdot \alpha))$$

(5.32)

with N being number of intervals over the fundamental period (at frequency f).

Conventional PWM is characterized by $\delta = 0$, while the optimized methods require T_s modulated so that it gets shorter periods at 0 and 60° but lengthier at 30° within each 60° sector. These ensure minimum angular errors, current distortion factor at low harmonics, and minimum torque oscillations. Random SVM represents a special type of frequency modulation.

Finally, note that the averaging principle used here does not provide any requirement on zero vector generation during t_0. Moreover, the sequence of the active and zero vectors within the sampling period is not unique and these degrees of freedom make the difference between space vector methods. The most well-known alternatives will be analyzed later in this chapter.

5.4 ADAPTIVE SVM: DC RIPPLE COMPENSATION

The presence of the DC voltage (V_{DC}) in Equation 5.30 compensates for the ripple of the DC bus voltage [24] (Figure 5.12). This ripple can be produced with insufficient filtering of the input rectifier power stage in a back-to-back converter structure.

Measuring the DC voltage at each sampling interval or with a larger sampling period will be appropriately compensated by Equation 5.30 for the effect of this ripple in the output voltage.

Figure 5.12 shows how the time intervals t_a, t_b are affected by the variation of the DC bus voltage in order to preserve the same harmonics on the load. The drawback of this method is that it reduces the maximum available voltage at the inverter output. The maximum available output voltage is achieved when the DC bus has the lower value within the ripple and the inverter operates at maximum modulation index of 0.866.

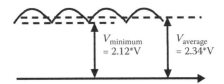

FIGURE 5.12 Reduction of the maximum output voltage by unfiltered DC bus with 6-step rectifier (*V* represents the grid side RMS voltage).

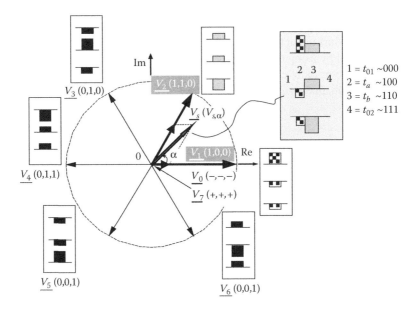

FIGURE 5.13 Generation of voltage space vector by SVM. (From Neacsu, D.O. Tutorial presented at IECON'01: The 27th Annual Conference of the IEEE Industrial Electronics Society 2001, IEEE Paper 0-7803-7108-9/01. © (2001) With permission of IEEE.)

5.5   LINK TO VECTOR CONTROL: DIFFERENT FORMS AND EXPRESSIONS OF TIME INTERVAL EQUATIONS IN (*D, Q*) COORDINATE SYSTEM

This SVM algorithm calculates the time intervals associated with each state based on the polar coordinates (V_s, α) of the desired space vector position. This is not very convenient for vector control methods for drives and for active filtering.

The result of the vector control methods is given in the coordinates (v_x, v_y) and these can also be expressed depending upon the polar coordinates (Equations 5.33 through 5.38) [25].

Sector 1

$$v_x = V_s \cdot \cos\alpha$$
$$v_y = V_s \cdot \sin\alpha$$

(5.33)

Sector 2

$$v_x = V_s \cdot \cos\left[\alpha + \frac{\pi}{3}\right]$$
$$v_y = V_s \cdot \sin\left[\alpha + \frac{\pi}{3}\right]$$

(5.34)

Sector 3

$$v_x = -V_s \cdot \cos\left[\frac{\pi}{3} - \alpha\right]$$
$$v_y = V_s \cdot \sin\left[\frac{\pi}{3} - \alpha\right]$$

(5.35)

Sector 4

$$v_x = -V_s \cdot \cos\alpha$$
$$v_y = -V_s \cdot \sin\alpha$$

(5.36)

Sector 5

$$v_x = -V_s \cdot \cos\left[\alpha + \frac{\pi}{3}\right]$$
$$v_y = -V_s \cdot \sin\left[\alpha + \frac{\pi}{3}\right]$$

(5.37)

Sector 6

$$v_x = V_s \cdot \cos\left[\frac{\pi}{3} - \alpha\right]$$

$$v_y = -V_s \cdot \sin\left[\frac{\pi}{3} - \alpha\right]$$

(5.38)

For the first sector, replacing Equation 5.33 in 5.30 yields:

$$\begin{cases} t_a = \dfrac{3}{2V_{dc}} \cdot T_S \cdot \left(v_x - \dfrac{1}{\sqrt{3}} v_y\right) \\[3mm] t_b = \dfrac{\sqrt{3}}{V_{dc}} \cdot T_S \cdot v_y \\[3mm] t_0 = T_S - t_a - t_b \end{cases}$$

(5.39)

For the second sector:

$$\begin{cases} v_x = V_s \cdot \left[\cos\alpha \cdot \dfrac{1}{2} - \sin\alpha \cdot \dfrac{\sqrt{3}}{2}\right] \\[3mm] v_y = V_s \cdot \left[\sin\alpha \cdot \dfrac{1}{2} + \cos\alpha \cdot \dfrac{\sqrt{3}}{2}\right] \end{cases} \Rightarrow \begin{cases} \sin\alpha = \dfrac{1}{2} \cdot \dfrac{v_y}{V_s} - \dfrac{\sqrt{3}}{2} \cdot \dfrac{v_x}{V_s} \\[3mm] \cos\alpha = \dfrac{1}{2} \cdot \dfrac{v_x}{V_s} + \dfrac{\sqrt{3}}{2} \cdot \dfrac{v_y}{V_s} \end{cases}$$

$$\Rightarrow \begin{cases} t_a = \dfrac{3}{2 \cdot V_{dc}} \cdot T_S \cdot \left(v_x + \dfrac{1}{\sqrt{3}} \cdot v_y\right) \\[3mm] t_b = \dfrac{\sqrt{3}}{V_{dc}} \cdot T_S \cdot \left(\dfrac{1}{2} \cdot v_y - \dfrac{\sqrt{3}}{2} \cdot v_x\right) \end{cases}$$

(5.40)

Similar calculus leads to the following results [25]:

Sector 1

$$t_a = \frac{3 \cdot T_s}{2 \cdot V_{dc}} \cdot \left(v_x - \frac{1}{\sqrt{3}} \cdot v_y\right)$$

$$t_b = \frac{\sqrt{3} \cdot T_s}{V_{dc}} \cdot v_y$$

(5.41)

Sector 2

$$t_a = \frac{3 \cdot T_s}{2 \cdot V_{dc}} \cdot \left(v_x + \frac{1}{\sqrt{3}} \cdot v_y \right)$$

$$t_b = \frac{\sqrt{3} \cdot T_s}{V_{dc}} \cdot \left(\frac{1}{2} \cdot v_y - \frac{\sqrt{3}}{2} \cdot v_x \right)$$

(5.42)

Sector 3

$$t_a = \frac{T_s \cdot \sqrt{3}}{V_{dc}} \cdot \left(v_y \right)$$

$$t_b = \frac{\sqrt{3} \cdot T_s}{2 \cdot V_{dc}} \cdot \left(-v_y - \sqrt{3} \cdot v_x \right)$$

(5.43)

Sector 4

$$t_a = \frac{3 \cdot T_s}{2 \cdot V_{dc}} \cdot \left(-v_x + \frac{1}{\sqrt{3}} \cdot v_y \right)$$

$$t_b = \frac{-\sqrt{3} \cdot T_s}{V_{dc}} \cdot v_y$$

(5.44)

Sector 5

$$t_a = \frac{\sqrt{3} \cdot T_s}{2 \cdot V_{dc}} \cdot \left(-v_x - 2 \cdot v_y \right)$$

$$t_b = \frac{\sqrt{3} \cdot T_s}{2 \cdot V_{dc}} \cdot \left(-v_y + v_x \right)$$

(5.45)

Sector 6

$$t_a = \frac{3 \cdot T_s}{2 \cdot V_{dc}} \cdot \left(v_x - \frac{2 \cdot \sqrt{3}}{3} \cdot v_y \right)$$

$$t_b = \frac{\sqrt{3} \cdot T_s}{2 \cdot V_{dc}} \cdot \left(v_y - \sqrt{3} \cdot v_x \right)$$

(5.46)

The time intervals allocated to the zero vectors remain

$$t_0 = T_S - t_a - t_b$$

(5.47)

Another mathematical form for the transformation of (v_x, v_y) coordinates into time intervals is provided by Equation 5.48 and is derived from (x, y)-type orthogonal coordinates of the active switching vectors denoted here as (V_x^2, V_y^2) and (V_x^1, V_y^1). The relationship for the time intervals yields:

$$t_1 = \frac{\left[V_y^2 \cdot v_x - V_x^2 \cdot v_y\right] \cdot T_s}{V_x^1 \cdot V_y^2 - V_x^2 \cdot V_y^1}$$

$$t_2 = \frac{\left[V_x^1 \cdot v_y - V_y^1 \cdot v_x\right] \cdot T_s}{V_x^1 \cdot V_y^2 - V_x^2 \cdot V_y^1}$$

(5.48)

Any of these forms expressing the time intervals represents only the first step for PWM implementation. Next, the time intervals corresponding to the inverter switching states need to be converted into a logical switching sequence applied to the gates of IGBTs. In order to implement this step, the switching reference function (also called modulation function) is defined.

5.6 DEFINITION OF SWITCHING REFERENCE FUNCTION

Once we know the time intervals for each state, we need to establish the sequence of these intervals. If we try to directly use the previous relationships, the definition of the switching sequence can be implemented only by the software. It is more advantageous to define a function called the switching reference function that represents the duty ratio of each inverter leg or the conduction time normalized to the sampling period for a given switch; this is a mathematical function varying between 0 and 1 centered around 0.5. This is also called the *modulation function*.

The switching reference function can be derived by algebraic operations between the time intervals previously calculated. For instance, on the first sector from Figure 5.13:

- The instant when S1 goes ON equals t01 from the beginning of the sampling period, and the value of the switching reference function for S1 will equal $[t_{01}/T_s]$.
- The instant when S3 goes ON equals a delay of $t_{01} + t_a$ from the beginning of the sampling period, and the value of the switching reference function for S3 will equal $[t_{01} + t_a/T_s]$.

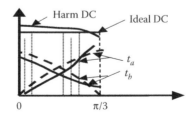

FIGURE 5.14 Space vector modulation with adaptive compensation.

- The instant when S5 goes ON equals a delay of $t_{01} + t_a + t_b$ from the beginning of the sampling period, and the value of the switching reference function for S5 will equal $[t_{01} + t_a/T_s]$.

The switching reference function is calculated at the sampling instants, and the real variation is a staircase waveform that can be interpolated as shown in Figure 5.14. The switching reference function has the same meaning as the reference used in sine-triangle comparison-based PWM methods. It is, therefore, simpler to use the mathematical definition of a continuous function not influenced by the sampling frequency rate. At this point, it is also easier to understand why the repetitive frequency used in PWM generation is called sampling frequency. It simply means to sample the switching reference function.

After successive derivations from Equations 5.38 through 5.43, one can calculate the switching reference function for all sectors. This is shown in Figure 5.15 for the case of regular SVM. The difference between this function and a pure cosine function is given by the following equation with a rich content in the third harmonic:

$$PW_{harm} = \begin{cases} 0.500 \cdot \sin\alpha, & \text{for } \alpha \in (0,60) \\ 0.866 \cdot \cos\alpha, & \text{for } \alpha \in (60,120) \\ -0.500 \cdot \sin\alpha, & \text{for } \alpha \in (120,180) \end{cases} \tag{5.49}$$

The third harmonic is present in the switching reference function but it is not present in the output phase or line voltages. It only represents the average of the $A-M$ voltage from Figure 5.16.

$$v_{AB} = v_{AM} - v_{BM} \Rightarrow 3\text{rd harmonic vanishes} \tag{5.50}$$

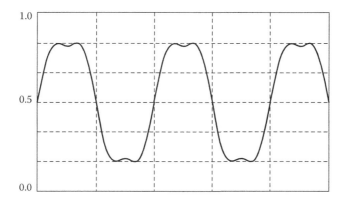

FIGURE 5.15 Switching reference function for the regular SVM. (From Neacsu, D.O. Tutorial presented at IECON'01: The 27th Annual Conference of the IEEE Industrial Electronics Society 2001, IEEE Paper 0-7803-7108-9/01. © (2001) With permission of IEEE.)

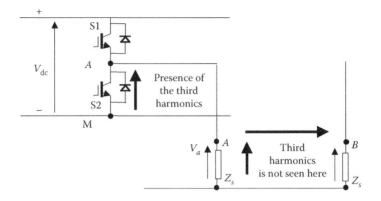

FIGURE 5.16   Third harmonics in the SVM generation.

The previous chapter showed the role and meaning of the injection of the third harmonics in the sinusoidal modulation. The amount of this harmonic injection has been shown to vary depending on the optimization objective:

- Maximization of the fundamental content (third harmonic with 0.25 of the magnitude of the sinusoidal reference)
- Minimization of the total content in harmonics (third harmonic with 0.16 of the magnitude of the sinusoidal reference)

The third harmonic in the SVM method is between the two values previously calculated at about 0.22. References [26–30] analyze the equivalence between regular, sampled sine-triangle methods and SVM and conclude that both methods are analogous and that conventional digital center-aligned PWM devices can be used for the implementation of the SVM algorithm.

5.7   DEFINITION OF SWITCHING SEQUENCE

5.7.1   CONTINUOUS REFERENCE FUNCTION: DIFFERENT METHODS

The number of switchings can be reduced with a special arrangement of the switching sequence so that only one switching on each inverter leg occurs during the transition from one state to the next. When using both available zero states from Figure 5.13, one zero state will start the sequence and the other will end it. For instance, the switching state sequence has to be _ _ _0 1 2 7 2 1 0_ _ _. The only remaining degree of freedom consists in the amount t_0 shared between the vectors V_0 and V_7.

The averaging theory used for SVM calculation does not define a way of sharing t_0 among the possible zero vector states. In the original method, t_0 was shared equally between the two zero vectors. But this is not the most optimal partition solution. The same low-sampling frequency algorithm is analyzed but with different sharing of the zero states. All the sectors and bisectrix symmetries are considered as well as the alternation of the zero-state vectors. Results for the sharing ratio are shown in

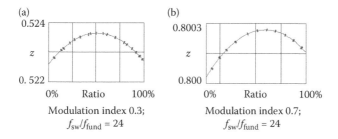

FIGURE 5.17 Effect of different sharing of the zero-states interval in content of fundamental (z). (From Neacsu, D.O. Tutorial presented at IECON'01: The 27th Annual Conference of the IEEE Industrial Electronics Society 2001, IEEE Paper 0-7803-7108-9/01. © (2001) With permission of IEEE.)

Figures 5.17 and 5.18 for a low number (equal to 24) of sampling intervals in the fundamental [31,32]. At larger sampling frequencies, the differences are smaller.

These results also show that high-performance SVM systems can be improved further by a different placement of the zero states within the sampling interval.

The two extreme situations for the sharing of the zero-state intervals are:

- *Method D-I-H* (direct-inverse-half) (Figure 5.19): Equal sharing of the zero vector intervals at each sampling interval ($t_0 = t_7$) [10,11] is shown in Figure 5.19 with a phase voltage waveform. Observe the trajectory of the tip of the current vector derived from this sequence.
- *Method D-I-O* (direct-inverse-one) (Figure 5.20): Use of only a zero-vector interval within each sampling period (e.g., $t_0 = 0$, $t_7 = T_s - t_a - t_b$).
- Both methods determine three switching on each sampling periods. Method D-I-O is often employed at high sampling frequencies, whereas in low frequencies, it produces even harmonics in the output phase voltage, as the waveform symmetries are no longer taken into account. The spectral differences between the voltages carried out by either Method D-I-H or Method D-I-O are small if the sampling frequency is large enough (Figure 5.21).

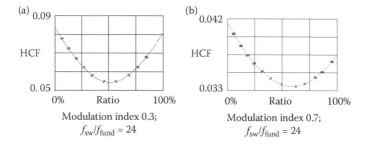

FIGURE 5.18 Effect of different sharing of the zero-states interval in HCF. (From Neacsu, D.O. 2001. Tutorial presented at IECON'01: The 27th Annual Conference of the IEEE Industrial Electronics Society 2001, IEEE Paper 0-7803-7108-9/01. © (2001) With permission of IEEE.)

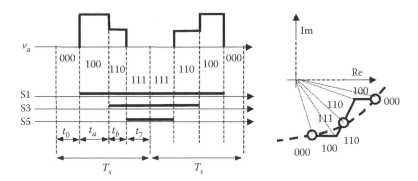

FIGURE 5.19 Pulse generation with method D-I-H.

- *Simple direct SVM* or *S-D-H method*: The simplest way to synthesize the output voltage vector is to turn on all the switches connected to the same DC link potential at the beginning of the switching cycle (sampling period) and to turn them off sequentially during the sampling interval. The zero-vector interval splits between V_0 and V_7 ($t_0 = t_7$). The drawback consists in switching all three inverter legs somewhere in the middle of the sampling interval in order to change from V_0 and V_7 (for instance, the switch sequence: ... 0127–0127–0127...). This method is similar to the usual sine-triangle comparison-based PWM (Figure 5.22).
- *Symmetrically generated SVM*: This modulation scheme is based on a symmetrical sequence within each sampling period. The phase voltages and switching signals are similar to the D-I-H method but the direct-inverse sequence is now inside the same sampling period (Figure 5.23). This method is similar to the center-aligned PWM devices and can be directly implemented in the existing PWM IC working on this basis.

Let us note here a very important difference between the *Symmetrically Generated SVM* and any of the other PWM algorithms described above. The use of the direct-inverse sequence produces harmonics at half the actual pulse frequency. This means

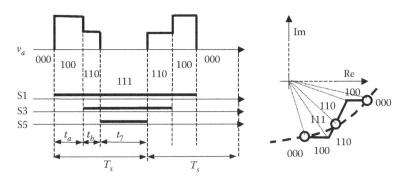

FIGURE 5.20 Pulse generation with method D-I-O.

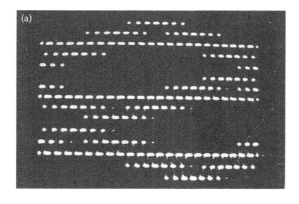

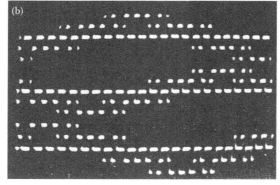

FIGURE 5.21   Three-phase voltage waveforms for (a) direct-inverse and (b) direct sequences at a low sampling frequency.

that the spectra are equivalent with that of center-aligned PWM which is measuring the sampling interval from the zero state to another zero state of the same type. Figure 5.24a presents spectra for a normal PWM generation, while Figure 5.24b for a PWM with direct-inverse sequence of pulses. We can thus see the appearance of higher harmonics at half the pulse frequency. It means that there is a price to pay for

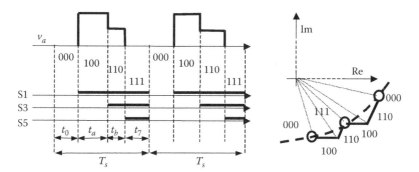

FIGURE 5.22   Pulse generation with Method S-D-H.

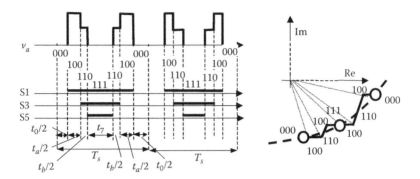

FIGURE 5.23 Pulse generation with Method S-G-S.

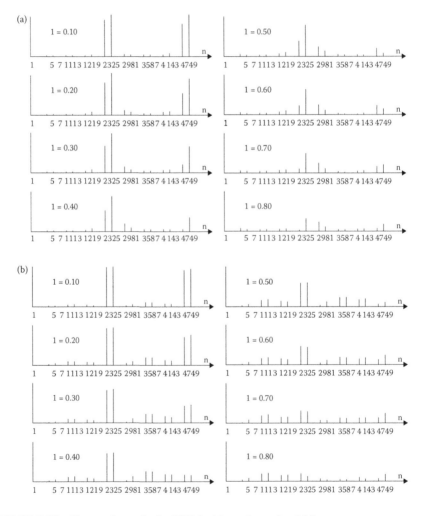

FIGURE 5.24 Harmonic results for PWM with a pulse ratio of 24.

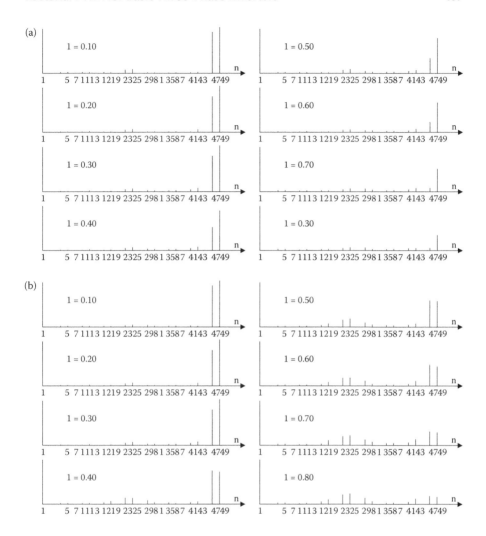

FIGURE 5.25   Spectra for PWM with a pulse ratio of 48.

the reduced count of switching instants. Figure 5.25 repeats the test for PWM with a pulse ratio of 48. It is therefore concluded that this type of PWM is better referred to as center-aligned PWM rather than quoting the two pulse sequence of direct-inverse.

5.7.2 DISCONTINUOUS REFERENCE FUNCTION FOR REDUCED SWITCHING LOSS

The previous chapter including the regular sampled PWM algorithms, demonstrated the advantages of discontinuous algorithms to reduce the inverter switching loss [33–37]. The same idea is now used with the SVM algorithm as support [38–42].

The averaging theory used for SVM calculation does not define the sequence of the switching states. Any possible sequence of states will satisfy the average relationship if the time intervals are calculated correctly. Any two neighboring vectors are different by only one switching in an inverter leg. Therefore, always select the zero vector that does not change the status of that zero vector. For instance, in the first sector, shown in Figure 5.13, the first leg does not switch when active vectors are changed (from 1 0 0 to 1 1 0 or vice versa). Selection of the homopolar vector is not unique; each time two zero vectors can be used. Always selecting 1 1 1 as the zero vector means that the first leg will not switch for the whole first sector (for instance, sequences ...–1 1 1–1 1 0–1 0 0–1 1 1–1 1 0–1 0 0–1 1 1–...). But this solution is not unique. In the first sector, the third leg does not switch when the active vectors are changed. This time, the lower switch will be ON during both active states (from 1 0 0 to 1 1 0 or vice versa). Always selecting 0 0 0 as the zero vector means that the third leg will not switch for the whole first sector (for instance, sequences ...–0 0 0–1 0 0–1 1 0–0 0 0–1 0 0–1 1 0–0 0 0...). The advantage of using any one of these solutions is the reduction in switching losses. There are two solutions to minimize the number of switching processes within each sector of the complex plane. This raises the question whether to change or not the selection of the zero vector and the whole optimized sequence when passing to the next sector.

If we do not ever change the selection of the zero vector and always select the zero vector as 0 0 0 or 1 1 1, we get one of the following solutions:

- Method DZ0: The null vector is always fixed as [0 0 0]
- Method DZ1: The null vector is always fixed as [1 1 1]

The switching reference function can be calculated as shown previously in Section 5.5 and it leads to the waveforms shown in Figure 5.26, with large intervals that have no switchings for each of the six IGBTs within the inverter. These functions are also not linear.

As can be seen, the switch that is not switched is either always on the low-side (for the selection of the 0 0 0 zero vector) or on the high-side (for the selection of the 1 1 1 zero vector). Three switches are ON for extended periods of time and this may create a problem in inverter bridges that use isolation circuits such as the bootstrap or charge pumps for their gate drivers. That is why these methods are not used in the industry.

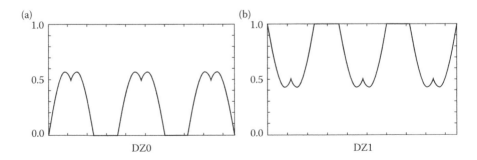

FIGURE 5.26 Switching reference function calculated for PWM generation with Methods DZ0 and DZ1.

To obtain a symmetrical stress from the power devices, the degree of freedom consists in alternating the zero vector at each 60° interval. As each phase can be kept unswitched for 120° consecutively, there is a degree of freedom in selecting where exactly the zero vector can be changed for sequences of 60°. For instance, the first leg can be kept unswitched from −60° to 60° (notice vectors 1 1 0 and 1 0 0 in Sector 1; 1 0 0 and 1 0 1 in Sector 6), but a 60° interval has to be chosen within this. Switching loss is approximately proportional to the magnitude of the current being switched and it would be better to avoid switching the inverter leg with the highest instantaneous current. In other words, the no-switching 60° interval can be selected at exactly the peak of the current.

How is the switching sequence selected? It can be based on four states in each sampling interval: a zero vector in the beginning, the first active state, the second active state, and another zero state at the end. Applying the principles explained earlier to such a switching sequence leads to results shown in Figure 5.27. The same phase voltages are obtained as in the case of *Method S-D-H*.

However, such a sequence violates the idea of having only one switch at a time and leads to four switches over a sampling interval. Another solution that respects this constraint and also takes advantage of the no-switch rule for 60° consecutively is different from conventional SVM. It uses a direct-inverse sequence:

$$|V_{A1} - V_{A2} - V_Z \mid V_Z - V_{A2} - V_{A1} \mid V_{A1} - V_{A2} - V_Z| \text{ (Figure 5.28)}$$

This selection of the switching sequence allows the maximum reduction of the number of switches.

There are several solutions accepted by the industry:

- *Method DD1*: The intervals of no switches coincide with the plane sectors. A null vector [1 1 1] is assigned for sectors 1, 3, and 5 and a null vector [0 0 0] is assigned for sectors 2, 4, and 6. The 60° interval without switches occurs right after the peak of the phase voltage. Generally, the current lags behind the phase voltage and the peak of the current fundamental settles in the next 60° after the peak voltage (Figure 5.29).

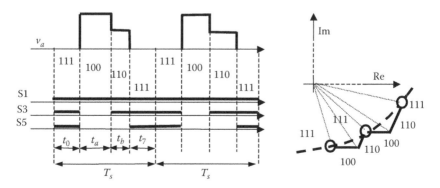

FIGURE 5.27 Switching sequence with four states taking advantage of no-switch rule on the first leg.

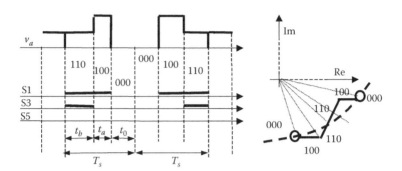

FIGURE 5.28 Modified space vector modulation.

- *Method DD2*: The 60° no-switch interval is spread equally around the peak of the fundamental of voltage. This method is very useful to reduce switching losses in grid-related applications, in which the power factor is unity and the peak of current is close to the peak of voltage (Figure 5.29).
- *Method DD3*: The 60° interval without switches can be spread equally around the measured peak of the phase current. Despite the difficulty of sensing the phase currents, this solution seems attractive, as it tracks the

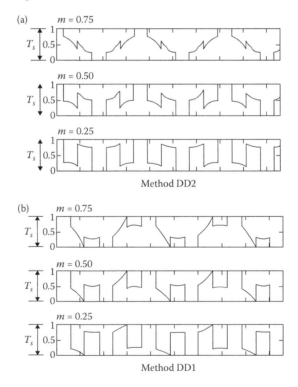

FIGURE 5.29 Pulse generation with methods DD1 and DD2 for different modulation indices.

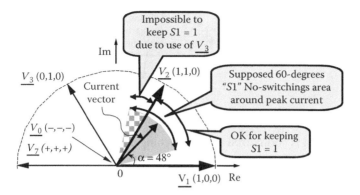

FIGURE 5.30 Vectorial discussion around the DD3 method. (From Neacsu, D.O. 2001. Tutorial presented at IECON'01: The 27th Annual Conference of the IEEE Industrial Electronics Society 2001, IEEE Paper 0-7803-7108-9/01. With permission.)

> maximum loss reduction. However, if the phase current and voltage out of phase are greater than 30°, the 60° no-switch region will overcome the vector sector (Figure 5.30).

If the voltage vector is, for instance, V_1 and the current vector is at 48°, equally spreading the no-switch region for the first inverter leg around the current vector leads to an area between 18° and 78°. For $\alpha > 60°$, keeping the first phase ON (1) is no longer possible as the Sector 2 is characterized by switches in the first phase due to the use of V_3 (0 1 0). There need not be any switches in the other two phases, but the other currents are less able to control the complexity, thereby compromising the merits of the method.

Let us analyze Figure 5.31 to understand how much power loss can be saved by using discontinuous PWM algorithms. Switching power loss for the discontinuous

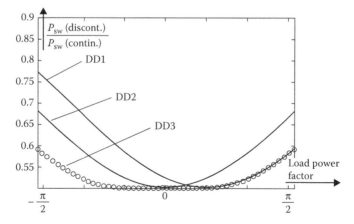

FIGURE 5.31 Switching loss for discontinuous PWM normalized to the switching loss for continuous PWM algorithm versus load power angle. (From Neacsu, D.O. The 27th Annual Conference of the IEEE Industrial Electronics Society 2001, IEEE Paper 0-7803-7108-9/01. © (2001) Printed with permission of IEEE.)

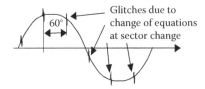

FIGURE 5.32 Output phase current when applying discontinuous PWM algorithms.

PWM algorithm is normalized to the switching power loss for the continuous PWM algorithm and shown in respect with the load power angle. The results are based on a computer-based analysis, considering the level of the load current at switching instant and the number of switching processes. The best case leads to 50% savings in switching loss [42–44].

This reduction in switching loss does not exactly come free. There is a small drawback as the discontinuous PWM methods can introduce oscillations around the points where the sector is changed. This is due to the different set of equations used within each sector to calculate the time intervals. The effects are clearer at low output fundamental frequencies and they result in increased loss in the load and may introduce instabilities of the feedback control system (Figure 5.32).

5.8 COMPARISON BETWEEN DIFFERENT VECTORIAL PWM

5.8.1 LOSS PERFORMANCE

The difference between the conduction loss among the SVM techniques is less than 3% of the total loss. Switching performance is presented in Table 5.1.

5.8.2 COMPARISON OF TOTAL HARMONIC DISTORTION/HCF

As shown in the previous chapter, the most important performance index for a power converter refers to the harmonic content in the output (input) current. This can be expressed by

TABLE 5.1
Switching Performance

Method	No. of Switchings within T_s	THD$_v$	No. of Switching States
Direct–Inverse (DIH)	3		4
Direct Inverse (DIO)	3		3
Simple direct (SDH)	6		4
Symmetric gen. (SGS)	6	Least	7
Direct–direct/000 (DZ0)	4		3
Direct–direct/111 (DZ1)	4		3
Direct–direct/sect (DD1)	4/2		3
Direct–direct/peak (DD2)	4/2		3
Direct–direct/mes (DD3)	4/2		3

$$HCF(\%) = \frac{100}{V_{(1)}} \cdot \sqrt{\sum_{n=5}^{\infty} \left[\frac{V_{(n)}}{n} \right]^2} \tag{5.51}$$

and represents a normalization of the harmonic content to the fundamental.

These PWM algorithms have different sequences during the sampling interval and the number of switching processes is different. For this reason, the definition of the switching frequency differs from method to method. A proper comparison must consider the same switching frequency even if reached by different means.

The best harmonic content is achieved for reduced loss algorithm operated at high-modulation indices. In low-modulation indices, the best harmonic content can be achieved from the conventional SVM. The approximate threshold for this kind of comparison lies at a modulation index of about 0.6.

Many other researchers or engineers present the current harmonic by normalization to the harmonics obtained by a six-step operation at the power stage. Figure 5.33 shows the difference between harmonics with an inverter obtained by a six-step operation and a PWM inverter. The operation with six pulses introduces larger harmonics at low frequency (Figure 5.34).

By normalization to this waveform, the monotony of the plots changes. Results for the most well-known methods derived from SVM are shown in Figure 5.35.

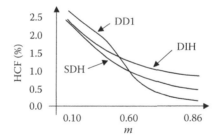

FIGURE 5.33   HCF for several modulation methods calculated for $f_1 = 50$ Hz, $f_{sw} = 3.6$ kHz.

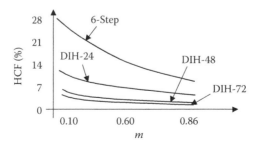

FIGURE 5.34   HCF for different number of sampling intervals over the fundamental.

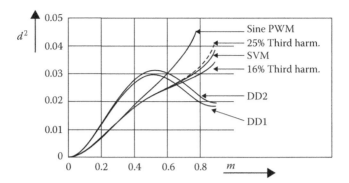

FIGURE 5.35   Distortion factor versus modulation index. (From Neacsu, D.O. Tutorial presented at IECON'01: The 27th Annual Conference of the IEEE Industrial Electronics Society 2001, IEEE Paper 0-7803-7108-9/01. © (2001) With permission of IEEE.)

5.9   OVERMODULATION FOR SVM

It has been shown that the maximum modulation index of the regular SVM algorithm is achieved when the circular trajectory with the largest radius becomes tangential to the external hexagon formed with the switching vectors (see Figure 5. 11). The linearity of the PWM ends at this point.

Many applications, however, require more voltage up to the six-step mode. Operation between these two limits is called *overmodulation* and was presented in Chapter 4. Let us see here how an *overmodulation* algorithm can be defined under SVM.

First, the operation is characterized by a trajectory of the averaged space vector along a circle of radius $m > 0.866$ as long as the circle arc is located within the hexagon and along the hexagon sides in the remaining portions. The same equations 5.30 are used for PWM generation when the tip of the averaged space vector is on the circular trajectory. For points that lie on the sides of the hexagon, there is no zero state ($t_0 = 0$) and the following equations are used for the active states:

$$t_a = T_s \cdot \frac{\sqrt{3} \cdot \cos \alpha - \sin \alpha}{\sqrt{3} \cdot \cos \alpha + \sin \alpha}$$

$$t_b = T_s \cdot \frac{2 \cdot \sin \alpha}{\sqrt{3} \cdot \cos \alpha + \sin \alpha} \tag{5.52}$$

At $m = 0.952$, the trajectory shows at the hexagon, and no portions move into the circular locus. In order to advance toward the six-step operation, *Operation Mode II* is defined. This time, the velocity of the tip of the averaged space vector is controlled. The higher the modulation index is expected to be, the higher is the velocity in the center of each hexagon side and the lower in the corners. This mode converges smoothly into a six-step operation when the trajectory is limited to six discrete positions in the complex plane. The structure of a pulse over each sampling interval is composed of only two active states. Zero states are never used.

Both operation modes are characterized by nonlinear transfer characteristics with addition of harmonics that jeopardize the harmonic performance. This is natural because the six-step operation has been already shown to have important harmonics.

5.10 VOLT-PER-HERTZ CONTROL OF PWM INVERTERS

Chapter 4 introduced the *volt-per-hertz* control associated with PWM techniques. It has also been shown that industry uses PWM with different numbers of pulses for different fundamental frequency intervals (Figure 5.36). A larger frequency ratio is therefore considered for low frequencies, whereas the power converter operates at high fundamental frequencies with a smaller frequency ratio. Extension of this method to SVM control is presented next.

SVM control of a gate turn-off (GTO) inverter with a switching frequency below 960 Hz is shown in Figure 5.36 [45]. The switching frequency is limited to GTO devices and has wide enough pulses to compensate for the voltage drop across the stator resistance. The following operation modes are accordingly obtained:

- At higher frequencies: An operation mode with a constant voltage is preferred in order not to exceed the induction machine nominal value. The voltage is kept constant between 40 and 80 Hz with an optimal SVM having 24 pulses.
- At lower frequencies: A PWM method is used with V_1/f = constant and a number of pulses on different frequency intervals: 24 pulses,... 20–40 Hz; 48 pulses,... 10–20 Hz; 96 pulses,... 5–10 Hz; 192 pulses,... <5 Hz.

The specifics of SVM for each operation mode are presented in Figure 5.37.

5.10.1 LOW-FREQUENCY OPERATION MODE

The relation V_1/f = constant can be written as V_1QT_s = constant, where Q is the number of pulses over the fundamental period (also called *frequency ratio*). For each frequency interval shown in Figure 5.36, the magnitude and RMS of the output phase voltage take values within a finite interval and the vectorial representation of this operation brings us inside a circular corona (for example, between V_{s1} and V_{s2} in Figure 5.38).

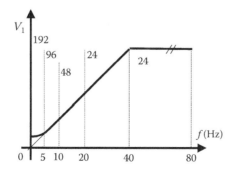

FIGURE 5.36 Volt-per-Hertz control.

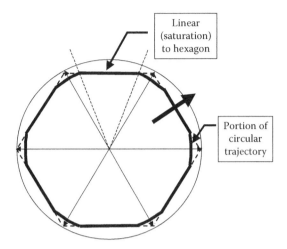

FIGURE 5.37 Operation modes.

The time allocated to the switching vectors neighboring a desired space vector position has been defined previously by Equation 5.30. These equations can be seen as $t_a, t_b = $ constant $V_s^* T_s$ for a given direction on the complex plane. The similarity between these forms of Equation 5.30 and the volt-per-hertz control condition ($V_1 QT_s =$ constant) implies a possible modification of the RMS value of the output voltage by adjusting the sampling interval T_s while both t_a, t_b are kept constant (Figure 5.39).

In order to decrease the frequency, the sampling period is increased by enlarging the zero-states' intervals. When the period doubles, the transition to the next domain occurs. This domain will be characterized by twice the number of pulses achieved by splitting the previous sampling period into two intervals. The use of the same time constants in all low-frequency operation modes is therefore possible, improving microcontroller and memory look-up table use. Figure 5.40 presents all possible positions of the tip of the space vector with the magnitude equaling the condition for transition from one domain to the next. Intermediary magnitude values are easily generated on the basis of the same time constants by enlarging the zero state intervals. This figure is also important as it shows the positions that need to have time

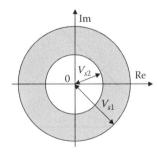

FIGURE 5.38 Circular corona.

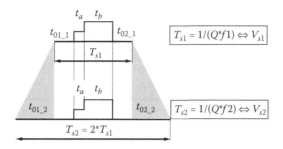

FIGURE 5.39 Pulse width changes over a circular corona in the complex plane.

constants stored in memory. All the other positions will use the same time constants (t_a, t_b). Considering all other symmetries of a three-phase system limits the total number of memory locations to

$$N_{ML} = 2 \cdot \left[1 + \frac{m}{12}\right] \quad (5.53)$$

5.10.2 High-Frequency Operation Mode

The operation of the induction machine in high fundamental-frequency range implies limitation of the voltage [46]. The same PWM pattern is supposed to be used for all these frequencies. Due to the nature of the high-frequency operation, the number of pulses in this pattern is limited (24 pulses in our example) and the harmonic results are not very good. It is a good practice to define an optimal PWM algorithm on the basis of harmonic elimination or global harmonic minimization for this domain.

Reducing this pattern to a 30° sector when considering the symmetries within a three-phase system shows a position on the real axis—one intermediary position and the last one on the 30° direction (for the pattern with 24 pulses over the fundamental cycle). There is a degree of freedom in neglecting Equation 5.30 and in defining a

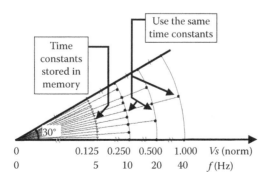

FIGURE 5.40 Operation within a 30° sector.

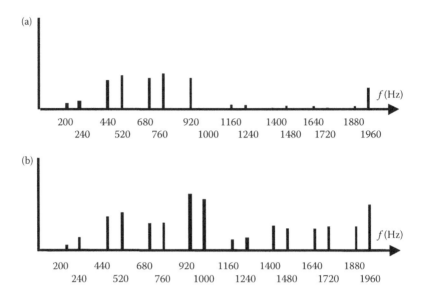

FIGURE 5.41 Spectra for different PWM patterns at 40 Hz. (a) Optimal PWM at fundamental frequency of 40 Hz. (b) Optimal SVM at fundamental frequency of 40 Hz.

new set of optimal time constants instead of calculating t_a, t_b on the basis of the optimization constraints. For instance, one can set the fundamental voltage at a desired value and limit both lowest harmonics as $\min(V_5^2 + V_7^2)$. This implies replacing the time constants calculated with Equation 5.30 by

$$t_a = 0.793 \cdot T_s$$
$$t_b = 0.152 \cdot T_s$$
$$\tag{5.54}$$

These values yield both fifth and seventh harmonics below 3% of the fundamental of the output voltage. These harmonic results are shown in Figure 5.41.

5.11 IMPROVING THE TRANSIENT RESPONSE IN HIGH-SPEED CONVERTERS

Numerous motor drives require operation at high speed, with fast transient response. This is achieved in control by applying as much voltage as possible within the actuator. Modern embedded controllers implement controller saturation and overmodulation operation in order to allow this increase in voltage.

In order to quantify this operation at highest voltage available, let us consider the induction machine drive with rotor in short-circuit as a symmetrical three-phase system with isolated neutral [47]. The stator voltage equation in complex variables yields:

$$\vec{v}_{s\lambda} = R_s \cdot \vec{i}_{s\lambda} + \frac{d\vec{\Psi}_{s\lambda}}{dt} + j \cdot \omega_\lambda \cdot \vec{\Psi}_{s\lambda} \tag{5.55}$$

with

$$\overrightarrow{\Psi}_{s\lambda} = L_s \cdot \vec{i}_{s\lambda} + L_m \cdot \vec{i}_{r\lambda} = L_{\sigma s} \cdot \vec{i}_{s\lambda} + L_m \cdot \left(\vec{i}_{r\lambda} + \vec{i}_{s\lambda} \right) \tag{5.56}$$

$$\vec{v}_{s\lambda} = R_s \cdot \vec{i}_{s\lambda} + \frac{\mathrm{d}}{\mathrm{d}t} \left(L_{\sigma s} \cdot \vec{i}_{s\lambda} + L_m \cdot \left(\vec{i}_{r\lambda} + \vec{i}_{s\lambda} \right) \right) + j \cdot \omega_\lambda \cdot \left(L_{\sigma s} \cdot \vec{i}_{s\lambda} + L_m \cdot \left(\vec{i}_{r\lambda} + \vec{i}_{s\lambda} \right) \right)$$
$$\tag{5.57}$$

We can consider a reference system with $\omega_\lambda = 0$.

$$\vec{v}_{s\lambda} = R_s \cdot \vec{i}_{s\lambda} + \frac{\mathrm{d}}{\mathrm{d}t} \left(L_{\sigma s} \cdot \vec{i}_{s\lambda} + L_m \cdot \left(\vec{i}_{r\lambda} + \vec{i}_{s\lambda} \right) \right)$$
$$= \left[R_s \cdot \vec{i}_{s\lambda} + \frac{\mathrm{d}}{\mathrm{d}t} L_{\sigma s} \cdot \vec{i}_{s\lambda} \right] + \frac{\mathrm{d}}{\mathrm{d}t} \left[L_m \cdot \left(\vec{i}_{r\lambda} + \vec{i}_{s\lambda} \right) \right]$$
$$= \left[R_s \cdot \vec{i}_{s\lambda} + \frac{\mathrm{d}}{\mathrm{d}t} L_{\sigma s} \cdot \vec{i}_{s\lambda} \right] + \vec{e} \tag{5.58}$$

The apparent power transferred between the inverter (actuator) and the electrical machine yields:

$$S = (3/2) \cdot \left(\vec{v}_{s\lambda} \cdot \vec{i}_{s\lambda}{}^* \right) = \left(\frac{3}{2} \right) \cdot \left[R_s \cdot \vec{i}_{s\lambda} + \frac{\mathrm{d}}{\mathrm{d}t} \left(L_{\sigma s} \cdot \vec{i}_{s\lambda} \right) + \vec{e} \right] \cdot \vec{i}_{s\lambda}{}^*$$
$$= \left(\frac{3}{2} \right) \cdot \left[\left(R_s \cdot \vec{i}_{s\lambda} + L_{\sigma s} \cdot \frac{\mathrm{d}}{\mathrm{d}t} \left(\vec{i}_{s\lambda} \right) \right) \cdot \vec{i}_{s\lambda}{}^* + \vec{e} \cdot \vec{i}_{s\lambda}{}^* \right] \tag{5.59}$$

In order to separate active and reactive (imaginary) power components, let us denote the phase shift between the stator voltage v_s and the back-emf voltage e with φ. This yields into

$$P_1 = \frac{V_{s\lambda} \cdot E}{\left(\dfrac{3}{2} \right) \cdot \left| \left(R_s + j \cdot \omega \cdot L_{\sigma s} \right) \right|} \cdot \sin \phi \tag{5.60}$$

$$Q_1 = \frac{V_{s\lambda} \cdot E \cdot \cos \phi - E \cdot E}{\left(\dfrac{3}{2} \right) \cdot \left| \left(R_s + j \cdot \omega \cdot L_{\sigma s} \right) \right|} \tag{5.61}$$

The active power P is depending strongly on the angle φ, while the reactive power Q is mostly influenced by the magnitude of stator voltage ($V_{s\lambda}$).

During a fast change in the speed reference, the load torque needs to increase quickly to a maximum for an increase of speed and to decrease quickly to a

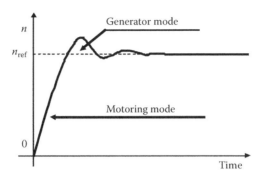

FIGURE 5.42   Explanation of the power transfer during transients.

minimum for a decrease in speed. During such a transient of the torque, there is a temporary transfer of reactive (imaginary) power between the mechanical system and the inverter (Figure 5.42). When the voltage applied ($V_{s\lambda}$) is higher, the amount of reactive energy (derived from power Q) transferred within the system yields lower. That is the inverter absorbs quicker the reactive energy from load. This yields into a faster transient.

This means that inverterized motor drives with higher voltage availability (or higher DC bus voltage) allow a faster transient of the drive. For a given voltage setup, the available applied voltage can be increased by control within certain limits. If such control is implemented within the embedded software, its speed depends on the sampling interval and the software arrangement. This usually yields in a fairly slow response. An alternative method [47,48] proposes to derive the additional voltage from the operation of the inverter with only two switches during the zero state intervals.

Observing the operation of the power converter with sinusoidal currents demonstrates that it is enough to control actively the current within two phases of the drive while the third current results from the current summation to zero within the neutral.

The proposed PWM algorithm is developed on the frame of a conventional *space vector modulation* algorithm while taking advantage of this property of currents. This controller uses two switches turned-on during a zero-state.

Depending on the value and direction of the third current, the uncontrolled inverter leg can go to the same DC potential with the other two producing a true zero voltage vector or can go on the other DC potential producing an active vector on the load. The phase current automatically determines the value of the voltage applied to the third phase of the load during the zero state.

Figure 5.43 shows the definition of the operation modes. Each state is defined with the conduction state (0 or 1) for each switch. The first set of three numbers correspond to the high- side switches, and the next set of three numbers correspond to the low-side switches. The uncontrolled switches are also shown in the figure as 0, while the other switches can take any value (0 or 1) according to the PWM algorithm. For more detail, all possible states are shown organized in Table 5.2. The same data is shown in Table 5.3 based on the phase of the current vector for the case of a PWM using a single zero state vector on each sampling interval

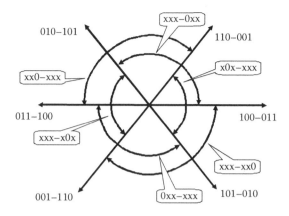

FIGURE 5.43 Definition of the inverter switching states.

α = angular coordinate of desired voltage vector within the 60° sector.
ϕ = phase shift between voltage vector and current vector.

For instance, the additional voltage ΔV contributed when operation takes place within the first sector (0° to 60°) is superimposed to the expected voltage vector. It can be calculated by averaging formulas.

If $\phi < 90 \Rightarrow I_a > 0 \Rightarrow \Delta V = 0$

If $90 < \phi < 150$

For $\alpha < \phi - 90 \Rightarrow I_a < 0 \Rightarrow \quad V = V\underline{1}$ and

$$\Delta \vec{V} = \frac{t_{000}}{T_s} \cdot \vec{V}_1 = \frac{1}{2} \cdot \left[1 - m \cdot \cos\left(\alpha - \frac{\pi}{6} \right) \right] \cdot \vec{V}_1 \tag{5.62}$$

TABLE 5.2
Voltage Vectors Generated During the Zero Vector States Based on Phase Current's Polarity

Sector 1: 0–60°	Zero vector 0-0-0	000-011	$I\underline{a} \geq \underline{0} \Rightarrow V = \underline{0}$	$I\underline{a} \leq \underline{0} \Rightarrow V = V_1$
	Zero vector 1-1-1	101-000	$I\underline{b} \geq \underline{0} \Rightarrow V = V_6$	$I\underline{b} \leq \underline{0} \Rightarrow V = \underline{0}$
Sector 2: 60–120°	Zero vector 0-0-0	000-011	$I\underline{a} \geq \underline{0} \Rightarrow V = \underline{0}$	$I\underline{a} \leq \underline{0} \Rightarrow V = V_1$
	Zero vector 1-1-1	110-000	$I\underline{c} \geq \underline{0} \Rightarrow V = V\underline{2}$	$I\underline{c} \leq \underline{0} \Rightarrow V = \underline{0}$
Sector 3: 120–180°	Zero vector 0-0-0	000-101	$I\underline{b} \geq \underline{0} \Rightarrow V = \underline{0}$	$I_b \leq \underline{0} \Rightarrow V = V_3$
	Zero vector 1-1-1	110-000	$I\underline{c} \geq \underline{0} \Rightarrow V = V\underline{2}$	$I\underline{c} \leq \underline{0} \Rightarrow V = \underline{0}$
Sector 4: 180–240°	Zero vector 0-0-0	000-101	$I_b \leq \underline{0} \Rightarrow V = \underline{0}$	$I_b \geq \underline{0} \Rightarrow V = V_3$
	Zero vector 1-1-1	011-000	$I\underline{a} \geq \underline{0} \Rightarrow V = V\underline{4}$	$I\underline{a} \leq \underline{0} \Rightarrow V = \underline{0}$
Sector 5: 240–300°	Zero vector 0-0-0	000-110	$I\underline{c} \geq \underline{0} \Rightarrow V = \underline{0}$	$I\underline{c} \leq \underline{0} \Rightarrow V = V\underline{5}$
	Zero vector 1-1-1	011-000	$I\underline{a} \geq \underline{0} \Rightarrow V = V\underline{4}$	$I\underline{a} \leq \underline{0} \Rightarrow V = \underline{0}$
Sector 6: 300–360°	Zero vector 0-0-0	000-110	$I\underline{c} \geq \underline{0} \Rightarrow V = \underline{0}$	$I\underline{c} \leq \underline{0} \Rightarrow V = V\underline{5}$
	Zero vector 1-1-1	101-000	$I\underline{b} \geq \underline{0} \Rightarrow V = V_6$	$I\underline{b} \leq \underline{0} \Rightarrow V = \underline{0}$

TABLE 5.3

Generation of Voltage Vectors Based on the Current Vector Phase

Sector		Zero Vector 0-0-0
0–60°	$\phi < 90 \Rightarrow$	$\Rightarrow I\underline{a} > 0 \Rightarrow V = 0$
	$90 < \phi < 150 \Rightarrow$	
	(a) For $\alpha < \phi - 90$	$\Rightarrow I\underline{a} < 0 \Rightarrow V = V\underline{1}$
	(b) For $60 > \alpha > \phi - 90$	$\Rightarrow I\underline{a} > 0 \Rightarrow V = 0$
	$\phi > 150 \Rightarrow$	$\Rightarrow I\underline{a} < 0 \Rightarrow V = V\underline{1}$
120–180°	$\phi < 90 \Rightarrow$	$\Rightarrow I\underline{b} > 0 \Rightarrow V = 0$
	$90 < \phi < 150 \Rightarrow$	
	(a) For $\alpha < \phi - 90$	$\Rightarrow I\underline{b} < 0 \Rightarrow V = V\underline{3}$
	(b) For $60 > \alpha > \phi - 90$	$\Rightarrow I\underline{b} > 0 \Rightarrow V = 0$
	$\phi > 150 \Rightarrow$	$\Rightarrow I\underline{b} < 0 \Rightarrow V = V\underline{3}$
240–300°	$\phi < 90 \Rightarrow$	$\Rightarrow I\underline{c} > 0 \Rightarrow V = 0$
	$90 < \phi < 150 \Rightarrow$	
	(a) For $\alpha < \phi - 90$	$\Rightarrow I\underline{c} < 0 \Rightarrow V = V\underline{5}$
	(b) For $60 > \alpha > \phi - 90$	$\Rightarrow I\underline{c} > 0 \Rightarrow V = 0$
	$\phi > 150 \Rightarrow$	$\Rightarrow I\underline{c} < 0 \Rightarrow V = V\underline{5}$

Sector		Zero Vector 1-1-1
60–120°	$\phi < 90 \Rightarrow$	$\Rightarrow I\underline{c} < 0 \Rightarrow V = 0$
	$90 < \phi < 150 \Rightarrow$	
	(a) For $\alpha < \phi - 90$	$\Rightarrow I\underline{c} > 0 \Rightarrow V = V\underline{2}$
	(b) For $60 > \alpha > \phi - 90$	$\Rightarrow I\underline{c} < 0 \Rightarrow V = 0$
	$\phi > 150 \Rightarrow$	$\Rightarrow I\underline{c} > 0 \Rightarrow V = V\underline{2}$
180–240°	$\phi < 90 \Rightarrow$	$\Rightarrow I\underline{a} < 0 \Rightarrow V = 0$
	$90 < \phi < 150 \Rightarrow$	
	(a) For $\alpha < \phi - 90$	$\Rightarrow I\underline{a} > 0 \Rightarrow V = V\underline{4}$
	(b) For $60 > \alpha > \phi - 90$	$\Rightarrow I\underline{a} < 0 \Rightarrow V = 0$
	$\phi > 150 \Rightarrow$	$\Rightarrow I\underline{a} > 0 \Rightarrow V = V\underline{4}$
300–360°	$\phi < 90 \Rightarrow$	$\Rightarrow I\underline{b} < 0 \Rightarrow V = 0$
	$90 < \phi < 150 \Rightarrow$	
	(a) For $\alpha < \phi - 90$	$\Rightarrow I\underline{b} > 0 \Rightarrow V = V\underline{6}$
	(b) For $60 > \alpha > \phi - 90$	$\Rightarrow I\underline{b} < 0 \Rightarrow V = 0$
	$\phi > 150 \Rightarrow$	$\Rightarrow I\underline{b} > 0 \Rightarrow V = V\underline{6}$

For $60 > \alpha > \phi - 90 \Rightarrow I_a > 0 \Rightarrow \quad \Delta V = 0$

If $\phi > 150 \Rightarrow I_a < 0 \quad \Rightarrow V = V\underline{1} \quad$ and

$$\Delta \vec{V} = \frac{t_{000}}{T_s} \cdot \vec{V}_1 = \frac{1}{2} \cdot \left[1 - m \cdot \cos\left(\alpha - \frac{\pi}{6} \right) \right] \cdot \vec{V}_1 \tag{5.63}$$

Exploiting these equations for all sectors yields the graphical representation from Figure 5.44 for the additional voltage. All this additional voltage is induced naturally without any special software algorithm except the new definition of the zero states.

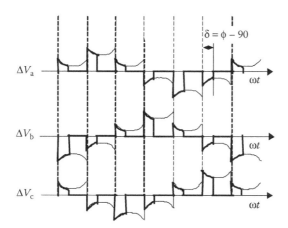

FIGURE 5.44 Additional voltage due to current-voltage phase shift.

Furthermore, the conventional operation of a space vector modulation algorithm is not modified at a current phase shift below 90°.

To quantify further this method, calculation is done for the steady-state operation at a given modulation index m and the same phase shift ϕ (constant) over the entire fundamental period of the generated voltage. The RMS of the additional voltage can be calculated as a function of the phase shift $\delta = \phi - 90$ and required modulation index m:

$$V_{RMS} = \sqrt{\frac{1}{2 \cdot \pi} \cdot \int_0^\delta \left[f(\alpha)^2 \right] \cdot \left[\frac{1}{9} + \frac{4}{9} + \frac{1}{9} + \frac{1}{9} + \frac{4}{9} + \frac{1}{9} \right] \cdot V_{dc}^2 d\alpha}$$

$$V_{RMS} = \sqrt{\frac{4}{18 \cdot \pi} \cdot V_{dc}^2 \cdot \int_0^\delta \left[1 - m \cos\left(\alpha - \frac{\pi}{6} \right) \right]^2 d\alpha}$$

(5.64)

$$V_{RMS} = V_{dc} \cdot \sqrt{\frac{4}{18 \cdot \pi} \cdot \left\{ \begin{array}{l} \delta - 2 \cdot m \cdot \left[\sin\left(\delta - \frac{\pi}{6} \right) + \frac{1}{2} \right] + \frac{1}{2} \cdot \delta \cdot m^2 \\ + \frac{1}{2} \cdot m^2 \cdot \left[\sin\left(2\delta - \frac{\pi}{3} \right) + \frac{\sqrt{3}}{2} \right] \end{array} \right\}}$$

(5.65)

This dependency is represented in Figure 5.45, and it represents an ideal maximum voltage. During the transient operation, the phase shift changes regularly and the additional voltage also changes. This helps understanding the limits and advantages of this method and assessing the usability in real industrial applications. The effects of using this modified PWM algorithm are shown in Figure 5.46.

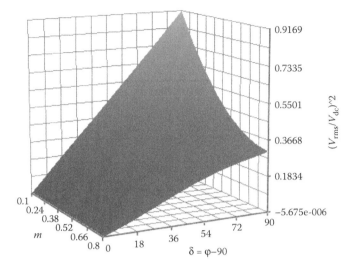

FIGURE 5.45 Additional voltage in respect to the current-voltage phase shift and modulation index.

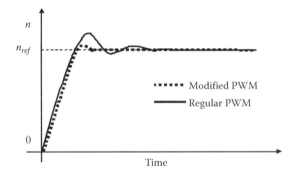

FIGURE 5.46 Results based on the mathematical model (speed as reference).

5.12 CONCLUSION

The space vector theory has been known for more than half a century, but only during the last 20 years, it has been consistently used in the control of three-phase converters. Details of PWM generation on the basis of vector representation as well as use of this concept in motor drives are the subjects of this chapter. All SVM variations are reviewed and the advantages of discontinuous reference functions are also outlined.

PROBLEMS

P5.1 Write relationship for Park/Clarke direct transforms and inverse transforms and prove the meaning of coefficients for power conservation and for magnitude conservation.

P5.2 Establish the mathematical expressions for waveforms in Figure 5.26. Consider each one as a phase function and write the mathematical expressions for the other two phases taking into account the phase shifts of 120° and 240°.

P5.3 Consider functions defined in P4.2. and define a new base in the vector space similar to the base defined in Equations 5.4 through 5.6. Then define a vector transform able to convert the $(d, q, 0)$ quasi-DC coordinates into phase voltages as shaped in Figure 2.22. Use a computer program (MathCAD, Matlab) to implement these relationships and run the program to plot control functions.

P5.4 Imagine a PWM pattern synchronized with the fundamental period and a count of 24 pulses on one period:
 • Draw the spatial distribution of the tip of the voltage vector.
 • For each discrete position thus determined, calculate the amount of time associated with each neighboring active vectors using Equations 5.29 and 5.30 for a modulation index of 0.55 defined as in Equation 5.31.
 • Consider the active vectors in the middle of the sampling interval and draw the time evolution of the voltage on the first phase during the first sector (*Method D-I-H*), using information from Figure 5.13.

P5.5 If a three-phase inverter is supplied from a three-phase central-point rectifier with very weak filtering or without filtering, calculate the maximum voltage we can have at the inverter output without overmodulation. Use Figure 5.11 and Equation 5.31 after definition of the minimum value of the rectified voltage (VDC). Consider the grid RMS voltage as 120 V per phase.

P5.6 Demonstrate Equations 5.41 through 5.46.

P5.7 Write Equation 5.48 for each sector, using the active switching vectors from Figure 5.13 and their ortho-normal coordinates in the complex plane.

P5.8 Use Figure 5.23 and define the instants for turning ON each of the six IGBT switches.
 • Write their expressions as a delay from the beginning of the sampling interval: For instance, $f(S1) = t_0/2$, $f(S2) = T_s - t_0/2$, and so on. Replace definitions from Equations 5.41 through 5.46 for each sector. Organize these results as a flowchart to be implemented in a microcontroller with a center-aligned PWM generator.
 • Write a computer program and draw the evolution of the calculated ON-time. Compare the results with Figure 5.15.

P5.9 Do the same as in P.5.7. with the switching sequence shown in Figure 5.28. Compare results with Figure 5.29.

P5.10 Explain an algorithm to take advantage of the best HCF from Figures 5.30 and Figure 5.32. How can we modify the switching reference function to ensure a smooth transition from one method to another?

REFERENCES

1. Park, R.H. 1929/1933. Two-reaction theory of synchronous machines. *AIEE Transactions*. No. 48, pp. 716–730 and no. 52, pp. 352–355.
2. Kron, G. 1942. The application of tensors to the analysis of rotating electrical machinery, Schenectady, NY, USA, General Electric Review.
3. Stanley, H.C. 1938. An analysis of the induction motor. *AIEE Trans*. 57, 751–755.
4. Kovacs, K.P. and Racz, I. 1959. Transiente Vorgange in Wechselstrommachinen. Budapest, Hungary, Akad Kiado.
5. Stepina, J. 1967. Raumzeiger als Grundlage der Theorie der Elktrischen Maschinen. *etz-A* 88(23), 584–588.
6. Serrano-Iribarnegaray, L. 1993. The modern space-phasor theory, Part I & Part II. *ETEP J.* 3(2), March/April, 3 May/June.
7. Murai, Y. and Tsunehiro, Y. 1983. Improved PWM method for induction motor drive inverters, IPEC, Tokyo, pp. 407–417.
8. Holtz, J., Lammert, P., and Lotzkat, W. 1986. High speed drive system with ultrasonic MOSFET-PWM-inverter and single-CHIP microprocessor control. *Conference Record 1986 IEEE Industry Applications Society Annual Meeting*, Part 1, Denver, Colorado, USA, pp. 12–17.
9. Van der Broeck, H.W., Skudelny, H.C., and Stanke, G. 1986. Analysis and realization of a pulse width modulator based on voltage space vectors. *IEEE-IAS Annual Meeting Conference Record*, Denver, USA, pp. 244–251.
10. Fukuda, S., Hasegawa, H., and Iwaiji, Y. 1988. PWM technique for inverter with sinusoidal output current. *19th Annual IEEE Power Electronics Specialists Conference*, Kyoto, Japan, vol. 1, 11–14 April, pp. 35–41.
11. Granado, J., Harley, R.G., and Diana, G. 1989. Understanding and designing a space vector pulse-width-modulator to control a three phase inverter. *Trans. SAIEE* 80, 29–37.
12. Halmos, P. 1986. *Finite-Dimensional Vector Spaces*. Springer-Verlag, New York.
13. Abrahamsen, J., Pedersen, J.K., and Blaabjerg, F. 1996. State-of-the-art of optimal efficiency control of low cost induction motor drives. *PEMC'96*, vol. 2, Budapest, Hungary, 2–4 September, pp. 163–170.
14. Murai, Y., Gohshi, Y., Matsui, K., and Hosono, I. 1992. High-frequency split zero-vector PWM with harmonic reduction for induction motor drive. *Trans. IA* 28(1), 105–112.
15. Murai, Y., Sugimoto, S., Iwasaki, H., and Tsunehiro, Y. 1983. Analysis of PWM inverter fed inductions motors for microprocessors. *Proceedings IEEE/IECON*, San Francisco, CA, USA, 10–14 November, pp. 58–63.
16. Fukuda, S., Iwaji, Y., and Hasegawa, H. 1990. PWM technique for inverter with sinusoidal output current. *IEEE Trans. PE* 5, 54–61.
17. Andersen, E.C. and Hann, A. 1993. Influence of the PWM control method on the performance of frequency inverter induction machine drives. *ETEP* 3(2), 151–160.
18. Holtz, J. 1992. Pulsewidth modulation—A survey. *IEEE Trans. IE* 39(5), 410–420.
19. Holtz, J. 1994. Pulsewidth modulation for electronic power conversion. *Proc. IEEE* 82(8), 1194–1212.
20. Handley, P.G. and Boys, J.T. 1991. Space vector modulation—An engineering review. PEVSD, London, UK, 87–91.
21. Neacsu, D. 2001. Space vector modulation. *Seminar IEEE-IECON*, San Jose, CA, USA, December.
22. Neacsu, D., Stincescu, R., Raducanu, I., and Donescu, V. 1994. Fuzzy logic control of a PWM V/f inverter-fed drive. *ICEM'94*, vol. 3, Paris, France, pp. 12–17.
23. Saetieo, S. and Torrey, D. 1998. Fuzzy logic control of a space vector PWM current regulator for three-phase power converter. *IEEE Trans. PE* 13(3), 419–426.

24. Enjeti, P. and Shireen, A. 1992. A new technique to reject DC-link voltage ripple for inverters operating on programmed PWM waveforms. *IEEE Trans. PE* 7(1), 171–180.

25. Schermann, M. and Schroedl, M. 1993. Methods of generating the voltage space vector by fast real-time PWM. *IEEE-PCC*, Yokohama, Japan, pp. 322–327.

26. Jacobina, C.B., Nogueira Lima, A.M., da Silva, E.R.C., Alves, R.N.C. and Seixas, P.F. 2001. Digital scalar pulse-width modulation: A simple approach to introduce non-sinusoidal modulating waveforms. *IEEE Trans. PE* 16(3), 351–359.

27. Lai, Y.S. and Bowes, S.R. 1996. A universal space vector modulation strategy based on regular-sampled pulse-width modulation [invertors]. *Proceedings IEEE IECON*, 22nd, vol. 1, pp. 120–126.

28. Bowes, S.R. and Lai, Y.S. 1997. The relationship between space-vector modulation and regular-sampled PWM. *IEEE Trans. IE* 44(5), 670–679.

29. Sun, J. and Grotstollen, H. 1996. Optimized space vector modulation and regular-sampled PWM: A re-examination. *Conference Record of the 1996 IEEE Thirty-First IAS Annual Meeting IAS '96*, San Diego, CA, USA, vol. 2, pp. 956–963, 6–10 October.

30. Holmes, D.G. 1992. A unified modulation algorithm for voltage and current source inverters based on AC–AC matrix converter theory. *IEEE Trans. IA* 28(1), 31–40.

31. Neacsu, D. and Lucanu, M. 1992. Output waveform optimization of the SVM inverters. *Proceedings of National Conference Electric Drives*, Iasi, Romania, 22–24, October, pp. B1–B6.

32. Holmes, D.G. 1995. The significance of zero space vector placement for carrier based PWM schemes. *Conference Record of the Thirtieth IAS Annual Meeting IEEE Industry Applications Conference, IAS '95*, Orlando, FL, USA, vol. 3, pp. 2451–2458, 8–12 October.

33. Kolar, J.W., Ertl, H, and Zach, F.C. 1989. Calculation of the passive and active component stress of three-phase PWM converter. *Proceedings EPE*, Aachen, Germany, pp. 1303–1311.

34. Alexander, D.R. and Williams, S.M. 1993. An optimal PWM algorithm implementation in a high performance 125 kVA inverter. *Conference Proceedings of the 1993 Eighth Annual Applied Power Electronics Conference and Exposition, APEC '93*, San Diego, CA, USA, pp. 771–777, 7–11 March.

35. van der Broeck, H.W. 1991. Analysis of the harmonics in voltage-fed inverter drives caused by PWM schemes with discontinuous switching operation. *Proceedings EPE '91*, pp. 3/261–3/266.

36. Houldsworth, J.A. and Grant, D.A. The use of harmonic distortion to increase the output voltage of a three-phase PWM inverter. *IEEE Trans. IA* IA-20, 1224–1228.

37. Hava, A.M., Kerkman, R.J., and Lipo, T.A. 1998. A high performance generalized discontinuous PWM algorithms. *IEEE Transactions on Industry, Applications*, vol. 34, no. 5, pp. 1059–1071, September–October.

38. Trzynadlowski, A.M. and Legowski, S. 1994. Minimum-loss vector PWM strategy for three-phase inverters. *IEEE Trans. PE* 9(1), 26–34.

39. Chung, D.W. and Sul, S.L. 1997. Minimum-loss PWM strategy for a 3-phase PWM rectifier. *Proceedings of the 28th Power Electronics Specialists Conference PESC*, St. Louis, MO, USA, vol. 2, pp. 1020–1026, 22–27 June.

40. Lai, R.S. and Ngo, K.D.T. A PWM method for reduction of switching loss in a full-bridge inverter. *IEEE Trans. PE* 10(3), 326–332.

41. Trzynadlowski, A.M., Kirlin, R.L., and Legowski, S. Space vector PWM technique with minimum switching losses and a variable pulse rate. *IEEE Trans. IE* 44(2), 173–181.

42. Faldella, E. and Rossi, C. 1994. High efficiency PWM techniques for digital control of DC/AC converters. *Conference Proceedings of the Ninth Annual Applied Power Electronics Conference and Exposition, APEC '94*, Orlando, FL, vol. 1, pp. 115–121, February 13–17.

43. Perruchoud, P.J.P. and Pinewski, P.J. 1996. Power losses for space vector modulation techniques. *IEEE-WPET*, Dearborn, MI, USA, pp. 167–173.
44. Ahmad, R.H., Karady, G.G., Blake, T.D., and Pinewski, P. 1997. Comparison in space vector modulation techniques based on performance indexes and hardware implementation. *23rd International Conference on Industrial Electronics, Control and Instrumentation, IECON 97*, New Orleans, LA, USA, vol. 2, pp. 682–687, 9–14, November.
45. Lucanu, M. and Neacsu, D. 1995. Optimal V/f control for space vector PWM three-phase inverters. *Eur. Trans. EPE* 5(2), 115–120.
46. Prasad, V.H., Borojevic, D., and Dubovsky, S. 1997. Comparison of high-frequency PWM algorithms for voltage source inverters. *Conference Proceedings of the 13th Annual Applied Power Electronics Conference and Exposition, APEC 1997*, Atlanta, GA, USA, pp. 857–863, 23–27 February.
47. Neacsu, D.O. 2007. Mathematical model for induction machine drives with modified PWM. *IEEE Int'l Symposium SCS, 2007*, pp.1–4.
48. Lucanu, M., and Neacsu, D. 1992. Optimization of the controller synthesis for three-phase inverters using space vector pulse width modulator. *The Transactions of the SAIEE*, vol. 83, no. 2, June 1992, pp.113–117.

6 Practical Aspects in Building Three-Phase Power Converters

6.1 SELECTION OF POWER DEVICES IN A THREE-PHASE INVERTER

Previous chapters have explained the operation of a three-phase converter and the need for pulse width modulation (PWM). This chapter investigates the three phase power converter as a system, outlining packaging and protection problems.

Power semiconductor devices for a three-phase power converter should be selected after determining the power converter ratings from the application requirements and taking into account the cooling and stress requirements for a given power level.

6.1.1 Motor Drives

When the power converter is used within a motor drive, its rating depends on the motor characteristics as it follows.

6.1.1.1 Load Characteristics

The application should provide information about the maximum torque required.

The power converter should take into account an increase of about 60% torque availability as an overload. Sometimes, this overload is considered within the rated torque with a derate of the nominal torque.

6.1.1.2 Maximum Current Available

The maximum phase current can be derived from the nominal power on the motor data.

6.1.1.3 Maximum Apparent Power

The power converter must be able to process the whole apparent power, including the active power that produces torque and circulates reactive power.

6.1.1.4 Maximum Active (Load) Power

The maximum power processed by the power converter can be calculated if the efficiency and cost of the motor at a fixed operation mode are known. This criterion is not very effective as both these vary highly with the mode of operation.

After the motor data has been investigated for power converter ratings of maximum phase voltage and currents, selection of power semiconductor switches should be made including additional margins required for overvoltage and overcurrent.

179

6.1.2 Grid Applications

The power is transferred from the grid mainly on fundamental frequency and the power semiconductor switches can be rated for their active power and a tolerable power factor. The fixed grid voltage automatically sets a fixed maximum voltage on the power semiconductor switches. Both current and voltage can be considered, respectively, with overvoltage and overcurrent. Modern snubberless converters do not require an overvoltage rating consistently larger than the operation voltage across the power semiconductor device.

Once we have a rough idea of the maximum levels of currents and voltage on power semiconductor switches, we have to check the cooling system. Appropriate switching power or energy losses for the required current level can be calculated based on the device model or estimated from device datasheets. For instance, *Powerex* insulated gate bipolar transistors (IGBTs) provide all switching energy losses estimated for different operation conditions. The switching losses are added to the conduction losses of each power semiconductor device to determine the cooling requirements. The datasheet also provides information about the junction-to-case thermal resistance for each semiconductor package.

The application should also provide information about the cooling system (air or coolant) and the temperature and pressure of the coolant agent. Simple equations determine the change in temperature under these cooling conditions, when the power loss is known. If the system has to work at a high temperature, an iterative process in a larger power device should be considered. There are some cases when the thermal requirement becomes more important than the maximum current. For instance, a 27 kW/300 V/90 A IGBT-based DC–DC power converter used in automotive applications may have the inlet coolant at 90°C. Selecting 100 A devices produces a junction temperature higher than recorded in the device datasheet.

Iterative design leads to selection of 300 A devices in order to overcome this thermal constraint. Higher current devices have lower thermal resistance because they have a different technology and can cope better in high temperature conditions. In many cases, this solution is cheaper than considering a more sophisticated cooling system with a lower thermal resistance.

Power semiconductor devices should be protected against extreme operating conditions and faults. Several protection requirements have become standard for power converters at any power level.

6.2 PROTECTION

6.2.1 Overcurrent

A very large value of the current can pass through a power semiconductor switch for diverse operational reasons. Let us try to understand the main sources of overcurrent and the means of electronic protection before the fuses burn.

Insulation breakdown or wrong connections can produce a short-circuit between two output wires (phases) (Figure 6.1). A simple shunt resistor followed by a linear optocoupler or a Hall-effect sensor on the DC bus can detect the unexpected peak of the current. Another protection method consists of using the same phase current

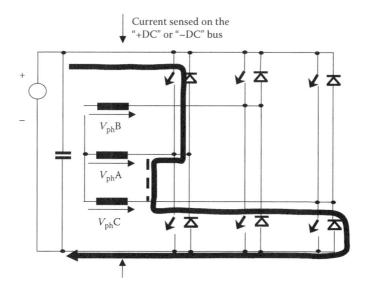

FIGURE 6.1 Short-circuit between two phases.

sensors that are used for current control. Each phase current is compared against two extreme thresholds for positive and negative levels.

Ground fault is another possible source of overcurrent and it may be caused by a motor insulation breakdown to the ground (Figure 6.2). This source of overcurrent can also be detected with one of the previous methods: sensing either the DC current or each phase current. It is important to note that many industrial power converters

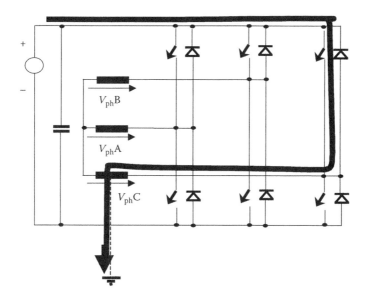

FIGURE 6.2 Ground fault.

controlled with vector-control methods do not measure all three-phase currents, but only two. They count on the symmetries of the three-phase system and calculate the third phase current as a difference between the sum of the other two currents and zero. Such a system cannot detect the ground fault if it occurs in the third phase.

Accordingly, a ground fault-protected system must sense all the three-phase currents.

Another source of undesired large currents arises from the shoot-through or cross-conduction fault. In certain conditions, the turn-on of an IGBT can produce a large positive (dv/dt) across the other IGBT on the same leg. Due to the Miller effect, this voltage variation can be accidentally transformed into gate current and turn-on the second IGBT. This would produce simultaneous conduction of both IGBT devices and short-circuit of the DC bus (Figure 6.3).

A general approach for protection consists of sensing voltage across each IGBT to prevent voltage build-up when the IGBT is in a controlled ON state. There are many circuit possibilities to implement this method: all of them sense the collector–emitter voltage and compare it with a fixed reference. Exceeding the reference shuts the gates off.

A simple circuit designated for this protection is shown in Figure 6.4 and it is called *Desat* protection. This name comes from the bipolar transistor's saturation, and basically this circuit verifies if the controlled power semiconductor is really in the normal ON state (or "*saturated*" for a bipolar transistor). The voltage drop on the switch is sensed and compared with a reference defined by a Zener diode.

If the transistor is unsaturated, the voltage drop is higher than the specified value, and the hysteresis comparator inhibits the control signal, turning-off the power semiconductor switch.

It is important to note that any of these three sources of overcurrent (line-to-line, ground, or shoot-through) can trigger the system protection in order to shut-off all six gates of a three-phase inverter. A quick shutdown generally produces a large voltage spike due to inductive components' tendency to keep the current circulating.

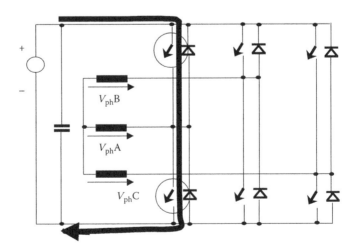

FIGURE 6.3   Shoot through.

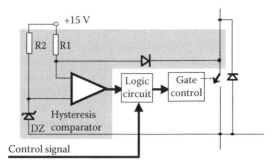

FIGURE 6.4   Desat protection.

This can be prevented by a soft shutdown of the IGBT under overcurrent, paying the price, though, of increased complexity of the control circuit.

The soft shutdown method shuts the gates off with a large gate resistor that is able to slow-down the turn-off waveforms. Synchronization of shutdown for all six gates implies additional complexity of the control circuitry. For this reason, the soft shutdown is used at low power levels where integrated circuits (IC) technology can easily accommodate the extra circuitry. *International Rectifier Corporation* has a nice series of high-voltage (600 and 1200 V) gate drivers (IR21xx and IR22xx) able to perform soft shutdown in the horsepower range.

6.2.2   FUSES

Processing power in high-voltage circuits implies also protection against overcurrent and, especially, short-circuits. Fuses represent the most known method of overcurrent protection. A fuse is a device able to break a high current through its own damage under the heat generated by that current. Figure 6.5 presents the structure of a fuse.

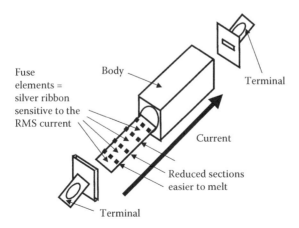

FIGURE 6.5   Basic structure of a fuse.

Current through the fuse produces heat, especially in the reduced sections. At high currents, these regions melt and the fuse is damaged. The rated current of the fuse is the maximum current carried continuously by the fuse without damage. This continuous current rating is defined with test procedures given in IEC269 or UL248 standards for ambient temperature, open air, and AC voltage at 50 or 60 Hz (grid). Another standard UL198L (DC Fuses for Industrial Use) provides the DC rating of the fuses in industrial applications.

An important parameter of a fuse is I^2t (squared RMS current multiplied by the clearing time) and it defines the fuse melting under a high fault current (Figure 6.6) [1]. The fuse can also melt when a lower current passes through the fuse for a longer time. This defines a dependency of the melting time that is inversely proportional to the applied current (Figure 6.7).

Selection of fuses for protection of a power converter depends on the voltage, total RMS current, the semiconductor device's rupturing I^2t value, device current di/dt, circuit inductance, ambient temperature, style of connection, and so on [1–3]. Diodes and other rectifier semiconductors are provided with datasheet information on a half-cycle surge rating characterized by the magnitude of a single sinusoidal half-cycle pulse at 50 or 60 Hz that the device can withstand. This value along with the half-cycle length (8.33 or 10 ms) is considered to calculate the semiconductor I^2t value.

Fuse selection is dependent on the total current containing both fundamental and harmonics and this is especially important in IGBT fusing. The high-switching frequency influences through the skin and proximity effects that are caused by nonuniform distribution of current density in the fuse elements. These effects are not very clearly known or defined, but they may justify premature opening of the fuse under the switching frequency components. A complete analysis of the power dissipation in a fuse under harmonics is presented in [4].

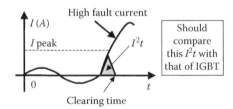

FIGURE 6.6   Fuse action on high fault current.

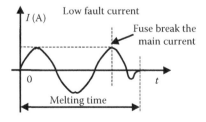

FIGURE 6.7   Melting of the fuse due to long-term current.

When a short-circuit occurs in an IGBT-based circuit, the collector–emitter voltage tends to increase immediately to a high value, which rapidly increases the internal power dissipation and failure of the device. Electronic protection circuits have been presented in the previous section. A fuse is used as protection when the electronic protection fails or is not used. The presence of the unprotected short-circuit in an IGBT can produce IGBT rupture, melting of the emitter connections or of the other circuit wires. If the IGBT rupture I^2t data is missing, a good practice is to calculate the I^2t for the copper bonding wire:

$$I^2 \cdot t = (100,000\ldots110,000) \cdot S^2 \tag{6.1}$$

where S is the wire section in mm^2.

This will ensure lower values than those experimentally defined for the IGBT case.

There are several possible distributions of fuses within a power converter. A complete solution includes protection on the DC bus, on phase currents, and on all IGBTs. This is not totally justified and a simpler or cheaper solution is generally satisfactory. The best compromise for the position of fuses within a three-phase inverter is shown in Figure 6.8.

When the fuse needs to break an inductive current within a DC circuit, the value of the circuit inductance determines the clearing time. The larger the inductance, the harder it is for the fuse to break the current. If a fuse is capable of suppressing a given amount of energy, then the DC voltage rating of a fuse is only valid for a specific time constant influenced by the amount of inductive component. For instance, a typical time constant for a capacitor bank, battery supply, and distribution circuits or UPS inverters is less than 10 ms, the DC motor armature has a typical time constant of 20–40 ms, and a traction system has a time constant of less than 100 ms.

Table 6.1 shows the entire UL (*Underwriters Laboratory*) classification of low-voltage fuses (Hundred of Volts). The "branch-circuit" rated class fuse is the most robust of the DC fuses. While the catalog number may vary from a manufacturer to another, these fuses fall into several classes, each having different performance characteristics and sizes. Such classes are: RK-1 and RK-5 (0.1–600 amps), T (1–1200

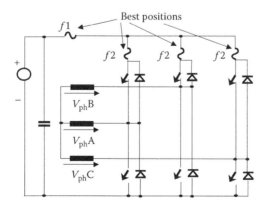

FIGURE 6.8   Best (reduced) placement of fuses.

amps), and CC (0.1–30 amps). Among them, the most popular, the RK-1 and RK-5 fuses fit in the fused, safety-switch disconnects available from all electrical supply houses and many home centers. These RK fuses are grouped in sizes by voltage with the shorter sizes having a 125-volt rating and longer sizes having a 250 or 600-volt DC rating. They are also grouped by current range (0.1–30, 35–60, 70–100, 110–200, 225–400, and 450–600 amps).

In North America, fuses for medium voltage applications (kV range) fall within one of the following categories:

- "E" fuses are generally used for transformers and general-purpose applications.
  - "E"-rated fuses for under 100 A must melt in 300 s at an RMS current within the range of 200–240% of the fuse's continuous current rating.
  - "E"-rated fuses above 100 A must melt in 600 s at an RMS current within the range of 240–264% of the fuse's continuous current rating.
- "R" fuses are used for motors with high starting currents and generally are back-up fuses that provide short-circuit protection for motor circuits.
  - R-rated fuses will melt in a range of 15–35 s at a current equal to 100 times the "R" rating.

The mounting of fuses on the busbar or printed-circuit-board should ensure a minimal electrical resistance of the contact, and under the assumption that the mounting support will sink heat from the fuse and not contribute to the fuse heating.

6.2.3 OVERTEMPERATURE

Power semiconductor devices are usually rated at a junction temperature between $-40°$ and $150°C$. Automotive, aerospace, or military applications require semiconductors rated below $-40°$ or above $150°$. For instance, power electronics for hybrid vehicles use the same cooling path as the engine providing a coolant at $90°$. MOSFET devices are available today at $175°$ or $200°$ junction temperature. The advent of silicon carbide-based power devices will make possible operation above $250°$.

A direct junction temperature measurement is very difficult. There are IGBT devices available in the market with junction temperature measurement based on a sensing diode on the same package, but they are expensive and, therefore, not widely used.

The most common approach to overtemperature protection consists of a temperature sensor on the cold plate or heatsink supporting the IGBT. If this thermocouple is mounted as close as possible to the IGBT, it provides a good reading of the device package's temperature. The measurement circuit is followed by analog processing. Unfortunately, the thermocoupler sensor is not linear and a linearization curve is needed if the temperature really needs measuring. This can usually be achieved by software using a piece-wise linearization method. Overtemperature protection does not really need this linearization or precise measurement, but requires only a comparison with a reference threshold in order to trigger the shutdown process.

Additional temperature monitoring is needed for the cooling system: this is either air-based or liquid-cooled. Some systems cooled with liquid also check the pressure.

6.2.4 OVERVOLTAGE

DC bus voltage and phase voltages are also monitored. If overvoltage on the DC bus is detected, the IGBTs within the three-phase inverter need to be shutdown. Power electronics systems may require monitoring of the phase voltages and shutdown in case of overvoltage. Circuits for voltage monitoring are based on resistive or trans-former `sensing of voltage followed by comparison with required thresholds. In high voltage, insulation with transformers from the power wires is required.

Except for these accidental overvoltage faults demanding fast action from the gate controller, overvoltage with slower transients is suppressed with devices called surge arrestors. This type of overvoltage can occur, for instance, at connection of a power electronics circuit to the power lines.

There are two classes of surge arresters: crowbar protection and clamping protection.

A crowbar device starts to conduct due to a quick change of its impedance when subjected to a large voltage. During this conduction interval, the voltage drop across the crowbar is limited to less than 15 V, allowing a large current to pass through it. It may be used in association with a dissipation resistor, a current-limiting device, or in series with a fuse that may blow due to the large current produced when the crowbar conducts. The energy is not dissipated on the crowbar device itself. The crowbar technique is also used in low-voltage DC/DC voltage regulators.

The most commonly used crowbar devices are air-gap protector, carbon-block protector, gas-discharge tube, and silicon-controlled rectifier (thyristor).

The second group of surge arrestors is composed of clamping devices. A clamp-ing device varies its internal resistance to limit the voltage transient by absorbing some of the transient energy. This is a serious limitation during application at large currents. Possible devices in this group are Zener diodes and metal oxide varis-tors (MOVs). MOV devices are mostly used in power converter applications. They have a voltage variable resistance and can support large currents during protection. However, they tend to degrade over time if high peak currents are repeated.

Varistor protective devices have peak current ratings from 20 A–70,000 A, peak energy ratings from 0.01 J–10,000 J, and mounting options to serve a wide range of applications. There are two major classes of products:

- *Multilayer Varistor* (MLV) family consists of compact surface mount devices with enhanced performance and filtering characteristics for circuit board level applications. They protect against electrostatic discharge, EFT and surge, offer low capacitance for high data rates and high capacitance for EMI filtering, and are widely used in computers, handheld devices, and automotive electronics.
- *Metal Oxide Varistor* (MOV) family is the most suitable to the circuits pre-sented in this thesis. They suppress higher energy voltage transients such as those generated by electrical load switching, they are offered in mounting options including bare disk, terminal connection, or radial and axial leaded packages.

MOV devices have a low power dissipation capability and they will be destroyed by repetitive transients. The MOV's lifetime should be carefully considered in accordance with the particular operation.

6.2.5 SNUBBER CIRCUITS

The transition of current between a power semiconductor switch and a diode has been explained in Chapter 2. Once current has been transferred from the turning off device to the turning-on device, the voltage starts to swing. This hard-switching induces a time interval during which both current and voltage are large in the turning-on device. As shown in Figures 2.2 and 2.7, this stress is more important for inductive loads due to the overvoltage produced by the current variation (di/dt).

6.2.5.1 Theory

Trajectories in the (I_C, V_{CE}) corresponding to the real operation are shown in Figure 6.9. They depend upon stray inductances, parasitic capacitances, and IGBT switching performances as di/dt, dv/dt. For instance, the IGBT package itself has a stray inductance of few tens of nanoHenries (nH) (for devices of the order of hundreds of amperes). The largest parasitic inductance is introduced by the DC bus connection of the inverter. It is very important to minimize the circuit parasitic inductances with a proper layout design [5,6].

Factors as di/dt, dv/dt can be partly adjusted through the gate circuit and the operation area can be minimized inside the datasheet safe operating area. The overvoltage produced by the recovering diode can also be limited by increasing the gate resistor.

It may be necessary to limit the slope of the current at turn-on of a power semiconductor device by inserting a series inductance. This is not usually the case in modern devices, but is required for some gate turn-off thyristors (GTO) or bipolar transistor-based inverters. Since the switching current adds up to the recovery current of the diode, sometimes an alternative solution consists of using a saturable inductor with a ferrite core in series with the diode. This inductor is supposed to take over all the voltage during recovery and it may reduce the recovery current.

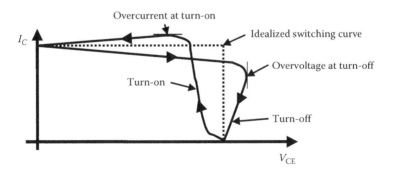

FIGURE 6.9 Trajectories during operation within a real circuit.

At turn-off, it may be necessary to limit the slope of the voltage. A better limitation along with power loss reduction can be achieved with *snubber circuits* [7] (Figure 6.10). When the snubber circuit is missing, the voltage will resonate due to the semiconductor parasitic capacitance and connections' inductance. The snubber circuit has to dump these oscillations.

For low-power applications, the parasitic inductance of the IGBT package and mounting on the bus bar are smaller than the inductance of the DC link. This is the case of discrete IGBTs or IGBT-based power modules used in applications above tens of amperes. Using power modules is definitely better as all connections are inside the package with extremely low parasitic inductances. The common solution consists of a simple decoupling capacitor across the entire inverter leg, providing a noninductive path for current transition (Figure 6.10a). High-frequency polypropylene film capacitors or other low equivalent series inductance-capacitors are especially designed for dual module IGBTs. They are mounted directly on the module terminals (Table 6.2) [8,9].

Depending on the estimated equivalent parasitic inductance, the decoupling capacitor can have values between 100 nF and 10 μF, usually 1 μF for each 100 A in the power semiconductor switch. A simplified calculation of the capacitor value can be made after neglecting the turn-off details within the semiconductor (Figure 6.11). The collector–emitter voltage is given by

$$V_{CE}(t) = V_{DC} - L_P \cdot \frac{di(t)}{dt} \Rightarrow \frac{di(t)}{dt} = \frac{1}{L_P} \cdot \left[V_{DC} - V_{CE}(t) \right] \tag{6.2}$$

$$V_{CE}(t) = V_{DC} + \frac{1}{C_s} \cdot \int i(t)dt \tag{6.3}$$

It yields:

$$\frac{di(t)}{dt} = -\frac{1}{C_s \cdot L_p} \cdot \int i(t)dt \text{ with the solution } i(t) = I_0 \cdot \cos\left[\frac{1}{\sqrt{C_s \cdot L_p}} \cdot t \right] \tag{6.4}$$

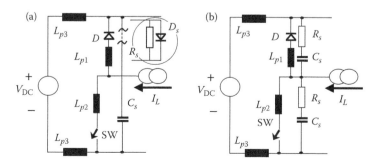

FIGURE 6.10 Snubber and switch equivalent circuit.

TABLE 6.1

Underwriters Laboratory (UL) Classification (Rating) of Low-Voltage Fuses

UL Class	Fuse Overload Characteristics	Rating [Amperes]	AC Voltage [Rating]	Available Ampere Ratings
L	Time-delay	200,000	600	200–6000
				601–4000
				200–2000
RK1	Time-delay	200,000	250	0.10–600
			600	
	fast-acting	200,000	250	1–600
			600	
RK5	Time-delay	200,000	250	0.10–600
			600	
T	Fast-acting	200,000	300	1–1200
			600	1–1200
J	Time-delay	200,000	600	1–600
	Fast-acting	200,000	600	1–600
CC	Time-delay	200,000	600	0.10–30
				0.25–30
	Fast-acting	200,000	600	0.10
CD	Time-delay	200,000	600	35–60
G	Time-delay	100,000	480	0.50–60
K5	Fast-acting	50,000	250	1–600
			600	
H	Renewable fuses fast-acting	10,000	250	1–600
			600	

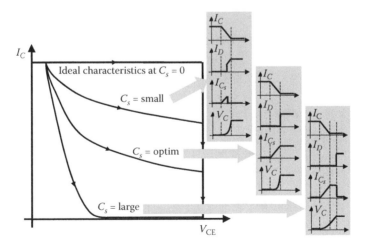

FIGURE 6.11   Discussion on the snubber capacitor value.

where I_0 is the load current at the moment of turn-off. Replacing this solution in Equation 6.3 yields:

$$v_{CE}(t) = V_{DC} + I_0 \cdot \sqrt{\frac{L_P}{C_s}} \cdot \sin \frac{t}{\sqrt{C_s \cdot L_P}} \qquad (6.5)$$

One can define a maximum desired voltage across the IGBT $[V_{MAX}]$ and replace for the maximum defined by previous equation.

$$V_{MAX} = V_{DC} + I_0 \cdot \sqrt{\frac{L_P}{C_s}} \Rightarrow C_s = L_P \cdot \left[\frac{I_0}{V_{MAX} - V_{DC}} \right]^2 \qquad (6.6)$$

Therefore, calculation of the needed capacitor value depends on the estimated value for the parasitic inductance.

Using the decoupling capacitor alone may not be the solution when the resonance between the DC link inductance and this capacitor produces a large bus ringing. An alternative solution is to insert a resistor-diode circuit in series with the capacitor. This will clamp the ringing. When the switch turns-off, the energy stored within L_{p3} is transferred to the capacitor C_s. The tendency of returning the energy to the bus inductance through oscillation is blocked by diode D_s. Moreover, the capacitor is decoupled during turn-on and the DC link parasitic inductance will smooth the turn-on transition and reduce the appropriate switching loss. The drawback of this approach is the additional inductance introduced in the circuit by the resistor–diode connection.

For high-power applications, the solution presented in Figure 6.10b is used. The snubber contains an R_s–C_s series network across each power semiconductor switch.

The C_s capacitance must be twice as large as the parasitic capacitance of the power semiconductor switch and its mounting. The R_s resistor is introduced to sustain the whole load current when C_s is discharged.

Accordingly, it yields $R_s = V_{DC}/I_L$

A second condition for R_s can be derived from the time constant for the discharging process. The snubber capacitor should discharge back to V_{DC} before the next turn-off moment, that is:

$$R_s = \frac{1}{6 \cdot C_s \cdot f_{sw}} \qquad (6.7)$$

The introduction of this resistor reduces the system efficiency due to inherent losses. The resistor loss at turn-off yields:

$$P_{Rs}(\text{off}) = \frac{1}{2} \cdot C_s \cdot \left[V_{pk}^2 - V_{DC}^2 \right] \cdot f_{sw} \qquad (6.8)$$

Losses at turn-on can be approximated as having the same value.

Using a resistor–diode assembly for dumping the voltage ring is another option. Advantages in this case are similar to those for clamping of the whole DC bus. The snubber capacitor is fully discharged during IGBT turn-on, whereas it is fully charged at turn-off. The losses in the snubber resistor are substantially higher in this case and can be expressed as

$$P_{Rs}(\text{off}) = \frac{1}{2} \cdot C_s \cdot V_{pk}^2 \cdot f_{sw} \qquad (6.9)$$

6.2.5.2 Component Selection

Snubber capacitors are subject to high peak and RMS currents as well as large dv/dt. The industry now provides capacitors especially built for this application. Snubber capacitors can be purchased as discrete components or as modules that allow connection of the snubber directly across the IGBT module terminal in order to minimize the terminal inductance. Table 6.2 presents different solutions for snubber capacitors provided by *Cornell-Dubilier* [9].

The snubber resistor should be selected to have the lowest possible inductance. Possible choices are carbon composite or metal film, but these are not easily available at high power. In this case, low inductance wire-wound resistors can be selected.

The diode in the snubber circuit experiences the same peak voltage as the snubber capacitor: the current is small in average but large in its peak. The blocking action of these diodes should be faster than the actual protected power semiconductor. Fast-switching diodes rated for the snubber capacitor voltage and circuit peak current should therefore be selected.

6.2.5.3 Undeland Snubber Circuit

Using Resistance-Capacitance-Diode (RCD) snubbers for both power semiconductor switches on one inverter leg requires many components and introduces large losses,

---

TABLE 6.2
Solutions for Special Snubber Capacitors

Code	Dielectric	Electrode	Voltage [V]	Capacitance [μF]	Max (dv/dt) [V/μs]
		Package type: Wrap and fill axial leads			
WPP	Polypropilene	Foil	250–1000	0.001–2.0	300–10,000
DPF	Polypropilene	Foil	250–2000	0.001–0.47	3000–10,000
940-1	Polypropilene	Double metalized	600–3000	0.1–4.7	100–2000
942-3	Polypropilene	Double metalized	600–2000	0.1–4.7	500–5000
		Package type: Dipped with radial leads			
CDx	Mica	Foil	500–1500	0.1–10 nF	>10,000
		Package type: Direct mount on IGBT module terminals			
SCD	Polypropilene	Double metalized	600–2000	0.1–10.0	100–2000

as demonstrated. A special snubber circuit has been proposed by Undeland (Figure 6.12) to minimize the number of components and to reduce the losses within the snubber. This circuit confines all losses in only one resistor, simplifying the energy recovery.

Capacitor C_{s2} separates the snubber circuit from the power stage during the intervals between switchings. At the end of each switching cycle, the excess energy within the inductance is discharged through D_{s2} and D_{s1} into the capacitor C_{s1}. The voltage across this capacitor tends to go above the DC bus voltage and the difference is dissipated on the snubber resistor R_s. This energy through R_s can be further recovered into the DC bus with regenerative snubbers.

6.2.5.4 Regenerative Snubber Circuits for Very Large Power

The higher the power within the power stage, the higher the losses in the snubber resistors associated with the six switches. For this reason, high-power converters are built with circuits able to recover something from this energy into the DC bus [10,11]. They are generally referred to as regenerative snubbers (Figure 6.13) [12–14].

It is worth noting that regenerative snubbers are useful in high-power converters equipped with slow-switching devices like gate turn-off thyristors (GTO) where losses are large. Such equipment is still in use in many places and some companies are currently producing GTO-based converters in multi-MW range. On the other hand, modern power semiconductor devices, for instance, IGBTs, are nowadays available in 1 kA range, and some of these devices do not need snubbering at all. Building snubberless power converters with IGBTs like Powerex MegaPack (300 V, 1000 A) makes this topic obsolete. However, due to historical reasons and due to the large number of GTO-based converters in use, regenerative snubbers are presented here.

6.2.5.5 Resonant Snubbers

The whole idea of using a snubber circuit can be reduced to controlling the slope of the current increase at turn-on and the slope of voltage at turn-off. The most minimal

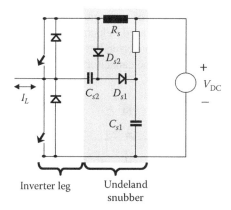

FIGURE 6.12   Undeland snubber with reduced losses.

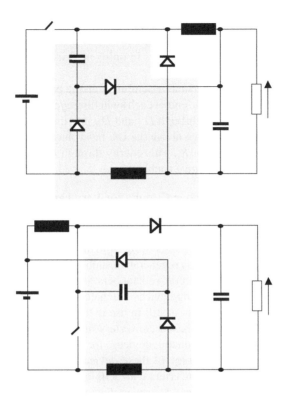

FIGURE 6.13 Circuit examples of regenerative snubbers.

solution has been shown to be a series inductor for turn-on and a parallel capacitor for turn-off. Complete solutions including resistance and diodes have been explained. Another concept is that of keeping all the transition losses out of the power semiconductor device by controlling its switching at zero current or zero voltage. This concept was first developed in the 1980s and is called resonant snubber. The simplest implementation of this concept consists of a circuit with a capacitor parallel to the power semiconductor device and an increased inductance in series (Figure 6.14).

This is represented as a buck converter, but it can also be a part of a converter leg. The capacitor might be the parasitic capacitor across a MOSFET device. The power

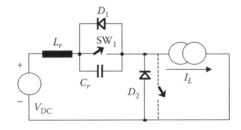

FIGURE 6.14 Principle of resonant snubber.

semiconductor switch S_{w1} will have transitions at zero voltage due to the resonance. It is controlled like regular switches within the buck or inverter leg operation.

Supposing S_{w1} is off, the load current I_L is passing through C_r and L_r charging the capacitor. Assuming a constant current I_L, the voltage across the resonant inductor stays zero while the capacitor voltage increases linearly.

$$V_{Cr} = \frac{I_L}{C_r} \cdot t \tag{6.10}$$

By difference, the voltage across D_2 decreases as

$$V_{D2} = V_{DC} - \frac{I_L}{C_r} \cdot t \tag{6.11}$$

Shortly, this diode turns-on and the charging time interval is defined as

$$t_1 = \frac{V_{DC}}{I_L} \cdot C_r \tag{6.12}$$

It is important to note that the slope of the voltage increase across the switch S_{w1} is ideally limited by resonance at I_L/C_r. The existence of the time interval t_1 does not considerably change the operation of the converter.

Next, the diode D_2 conducts a part of the load current, while the rest of the current circulates through the series resonant circuit L_r–C_r. The voltage across the capacitor C_r is the solution of the differential equation:

$$L_r \cdot C_r \cdot \frac{d^2 v_{C_r}(t)}{dt^2} + v_{C_r}(t) = V_{DC} \tag{6.13}$$

with $v_{C_r} = V_{DC}$ as initial condition.

The expression of the capacitor voltage yields:

$$v_{C_r}(t) = V_{DC} + \sqrt{\frac{L_r}{C_r}} \cdot I_L \cdot \sin\left[\frac{1}{\sqrt{L_r \cdot C_r}} \cdot (t - t_1)\right] \tag{6.14}$$

This shows an increase of the capacitor voltage after turning-on the diode D_2 and then a decrease toward zero according to the resonant swing. The capacitor voltage crosses zero only if:

$$V_{DC} \leq \sqrt{\frac{L_r}{C_r}} \cdot I_L \tag{6.15}$$

which is a very strong constraint for sinusoidal inverters. For small load currents, the voltage across capacitor will not cross zero. The current variation through L_r and C_r is a cosine function during this time. The moment of time corresponding to zero capacitor voltage is given by

$$\Delta t_x = \sqrt{L_r \cdot C_r} \cdot \left[\pi \arcsin \frac{V_{DC}}{Z_r \cdot I_0} \right] \tag{6.16}$$

After this moment, diode D_1 turns-on and takes over the L_r current and the C_r is no longer conducting current. The voltage across the inductor L_r is clamped at V_{DC} and its current goes linearly to zero. Throughout this interval of current decrease, the voltage across S_{w1} is kept at zero, and any turn-on command produces commutation at zero, voltage after the L_r current goes to zero. The time associated with this event is given by

$$\Delta t_y = \sqrt{L_r \cdot C_r} \cdot \frac{I_L \cdot \sqrt{\dfrac{L_r}{C_r}}}{V_{DC}} \cdot \sqrt{1 - \left(\frac{V_{DC}}{I_L \cdot \sqrt{\dfrac{L_r}{C_r}}} \right)^2} \tag{6.17}$$

The duration of the OFF interval is thus limited by parameters of the resonant circuit:

$$\Delta t_x \leq T_{off} \leq \Delta t_y \tag{6.18}$$

It can be noticed that power semiconductor devices are switched at zero voltage without switching losses. After S_{w1} turns-on, current through S_{w1} increases slowly due to L_r under a constant voltage V_{DC}. Diode D_2 stays in conduction for another short time interval under the same equivalent circuit derived previously during the D_1 conduction. This interval ends when the current through D_2 gets to zero. This current equals the difference between the load constant current IL and the linearly increasing current through S_{w1}.

A complete solution is presented in Figure 6.15 for a three-phase inverter. The capacitors are distributed in parallel with each switch, while the inductance is placed on the DC bus and increased from the value of the parasitic inductance. Modern MOSFET-based inverters can take advantage of the MOSFET's inherent parallel capacitance. After the parasitic inductance is estimated, additional inductance may become necessary to achieve the desired resonant frequency. The resonant frequency influences the voltage swing slope and the delay to zero crossing.

The early 1990s brought the explosive development of IGBTs and this concept has been widely developed in what we know today as resonant converters. A special chapter is later dedicated to this topic.

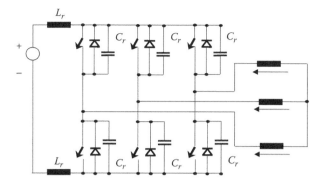

FIGURE 6.15 Distributed resonant snubber.

6.2.5.6 Active Snubbering

Voltage overshoot protection can also be achieved by including an additional stage in the gate driver (Figure 6.16) [15]. At turn-off, the protection transistor Q_p is turned-on and the gate is discharged through it. When the IGBT collector voltage reaches the breakdown voltage of the Zener diode, a current flows through the gate of Q_p and turns it off. The remaining current flows through R_{off}, slowing down the dv/dt rate. An additional benefit of this method is that switching power is reduced by half.

6.2.6 GATE DRIVER FAULTS

Another possible fault can occur at the gate-driver level. A faulty operation of the gate driver leading to absence of the control pulses at the IGBT gate should be detected and all the six gate drivers of the inverter turned-off. This is generally processed through the control device, a field programmable gate array (FPGA) or a digital signal processor circuit.

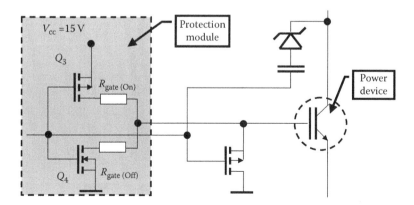

FIGURE 6.16 Active voltage overshoot protection.

6.3 SYSTEM PROTECTION MANAGEMENT

Complex systems including multiple power converters, sources, or loads have a protection system that sets several levels of priority for communication between them. This is also discussed in Chapter 7 [16].

6.4 REDUCTION OF COMMON MODE EMI THROUGH INVERTER TECHNIQUES

Chapter 1 has shown the importance of preventing common- and differential-mode electro-magnetic inference (EMI) in switching power converters and the appropriate standards have been described. Special EMI filters are commercially available for currents up to 100 A in grid-connected applications. They are based on higher order passive filters especially calibrated to limit EMI according to standards.

Let us now take a look at some circuit solutions for the common-mode EMI reduction. Three-phase inverters in which the neutral is not connected experience a continuous variation of the neutral voltage with respect to earth. This is illustrated in Figure 6.17 for a pulse width modulation (PWM) algorithm that represents a sequence of active and zero states already known for the three-phase inverter.

Each state of the inverter operation produces a different level of neutral voltage as shown in Table 6.3. The largest neutral voltage change (step) occurs when using zero states. A possible minimization of the common-mode voltage and ground current can be achieved by avoiding zero states within the PWM generation [17,18]. The drawback of such a solution is in increasing the ripple of the motor currents and limiting the maximum modulation index.

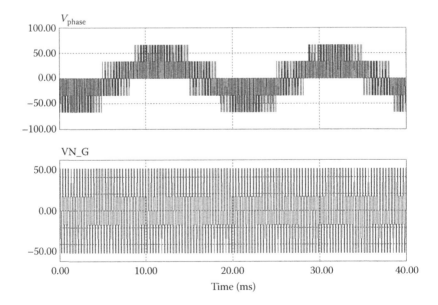

FIGURE 6.17 Variation of the neutral point voltage in respect with the middle point of the DC link for a Sinusoidal PWM at 3 kHz and $V_{DC} = 100$ V.

TABLE 6.3
Neutral Voltage for Each State of Inverter Operation

[1 0 0]	$-0.16 * V_{DC}$
[1 1 0]	$0.16 * V_{DC}$
[0 1 0]	$-0.16 * V_{DC}$
[0 1 1]	$0.16 * V_{DC}$
[0 0 1]	$-0.16 * V_{DC}$
[1 0 1]	$0.16 * V_{DC}$
[1 1 1]	$0.50 * V_{DC}$
[0 0 0]	$-0.50 * V_{DC}$

The parasitic coupling between the neutral point of the load and ground creates a path for the common-mode current flow. Note that the slope of neutral point voltage variation follows the voltage variation across the power semiconductor switches. The faster are these switchings, the larger is the current to ground. Figure 6.18 shows the capacitor path of the common-mode current.

Capacitor C_g can be the machine's stray capacitance or the distributed parasitic capacitance to ground. These common-mode currents create EMI problems and can produce damage to the electrical machine through bearing current, shaft voltage, insulation breakdown, or current flowing through the stray capacitors between motor and frame. These currents show components within the range of 100 kHz to tens of MHz and cannot completely be removed with ordinary chokes or EMI filters (like *baluns*).

A prior solution considered a common-mode transformer with an additional winding shorted by a resistor (Figure 6.18) [19–21]. The neutral point voltage is detected with an *RC* three-phase network. In this solution, care has to be taken to choose the appropriate *R* and *C* components, as they appear in parallel with each load phase. One improvement is to create the neutral voltage with an iron core transformer that offers very large impedance in parallel with each load phase (Figure 6.19). The resulting current circulates through the fourth winding of a four-winding ferrite-core common-mode inductor [22]. This is used for common-mode current cancelation. Furthermore, an *RC* circuit is used (Figure 6.19) to limit the power dissipation, as only the edges of the common mode voltage are addressed in order to minimize their slope.

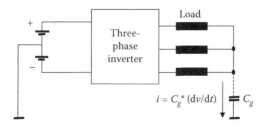

FIGURE 6.18 Common mode current.

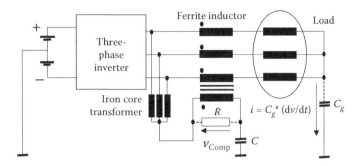

FIGURE 6.19 Common mode transformer.

This group of methods has proven inefficient in withdrawing the aperiodic ground-current. Elimination of both oscillatory and aperiodic ground currents (common-mode voltage) have been attempted with active circuits [21,23]. They can be used in the horsepower range in which high-frequency transistors are available for common-mode voltage control. One of these solutions is shown in Figure 6.20 [24]. The common-mode voltage at the inverter output is reconstructed with a set of small capacitors C_x and used to control an inverter leg. This adds a compensating voltage at the inverter outputs through the transformer T_r. This completely cancels the common-mode voltage on the load.

The implementation issues of this method relate to the choice of the transistors in the active circuitry. Transistors are operated in the active region following emitters and should have a wide frequency bandwidth and low output impedance to eliminate any influence of the output current in the compensating voltage. The high-frequency bandwidth ensures that the compensating voltage precisely follows the slopes of the inverter output voltages. The power dissipation within these transistors is very small (~0.5%) as the transistors carry only the transient part of the load voltage.

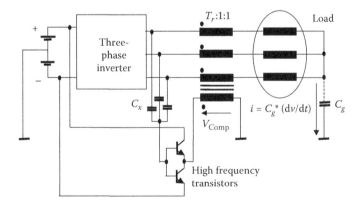

FIGURE 6.20 Active control with high frequency bandwidth transistors.

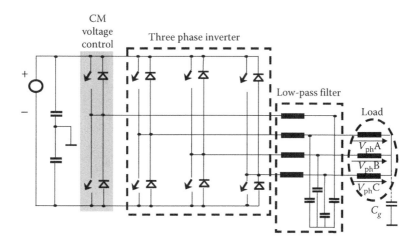

FIGURE 6.21 Four leg inverter.

At high-power levels, none of these approaches based on active common voltage canceling is convenient [25]. The alternative solution consists in using a fourth converter leg (Figure 6.21).

The power converter becomes a converter with four identical legs followed by LC low-pass filters. For balanced systems, the fourth leg can be derated with respect to the conventional inverter. The role of this additional leg is to complement the neutral voltage so that the instantaneous sum of all pole voltages is zero and no common-mode current is created. The drawback is that it is not practical to add a fourth load phase for the compensating current. A filter system with four phases can be used to fictitiously create the fourth phase and cancel common-mode voltage at the neutral point. If the load is perfectly symmetrical, this idea works perfectly. It is limited only by the frequency characteristics of the transfer function through the passive components used in filter and load.

Summation of voltage effects is therefore created through the low-pass filters LC when the fourth leg voltage is generated by reversing the information from Table 6.2, that is, to have always two switches tied to the positive DC rail and two switches to the negative rail. However, zero states cannot be used within this approach. Other PWM algorithms can however be defined without the use of zero states. A possible solution is to create the effect of zero state by employing two opposite vectors. For instance, if the last active state before the zero state was [1 0 0], we create the effect of the zero state by using the active states [1 0 0] and [0 1 1], each for half of the time desired for the zero state. Figure 6.22 illustrates this principle for a single pulse within the PWM algorithm. The extended time intervals associated with the active states produce more ripple on the load phase currents. In other words, a proper selection of the PWM algorithm can help in reducing the common-mode voltages at the price of increased ripple on the load.

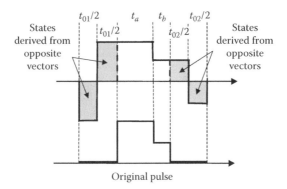

FIGURE 6.22 PWM without zero states.

6.5 TYPICAL BUILDING STRUCTURES OF THE CONVENTIONAL INVERTER DEPENDING ON THE POWER LEVEL

As shown in Chapter 1, the same circuit topology can be used at 10 or 1000 A, but building the appropriate power converters differs with the power level. In order to understand constraints for packaging power converters at different power levels, let us start with a review of power semiconductor packages.

6.5.1 PACKAGES FOR POWER SEMICONDUCTOR DEVICES

Figure 6.23 illustrates different packages used for IGBT devices. For currents of tens of amperes through-hole packages, such as TO-220 and TO-247, are preferred for power semiconductor switches. They are used with printed circuit boards (PCBs) to build power converters below 40 A. This direction toward use of PCBs has been

FIGURE 6.23 IGBT packages.

imposed by power converter manufacturers for cost reasons and to take advantage of the existing PCB-automation tools. These packages benefit from putting both the power semiconductor switch and the associated diode within the same package and offering it at very low cost per ampere. For instance, a 20 A/600 V IGBT/diode can be found under $1.50 in a single component order, and a 60 A/600 V IGBT for less than $5.00 for a single component order (digikey data shown for Fall 2012).

For low- and medium-power applications, IGBTs are packaged in dual (inverter leg) or six-pack assemblies. Unfortunately, the packaging is not consistent from one manufacturer to another and it is similar to the former bipolar Darlington power modules.

Modern power modules also include control and protection circuitry within the same package in order to simplify the inverter building and reduce costs of auxiliary parts. There is no standard for these intelligent power modules and they are not interchangeable as characteristics, control, or protection. This becomes a serious limitation to paralleling such modules. Single inline package (SIP) and dual inline package (DIP) modules are another alternative for power modules used in low power appliances. Given the wide emergence of these devices over the last years, a special chapter is later on dedicated to their presentation.

Different manufacturers have tried during the last years to provide standard packages, especially for the low-power market where power converter manufacturing is based on more automation. The EconoPACK (Figure 6.24) and EconoPIM modules are dedicated packages below 20 kW (1200 V, 100 A) and contain a full bridge with through-hole terminals able to connect the control circuitry from a PCB. Above 100 kW, integrated hybrid modules (IHM) modules are used.

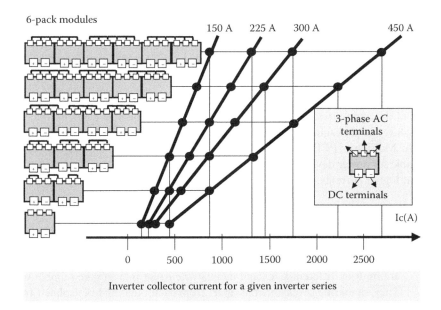

FIGURE 6.24 Packaging for Parallel applications: EconoPACK+.

Because a power converter manufacturer generally has to address a wide power range and provide very large volumes at lower power levels, a new approach has won market share during the last few years. The packaging has been changed to accommodate easy paralleling of power modules to define a very large power range that can be easily manufactured. The major features are:

- Define a flow-through concept by separating the power DC terminals on one side and phase output terminals on the other.
- Parallel the three legs of a six-pack IGBT if and when necessary.
- Use the same housing for dies that support currents from 150 to 450 A in order to achieve easy scaling of heatsinks, bus bars, and drivers.

At higher currents, the IGBT modules are mounted directly on heatsinks or cold plates while the electrical terminals are connected through screws on top to special structures called bus bars.

In the medium-power range, there is always a temptation to save money by paralleling multiple low-power IGBTs packaged in TO-220 or TO-247 packages. However, such an approach loses the advantages of the PCB mounting and requires high-current wiring of all semiconductor power terminals.

All of these higher power modules are more expensive. For instance, a 300 A/600 V dual (half-bridge) IGBT can be purchased under $150 in a single component order, a 600 A/1200 V dual IGBT under $400 in a single component order; while the largest in family, the 1400 A/1200 V dual IGBT can be found under $1000 (digikey data shown for Fall 2012). It is important to understand that the cost of a module is mostly dependent on the mechanical packaging and not on the size of the semiconductor die that is inside. This is why the cost becomes advantageous if the package accommodates the largest semiconductor die that it can.

In the very high-power range, IGBTs are packaged as discrete devices only.

6.5.2 CONVERTER PACKAGING

Once the power semiconductor devices have been selected and the size and terminals of the appropriate module have been understood, the next element to look at is the converter packaging. It has been mentioned that PCBs are the best solution below 40 A. Multilayer PCBs allow large currents on different isolated layers. They are suitable for power devices with through-hole terminals.

At higher currents, there are two options for power distribution:

- High current (heavy-gauge) wires: Heavy-gauge wires can be used at reasonable power levels but they introduce difficult routing and bending within the converter enclosure.
- Copper bus bars with tapped holes for cable connection. Copper bus bars are built in different sizes and can carry current in a simple, reliable way. They should be several inches apart from each other and be isolated from the cabinet by fiberglass reinforced plastic spacers.

An alternative solution has been recently introduced that uses laminated bus bars built of a multilayer structure of copper and dielectric insulator. They were first used in computer and telecommunications systems, but were introduced recently in medium- and high-power converters (Figure 6.25). The advantages of this technology is better cooling, lower resistance than wires (lower voltage drop), minimized stray inductance (lower voltage overshoots), and the possibility to use different copper layers in the laminated package for different purposes. For instance, a direct comparison of a connection with twisted wires and one with a laminated bus bar shows half DC resistance of the new solution (0.006 versus 0.0032 Ω), and a substantial decrease in the high frequency impedance at 1 MHz from 0.078 to 0.019 V. Using each layer for another function highly improves packaging of power converters for modern requirements up to 1000 A or 5000 V [26].

Due to their ruggedness, they can also be used as mounting platforms for auxiliary components, such as protection circuitry breakers or snubbers [27]. Moreover, special structures are built for IGBT devices or modules to accommodate their terminals (Figure 6.25).

6.5.3 Enclosures

After the electronics has been packaged and assembled together, the entire equipment should be placed inside an enclosure. Enclosures are defined by NEMA standard ICS 1-110 [28] as

- Type 1 is a general purpose enclosure, useable indoor, aiming at personnel protection and water dripping inside the electronic circuitry.
- Type 4 is a watertight, dustproof, nonventilated indoor or outdoor enclosure.
- Type 12 is a dustproof, driptight, indoor enclosure.

Usually, enclosures are built with steel of Gauge 10-12, occasionally with Gauge 14. All doors should have safety closing. Freestanding enclosures should be under 90 inches.

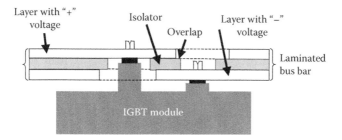

FIGURE 6.25 Possible use of a laminated bus bar at IGBT module connection.

The IP Code (*Ingress Protection Rating*) [29] provides another classification and rates the degree of protection provided against the intrusion of solid objects (including body parts like hands and fingers), dust, accidental contact, and water in *mechanical casings* and with electrical enclosures. IP codes have the format IP*xx*, where the *xx* represent numerals from the coding scheme. The first number in the sequence signifies the degree of protection against the entry of foreign solid objects (0 = not evaluated; 1 = for greater than 50 mm diameter; 2 = for greater than 12.5 mm diameter; 3 = for greater than 2.5 mm diameter; 4 = for greater than 1 mm diameter; 5 = dust protected; 6 = dusttight). The second number signifies the degree of protection against the entry of moisture and may be anything from 0 through 8 (0 = Not evaluated; 1 = Dripping water: from vertical; 2 = Dripping water: at 15° tilt; 3 = Spraying water; 4 = Splashing water; 5 = Jetting water; 6 = Powerful jetting water; 7 = Temporary immersion; 8 = Continuous immersion). Test procedures generally accompany the actual rating for a better understanding of the enclosure's capabilities.

Enclosure cooling is done by radiation and convection. Reference [28] provides empirical relationships for estimation of temperature rising, for an ambient under 50°C:

Radiation

$$T_{rise}[°C] = \left(\frac{426 \cdot W[\text{watt/sq. inch}]}{\eta} \right)^{0.84} \tag{6.19}$$

Convection

$$T_{rise}[°C] = \left(714 \cdot W[\text{watt/sq. inch}] \right)^{0.80} \tag{6.20}$$

In the first cases, η = emissivity (Table 6.4).

TABLE 6.4
Radiation Emissivities for Certain Materials

Material	Emissivity
Polished silver	0.02
Polished aluminum	0.05
Polished brass	0.60
Copper	0.15
Oxidized steel	0.70
Cast iron	0.25
AL paint	0.55
Black gloss paint	0.90
White lacquer	0.95
White enamel	0.95

6.6 AUXILIARY POWER

6.6.1 REQUIREMENTS

The numerous gate driver circuits as well as the current and voltage sensing circuitry need isolated power supplies. If the gate drivers for lower power converters (HP range) can be simply supplied through bootstrap power supplies, dedicated power supplies are necessary at higher power levels.

The number of these isolated power supply channels can be very high. For instance, a conventional three-phase inverter may require a supply of 6 gate drivers, a bipolar voltage for the phase current measurement circuits, one for the measurement of the DC bus voltage, a 5 V or 3.3 V for the digital control circuitry, and possibly some voltages for isolated communication channels. The internal power distribution system hence becomes very complex.

In order to reduce the component count and to maintain the system's efficiency quite low, the flyback converter topology is mostly used. It is the simpler possible one-switch converter, allowing generation of multiple secondary side voltages. Since the regulation is in general done on only one channel of secondary voltage, the different supply voltages may have weaker regulation than expected. If this is the case, the regulation can be improved further with local nonisolated voltage regulators.

The control of the flyback power supply as well as the control of any other low voltage PWM type voltage regulator can be achieved with a family of integrated circuits that follows the original design of Bob Mammano.

6.6.2 IC FOR POWER SUPPLIES

The introduction of the PWM control IC is an excellent example of disruptive innovation. This integrated circuit has been invented by Bob Mammano [30] in 1975, and introduced to market in 1976 by *Silicon General Company,* as SG1524. In 1970s, similar PWM Control ICs (Figure 6.26) were developed by multiple corporations like Motorola MC3420, Texas Instruments TL454, Signetics NE5560, Ferranti ZN1066, Fuji FA553 [32].

The IC is based on a mixed-mode IC technology with a very simple structure, well-known today. Its apparition coincides with the advent of switching power supplies in late '70 s and it did satisfy a clear market need, able to add value to the customers. It also helped the creation of a new market through an interesting paradigm (*vendors competing customers*). It did enable new power supply technologies and subsequently a more incremental development. Circuits like current mode ICs, cycle-by-cycle current limiting protection, single-ended, push-pull supplies, LDOs, hot-swap, soft-switching, so on, have benefited from the original PWM chip idea.

Today, the power control IC industry has a market of more than $5 billion.

As the technology substrate has developed, additional features were incorporated along the conventional PWM control with current regulation. Some of such features are next listed:

- Advanced soft-start
- Quasi-resonant flyback

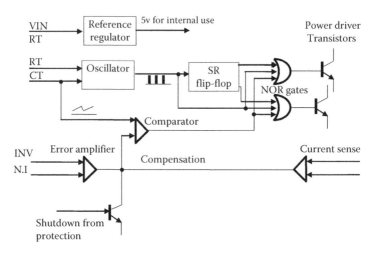

FIGURE 6.26   Structure of the PWM control IC.

- Valley switching for low EMI
- Stand-by power requirements
- No-load power requirements (like < 300 mW no-load power)
- Current-mode control
- Multimode power saving with automatic switching between operation modes
- Pulse skipping, or pulse density modulation
- Burst operation, up to hard-switching
- Green mode status indicator (can disable PFC)
- Both line and load over-voltage protection
- Bounded frequency range, with frequency foldback

A derivative product was the PWM control chip for the flyback power supplies. It is the most used class of ICs for isolated power supplies, featuring the entire controller in a 8-pin package. The history of the original *Unitrode* products was marked by the following milestones:

- 1980s—Introduced as UC3842: simple structure, 144 transistors, in 7.5 mm technology, sold for $1.75.
- 1990s—Feature improvement: sold as UCC3802, 478 transistors, in 3.0 mm technology, sold for $0.85.
- 2000s—More features: sold as UCC38600, 1158 transistors, in 0.5 mm technology, sold for $0.45.

The application circuitry is illustrated in Figure 6.27 for the case of a single secondary. Similar designs can consider the generation of multiple secondary voltages while using a single control IC.

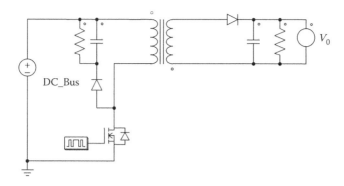

FIGURE 6.27 Single-switch flyback power supply.

6.6.3 OPERATION OF A FLYBACK POWER CONVERTER

Depending on the conduction state of the power MOSFET, there are two operational states of the circuit. When the switch is ON, the secondary side diode becomes reverse polarized and turns-off. The MOSFET device allows a current circulation through the primary winding, and the inductor core flux increases linearly from its initial value.

$$\phi(t) = \phi(0) + \frac{V_{DC}}{N_1} \cdot t \tag{6.21}$$

After a conduction time interval t_{on}, the power MOSFET turns-off and the energy stored in the core causes the current to flow in the secondary side, through the diode D. The voltage across the secondary winding becomes $[-V_0]$, and the flux decreases linearly:

$$\phi_{max} = \phi(0) + \frac{V_{DC}}{N_1} \cdot t_{on} \tag{6.22}$$

$$\phi(t) = \phi_{max} - \frac{V_0}{N_2} \cdot (t - t_{on}) \tag{6.23}$$

For a steady-state operation, the final value should be equal to the initial value for the next cycle:

$$\phi_{fin} = \phi_{max} - \frac{V_0}{N_2} \cdot (T - t_{on}) = \phi(0) + \frac{V_{DC}}{N_1} \cdot t_{on} - \frac{V_0}{N_2} \cdot (T - t_{on}) = \phi(0) \tag{6.24}$$

Hence:

$$\frac{V_{DC}}{N_1} \cdot t_{on} = \frac{V_0}{N_2} \cdot (T - t_{on}) \Rightarrow \frac{V_0}{V_{DC}} = \frac{N_2}{N_1} \cdot \frac{D}{1 - D} \tag{6.25}$$

The duty cycle yields:

$$D_{\text{CCM}} = \frac{V_0}{\dfrac{N_2}{N_1} \cdot V_{\text{DC}} + V_0}$$

(6.26)

This calculation has followed a similar approach with a buck-boost converter. However, the operation of a flyback power converter may be influenced by discontinuous mode of operation (DCM), when the equations become more complex. The output voltage will also depend on the load in such case (Figure 6.28).

While operated in DCM, the entire energy is transferred to the load and the duty cycle depends on the load energy, input voltage, and inductance value:

$$D_{\text{DCM}} = \sqrt{\frac{2 \cdot P_{\text{DCM}} \cdot L \cdot f_{\text{sw}}}{V_{\text{DC}}^2}}$$

(6.27)

where V_{DC} is the input voltage.

It is important to note that the control of energy is assured during the ON-time of the primary-side MOSFET, while the energy is delivered to load during the OFF-time. This means there is no control possible during the time the energy is actually transferred to load. If the converter operates in CCM, the energy accumulated in inductor will be delivered to the load uncontrolled over several cycles, before an action can be done by control. This behavior is also seen in the small signal model as a *right-half-plane zero*. The phase decreases with increasing gain, and this must be considered when defining the control-loop compensation. The general rule for

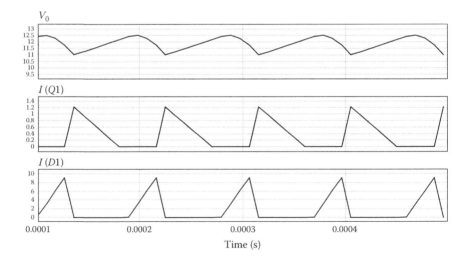

FIGURE 6.28 Typical waveforms for the operation of a flyback power converter in discontinuous conduction mode.

converters with a *right-half-plane zero* is to design at the lowest input line voltage and at the maximum load current, restricting the bandwidth of the feedback loop to about one-fifth the *right-half-plane zero* frequency. The *right-half-plane zero* frequency for CCM yields as

$$f_{RHPZ} = \frac{(1 - D)^2 \cdot V_o}{2 \cdot \pi \cdot L \cdot D \cdot I_{out} \cdot \left(\dfrac{N_2}{N_1}\right)^2} \tag{6.28}$$

The same behavior of *right-half-plane zero* is also seen in DCM. However, this is usually not a problem anymore as the frequency moves above half of the switching/sampling frequency.

A final comment about the control system relates to the actual use of the flyback power supply within a multiphase power converter. In such application, there are multiple secondaries of the flyback transformer required to power multiple low voltage circuits with galvanic isolation. These loads have variable currents and it is virtually impossible to regulate all voltages at a while. The compromise is usually to set the feedback loop after the most restrictive secondary that is usually the supply of digital circuitry. Additional solutions require small on-board 3-pin voltage regulators on each secondary for precise regulation without isolation. However, the gate drivers and some current, temperature and voltage sensing devices (connected in differential mode) do not require accurate regulation of voltage.

The latest flyback control ICs benefit from the following features [31,32]:

- Operation with variable frequency to maintain discontinuous or boundary (also called transition mode) conduction modes with advantages in
  - Efficiency optimization (operation with diode turn-off by zero current is virtually lossless, with no diode reverse recovery, and it improves efficiency)
  - Smaller size of the flyback transformer given the requirement for a smaller inductance
  - Small-signal modeling, with the plant model reduced at a first order model for either voltage-control mode or current control mode
- Operation in "green mode", where the PWM sequence is shut-down at low load, with a burst mode operation
- Special protection features like over-voltage detection, maximum ON-time programming, fast latch-up fault recovery, and thermal shutdown

There is disadvantage of higher peak current when operated in DCM or BCM, with bad influence in EMI, MOSFET conduction loss, over-voltage protection. However, the cost difference of using a higher current MOSFET is minimal and the problems are thus solved.

Despite being the most critical design component, the actual design and construction of a flyback transformer is beyond the scope of this book as flyback transformer can be acquired off-the-shelves or by order with system-level requirements.

6.7 CONCLUSION

This chapter presents details related to the building of a three-phase power converter. Information about three-phase power converters can easily be found in many textbooks, but the way the converter is built and protected is also important. Modern techniques have been shown to improve performance criteria such as efficiency, power density, and input or output harmonics.

PROBLEMS

P6.1 A 24V/120V boost converter is built with an IGBT and a diode switched at 20 kHz. The IGBT is protected with a snubber capacitor. The parasitic inductance of the circuit is 10 nH, the maximum input current is 100 A and a voltage increase of maximum 5% is allowed. Estimate the required snubber capacitance.

P6.2 Select a resistor to form an RC snubber circuit for the previous converter. Calculate losses within the resistor.

P6.3 Consider a single-phase IGBT inverter with snubber circuits across each IGBT. The DC voltage is 270 V, switching frequency is 16 kHz, maximum current is 120 A and a voltage overshoot of 10% is allowed. The bus parasitic inductance has been estimated at 20 nH. Define the values for the resistor and capacitor and estimate resistor power losses.

P6.4 Explain how an *RC* network connected in parallel with the load would serve as a turn-off snubber for all 4 IGBTs in the previous problem. Calculate values of components within such network and estimate power losses. Why the solution of previous problem is more preferable?

P6.5 Rewrite the SVM time equations for

P6.6 The common mode current is produced by derivative of the neutral voltage. This current is lower for PWM algorithms that do not produce large variations of the neutral point voltage. Considering Table 6.3 along with the switching sequences considered in the previous Chapter for the SVM algorithms, determine which state sequence produces the lowest peak-to-peak common-mode current.

P6.7 Draw for each sequence the qualitative evolution of the common mode current. What is the frequency of the most important component?

P6.8 A buck converter built with an IGBT and a diode is switched at a frequency f_{sw} with a constant duty cycle producing an on-state loss of 100 W and a switching loss given by $0.01 * f_{sw}$. The maximum junction temperature of the IGBT is 150°C and the junction-to-case thermal resistance is 2°C/W. The cooling system maintains a quasi-constant case temperature at 60°C. What is the maximum allowable switching frequency.

P6.9 Consider the same IGBT mounted on a heatsink while the ambient temperature is 27°C. Consider a switching frequency of 16 kHz and calculate what is the maximum heatsink thermal resistance.

REFERENCES

1. Mohan, N., Undeland, T., and Robbins, W. 2002. *Power Electronics.* 3rd edition. John Wiley and Sons, New York.
2. Cline, C. 1995. Fuse protection of DC systems. *Annual Meeting of the American Power Conference.*
3. Anonymous 2002. *Semiconductor Fuse—Application Guide.* Ferraz-Shawmut Corporation, January.
4. Anonymous 2002. *Introduction to Power Electronics and Protection Methods.* Ferraz-Shawmut Corporation, January.
5. Severns, R. 1997. Design of snubbers for power circuits. *PCIM.*
6. Zhang, Y., Soghani, S., and Chokhawala, R., 1995. Snubber considerations for IGBT applications. IRF Design Tip Documentation
7. Severns, R. 2008. Snubber Circuits for Power Electronics, Ed.
8. Iov, F., Blaabjerg, F., and Ries, K. 2003. Prediction of harmonic power losses in fuses located in DC-link circuit of an inverter. *IEEE Trans. IA* 39(1), 2–9.
9. Anonymous 2004. *Cornier-Dubilier Databook. Documentation.* Chicago, IL
10. Thiyagarajah, K., Ranganathan, V.T., and Ramakrishna, R.M. 1991. A high frequency IGBT PWM rectifier/inverter system for AC motor drives operating from single phase supply. *IEEE Trans. PE* 6(4), 576–584.
11. Deacon, J.H., Van Wyk, J., Schoeman, J. 1999. An evaluation of resonant snubbers applied to GTO converters. *IEEE Trans. IA* 23(2), 292–297.
12. Steyn, C. and Van Wyk, J. 1989. Optimum nonlinear turn-off snubbers: Design and applications. *IEEE Trans. IA*, 25(2), 298–304.
13. Swanepoel, P.H. and Van Wyk, J.D. 1994. Analysis and optimization of regenerative linear snubbers applied to switches with voltage and current tails. *IEEE Trans. PE* 9(4), 433–442.
14. Steyn, C. 1989. Analysis and optimization of regenerative snubbers. *IEEE Trans. PE* 4(3), 362–370.
15. Heath, D. and Wood, P. 1999. Overshoot voltage reduction using IGBT modules with special drivers. IRF Design Tip Documentation.
16. Donescu, V. 2001. Fault detection and management system broadcasts motor drive faults. *PCIM*, June.
17. Cacciato, M., Consoli, A., Scarcella, G., and Testa, A. 1998. Continuous PWM to square-wave inverter control with low common mode emissions. *IEEE PESC*, pp. 871–877.
18. Holmes, DG. 1996. The significance of zero space vector placement for carrier based PWM schemes. *IEEE Trans. IA* 32(5), 1122–1129.
19. Ogasawara, S. and Akagi, H. 1996. Modeling and damping of high frequency leakage currents in PWM inverter-fed AC motor. *IEEE Trans. IA* 32(5), 1105–1114.
20. Swamy, M.M., Yamada, K., and Kume, T.J. 2001. Common mode current attenuation techniques for use with PWM drives. *IEEE Trans. PE* 16(2), 248–255.
21. Ogasawara, S., Ayano, H., and Akagi, H. 1998. An active circuit for cancellation of common mode voltage generated by a PWM inverter. *IEEE Trans. PE* 3(5), 835–841.
22. Shimizu, T., Kimura, G. 1996. High frequency leakage current reduction based on a common mode voltage compensation circuit. *IEEE PESC*, pp. 1961–1967.
23. Pelly, B. 2002. Active common mode filter connected to the AC line. Patent application No. 20020171473, November 2002.
24. Oriti, G., Julian, L., and Lipo, T.A. 1997. An inverter/motor drive with common mode voltage elimination. *IEEE IAS, Conference Record*, vol. I, pp. 587–592.
25. Julian, L., Oriti, G., and Lipo, T.A. 1999. Elimination of common mode voltage in three-phase sinusoidal power converters. *IEEE Trans.* 14(5), 82–989.
26. Whistler, R.J. 1999. Laminated bus bars eliminate unmanageable cabling in high power systems cabinets. PCIM Magazine, June 1999, pp. 1–4.

27. Dimino, C.A., Dodballapur, R., Pomea, J.A. 1994. A low inductance simplified snubber power inverter implementation. *Proceedings of the HFPC*, April.
28. Sueker, K. 2005. Power electronics design. *Newnes Ed.*
29. Bisenius, W.S. 2012. Ingress protection: The system of tests and meaning of codes. *Compliance Engineering Magazine*, www.ce-mag.com.
30. Mammano, B. 2007. A historical perspective of the power electronics industry, plenary session. *APEC*.
31. Picard, J. 2010. Under the hood of flyback SMPS designs. *2010 Texas Instruments Power Supply Design Seminar, SEM1900*, Topic 1, TI Literature Number: SLUP261, 2010.
32. Fujii, M., Maruyama, H. and Boku, K. 2002. FA5553/FA5547 series of PWM control power supply IC with multi-functionality and low standby power. *Fuji Electr. Rev.* 54(2), 68–72.
33. Pelly, B. 1994. Choosing between multiple discretes and high current modules. IRF Design Tip Documentation.

7 Thermal Management and Reliability

7.1 THERMAL MANAGEMENT

7.1.1 THEORY

The most important criterion in packaging consists of thermal management. It is not only the avoidance of the maximum temperature rating for device protection, but also the desire to operate at lower temperature. The power semiconductor devices operate better at lower temperature, with performance maintained within the specified data. Moreover, the lifetime of a power semiconductor device depends strongly on the temperature it operates at. A rule of thumb for silicon devices is that failure rates often double for every 10–15°C rise in operating temperature beyond 50°C [1].

After the power loss has been calculated at device level, the cooling system can be dimensioned. Thermal calculations are subsequently performed at all levels:

- At device level, for heat transfer from the semiconductor die to the external heat-sink and/or the selection of the device's package;
- At converter level, for selection of the heat-sinks or cold-plates and the calculation of the radiated ad conducted heat toward the exterior;
- At equipment level, for selection of the enclosure.

All power semiconductors dissipate their switching and conduction losses and these should be removed as fast as possible. As this power is mainly removed through a contact surface with a cooling system, the whole size of the power converter and the power density within the equipment depend on the quality of the thermal transfer through the selected cooling system. Modern power converters expect a power removal of up to 200–500 W/cm²—that represents about half of the mean power density of the sun's surface.

A model for a typical thermal circuit of a power semiconductor device is presented in Figure 7.1 and the change in temperature from the junction to the cooling agent $(T_j - T_s)$ under a given power dissipation P is provided by the following relationship:

$$T_j - T_s = P \cdot \left[R_{thjc} + R_{thcs} + R_{thss} \right] \tag{7.1}$$

where R_{thjc} represents the junction-to-case thermal resistance, R_{thcs} represents the case-to-cooling thermal resistance and R_{thss} represents the thermal resistance of the cooling system from the cooling agent (air, water, liquid) to the surface. It is,

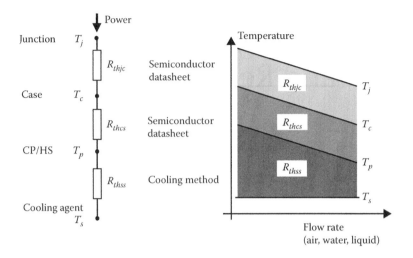

FIGURE 7.1 Equivalent model of a thermal system for a semiconductor device.

therefore, obvious that a lower equivalent resistance will keep the junction closer to the temperature of the cooling agent. The first two terms are provided by the semiconductor device datasheet and they are the same for a given device. The thermal resistance of the silicon itself is only 2–5% of the total thermal resistance, and the thermal properties of the cooling system are more important. Among these, the dependency of the thermal resistance on the thermal conductivity variation with temperature is very important for thermal modeling. Fortunately, this dependency for materials like Aluminum and Copper is minimal, and we can assume a quasi-constant thermal resistance for the entire operation range.

The second issue with Equation 7.01 relates to the interpretation of temperatures [2]. Thermal analysis in three-dimensional systems has pointed out the temperature variation on every geometrical direction, even on a state of equilibrium. The temperature values of Equation 7.01 are an average approximation of the reality, and the entire theory herein presented is neglecting the actual temperature distribution. Equation 7.1 can also be seen as an one-dimensional analysis.

The solution for improvement for a given structure is to go to higher ratings in order to minimize the thermal resistance. The last term in Equation 7.1 corresponds to the cooling system, and that depends on the method chosen, the cooling agent, the material and shape of the heat-sink or cold-plate, the flow rate of the cooling agent, and so on. Let us analyze the options we have.

Figure 7.1 shows that the higher the flow rate, the smaller the thermal resistance and the better the cooling. However, the cooling device itself and the connecting pipes, heat-sink, or cold-plate limit the flow rate of the cooling agent. Secondly, let us note the multitude of choices for cooling systems and their selection depends on the system requirements and cost (Figure 7.2).

The cooling agent can be air, water, or a special agent with a larger heat capacity. For example, removing the same power dissipation of 244 W from a three-phase

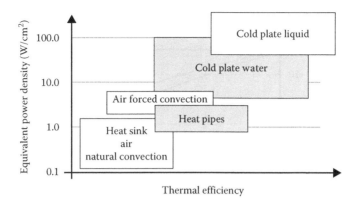

FIGURE 7.2 Thermal efficiency of different methods used for cooling.

converter produces an increase of 40°C on an air-cooled heat sink and only 2.8°C rise in a water-cooled system. Liquids with better heat capacity are based on different glycol solutions [3].

The simpler solution consists of air cooling. A *heat spreader* is therein necessary to interface between the device/heat-sink and the ambient. This can be a simple copper or aluminum plate, or a more complex heat-sink structure. For example [5], the temperature drop across a 1 mm thick copper plate is approximately 0.25°C at a heat flux of 10 W/cm², and it increases to 2.5°C at 100 W/cm² or 12.5°C at 500 W/cm². *Heat spreaders* are quite effective at low heat fluxes and may become unattractive at higher heat fluxes. This conclusion made room for liquid cooling (Figure 7.2).

Historically, the liquid cooling was not very attractive near electronic equipment. As the power electronics matured, the new development efforts moved from the power electronics converter to the auxiliary equipment, including the heat removal devices. First liquid-cooled systems were introduced in 1982 to large-scale integrated circuits [5]. The emergence of insulated gate bipolar transistor (IGBT) devices in early 1990s has also boasted the use of liquid cooled cold-plates in power electronic equipment. Today, the liquid cooling is the method of choice for medium and high power converters.

Most liquid cooling systems are water based and need to be protected from freezing (Figure 7.3) and this needs to target temperatures where the first ice crystals are formed and the flow is jeopardized. For this reason, *ethylene glycol* was used in solutions up to 40%. The recent focus on environment recommended the use of propylene glycol instead of ethylene glycol (Table 7.2). Either type of glycol needs to be pure enough in the sense it does not contain other additives like those used in automotive applications.

Once the method has been selected, the type of cold-plate or heat-sink is the next area of focus [5]. Different materials like aluminum or copper are used in manufacturing (Table 7.1) and their shapes can be different to facilitate the easy transfer of thermal energy. Materials and shapes of thermal devices able to handle these requirements are designed and selected based on knowledge of physical laws of thermal conduction, thermal radiation, and nature of forced convection or phase convection. The final criterion here is the cost of the device, as a more complex mechanical structure able to remove more heat will also cost more.

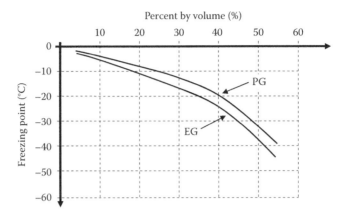

FIGURE 7.3 Freezing point of various cooling agents (EG = ethylene glycol, PG = propylene glycol).

7.1.2 TRANSIENT THERMAL IMPEDANCE

All the previous analyses have focused on steady-state thermal aspects when the average loss of power is known. In many applications, power semiconductor devices are stressed by transient over-currents with large instantaneous dissipation.

Modeling transient thermal impedance requires definition of a new parameter, *heat capacity*. This represents the rate of change of the heat energy with respect to the material temperature. The heat capacity per volume yields:

$$\frac{dQ}{dT} = C_v \tag{7.2}$$

The transient behavior of the junction temperature is related to the time-dependent heat diffusion equation with a simplified solution given by the analog equivalent circuit shown in Figure 7.4, and often called *the Cauer model*. The equivalent of the heat capacity is a capacitor able to slow down the junction temperature variation

TABLE 7.1

Thermal Conductivity of Different Materials Used for Heat Extraction

Material	Thermal Conductivity (Cal-g/cm²/s/°C)
Water	1.00
Steel	0.11
Aluminum	0.50
Copper	0.92

TABLE 7.2
Liquid Thermal Conductivity

Fluid	Thermal Conductivity (W/m K)
Water (fresh)	0.609
Ethylene glycol	0.258
Propylene glycol	0.147

when a step change in power is applied. The ground potential represents the ambient temperature, and the heat capacities are connected from each node to the ambient (ground potential). This model can easily be implemented on any software for simulation of electrical circuits. As an alternate option, *the Foster model* is considering all thermal capacitors connected in parallel with equivalent thermal resistances. More recent computer models consider three-dimensional modeling of the heat transfer.

Identical to the analog circuitry, a thermal time constant can be defined as

$$\tau_{th} = \frac{\pi}{4} \cdot R_{th} \cdot C_{th} \tag{7.3}$$

Packaging materials and structures are always designed to minimize the thermal resistance as the loss power is transferred usually in average. For this reason, the thermal time constant and the power transient capability of a device are limited. However, it has been proven that power semiconductor devices can withstand large overload capabilities that exceed their average power ratings. Completing Figure 7.1 with the transient model yields the equivalent circuit shown in Figure 7.5. The temperature evolution in space when a pulse power is applied is also shown.

Transient thermal models are very useful in thermal analysis of power converters switched at high frequency with a variable duty cycle. The cooling system should sustain pulses of power with considerable thermal dynamics.

As an example, let us consider a three-phase power converter that supplies intermittently a motor fan. The steady-state operation of electronics has depicted 240 W of power loss. The fan works for 20 s at each 60 s. A very simple Cu plate used for

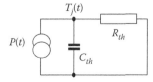

FIGURE 7.4   Transient thermal model.

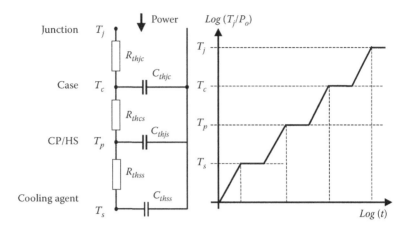

FIGURE 7.5   Transient equivalent circuit and temperature evolution in space.

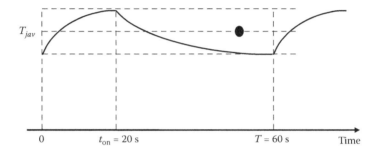

FIGURE 7.6   Transient thermal behavior of the considered example.

cooling will allow a steady-state temperature increase of $\Delta T_{jav} = 32$ K. The simulation results for the complete thermal dynamics are shown in Figure 7.6.

7.2   THEORY OF RELIABILITY AND LIFETIME—DEFINITIONS

From a mathematical perspective, *reliability* represents the probability that a product performs its intended function without failure under specified environment conditions, for certain time. Reliability can also be shown as a probability distribution of cumulative failures. The analytical expression yields as

$$R(t) = \frac{Number\_of\_components\_surviving\_at\_a\_moment\_t}{Number\_of\_components\_surviving\_at\_t = 0} \quad (7.4)$$

Over the years, the term reliability got a wider connotation, as it is mostly used beyond its analytical meaning, to express the quality of being reliable, that is the extent to which an experiment, test, or measuring procedure yields the same results on repeated trials.

The *failure rate* can be expressed as $f(t)$:

$$f(t) = \frac{Number\_of\_components\_failing\_per\_time\_unit\_at\_a\_moment\_t}{Number\_of\_components\_surviving\_at\_a\_moment\_t}$$

(7.5)

For a constant failure rate:

$$R(t) = \exp(-f \cdot t)$$

(7.6)

This definition is not very friendly as it deals with very small numbers. As an alternative, we may use the failure rate in a specified time interval of 1×10^9. This allows us to define *FIT* (*failures-in-time*) as

$$\text{FIT} = \frac{1 \cdot \text{Failure}}{1 \cdot 10^9}$$

(7.7)

In equipment where high reliability is a must, failure rate of the semiconductor devices usually range from 10 to 100 FIT (1 FIT = 10^{-9}/h) [44].

Mean time between failures (MTBF) is the predicted time between inherent failures of a system during operation. This means:

$$\text{MTBF} = \frac{1}{f}$$

(7.8)

Lifetime of an electronic component is specified as the time at which the electrical parameters have drifted out of some specified limits mainly due to wearout (Figure 7.7). In reality, during the time interval with a constant failure rate, failures due to improper use of devices or due to accidental ambient factors may happen and they are generally identified as *random failures*. As a rule of thumb, if the failure rate during the usage interval (interval with random failures) is above 0.1%, one should revise carefully the design of that system.

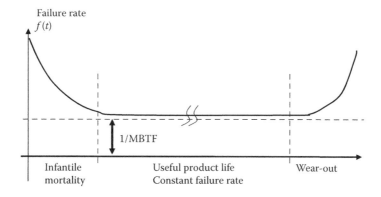

FIGURE 7.7 Bathtub curve for lifetime of a power electronics equipment.

All of these measures for reliability have a statistical character. Data can be depicted from numerous experimental or field test data or determined by accelerated power and temperature cycling tests.

Reliability of power electronics equipment has become a priority for any system developer [6]. Once the knowledge about the power electronics equipment perfected, the studies of reliability became a priority. As of 2012, reliability is one of the most dynamic fields of interest in power electronics. While studies are dedicated to different application fields (integrated circuits [7], energy distribution networks [8], wind energy [9,10], distributed generation [11], photo-voltaic systems [12], high-power converters [13]), this Chapter provides a general introduction to reliability of power electronic equipment.

In order to improve device lifetime and reliability, efforts are made at all phases of device's life:

- During the manufacturing process, attention is paid to quality. *Statistical process control* (SPC) is the application of statistical methods to the monitoring and control of a process to ensure that it operates at its full potential to produce conforming product. This is necessary in order to produce as much conforming or reliable devices and systems with the least possible waste.

- During operation, the ambient conditions are severely monitored and the control system adapts depending on the external factors. *Advanced control systems* are designed to withstand a variety of environmental conditions or variation of parameters, generally considered as uncertain parameters or disturbances. In this respect, modern computer-based simulation and optimization tools allow design under complex criteria. Probably the most important example of a robust control technique [14] is H-infinity loop-shaping, which was developed by Duncan McFarlane and Keith Glover of Cambridge University. The H-infinity loop-shaping method minimizes the sensitivity of a system over its frequency spectrum, and this guarantees that the system will not greatly deviate from expected trajectories when disturbances enter the system. Another example is LQG/LTR, which was developed to overcome the robustness problems of LQG control.

- Given the complexity of operation and failure constraints, modern control systems are joined by expert systems. Design for reliability (DFR) includes an advanced fault management system able to decide on the operation of the power electronic equipment when special situations or random faults occur. Expert systems are able to predict failure, to provide scenarios for getting out of failure mode, and/or operation optimization for minimization of risks. Fault management system is a simple example of an expert system targeting reliability improvement. A particular case is the design with redundancy. This means that if one part of the system fails, there is an alternate success path, such as a backup system. An UPS system might use two batteries. If one battery fails or gets discharged, the UPS still operates using the other battery. Redundancy significantly increases system reliability, and is often the only viable means of doing so.

- Accurate lifetime prediction should help scheduling the replacement of each individual component and the periodic maintenance. This is usually achieved with *accelerated test plans.* Because reliability is a probability, even highly reliable systems have some chance of failure. A single test is not able to generate enough statistical data. Multiple tests or long-duration tests are usually very expensive. In order to secure the statistical information necessary for failure and reliability analysis, a set of experiments is specifically designed with the method of *design of experiments,* which has been developed by people like *G. Taguchi.* The method consists of introducing noise factors into experiments in order to quantify different effects and to use a signal-to-noise metric. The *design of experiments* method is heavily used in manufacturing for calibration of the robust design process. Different sets of tests are required for assessing the lifetime of various power devices and systems. The purpose of accelerated life testing is to induce field failure in the laboratory at a much faster rate by providing a harsher, but representative environment. In such a test, the device or system is expected to fail in the laboratory just as it would have failed in the field (but in much less time), based on destructive energy equivalence. This helps to discover failure modes or to predict the normal field life from the high-stress life. Establishing the appropriate accelerated test plan is a science by itself.

7.3   FAILURE AND LIFETIME

7.3.1   SYSTEM FAILURE RATE

The failure rate of the entire system is calculated from the failure rates for each individual component. In most systems, like it is the case of power electronics circuits, all components are considered as being important. From reliability point of view, this means that the components are considered as being "*in series*" within the system. In a *series system* that includes n components, the overall failure rate λ_{SYSTEM} is given by

$$\lambda_{\text{SYSTEM}} = \sum_{i=1}^{N} \lambda_i \qquad (7.9)$$

where λ_i corresponds to the individual failure rates of the elements [15].

7.3.2   COMPONENT FAILURE RATE

Even for components manufactured under identical conditions, the failure rates in the field can vary by a factor of 10 depending simply on the conditions the device was used [44]. Each individual component is considered to have a *constant failure rate* (λ), which is weighed (derated) through a series of stress factors.

$$\lambda = \lambda_{\text{base}} \cdot \prod_{i=1}^{m} \pi_i = \lambda_{\text{base}} \cdot \pi_T \cdot \pi_S \cdot \pi_A \cdot \pi_R \cdot \pi_E \cdot \pi_C \cdot \pi_Q \qquad (7.10)$$

The stress factors (π_i) refer to temperature, voltage stress, circuit/application, rating, environment, construction factor, and product quality, respectively. They have the meaning of derating the datasheet information for the constant failure rate depending on manufacturing, environmental conditions, and circuit use of devices.

As an example, let us consider a silicon rectifier diode [44] with a maximum rated current of 1 A, at an ambient temperature of 30°C, and a rated maximum junction temperature of $T_{j\max} = 150°C$. The rectifier has an actual operation at a current of 0.5 A, at 40% of the rated voltage, at ambient temperature, on a system fixed to the ground, with a junction temperature of $T_j = 100°C$. The datasheet also provides the base failure rate of $\lambda_{base} = 0.0010/10^6$ h. The manufacturer is also providing the following weigh factors:

$$\pi_E = 6.0 \text{ for a fixed application (on the ground)}$$

$$\pi_Q = 2.4 \text{ and } \pi_C = 1.0 \text{ for the specific production line}$$

$$\pi_T = 8.0 \text{ when operated at } T_j = 100°C$$

$$\pi_S = 0.11 \text{ at 40\% of the rated voltage}$$

Using the above values, λ_P is calculated as follows:

$$\lambda = \lambda_{base} \cdot \prod_{i=1}^{m} \pi_i = 0.0010 \cdot (8.0 \cdot 0.11 \cdot 1 \cdot 1 \cdot 6.0 \cdot 1 \cdot 2.4) = 0.0126/10^6 \quad (7.11)$$

In most cases, the steady-state temperature is the only factor incorporated in the model (π_T). For instance, an IGBT device operated always at a junction temperature of 125°C will have a shorter lifetime than the same device operated always at a junction temperature of 75°C. This dependency is illustrated by the *Aarhenius law*:

$$\lambda_T = \lambda_{ref} \cdot \exp\left(-\frac{E_{dev}}{k \cdot T}\right) \quad (7.12)$$

where E_{dev} represents the *thermal activation energy* of a particular failure, and K is the Boltzman constant $= 8.617 \times 10^{-5}$ eV/K. This can be shown in a similar form as

$$\pi_T = \pi_{ref} \cdot \exp\left(-\frac{E_{dev}}{k \cdot T}\right) \quad (7.13)$$

The *thermal activation energy* (or simply called *activation energy*) concept comes from chemistry where it represents the minimum amount of energy required to convert a normal stable molecule into a reactive molecule. The *activation energy* levels for different failure modes of power semiconductors are in the range 0.1... 2.0 [eV].

Since different wear-out mechanisms are possible (see later on, Section 7.3.4.), each one can be provided a failure rate characterized with a different activation energy and the equivalent activation energy value for the entire system is calculated based on the dominant failure [16] or based on a weighed activation energy. Since the energy activation and the weigh factors depend mainly on the manufacturing of power semiconductor device, it is very difficult to maintain a single set of weighs. This is why the use of the *activation energy method* for reliability calculation has its pros and cons.

It is possible to neglect the variation due to factors other than the steady-state temperature since this is the most important stress factor. Analogous to the *Aarhenius equation*, other relationships are available to characterize other stress factors. Furthermore, other common ways to determine the life stress relationship are: the Eyring Model, the Inverse Power Law Model, the Temperature-Humidity Model, the Temperature Nonthermal Model.

7.3.3 FAILURE RATE FOR DIVERSE COMPONENTS USED IN POWER ELECTRONICS

The base failure rate, reliability, or MTBF for each individual component is provided by manufacturer or by industry standards. The industry standards are more generic, and do not take into account the specifics of each production line. The manufacturer provided reliability data is based on measurements and accelerated life tests. They provide base values for each product that still needs to be weighed with the actual operation conditions (environment and circuit).

Table 7.3 shows example of data depicted from the U.S. military standard MIL-HDBK-217. As will be shown later on, this type of standards regulation does not provide data about high power switching semiconductor devices, and data needs to be adapted from conventional transistors or modules.

All of these components are present in different power electronic structures. In medium and high power converters, the power semiconductors are more important.

TABLE 7.3

Example of Failure Rates for Electronic Components

Description	Type	MTBF (Years)
Connector	Per pin	2283
Semiconductors	Si diode	2283
	Si transistor	1426
Resistor	Carbon	11,415
	Wire-wound	4566
	Film	2283
Capacitor	Electrolytic	76
	Tantalum	114
	Paper	228
	Ceramic	456
	Plastic film	5707
Magnetics	Power inductor/copper	2283
	Transformer/copper	570

As shown before, the failure of the entire system is given by the failure rates for each component. Hence, the most critical component is the DC link electrolytic capacitor.

Capacitor lifetime is limited by electrochemical degradation that proceeds in predictable manner, accelerated by environmental factors like temperature and voltage stress. Historically, the lifetime of electrolytic capacitors has progressed from 1000 h at 65°C, 40 years ago, to 15,000 h at 105°C today [17]. The predicted lifetime at a given temperature yields with the *Arrhenius* law equations used by all commercial capacitor vendors. For instance, a capacitor rated with 10,000 at 105°C, could survive 160,000 at 65°C.

The reliability of the power semiconductor module depends on many different physical and chemical processes. It can be demonstrated that power semiconductor module could contribute better performance for the system reliability than individual components [18]. Moreover, a power module provides a better thermal design and an excellent layout, both with effects on the system reliability. Using a power module supplied from the manufacturer rather than using individual components is recommended for the inverter application.

7.3.4 FAILURE MODES FOR A POWER SEMICONDUCTOR DEVICE

As the most part of the R&D is dedicated to reliability calculation and improvement for the power semiconductor devices, they will be presented in more detail in what follows.

There are two major categories of failure in power electronic equipment:

- Random (accidental) failures
- Wear-out mechanisms

Manufacturers of power semiconductor devices consider reliability within the new technologies [19]. The safe-operating-area needs to be enlarged so that devices are extremely rugged, capable of withstanding current, voltage and temperature conditions well in excess of their nominal continuous ratings.

7.3.5 WEAR-OUT MECHANISMS IN POWER SEMICONDUCTORS

The most observed failures due to wear-out mechanisms for a power semiconductor device are package related and they are accelerated by thermo-mechanical stress [20].

Considering operation of power electronics equipment for a long enough time interval will show changes in parameters. These changes can be globally called wear-out mechanisms. They occur both in the semiconductor die and the mechanical packaging [21].

Examples of wear-out mechanisms in the packaging and assembly of power semiconductors:

- Bond wire
  - Bond wire lift-off
  - Bond wire heel crack

- Damage of the AL metallization
- Electromigration in the bonding wires
- Solder joints degradation
- Aging of the direct copper bond
  - Cracks and void formation
  - Delamination of copper metallization
- Die fracture due to other pre-existing defects

Examples of wear-out mechanisms in semiconductor die (mostly similar to those studied in microelectronics) [22,23]:

- Ionic contamination
- Hot carrier injection
- Slow trapping
- Gate–oxide breakdown
- Negative bias temperature inversion

The thermo-mechanical wear-out processes are the most important [24]. Any of these wear-out mechanisms can be analyzed individually with the *thermal activation energy* concept.

The ageing of devices (wearout) is accelerated by operation under application-related stress factors like the repetitive protected short-circuit [25]. A *critical energy* has been defined in [25] to account for the cumulative degradation effect of repetitive short-circuits while enough time was allowed between short-circuits for device cooling. It has demonstrated that it takes around 10,000 protected short-circuits to reach failure of a conventional 600 V IGBT [25].

The repetitive short-circuit events produce a thermal cycling with a large temperature difference within the IGBT. This thermal cycling introduces repetitive compressive and tensile stress in the emitter metallization film due to the CTE mismatch between aluminum metallization and the silicon chip. This stress leads to high plastic deformation with effects in increasing of the metallization resistance and weakening of the bond wire contacts (resistance produces more dissipation, hence increase in local temperature, and fatigue follows).

After understanding all the possible wear-out mechanisms and what it takes for them to occur, the design engineers are setting condition monitoring systems able to tell at each moment the actual status of the component or subsystem and how many hours are still available before the next maintenance checkup. Such systems are well known from laptop computers, where the battery status is monitored based on the amount of hours it is used, and the predicted lifetime is at any time reported to the user. Similar systems are to be implemented for IGBT-based equipment and their studies are in a very embryonic state [26]. The idea is to monitor and measure in-circuit the most important device parameters (gate threshold voltage, leakage current, voltage drop in conduction) along with the operating temperature and to assess the performance degradation and ageing of the IGBT device from such information.

7.4   LIFETIME CALCULATION AND MODELING

7.4.1   Problem Setting

Lifetime calculation for given equipment comes down to applying the analytical formula onto the lifetime rates for each individual component. The modeling and/or calculation of lifetime rate for each component is done usually independent of the final equipment. The lifetime of each component is investigated by analyzing all the possible failures. A fault tree is built sometimes based on experimental (field) evidence [27]. Each individual failure mechanism is isolated and further investigated by appropriate testing or by physical modeling (Figure 7.8).

If the failure rate for an individual failure mode needs to be determined by experiment, a set of accelerated tests is adopted since most failure mechanisms are based on accumulated degradation (wearout) that occurs after long time in practice.

The second possible approach for determination of the failure rate for a certain failure mode is based on physical description of the system. Equations for operation under specified circuit and environmental conditions are required to qualify the production of failure. At this moment, the method of physics of failure is more developed for semiconductor materials (microelectronics) and under major investigation for thermo-mechanical failures (packaging).

Let us assume some equipment has failed at a given moment after hours of proper operation. The obvious question is *"what happened?"* and the first reaction would be to blame an external event. If no obvious external event has occurred (there was no mechanical, thermal or electrical stress unexpectedly coming from outside the proper operation of each device/component within the equipment), the investigation should go back toward the manufacturing methods and the technology inside each device. Each device or component within that equipment is subjected to multiple failure mechanisms, each mechanism being triggered by its own accumulated degradation. The multiple mechanism model extrapolates independent *acceleration factors* for each component's mechanism of concern based on the component's stress states. The concept of *acceleration factor* and the thermal dependency of each accumulated degradation mechanism will be studied in the next section (Figure 7.9).

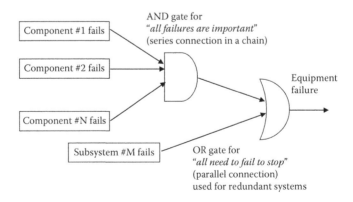

FIGURE 7.8   Example of fault tree for a power converter.

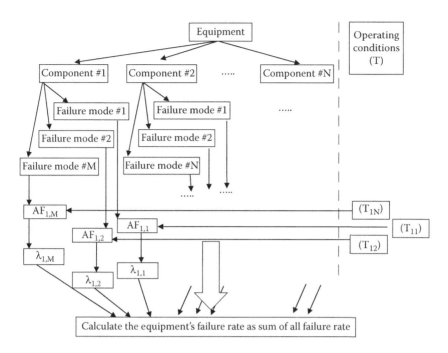

FIGURE 7.9 Multiple mechanism model.

7.4.2 ACCELERATED TESTS FOR ELECTRONIC EQUIPMENT

7.4.2.1 Using the Activation Energy Method

Since the expected lifetime of power electronics equipment is between 10 and 30 years, it is not practical to gather data from actual life experiments. Some technologies or newer generation of devices were not even imagined 30 years ago. For this reason, the lifetime is estimated by accelerating the stress factors. The qualification tests last for a shorter time, and the statistical model of lifetime prediction can be determined more easily.

Given the multiple stress factors from Equation 7.09, it is practical to limit the qualification tests to temperature as the main stress factor. Accelerated tests are therefore designed for temperature as the unique stress factor. Reconsidering the *Aarhenius law*:

$$\lambda_T = \lambda_{\text{ref}} \cdot \exp\left(-\frac{E_{\text{dev}}}{k \cdot T}\right) \tag{7.14}$$

Helps depicting the *time to the failure* as

$$\tau = A \cdot \exp\left(\frac{E_{\text{dev}}}{k \cdot T}\right) \tag{7.15}$$

Therefore, the actual *time-to-failure/temperature* dependency belongs to the same characteristics with the *accelerated-time-to-failure/test-temperature* point. By selecting a test stress temperature T_s, and measuring the *test-time-to-failure* τ_s, one can get information about the actual *time-to-failure* τ_0 when operated at a lower actual temperature T_0 (Figure 7.10). Such tests are also referred to as *HAST (highly accelerated stress tests)*.

It is common to define an *acceleration factor*:

$$AF_T = \frac{\tau_0}{\tau_s} = e^{\left(\frac{E_{dev}}{K}\left(\frac{1}{T_0} - \frac{1}{T_s}\right)\right)} \tag{7.16}$$

Let us consider an example. Several samples for a power converter product have been tested at 125°C and the statistical *time-to-failure* was determined as 1000 h for a known failure mode with the activation energy of 0.5 eV. Measurements for the consistent operation of the converter in the field show operation at a temperature of a 50°C. The acceleration factor yields:

$$AF_T = e^{\left(\frac{0.5 \cdot eV}{8.617 \cdot 10^{-5} \cdot eV}\left(\frac{1}{273+50} - \frac{1}{273+125}\right)\right)} = e^{\left(\frac{0.5 \cdot eV}{8.617 \cdot 10^{-5} \cdot eV}\left(\frac{75}{323 \cdot 398}\right)\right)} = 26.3 \tag{7.17}$$

Hence, the prediction for the *time-to-failure* within the actual operation conditions yields:

$$\text{Time-to-failure} = 1000 \cdot 26.3 = 26,300 \, h = 3 \, \text{years} \tag{7.18}$$

If there are multiple known failure modes of the same device, a generic (weighed or averaged) thermal activation energy can be considered. Alternatively, a more accurate modeling would consider all the possible failure mechanisms, each one with its own thermal activation energy and its own accelerated test plan. The lifetime has to be calculated from results pertaining to all failure modes (Figure 7.9). The failure rate in FITs (number fails in 10^9 device hours) yields:

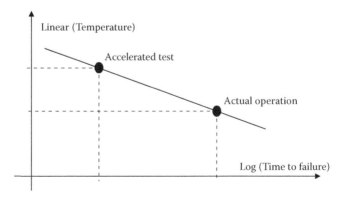

FIGURE 7.10   Accelerated tests for thermal related wear-out.

$$\text{FIT} = \sum_{i=1}^{\beta} \left[\frac{x_i}{\left(\sum_{j=1}^{k} \text{TDH}_j \cdot \text{AF}_{ij} \right)} \right] \cdot \frac{M \cdot 10^9}{\sum_{i=1}^{\beta} x_i} \qquad (7.19)$$

where:

- β represents the number of distinct failure mechanisms considered for the study.
- k represents the number of different lifetime tests combined in the calculation of each failure (i).
- x_i represents the total number of failures for a specific failure mechanism (one of β).
- TDH_j represents the number of test hours before a certain failure (i).
- AF_{ij} represents the acceleration factor for the failure (i) associated with the test (j).
- M represents a factor determined by the confidence level for the estimation (so-called *Chi square confidence value*).

This method works both ways:

- When we want to predict failure rate for each component, independent of the operation, just for simulated accelerated tests, the direction is from accelerated tests to the nominal value (*benchmark value*).
- Otherwise, we may know a value for the nominal lifetime, and use the actual operating conditions to predict the actual lifetime considering within the acceleration factors the effect of the operating conditions.

The accelerated test plans can be designed for voltage stress or humidity stress. When both the temperature and voltage stress factors are used, the acceleration factor of the test should be calculated as a product $A = \text{AF}_T * \text{AF}_V$, where AF_T is the thermal acceleration factor, and AF_V is the voltage related acceleration factor. This is provided by the exponential law:

$$\text{AF}_v = \exp\left[\gamma \cdot \left(V_A - V_0 \right) \right] \qquad (7.20)$$

However, the voltage and/or humidity derate are not very much used in practice for power electronics equipment.

7.4.2.2 Temperature Cycling

Another accelerated test procedure refers to *temperature cycling*. This time, the effect of *temperature change* on the equipment's lifetime is calculated. Each time the equipment powers up from ambient temperature to the actual operation, there is a gradient of temperature than influences the *time-to-failure*. This test is very important for the validation of the packaging and overall assembly and it does not require electrical power in the circuit. The device is heated and cooled by an external heat source, like an oven, to produce the variation in ΔT_{jc}. Materials have different thermal expansion coefficients and the temperature cycling induces certain failures due to the thermo-mechanical expansion. Such failures can be package cracking, die cracking, wire-bond lift-off, and an increase in contact resistance.

The *Coffin-Manson model* is generally used to designing this acceleration test. The *time-to-failure* τ is proportional to $[\Delta T]^n$.

$$\tau = A \cdot \left[\Delta T\right]^n \tag{7.21}$$

where A is a constant. The *acceleration factor* yields:

$$\text{AF} = \left[\frac{\Delta T_s}{\Delta T_0}\right]^n \tag{7.22}$$

where ΔT_s is the stress temperature difference and ΔT_o is the actual operation temperature difference.

Let us consider an example. A *temperature cycling test* was designed with a temperature change from $-65°C$ to $125°C$. The statistically determined *time-to-failure* was 1000 cycles (the components failed after around 1000 cycles). The manufacturer of the component provides information about the modeling of the package with $n = 6$. What is the time-to-failure for the same system performing 100 cycles per day, with an actual temperature cycle between the ambient temperature of $25°C$ and the steady-state operation at $90°C$?

First, we will determine the acceleration factor:

$$\text{AF} = \left[\frac{125 + 65}{90 - 25}\right]^6 = 623.794 \tag{7.23}$$

The expected time-to-failure under actual operation conditions yields:

$$\tau_{\text{actual}} = \frac{623.794 \cdot 1000[\text{cycles}]}{100 \cdot 365} = 17.09 \text{ years} \tag{7.24}$$

7.4.2.3 Accelerated Tests for Power Cycling

Numerous motor drive applications require an operation with frequent acceleration and stops. This is the case of traction power converters for urban trains, elevators and so on. A short-distance train travels for only a few minutes between stops and is expected to operate 30 years in the field. This corresponds to a lifetime of 100,000–135,000 h and 5–10 million travel-stop cycles [32]. Other applications subjected to similar operation include different machine-tools, or home appliance equipment [31]. This operation with large gradients of temperature derived from frequently heating and cooling the power semiconductor devices favors specific wear-out mechanisms. Given the large spectrum of possible load profiles (like different distances between stops, and so on), it is virtually impossible to define a unique test profile. Field reports show that the use of IGBT devices in this operation mode with frequent stops, eventually led to wearout and failure related to

- Disconnection of the aluminum bonding wires from the silicon chips (*bond-wire lift-off*).
- Increase in module R_{thjc}.

In order to qualify the power electronics equipment for this type of operation, accelerated tests for power cycling have been designed [32]. These tests involve operation of the device with electrical power (Figure 7.11). The power semiconductor device is set-up for conduction under a permanent gate control. The collector current is externally switched on and off, and a very large value of the current is adopted. The junction temperature is closely monitored with the goal to emulate a very large temperature difference. The greater the variation in ΔT_{j-c}, the greater the thermal stress, and the shorter is the power semiconductor lifetime. The same reasoning as explained in Figure 7.10 is adopted for the design of experiment (Figure 7.12).

Obviously, there are different variations of the test as different parameters and measurement methods differ [32–35].

A very different approach for the power cycling tests is specified in Section 8.2.6 of the IEC standard 60747-9 of 2001. It is named the "*Intermittent operating life test*" and is a power cycling test based on controlling the gate of an IGBT to turn on and off for cycling a very large load current. Most of the power cycling tests proposed by industry and academia in the past meets the requirements of the IEC test. However, they do not achieve power cycling by gate control. Hence, the difference between the two methods is that the IEC approach includes switching loss in the device while the switching loss is not present in tests where the device is held permanently on. This is not expected to significantly affect the test results.

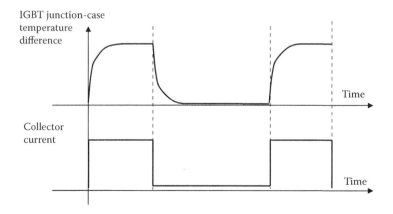

FIGURE 7.11   Principle power cycling.

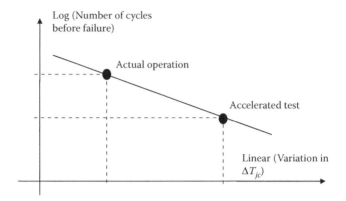

FIGURE 7.12   Accelerated tests with power cycling.

Once the test procedure has been decided on, the engineer needs to define the failure criteria. Usually, this is defined as one or multiple of the following:

- Increase of the conduction voltage drop by 5–20%
- Increase of the thermal resistance R_{thjc} by 20%
- Increase of the Gate-Emitter threshold voltage by 20%
- Increase of the leakage current by 20%
- Increase of the steady-state gate current by 20%

7.4.3   Modeling with Physics of Failure

There is a tremendous recent effort for the development of accurate failure models based on the actual physical operation of the device. The following steps are generally used in this respect:

- Identify the potential failure mechanisms.
- Expose the power semiconductor device to highly accelerated stress to find the dominant root-cause.
- Identify the dominant failure as the weakest link.
- Model the dominant mechanism.
- Combine the data.
- Develop an analytical model for the dominant failure.

7.5   STANDARDS AND SOFTWARE TOOLS

7.5.1   Standards

There are different standards that provide guidance for calculation of the reliability performance of electronic components. The most quoted reference is the U.S. Military Handbook for "*Reliability Prediction of Electronic Equipment*" (MIL-HDBK-217) [36]. The prediction based on MIL-HDBK-217 considers systems as being in a series that means any device failure is as important. A similar effort is assembled in the 3rd issue of *Telcordia SR-332* (entitled *"Reliability Prediction Procedure for Electronic Equipment"*) released in January 2011 [37]. The *Telcordia* model for reliability of electronics equipment uses a black-box technique. The failure rate is calculated from a steady state failure rate, weighed by quality, stress and temperature ratios:

$$\lambda_{BB} = \lambda_G \cdot \pi_Q \cdot \pi_S \cdot \pi_T \tag{7.25}$$

Other standards are released by NERC and NIST for the U.S. power systems. Finally, *International Electrotechnical Commission* has elaborated two standards on the field [8]:

- The IEC 61709 standard was published in 1996, it is derived from a manufacturer standard and it is very similar to MIL standard. The failure rates and the influence factors are extrapolated from field data of items operating in different environmental conditions. They are referred to the equipment manufactured in the first half of 1990s.
- The IEC62380 standard was released in 2004, and compiles data collected in between 1999 and 2001 from the telecommunication sector. However, the standard is applicable to both ground and airborne equipment.

All of these standards provide models for conventional base components like resistors, capacitors, inductors, transformers, transistors and diodes. They do not include models for newer power devices like IGBTs, GTO, IGCT, and the like. There is actually no standard available for models pertaining to these power devices.

The *European Power Supply Manufacturers Association* has released in 2005 a document entitled *"Guidelines to Understanding Reliability Prediction"* [38]. The report presents results for an interview of 16 EU companies on the methods they use for reliability prediction. The results are very disappointing. For example the *MTTF* at room temperature (25°C) of a small 1 W DC-DC converter with 10 components ranged from 95 years to 11,895 years—a variation of 125:1! When it was agreed that the major reason for such variation was an inappropriate assumption to consider the potted product a *"hybrid assembly,"* re-calculation showed the variation to range from 1205 years to 11,895 years—a variation of 9.9:1.

We should expect in the near future, more effort on defining standards and procedures for reliability prediction. In this respect, let us note that on December 31, 2012, the *North American Electric Reliability Corporation* (*"NERC"*) filed with the *Federal Energy Regulatory Commission* a new reliability standard (EOP-004-2) that impacts the reliability of the power grid.

7.5.2 SOFTWARE TOOLS

7.5.2.1 Tools Derived from Theory of Reliability

Information from these standards is used to elaborate software tools for reliability prediction and several examples of software tools are herein briefly revised. These solutions consider the electronic equipment in a kind of static operation, under a constant failure rate, without modeling the change of parameters in time (*wearout*). They simply convey the standards' requirements into a software format. Since these solutions are provided by small companies, they may appear and disappear inadvertently, so this list is provided for understanding the current capabilities rather than an ultimate solution.

- *DfR Solutions*—Offers a software tool called *SHERLOCK* for reliability prediction within electronic PCB (printed-circuit-boards). The software is able to import bill of materials from a SPICE simulation, to build the FEA, and to define the product lifecycle in respect to vibration, shock, and temperature variation. It also has a section able to perform certain physics of failure analysis for solder joint fatigue, plated through-hole fatigue, ceramic capacitor breakdown, and IC wearout. During the years 2007–2011, a library of over 250 root-cause investigations were reported.
- *T-Cube Systems*—Offers *RELCALC* that is a direct implementation of MIL-217 standard. Similar tools are offered for more than 25 years.
- *Relex Scandinavia*—Offers a similar package called *Relex Reliability Studio*, since 1986.

7.5.2.2 Tools Derived from Microelectronics

On a different line of thought, wear-out models can be developed [39] for microelectronic devices within programs like:

- FaRBS [40] (the name comes from *Failure Rate Based Spice*) = It is able to estimate failure rates within microelectronic devices based on physics of failure and the method of sum of failures.
- BSIMProPlus of Cadence [41].
- RelXpert (formerly BTABERT) by *Celestry* provides a SPICE simulation of both the HCI and NBTI degradation for IC technologies below 0.13 μm.

These programs allow the modeling of the change with time of parameters like threshold voltage, transconductance, gate charge, leakage current, I-V characteristics (including the voltage drop on device). It is worthwhile to also mention the AgeMOS model [42] developed by Cadence, that includes device degradation due to hot carrier injection and negative bias temperature instability.

In 2003, *IBM Corporation* has elaborated a program called *RAMP*, able to model the wearout of semiconductor devices like microprocessor (microelectronics). It did use a *competing risk model*, incorporating failures like electromigration, stress migration, time-dependent breakdown, temperature cycling, hot carrier injection, and negative bias temperature inversion.

7.5.2.3 Power Electronics Specifics

Even from mid-1990s, efforts for computer-based estimation of reliability for power electronics equipment were reported [43]. Since there was not a program able to do both circuit simulation and reliability calculation based on the component's constant failure rate weighed by the actual operation point, researchers have tried to pass data in between programs in order to achieve this goal [27,39].

As of 2012, a complete program for the prediction of reliability in power electronics is still not available. The knowledge from microelectronics and packaging technologies are complementary and we should expect a solution able to model both types of failures under the same model.

Due to the importance of thermo-mechanical failures caused by the mismatch in the Thermal Expansion Coefficient of the different joined materials within an IGBT, such failures are usually considered as major and followed alone for the estimation of lifetime. They usually lead toward bondwire liftoff or solder layer degradation.

Semiconductor failures like electromigration, time-dependent breakdown, hot carrier injection, negative bias temperature inversion, or even the gate oxide breakdown or passivation are well known in the microelectronics and their activation energy seems to be low enough to prevail also in power semiconductor devices. Their occurrence would change the IGBT's parameters and accelerate the thermo-mechanical issues. This is probably why the power semiconductor companies direct their reliability efforts into two directions:

- Hot-carrier injection and the sistership phenomena of radiation hardening
- Minimization of technological parameter variation for sustaining a more robust design

Most of the so-called "Hi-Rel" power semiconductor devices are sold in the same packages as conventional IGBTs or MOSFETs. At most, the hermetic packages are considered to alleviate the occasional exposure to special environmental conditions like dust, or humidity.

In conclusion, the simultaneous occurrence of wearout in both microelectronics and packaging should be ideally considered within an advanced model. Applying the competing risk model helps us to reduce the analysis to microelectronics (semiconductor) failures for low power devices (like power supplies MOSFETs), and to thermo-mechanical packaging failures in high power devices like IGBT and IGCT.

Finally, let us note that any of these tools is not complete as nobody did merge them with conventional circuit simulators. This is one step the future should offer. For instance, there is no tool able to compute in a single step the answer to a question like "if in this converter I change the gate resistor from 10 Ohm to 20 Ohm, the MTBF goes from 100,000 h to 145,000 h."

7.6 FACTORY RELIABILITY TESTING OF SEMICONDUCTORS

All power semiconductor manufacturers pass devices through a series of reliability tests according to their own methodologies [2,44]. Such tests can subscribe to two major classes:

- Initial burn-in tests to minimize the infantile mortality [45]
- Qualification tests based on samples

Even if methods differ from manufacturer to manufacturer, the qualification tests aim at meeting requirements from standards for discrete semiconductor devices like JIS C 7021 (Japan) or IEC60747, IEC60068, and IEC60749 (Europe). Such tests are intended to "*pass*" or "*fail*" the products based on certain *failure criteria*. These tests are accelerated test procedures established for the power device only, and are different from tests shown in Section 7.4.2 that are intended to simulate the actual use of the electronic equipment. They are based on production samples.

These qualification tests can be classified in three groups:

- Chip related tests (high temperature reverse bias test, high temperature gate stress, and so on)
- Stability of the package under specified operation and storage temperature (steady-state and transient)
- Mechanical integrity of the package (vibration and shock)

7.7   DESIGN FOR RELIABILITY

DFR represents an emerging discipline that refers to the process of designing reliability into products. During system design, the top-level reliability requirements are then allocated to subsystems by design and reliability engineers working together. As shown before, the reliability models use block diagrams and fault trees to provide a graphical means of evaluating the relationship between different parts of the system. These models incorporate predictions based on parts-count failure rates taken from historical data. While the predictions are often not accurate in an absolute sense, they are valuable to assess relative differences in design alternatives.

As an alternative, the *physics of failure* represents a design technique that relies on understanding the physical processes of stress, strength, and failure at a very detailed component level. This helps the redesign of each component to reduce the probability of failure. Most typically such evaluation of the physics of failure comes down to the test and introduction of new materials. The issues related to the reliability of the highly-integrated power modules as well as of various passive components represent a very active and promising field of research, requiring a multiphysics approach to thermal-mechanical strain control for controlled reliability and full capacity utilization.

The movement of the design effort from circuit design and validation toward proving sustainability and reliability of the design has opened up new R&D directions:

- Analysis of the suitability of a device or system for a purpose, with respect to time
- Determination of the capacity of a device or system to perform as designed, within conventional context or while taking advantage of new materials or knowledge
- Calculation of the resistance to operation under extreme conditions or in failure mode of a device or system, as well as the ability of a device or system to fail without catastrophic consequences
- Calculation of the probability that a functional unit will perform its required function for a specified interval under stated conditions

Pure *reliability engineering* relies heavily on mathematical models such as reliability prediction, Weibull analysis, thermal management, reliability testing, and accelerated life testing. Because of the large number of reliability techniques, their expense, and the varying degrees of reliability required for different situations, a more systematic and/or academic approach is recently considered.

The issues related to the reliability of the highly-integrated power modules as well as of various passive components represent a very active and promising field of research, requiring a multiphysics approach to thermal-mechanical strain control for controlled reliability and full capacity utilization.

A structural introspection outlines possible topics of research around a power module:

- Influence of semiconductor fabrication defects and defect migration under operation
- Heat-fatigue phenomenon in semiconductor modules, under thermal cycle and power cycle
- Stress at the bond wire–bond pad interface within a power semiconductor device is able to recommend the proper shape of the connection and to model the fatigue
- Stress at insulation substrate-base plate

With the increasing interest in reliability, there are numerous recent reports of cases when the reliability concerns were considered in the design phase. For instance, [46] reports a method for the on-line manipulation of the switching frequency and current limit to regulate the losses in order to prevent over-temperature and hence the power cycling failures in IGBT power modules.

7.8 CONCLUSION

This chapter makes an introduction to the thermal and reliability aspects related to the design of medium or high-power power electronics equipment. Thermal calculation is well known and was previously used to properly size the cooling systems.

The recent years has shown an increasing interest in reliability prediction. The theory of reliability is herein revisited and the peculiar aspects of applying this theory to power electronics equipment are revealed. Faults can occur through degradation of performance in both semiconductor die and thermo-mechanical setup. The most important failure mechanisms need to be qualified with experimental tests or by investigation of the physics of failure and analytical modeling. The experimental tests need to be designed based on theories of design of experiments, for either temperature cycling or power cycling. The analytical modeling needs a greater understanding of the physics of phenomena occurring during failure. In either case, monitoring of the actual operation characteristics helps accurate estimation of the lifetime. This helps anticipating failures and proper service or maintenance scheduling.

Finally, principles of DFR help us to consider reliability into early phases of design, in order to increase the product robustness and lifetime.

The entire field of reliability prognosis and estimation is under major investigation and we may expect to see more standards and software tools in the near future.

REFERENCES

1. Sparrow, E.M., Chevalier, P.W., and Abraham, J.P. 2006. The design of cold plates for the thermal management of electronic equipment. *Heat Transfer Eng.* 27(7), 6–16.
2. Lutz, J., Sclangenotto, H., Scheuermann, U., and De Doncker, R. 2011. *Semiconductor Power Devices—Physics, Characteristics, Reliability.* Springer, Heidelberg/Dordrecht/London/New York.
3. Anon. 2002. Thermal management products. Application Note, Ferraz-Shawmut.
4. Sueker, K.H. 2005. *Power Electronics Design—A Practitioner's Guide,* Ed. Newness/Elsevier, Amsterdam/Boston/ Heidelberg/London/ New York/Oxford/Paris/ San Diego/San Francisco/ Singapore/ Sydney/Tokyo.
5. Kandlikar, S.G. and Hayner, II, C.N. 2009. Liquid cooled cold plates for industrial high-power electronic devices—Thermal design and manufacturing considerations. *Heat Transfer Eng.* 30(12), 918–930.
6. Blaabjerg, F. 2012. Power electronics and reliability in renewable energy systems. *IEEE IECON Keynote Speaker.*
7. Wyrwas, E.J. and Bernstein, JB. 2011. Quantitatively analyzing the performance of integrated circuits and their reliability. *IEEE Instrumentation and measurement Magazine,* February 2011, pp. 11–31.
8. Anon. 2007. Advanced power converters for universal and flexible power management in future electricity networks, Report D6.1, Project co-funded by the European Community under the Sixth Framework Programme.
9. Tavner, P. 2010. Reliability and availability of wind turbine electrical & electronic components. Durham University Document.
10. OKeefe, M. 2009. Thermal stress and reliability for advanced power electronics and electric machines. *NREL Presentation at 2009 U.S.DOE Hydrogen Program and Vehicle Technologies Program Annual Merit Review and Peer Evaluation Meeting.*
11. Waseem, I., Pipattanasomporn, M., and Rahman, S. 2009. Reliability benefits of distributed generation as a backup source. *IEEE Power and Energy Society General Meeting, PES'09,* pp. 1–9, http://ieeexplore.ieee.org/xpl/mostRecentIssue.jsp?punumber = 5230481.
12. Kaplar, R., Marinella, M., and Brock R. 2011. Stress testing of semiconductor switches for PV inverter applications at Sandia national laboratories. *Utility-Scale Grid-Tied PV Inverter Reliability Workshop,* Sandia Laboratories.
13. Wolfgang, E., Amigues, L., Seliger, N., and Lugert, G. 2008. Building-in reliability into power electronic systems. *ECPE Workshop,* pp. 246–252.
14. Kozola, S. 2008. Reliability analysis and robust design using MATLAB® products. *LMCS Journée nationale pour la modélisation et la simulation 0D/1D,* 17th April.
15. Calleja, H., Chan, F. 2010. Reliability: A neglected topic in the power electronics curricula. *J. Power Electron.* 10(6), 660–666.
16. Lall, P. 1996. Tutorial: Temperature as an input to microelectronics—Reliability models. *IEEE Trans. Reliabil.* 45(1), 3–9.
17. Parler, S. 2000. Reliability of CDE electrolytic capacitors. *CDE Internet Paper.*
18. Liu, J. and Henze, N. 2009. Reliability consideration of low-power grid-tied inverter for photovoltaic application. *24th European Photovoltaic Solar Energy Conference and Exhibition,* Hamburg/Germany, September.
19. Blake, C., McDonald, T., Kinzer, D., Cao, J., Kwan, A., and Arzumanyan A. 2005. Evaluating the reliability of power MOSFETs. *Power Electron. Technol.* 40–44, November.

20. Lu, H., Bailey, C., and Yin, C. 2009. Design for reliability of power electronics modules. *Microelectron. Reliabil.* 49, 1250–1255.
21. Ye, H., Lin, M., and Basaran, C. 2002. Failure modes and FEM analysis of power electronic packaging. *Finite Elements Anal Design* 8, 601–612.
22. Li, X., Qin, J., and Bernstein, J.B. 2008. Compact modeling of MOSFET wearout mechanisms for circuit-reliability simulation. *IEEE Trans. Device Mater. Reliabil.* 8(1), 98–121.
23. Shen, C.C., Hefner, Jr, A., Berning, D.W., and Bernstein, JB. 2000. Failure dynamics of the IGBT during turn-off for unclamped inductive loading conditions. *IEEE Trans. Industry Appl.* 36(2), 614–624.
24. Patil, N., Das, D., Goebel, K., and Pecht, M. 2008. Failure precursors for insulated gate bipolar transistors. *2008 Int'l Conference on Prognosis and Health Management.*
25. Arab, M., Khatir, Z., and Lefebvre, S. 2008. Investigations on ageing of IGBTs under repetitive short-circuit operations. *Power Electron. Europe* 5, 22–24.
26. Yang, S., Xiang, D., Bryant, A., Mawby, P., Ran L., and Tavner P. 2010. Condition monitoring for device reliability in power electronic converters: A review. *IEEE Trans. Power Electron.* 25(11), 2734–2752.
27. Smith, M.A. and Atcitty, S. 2009. Power electronics reliability analysis. *SANDIA Report SAND2009-8377*, Printed December.
28. Anon. 2010. *Power Semiconductor Reliability Handbook.* Alpha and Omega Semiconductor, May.
29. Celaya, J.R., Patil, N., Saha, S., Wysocki, P., Goebel, K. 2009. Towards accelerated aging methodologies and health management of power MOSFETs (Technical Brief). *Annual Conference of the Prognostics and Health Management Society.*
30. Chamund, D. and Newcombe, D. 2010. *AN 5945 IGBT Module Reliability.* Dynex Semiconductor, pp. 1–8, October.
31. Neacsu, D.O. 2010. Towards an all-semiconductor power converter solution for the appliance market. *IEEE Industrial Electronics Conference IECON*, Glendale, AZ, USA, November.
32. Beutel, A.A. 2006. A novel test method for minimising energy costs in IGBT power cycling studies, PhD Thesis, University of the Witwatersrand, South Africa.
33. Schuler, S. and Scheuermann, U. 2010. Impact of test control strategy on power cycling lifetime. *PCIM.*
34. Amro, R.A. 2009. Packaging and interconnection technologies of power devices, challenges and future trends, world academy of science. *Eng. Technol.* 49, 691–694.
35. Amro, RA. 2006. Power cycling capability of advanced packaging and interconnection technologies at high temperature swings. PhD Thesis, Chemnitz University of Technology.
36. Anon. 1990. *MIL-HDBK-217F = Military Handbook—Reliability Prediction of Electronic Equipment*, Original copy of 1990 (with subsequent revisions).
37. Anon. 2011. SR-332 = Reliability prediction procedure for electronic equipment. Telcordia (www.tecordia.com), January.
38. Anon. 2005. Reliability guidelines to understanding reliability prediction. European Power Supply Manufacturers Association, 24 June.
39. White, M. and Bernstein, J.B. 2008. Microelectronics reliability: Physics-of-failure based modeling and lifetime evaluation. NASA JPL Publication 08-5 2/08.
40. Bernstein, J.B., Gurfinkel, M., Li, X., Walters, J., Shapira, Y., and Talmor, M. 2006. Electronic circuit reliability modeling. *Microelectron. Reliabil.* 46, 1957–1979.
41. Anon. 2003. BSIMPro, Cadence datasheet.
42. Anon. Cadence virtuoso UltraSim full chip simulator datasheet. Cadence Design Systems. Available from: http://www.cadence.com/.

43. Kamas, LA. and Sanders, S. 1996. Power electronic circuit reliability analysis incorporating parallel simulations. *IEEE Workshop, Computers in Power Electronics*, Portland, OR, USA, pp. 45–51.

44. Anon. 1998. Semiconductor device reliability. Mitsubishi high power semiconductors. *Application Note*, August.

45. Gerstle, D., and Lee, P. 2005. Impact of burn-in on power supply reliability. *Power Electron. Technol.* 20–25, September.

46. Murdock, D.A., Torres, J.E.R., Connors, J.J., Lorenz, R.D. 2006. Active thermal control of power electronic modules. *IEEE Trans. Industry Appl.* 42(2), 552–558.

8 Implementation of Pulse Width Modulation Algorithms

8.1 ANALOG PULSE WIDTH MODULATION CONTROLLERS

The pulse width modulation (PWM) controller represents the final module within the feedback control path (Figure 8.1). It is therefore suitable to implement this module in the same technology as the control module. If the feedback control is achieved with analog circuits for high-speed, closed-loop control, the PWM controller should also be analog.

The basic role of the PWM module is to convert a reference signal to a train of pulses with a duty cycle variable upon the reference [17,18,19,26,29]. In a three-phase system, the reference is represented by a set of three-phase variables, normally symmetrical, and the PWM pattern is delivered for the six switches of the three-phase inverter.

The simplest implementation of the PWM controller separates PWM generation for each inverter leg. The conversion from reference to the upper switch control signals is achieved by comparing the reference with a triangular signal that has constant magnitude and frequency. This has been shown in Chapter 3. Figure 8.2 illustrates the simplest possible PWM generator with a simple operational amplifier. This operational amplifier provides rail-to-rail output.

The negative input of the operational amplifier sees a triangular waveform generated by charging and discharging the capacitor C from the output voltage. The positive input of the operational amplifier sees a voltage derived from the positive feedback through R4 and R3 and the input voltage VREF. Basically this side operates as a *Schmitt* trigger comparator, and the input voltage VREF controls the output pulse duty cycle. The pulse width is accordingly modified around half of the switching period depending on the polarity and value of the input voltage. Finally, the low-side switch control is achieved by reversing the control signal for the high side.

This solution does not provide a synchronization of the generated PWM with an external signal, and the period of the generated signal has its own variations due to supply voltage and temperature. An improved solution uses the integrated circuit (IC) timer 555 with input for synchronization and analog reference. The command pulses can be achieved with the same circuit timer 555, or both the command pulses and PWM can be generated within the same device, the dual timer 556. The 555 used to generate the command pulses can be applied as input to all three phases. The PWM is generated on each phase with an additional 555 circuit (Figure 8.3).

Alternative solutions use the same triangular signal for all three channels and they are based on accurate voltage oscillators (Figure 8.4).

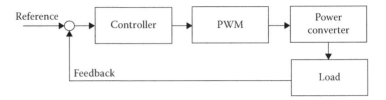

FIGURE 8.1 Schematics of the feedback control loop including a power converter.

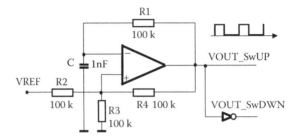

FIGURE 8.2 Simple PWM generator based on operational amplifier.

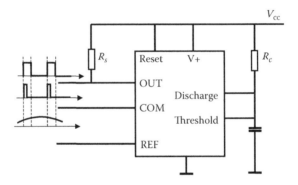

FIGURE 8.3 PWM generator with 555 oscillator.

There are many similar solutions for analog implementation of the PWM controller; all of them work on the principles explained here. The most advanced solutions also include *Shutdown* pins for each channel in order to discontinue the PWM generation when a fault occurs. Moreover, modern requirements may differ with respect to the power delivered within the PWM signals.

Designed primarily for power supply control, TL5001 from Texas Instruments follows this type of structure, with a dead-time generator and an open collector output transistor able to control the final power-driver stage. Additional features such as under-voltage lockout and short-circuit protection are included.

Given the limited flexibility in changing PWM parameters, these analog-based solutions are hardly used. They still remain a valuable choice in high-frequency

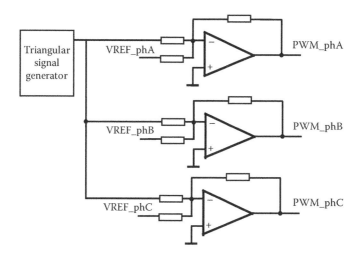

FIGURE 8.4   Schematics of a three-phase PWM generator with unique triangular generator.

servo-drives where the whole controller needs to be implemented in analog due to the extremely high bandwidth required for the control loops. Modern alternatives include high-speed field programmed gate arrays (FPGA) [16] or application-specific integrated circuit (ASIC) devices with predefined control loops and a PWM generator.

8.2 MIXED-MODE MOTOR CONTROLLER ICs

Motor drives in the horsepower range can be controlled with ICs without too many external components. This became a standard requirement for appliances or servo-drives when costs had to be reduced for commercial purposes.

There are many excellent mixed-mode IC-technology solutions in the market. They are designed to control low-voltage motor drives with applications in the automotive and consumer sectors. These controllers incorporate analog controllers, PWM generators with protection and deadtime circuitry, and gate drivers. The most advanced solutions also include power supply for the high-side MOSFET transistor instead of a conventional bootstrap external supply. In several automotive applications, the gate driver needs a separate power supply as the DC bus voltage may decrease below the limit required for proper gate control. For instance, A3948 [1] from Allegro Microsystems includes a boost inductor with three pairs of drivers to maintain gate control voltage.

Control systems are grouped around two application classes: the general brushless DC (BLDC) motor able to provide continuous rotation and the stepper motor able to provide start/stop operation or great positioning. Both provide integrated control solutions in mixed-mode IC technology.

The first generation of step-by-step motor incorporated the PWM current controller and the power stage within the same design, as in the Allegro A3966, in which the required phase reference currents are set by an external reference. Further developments in technology pushed for a complete solution, including a digital-to-analog controller to set current levels (A3967–A3977). The digital interface in these models allows communication with higher hierarchical levels. Further R&D is dedicated to specific applications and it includes transmission of fault conditions over the serial or parallel digital communication interface.

All these solutions are at the edge of IC technology and the future may see new products dedicated to higher power levels or with more digital and analog functions for better protection and control. For the moment, these solutions are limited to the low-voltage converter bus and small power motors.

8.3 DIGITAL STRUCTURES WITH COUNTERS: FPGA IMPLEMENTATION

8.3.1 PRINCIPLE OF DIGITAL PWM CONTROLLERS

The same implementation of a timer followed up by a comparison with the reference signal can be implemented in digital with counters and compare units. Modern microcontrollers have incorporated compare units along their timers/counters that make straightforward the PWM implementation. A single-channel PWM can be implemented with a counter, as in Figure 8.5.

PWM signals for the six switches within a three-phase inverter can be generated in several ways with counters. Generally, one signal is generated for each phase and a complementary pair of signals for the low side or the high side is achieved by a logical inversion. The designer must pay attention to the polarity within the gate driver in order to send the proper control signal to the controlled insulated gate bipolar transistor (IGBT)/MOSFET. As a rule of thumb, the gate drivers used or built in the U.S. generally do not invert the control signal polarity, whereas those made in Europe or Japan do change the control signal polarity. This is why all PWM from microcontrollers initially designed for the Japanese or European markets have negative outputs, expecting the gate driver to reverse the signal polarity.

Finally, the design engineer needs to verify if the digital system can build the deadtime by itself or whether an external deadtime generator is necessary. Sometimes, the gate driver itself is able to generate fixed values of deadtime.

PWM can be generated with counters for a three-phase inverter in one of the following topologies:

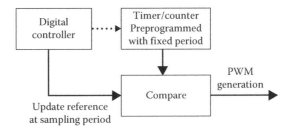

FIGURE 8.5 Principle of counter/timer use in single-channel PWM generation.

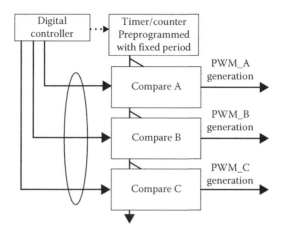

FIGURE 8.6   Three-phase PWM generation with a single counter.

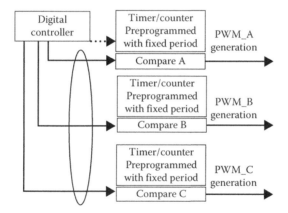

FIGURE 8.7   Three-phase PWM generation with 3 counters and 3 compare.

- A single common counter and three compare units, one for each phase, using the same counting device (Figure 8.6)
- A counter and a compare unit for each individual phase (Figure 8.7)

Depending on the internal structure of each timer, generation of the pulse width can be different. Figure 8.8a shows a left-aligned variable pulse width control, Figure 8.8b a right-aligned PWM, and Figure 8.8c a center-aligned PWM. The last one is a logical result of using the first two solutions back-to-back. Commercially available PWM generator circuits usually implement the center-aligned approach.

Some of the digital PWM ICs have been available in the market since the late 1980 s. Due to simplicity in designing with FPGA and ASIC, these PWM circuits no longer represent an appealing solution, but their principle of implementation can be used as a starting point for modern FPGA [16] or ASIC solutions.

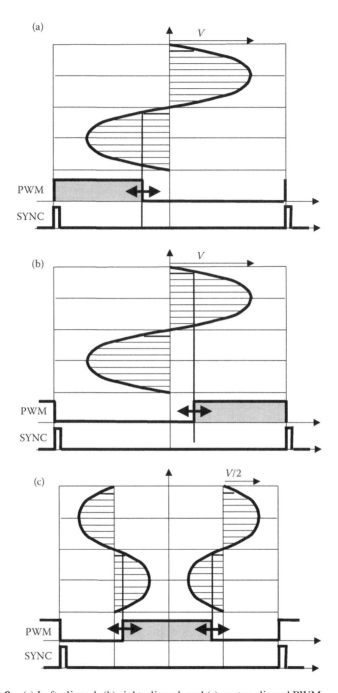

FIGURE 8.8   (a) Left-aligned, (b) right-aligned, and (c) center-aligned PWM.

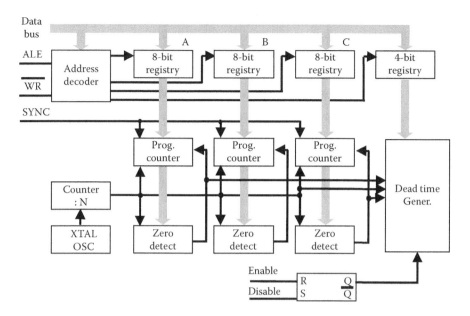

FIGURE 8.9 Block diagram of SLE 4520.

8.3.2 Bus Compatible Digital PWM Interfaces

Figure 8.9 shows the block diagram of the Siemens SLE4520 circuit. A similar solution has been implemented by Dynex.

The SLE4520 circuit is designed as an external interface for any 8-bit microcontroller and its operation can be programmed from microcontroller or microprocessor through a data bus and a control signal bus. The time constant or pulse width associated to each phase can be stored within an 8-bit registry and loaded into counters at each SYNC signal. The programmable counters count down to zero when the state of the switching signals for the inverter control change. The deadtime constant can be programmed within a 4-bit registry. When a fault occurs in the power stage, the emergency shutdown pin is activated and it cancels out the PWM signals through the RS flip-flop.

Similar solutions for digital devices able to generate PWM with or without deadtime are developed to directly interface on the data and address bus of modern microprocessors or digital signal processors (DSPs) [20,23–25]. The same idea is actually used in custom-made FPGA or ASIC devices, but it is worthwhile quoting more examples of digital devices available for PWM generation from a processor bus.

The IXDP610 from *IXYS* can provide a single-channel pair PWM control for a switching power converter bridge [21,22]. It has a complete digitally programmable interface from an 8-bit microprocessor. This has a digital comparator, comparing a counting timer with a preprogrammed constant followed by a deadtime generator. The IXDP610 is able to control power PWM devices that have switching frequencies between zero and 390 kHz, 7-bit or 8-bit resolution, and up to 11% deadtime. The output is able to drive directly 20 mA and it is suitable for opto-couplers, gate driver circuits, or low-power power modules. It also features a pulse-by-pulse shutdown protection against over-current, over-voltage or over-temperature, controlled by a logic signal from protection sensors.

8.3.3   FPGA IMPLEMENTATION OF SPACE VECTOR MODULATION CONTROLLERS

Space vector modulation (SVM) represents a special case in which a digital method is required to calculate time intervals to be loaded in counters/timers, followed up by another method able to define the switching sequence. There are numerous solutions possible for digital hardware implementation. One of them is next presented [2] and it considers an SVM generation with only three states during each sampling period (*active1—active2—zero state*). The following interval will have a reversed sequence (*active2—active1—zero state*), in which the zero state is always selected with only one switching difference from the last active state.

The memory look-up table can be reduced if we take into consideration the symmetries of the three-phase system as well as the symmetries around the bisectrix of each generalized 60° sector. The most significant bits (MSB) of the angular coordinate reflect the sector number and they are used to select the proper switching sequence in the second memory look-up table. The fourth MSB is used to define the position within each sector and also to establish the sequence of the active states. The last four bits of the angular coordinate are used to read the t_a, t_b time constants needed for SVM generation. This memory look-up table stores these constants for an interval of 30° only (2^4 increments within 30°). Each time constant is defined on eight bits and this resolution is considered as enough for this PWM application.

An external periodic signal is used as the system clock and a frequency divider counts the sampling interval period. The definition of the pulse width is limited by having only $2^8 = 256$ points for a sampling interval, as shown in the definition of the memory look-up table. The overflow of the sampling period counter changes the state of a flip-flop D. The outputs of this flip-flop control the sequence of the time intervals t_a and t_b through two OR gates. Finally, a "switch" module is used to designate the meaning of two signals A1 and A2 corresponding to the different meanings of t_a and t_b in the first half of the 60° sector and in the second half of the 60° sector. For instance, generation of a position at 23° and a modulation index of 0.6 leads to $t_a^1 = t_a(V_1) = 0.36\,T_S$ and $t_b^1 = t_b(V_2) = 0.23\,T_S$. By symmetry, generation of a vector at 37° with the same modulation index leads to $t_b^1 = t_a(V_1) = 0.23\,T_S$ and $t_a^1 = t_b(V_2) = 0.36\,T_S$. The control switching sequence is therefore decoded with A1 and A2 and the fourth MSB with another memory look-up table.

Using two memory look-up tables is not convenient, especially because they are of different sizes. The switching sequence can be defined with a combinatorial circuit and a series of multiplex circuits (Figure 8.10). Observing all the possible switching sequences, the synthesis can be developed as shown in Table 8.1.

Figure 8.11 shows the logic decoder used instead of the memory look-up table and is built up of a divider of six Johnson counters *[D0-D1-D2]*. The outputs of these

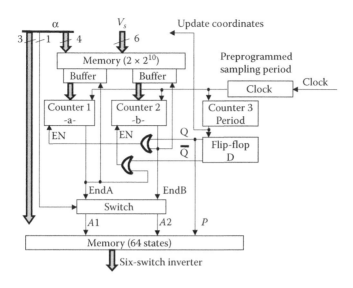

FIGURE 8.10   Principle of pure digital implementation of an SVM algorithm.

counters can be used as addresses for the multiplex circuits having P, $A1$, and $A2$ as inputs. Table 8.2 illustrates this decoding of the end of counting signals.

Other digital or FPGA syntheses of the SVM algorithm can be found in [3–7]. Many of them implement the SVM algorithm in a straightforward manner with calculation of Equation 5.30 followed by switching sequence decoding. The latter module can be changed from one SVM algorithm to another, for instance, in order to reduce losses (see Chapter 5).

The final stage before firing the gates of the power semiconductor devices is represented by the deadtime generator. The role and calculation of the deadtime interval have been explained in Chapter 3. The deadtime generator is usually implemented together with the PWM algorithm, but special circuits for deadtime generation also exist.

TABLE 8.1
Defining the Switching Sequence from the End of Counting Signals

| | Initial Zero Vector 000 | | | | Initial Zero Vector 111 | | |
Sector	Signal B	Signal A1	Signal A2	Sector	Signal B	Signal A1	Signal A2
1	S1	S3	S5	1	S6	S4	S2
2	S3	S1	S5	2	S6	S2	S4
3	S3	S5	S1	3	S2	S6	S4
4	S5	S3	S1	4	S2	S4	S6
5	S5	S1	S3	5	S4	S2	S6
6	S1	S5	S3	6	S4	S6	S2

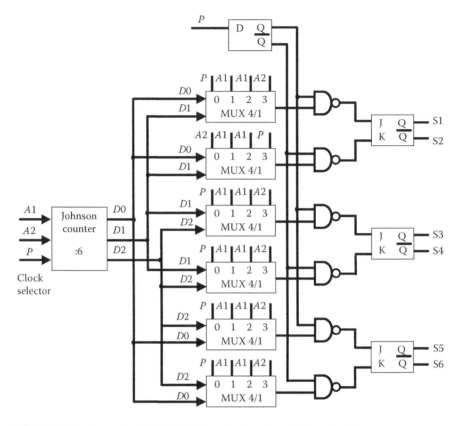

FIGURE 8.11   Example of SVM digital synthesis with multiplex circuits.

TABLE 8.2
Proposed implementation

S	Address D1-D0	S1 Turn-on	S2 Turn-on	Address D0-D2	S3 Turn-on	S4 Turn-on	Address D2-D1	S5 Turn-on	S6 Turn-on
1	00	P	A2	00	A1	A1	00	A2	P
2	01	A1	A1	00	P	A2	10	A2	P
3	11	A2	B	01	P	A2	10	A1	A1
4	11	A2	B	11	A1	A1	11	P	A2
5	10	A1	A1	11	A2	P	01	P	A2
6	11	B	A2	10	A2	P	01	A1	A1

8.3.4 DEADTIME DIGITAL CONTROLLERS

A solution for deadtime implementation with counters is shown in Figure 8.12 and it corresponds to *IXYS* circuits IXDP630/631PI [21,22]. The deadtime is always eight clock periods and the clock can be an external crystal or an RC oscillator circuit. Controlling the clock period, one can adjust the deadtime interval. The same IC is able to generate deadtime on all three phases separately. An additional "output enable" signal is able to shutdown all or each output. The operation of this circuit can be understood from Figure 8.12. Each positive edge of the control signal SIN delays the positive edge of the gate signal *S_High* and each negative edge of the control signal SIN delays the positive edge of the gate signal *S_Low*. Negative edges of *S_High* and *S_Low* directly follow the appropriate edges of the control signals.

8.4 MARKETS FOR GENERAL-PURPOSE AND DEDICATED DIGITAL PROCESSORS

8.4.1 HISTORY OF USING MICROPROCESSORS/MICROCONTROLLERS IN POWER CONVERTER CONTROL

Chapter 1 presented the main control functions required for implementation on the digital control system platform. There are two directions of development in the digital world able to implement all these functions.

The first direction focuses on architectures of general-use microprocessors. These devices achieve incredibly fast running of the control code, and include many functions implemented in the software. The evolution of microprocessors in the last 25 years started with families of *bit-slice* devices, such as INTEL3000 or AMD2900; 8-bit microprocessors such as INTEL I8080, ZILOG Z80, and I8085; 16-bit microprocessors such as I8086, Z8000, Motorola 68000, and Texas Instruments 16008/16032; and 32-bit microprocessors scuh as 68020 or I80385. They were upgraded in the late 1980 s with arithmetic coprocessors used in tandem, such as INTEL 80826/80287,

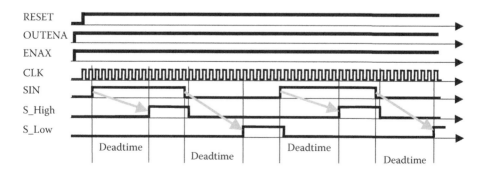

FIGURE 8.12 Time diagram of the deadtime generator circuit IXDP630/631PI.

National Semiconductor NS 32016/32081. This evolutionary step also included other LSI circuits:

- High-speed hardware multipliers able to calculate 24×24 bits in fixed point in 200 ns, and 16×16 in 34 ns (for instance, MPY016 K-TRW)
- Arithmetic modules for calculation in variable points up to 80 bits (e.g., AMD 511, 9512), which work as slave coprocessors to the main microprocessor device
- Multichannel acquisition systems (e.g., AD162, AD364, DAS1150) or dedicated interfaces to existing microcomputers in digital systems, such as INTEL, PROLOG, MOSTEK, TI, or DEC, and so on
- Fast RAM memories, with double port access and huge capacity
- Special digital circuits used to fast interface to microprocessor buses
- Circuits dedicated to generation of the multichannel PWM for control of power converters
- Complex digital circuits containing ROM-I/O-Timer peripherals dedicated to extension of existing microprocessors (e.g., MC6846 used in conjunction with Motorola 6800 device)

In order to fully understand this evolutionary step, let us take an example. *Zilog* Z80 was a quite widely used 8-bit microprocessor family in the 1980 s. The clock speed was limited to 4 MHz and the peripheral functions were achieved on separate chips from the main device. A special device called PIO-Z80 was dedicated to the parallel I/O, another one to the serial communication SIO-Z80; a special memory access unit DMAC-Z80 and a timer/counter circuit CTC-Z80 completed this family. A fully operational microsystem was required to contain all these ICs on a common printed-circuit board with all the data, control, and address buses accessible from outside. Once this board was realized and tested, coding was done directly on assembly language without any real-time debugger or the possibility to visualize the operation through a monitor program.

All these digital solutions of the late 1980 s or early 1990 s had, however, a quite limited processing speed. Some of them are still in use due to their generality and simplicity. An alternative able to increase the general operation speed considers off-line processing and storage of the control results in a large memory look-up table. The best example is offered by implementation of different PWM algorithms through memory look-up tables. Initially, there was a substantial limitation due to the inherent limited speed of access to these memory look-up tables. Modern ROM devices can achieve access times sufficient for the switching of modern power converters. We will come back to these solutions few years later with the advent of FPGA devices.

The second major direction of development was in modern microcontrollers designed for industrial applications and including dedicated peripherals within the same silicon die. These microcontrollers incorporated on the same chip more memory and I/O interfaces along with their MUX, A/D converters, timers, PWM channels, and so on. Several examples that met with success in the beginning of the 1990 s are still in use today:

- INTEL 8051 (with its versions 8031, 8751, 8052, 8032, and different manufacturers like *Siemens* or *Philips)* contains:
  - An 8-bit microcomputer with its own clock generator
  - A dedicated module for Boolean processing
  - kB of ROM (8031 without memory, 8751 equipped with EPROM, and 8052 with 8 kB and a BASIC interpreter)
  - 128 bytes RAM (with 256 at 8052)
  - Two 16-bit timer/counter circuits (8052 equipped with three such counters) used especially for PWM generation
  - Four programmable I/O, bidirectional, with option for each bit to read, write and program
  - Serial I/O channels
  - Two external interrupt inputs, that can be enabled or disabled by software (8052 equipped with six interrupts including two external inputs)
  - Dedicated registers for arithmetic operations, stack management, I/O latches
  - A strong assembly language with 111 instructions, enhanced with interpreter for BASIC or PL/M
- INTEL 80186 microcontroller created after the 80196 microprocessor with an architecture similar to the 8086-2 but also including:
  - A clock generator at 8 MHz
  - Two high-speed Direct Memory Access (DMA) channels
  - Programmable interrupt controller with five levels of priorities
  - Wait state generator
  - Dedicated bus controller
  - Possible coprocessor interface
  - Software compatible with 8086 and option for programming in ASM86, PL/M86, PASCAL 86, FORTRAN86, LINK86, C, and so on
- INTEL 8096 16-bits microcontroller with enhanced interfaces to analog and digital systems. Different versions include:
  - kB or no memory
  - Four or eight 10-bits A/D channels with a conversion time of 42 μs
  - Five I/O ports
  - Serial output port
  - One external interrupt
  - PWM output
  - Four channels for fast capture and event timing with the existing counter
  - Six fast outputs programmable to generate pulses at fixed time intervals with a resolution of 2 ms
  - Two timers of 16 bits each, one for real-time synchronization and the second for countdown of external events
  - Watchdog circuit
  - Assembly language with interpreter for PL/M
- Additional microcontroller examples from the same historical class are Motorola MK 68200, MK3870, NEC μPD 78312, INTEL 8748, Philips PCB8049, MAB 4048, and so on.

8.4.2 DSPs Used in Power Converter Control

All these efforts to add more functions to microcontroller chips have not been enough for modern applications, and semiconductor engineers have moved forward to DSP devices [20,23–25]. DSPs have emerged in the communications business for fast processing of information from data and signals. They are characterized by very quick clock rates and additional processing circuitry within the same chip. It is normal to expect a 16×16 multiplication calculated within 100 ns on any of these devices.

Lately, the use of DSPs in power conversion control has been widened by dedicated motor control DSPs equipped with special peripherals designed for multiphase converter control. These devices are generic, are called *Motor Control DSPs,* and they feature three-phase power conversion PWM control useable in both machine drive and three-phase grid-tied applications. The main DSP producers in the world are—in an arbitrary order—Texas Instruments, Motorola, and Analog Devices. A brief description of the features of each type of DSP and its specifics in controlling three-phase power conversion follows.

The largest DSP market share belongs to Texas Instruments. Their DSPs have passed through a series of iterations and versions in the last 10 years. The TMS320 family of DSPs [23–25] is made on a Harvard architecture that allows execution of several operations simultaneously for increased running speed.

The first use of the DSP to control power converters seems to be based on TMS32010/32020, a general use DSP. Its features include:

- Working on 16 bits with an instruction cycle of 160 ns
- A small RAM memory of 144×16 bits
- A parallel 16-bit module for a 160 ns multiplication
- Eight I/O ports of 16 bits
- An external interrupt
- Special registers for data processing
- A 16-bit timer
- Interface for memory access in multiprocessor mode

In the early 1990s, Texas Instruments had developed a set of peripherals for motor drive control under the name *Event Manager.* This was their first generation motor control DSPs, TMS320F(C)24x, and it was optimized in the second generation TMS320F(C)240x that was released in the late 1990s. In 2002, TI released the third generation motor control DSPs, TMS320F(C)280x, which had a very high-speed Central Processing Unit (CPU), able to run programs at a clock frequency of 150 MHz. Section 8.5 will present in detail all hardware and software possibilities for PWM implementation on a TI DSP using the Event Manager.

Analog Devices has another good DSP program with motor control features. Several architecture solutions have been tried and developed at *Analog Devices* and the most remarkable probably refers to the development of a motor controlled coprocessor. The first solution, called ADMC201, was subsequently developed into ADMC330, ADMC401, and finally incorporated as a peripheral of a DSP circuit within the same package. Section 8.7 will reiterate the historical importance of this architecture development.

The DSP family ADSP-219x represents another solution from *Analog Devices* with many applications in power electronics [20]. This 16-bit fixed-point DSP core is able to perform 160 MIPS in its modern versions. Other power converter control-related peripherals include:

- On-chip RAM 8 k Words shared between program RAM and data RAM
- External Memory Interface with dedicated Memory DMA Controller for data and instruction transfer
- Eight-channel, 14-bit analog-to-digital converter system with up to 20 MSPS sampling rate
- Three-phase 16-bit center-based PWM generation unit with 12.5 ns resolution at 160 MHz core clock
- Dedicated 32-bit encoder interface with companion encoder event timer
- Dual 16-bit auxiliary PWM outputs
- Three programmable 32-bit timers
- Sixteen general-purpose I/O pins
- SPI and synchronous serial interface

Other DSPs of the same generation are NEC7720, Fujitsu MB8784, and STC-DSP128.

The large number of customers and the extensive use of these DSPs in the world have encouraged development of software libraries dedicated to DSP applications. Programmable in both assembly language and C language, these devices helped develop a software culture and a style for programming proper to motor-drive applications. Recent efforts will probably provide code modularity and automatic software validation. Conventional development tools have been extended *in the loop viewers* and automatic programming tools. Companies like MathWorks have produced systems able to include the DSP system in a PC-based simulation program for quick development and debugging of new codes dedicated to power converter control.

8.4.3   PARALLEL PROCESSING IN MULTIPROCESSOR STRUCTURES

In parallel with the tremendous development in DSP devices, engineers have tried to use digital systems with parallel processing for control of power converters. These systems allow more processing power even if they do not directly benefit from special peripherals. The whole control algorithm is therefore brought down to the time scale of the PWM algorithm.

The software operation in a multiprocessor system is based on a waiting list for processes, special set of instructions for creation of parallel tasks, a minimal context for fast communication between processes, several timers, and an evolved interrupt system. Selection of the communication protocol between parallel microsystems is very important and there are several options available: series transfer, parallel transfer through handshaking, DMA transfer, FIFO transfer, double-port RAM memory transfer, and multiple bus architecture.

A device in this category, INMOS Transputer 414, uses for execution reduced instruction set controller processors able to run with 32 bits at 50 ns and to communicate internally at high speed.

The same tendency to use parallel hardware is seen in the development of software platforms. The simplest microprocessors need to model numerous concurrent events sequentially. Assembly language is the quickest way to implement these models, though programs written in assembly language often have sequences that are redundant and too detailed. To simplify this process, many engineers use high-level programming languages like C, PL/M, PASCAL, Visual BASIC, and so on. Programming is therefore reduced to establishing clear tasks and writing special modules for each task. The main program is then grouping these tasks based on priority levels.

The sequential processing of tasks does not extract all benefits from the parallel processor hardware. The first programming language that consistently expressed parallelism in execution was APL [8]. However, the task was computed through computing power distribution, since, at that time, real parallel hardware was not available. The programming language ADA represents an important evolutionary step, especially in conjunction with the specialized 32-bits microprocessor iAPX432. The first truly parallel evaluation of tasks was carried out with the programming language OCCAM, which was able to achieve synchronization and communication between tasks on different hardware. OCCAM is the programming language for *transputer,* but it can be adapted to any other computer with minimal modifications.

OCCAM shares tasks in individual processes that run separately by communicating between them. The whole program can be seen as a network of interconnected processes. The synchronization between processes requires that all of them run in the time required by the slowest process and that results are reported at each step. *Transputer* devices simulate this concurrency through software *time-slicing,* followed by communication through the four serial communication channels. This structure can be extended further, if necessary.

Despite the computing power of these parallel architectures, they are not used in large series production of power converters. Manufacturing operations require a reduced number of terminals to be soldered and a cheap semiconductor solution. Furthermore, firmware considerations also encourage the use of a simpler software-based solution.

8.5   SOFTWARE IMPLEMENTATION IN LOW-COST MICROCONTROLLERS

8.5.1   SOFTWARE MANIPULATION OF COUNTER TIMING

Many microcontrollers do not benefit from a dedicated motor control peripheral and they can implement PWM algorithms using software. A minimal timer is still required for real-time accounting. The software has to calculate the time intervals necessary for the next sampling interval and to sequentially program the timer for each state. Obviously, the resulting PWM cannot have a very large PWM frequency, given the real-time requirements, but it may satisfy many low-cost applications.

Consider an implementation of the conventional SVM algorithm using a single timer circuit and a parallel interface for inverter control. This digital structure imposes some requirements in designing the software controller:

- Each time constant must be calculated by the software on the basis of general SVM relationships. This will limit the maximum frequency of the fundamental waveform able to be generated by this system.
- The timer channel should produce an interrupt at the beginning of each time interval followed by a software compensation for all delays produced by this approach.
- A parallel interface is used to generate the PWM output signals on each interrupt produced by the counter.

A time diagram for this approach is presented in Figure 8.13. Due to the pure software implementation, the switching frequency is limited in the support processor and not too many other features can be run on the same platform. The advantage, however, is an extremely reduced list of hardware requirements: a simple timer and software processing of the comparison result is enough.

This solution is not very practical, though, and another solution is presented in Figure 8.14, in which four timer channels are used for synthesis of each sampling interval. The end of each timing interval produces the start of the following counter and the timing of the next inverter state. The same software structure is required to calculate the PWM.

8.5.2 CALCULATION OF TIME INTERVAL CONSTANTS

The second major requirement to implement a PWM algorithm is that the constants to be loaded within timers for counting the different state intervals must be determined very quickly.

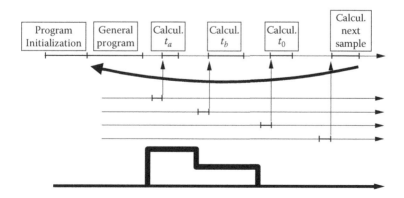

FIGURE 8.13   Time diagram of a pure software implementation of SVM.

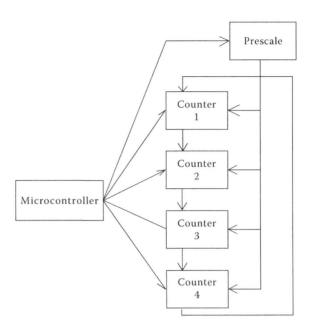

FIGURE 8.14   SVM software implementation based on four counters.

Chapter 5 showed the symmetries in the generation of an SVM algorithm, symmetries after each 60° interval and around the bisectrix of each interval. This can help in reducing the size of the look-up table required to implement the SIN/COS function. Many DSP or microcontroller devices already have in their ROM memory a brief SIN look-up table, usually with 256 or 512 points over a complete cycle. If a higher resolution is required, a new look-up table can be generated by a computational software package like MATLAB or MATHCAD and incorporated with the control program. The number of points within the SIN look-up table determines the resolution in defining each pulse width. This is important for high-switching frequency applications with advanced controlllers. Because memory is not generally a serious constraint, incorporation of a high-resolution SIN look-up table is not a problem in modern microcontrollers.

A totally different approach that is able to take advantage of the computing speed of modern controllers while saving memory requirements is to approximate the reference signals by interpolation. Different mathematical approaches for interpolation are available and they can be used to extend the range of a SIN look-up table from what is already installed in the microcontroller or DSP ROM memory to what the current requirements are to generate the PWM.

An alternative solution to achieving interpolation is fuzzy logic. The theory of fuzzy logic was first proposed in a paper of Professor Lotfi Zadeh in 1965 [9]. For many years this theory was not of interest until, in the 1970s, a series of books extensively presented the mathematical aspects of fuzzy logic [10,11]. In the early 1990s, many papers tried to implement concepts of fuzzy logic in the control of power converters, especially in replacing conventional PI controllers with fuzzy logic controllers. But it did not really succeed in power converter applications, mainly due to existence of other conventional nonlinear control methods and some difficulties in implementation.

A special feature of the fuzzy logic theory was outlined at the 1992 IEEE conference for fuzzy logic systems, where it was repositioned as a logic of interpolate thinking.

This section presents an example of the use of fuzzy logic to implement the interpolation concept, which generates an SVM from a SIN look-up table with a reduced number of points [12,13].

Consider a PWM model with 24 points over the fundamental cycle. The zero states of the model are shared equally during each sampling interval. To simplify the explanation the demonstration is limited to a generalized 60° sector. The fuzzy logic-based control relations can then be expressed as

$$\text{If } \alpha = i * \frac{\pi}{12}, \text{ then } K_a = K_i \text{ with } i = 0,\ldots,4. \tag{8.1}$$

This says that for a reduced number of five angular coordinates over the 60° sector, we know precisely what the time constants for each active vector are. The whole set of angular coordinates is therefore described by five fuzzy subsets. Figure 8.15 presents the membership functions for each variable. This choice of the membership functions along with a linear defuzzification method presents less distortions when equivalence with an identical crisp system is sought. Besides, this is very simple to implement. The effect of different approximation approaches in the output phase voltage is shown in Figure 8.16.

The linear defuzzification approach produces a very good interpolative approximation, each harmonic of the output phase voltage having an error below 10^{-5} of

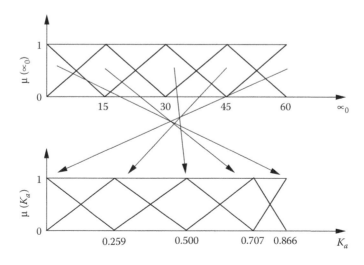

FIGURE 8.15 Fuzzy subsets for the proposed method.

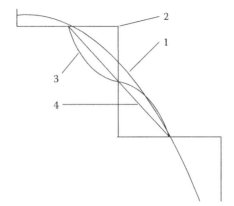

FIGURE 8.16 Effect of different approximation methods in the output phase voltages: (1) sinusoidal PWM; (2) staircase PWM; (3) center-of-gravity defuzzification; and (4) linear defuzzification.

the precise calculation method. This is also reflected in the total harmonic distortion (THD) calculation for the output phase voltage. Figure 8.17 shows the similarity between the two methods.

A software implementation algorithm for this method is presented next along with a numerical example:

1. Calculate the polar coordinate of the next desired vector position (V_s, α).

Example: (V_s, α) = (0.45, 212°)

2. Define α_0 as the angular coordinate within the generalized sector.

$$\alpha_0 = 212° - (\text{INT}(212/60))60 = 32°$$

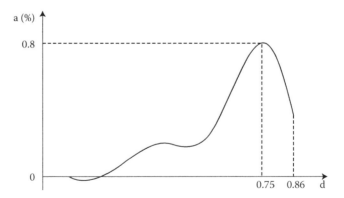

FIGURE 8.17 Relative differences between the two methods.

3. Fuzzy estimation of the time intervals allocated to each active vector K_a, K_b based on α_0.

Define the sector number:

$$I = INT(\alpha/15) + 1 = 3$$

Calculate the membership degree of α_0 to the left subset:

$$\mu_1(\alpha_0) = (15 * 1 - \alpha_0)/15 = 0.86$$

Calculate the membership degree of α_0 to the right subset:

$$\mu_2(\alpha_0) = 1 - \mu_1(\alpha_0) = 0.14$$

Calculate the time interval by linear defuzzification:

$$K_a = (\mu_1(\alpha_0) * K_I + \mu_2(\alpha_0) * K_I + 1) = 0.86 * 0.5 + 0.14 * 0.707 = 0.528$$

Calculate the second time interval by linear defuzzification:

$$K_b = (\mu_1(\alpha_0) * K_{6-I} + \mu_2(\alpha_0) * K_{5-I}) = 0.86 * 0.5 + 0.14 * 0.259 = 0.467$$

4. Determine the actual time constants by multiplication with the sampling interval period.
5. Calculate $t_{01} = 0.5 * (T - t_a - t_b)$.
6. Select the appropriate switching pattern ($t_{01} \rightarrow t_a \rightarrow t_b$ or $t_{01} \rightarrow t_b \rightarrow t_a$).
7. Timer programming.

This fuzzy logic-based approach can be extended to V/Hz control of three-phase induction motor drives. The V/Hz characteristic is not linear for the whole range, but requires some nonlinearity in the low-frequency range in order to compensate for the voltage drop on the stator resistances. This nonlinearity can also be mapped with a fuzzy logic-based variable. The resulting controller has two inputs and can be implemented based on the rule table shown in Table 8.3. The linquistic degrees are general without any relationship to their language sense. Each rule can be interpreted as:

IF α is NS (negative small) AND f is PS (positive small), THEN t_a is NS (negative small).

The membership functions are considered triangular, symmetrical, with prototypes provided by the time constants calculated for the 24-pulse PWM case. The resulting control surface is shown in Figure 8.18, comparing the ideal and approximate cases.

8.6   MICROCONTROLLERS WITH POWER CONVERTER INTERFACES

Given the large market for low-power motor drives with applications in appliances and servo-drives, some manufacturers have developed microcontrollers dedicated to motor control. The pressure to make these devices low cost is extremely high and giving up features not necessary in basic inverter control applications could make

TABLE 8.3
Rule Table for The Fuzzy Logic Controller

\αf\	Negative Big (NB)	Negative Small (NS)	Positive Small (PS)	Positive Big (PB)
Negative big (NB)	Positive big (PB)	Positive big (PB)	Positive medium (PM)	Negative big (NB)
Negative medium (NM)	Positive big (PB)	Positive medium (PM)	Zero (Z)	Negative big (NB)
Negative small (NS)	Positive small (PS)	Zero (Z)	Negative small (NS)	Negative big (NB)
Zero (Z)	Positive small (PS)	Zero (Z)	Negative medium (NM)	Negative big (NB)
Positive small (PS)	Zero (Z)	Negative small (NS)	Negative medium (NM)	Negative big (NB)
Positive medium (PM)	Zero (Z)	Negative small (NS)	Negative medium (NM)	Negative big (NB)
Positive Big (PB)	Zero (Z)	Negative small (NS)	Negative medium (NM)	Negative big (NB)

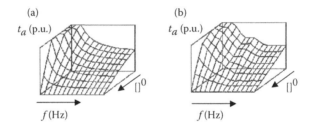

(a) t_a (p.u.) f (Hz) $[]^0$ (b) t_a (p.u.) f (Hz) $[]^0$

FIGURE 8.18 Controller surfaces: (a) ideal and (b) approximative.

them very competitive. Along with a simple, low-cost 8-bit central processing unit, these devices benefit from dedicated hardware interfaces for PWM generation and A/D conversion and data acquisition. Different communication interfaces complete the internal structure, which is optimized for motor drive applications.

A good example within this category is the NEC μPD78098x [14]. This class of microcontrollers has seven channels of programmable timer/counters for event management along with a low-cost central processing unit running at 8 MHz. Using a 10-bit timer for PWM generation, it has dedicated circuitry able to generate three pairs of output signals (inverter control) and a programmable 8-bit deadtime generator. Eight 10-bit A/D conversion channels are also available for power converter feedback control. A UART communication interface may place this device as a lower level controller in a hierarchical structure.

Because of the limited clock frequency (8 MHz), the PWM cannot be generated with a carrier frequency of 15.6 kHz (64 μs) or higher when the timer is programmed to an 8-bit resolution or on 3.9 kHz (256 µs) for the 10-bit resolution. The clock frequency is scaled for different switching frequencies below these values. For higher switching frequencies, the 8-bit or 10-bit resolution cannot be achieved, and the timer should work with multiple interrupts from the same timer and should divide the frequency by an appropriate value.

Motorola provides a similar device that can run at up to 32 MHz, and that has a fault-tolerant PWM controller for motor drive applications. The Motorola 68HC08 family also includes a deadtime generator and an interesting deadtime compensation capability [27,28]. These are created around the so-called *Timer Interface Module*, which also provides capture features. Finally, this device includes USB, CAN, and SPI interfaces with plans for a local interconnect network. As in many modern microcontrollers or DSPs, it has an in-circuit flash option.

8.7 MOTOR CONTROL COPROCESSORS

Any digital feedback control solution can be implemented into a general-purpose processing unit, but the particular control and measurement of a power electronics system requires special peripherals. Current research is aimed at improving the performance of power-dedicated peripherals, and placing it along with the main computing unit.

Another direction of research is toward application-specific peripherals outside of the main processing unit in the form of a digital coprocessor [15].

Analog Devices' ADMC201/./331 is the most well-known commercial solution using this approach. Developed for motor control applications, it has modular blocks that implement the vector control method within the same programmable digital system. Developed initially as a partner for ADSP2105, it is used in conjunction with other DSPs or microcontrollers as well.

The latest descendent of this family has been incorporated along a 26 MIPS fixed-point central processing unit within the ADMC401. It includes:

• Eight-channel simultaneous sampling A/D converters
• Three-phase 16-bit PWM generator
• Two 8-bit auxiliary PWM outputs
• Park direct and reverse digital transform based on angular coordinates

ADMC331 includes predefined hardware mathematical functions used in motor control, such as vector transforms based on angular coordinates. Hardware based vector transforms provide substantial time saving in a three-phase system control. The feedback control loops can run at higher rates implementing control loops with larger bandwidth.

More recently, International Rectifier developed a similar device (Accelerator TM) able to provide a feedback control loop with a bandwidth of 5 kHz by using FPGA support [15]. Again, all elements of a vector control algorithm were included in VHDL models and implemented within a flexible FPGA structure, providing fast parallel processing of the field orientation algorithm. The system makes possible code modularity and portability for specific applications. This digital system is designed to sit on the DC power bus and to directly control the power device's gate circuitry. Isolation of the communication interface with the higher hierarchical level is provided.

8.8 USING THE EVENT MANAGER WITHIN TEXAS INSTRUMENT'S DSPs

8.8.1 EVENT MANAGER STRUCTURE

Texas Instrument's has manufactured a set of DSP devices dedicated to motor control applications. The particular peripheral module of these DSPs is called *Event Manager* and it has all functions necessary for motor control. There are slight differences between the *Event Manager* included in the '24x device, '24xx device, and '280x device, but all are meant to help power converter control. We refer to the *Event Manager* from the '24xx devices only noting slight differences where possible. The '24xx series was the most used in early 2000s.

There are two *Event Managers* within the '24xx device, each including:

- Two *general-purpose timers* that are used to implement different PWM algorithms, quadrature encoder or generation of the sampling period for different control systems. They can work on the DSP clock or on an external clock and can have programmable periods and counting directions. There are local programmable compare modules able to detect a time moment and to release an interrupt or to set an output. An interesting feature allows starting A/D conversion synchronized with a timer operation.
- Three *compare units* that direct PWM generation by comparing timer values and predefined time constants. Each compare unit has two associated PWM outputs, working on a time base provided by one of the timers. Outputs of the compare units can be programmed easily for different purposes, including carrier-based PWM generation or hardware SVM generation.
- PWM circuitry that include:
  - A hardware SVM state machine used for hardware implementation of one version of the SVM algorithm
  - Symmetrical or asymmetrical PWM generators
  - Programmable deadtime generator
  - Programmable output logic
    - Three *capture units* that have two-level deep memory FIFO stacks, able to provide a time stamp for external events, useful in synchronization of events.
    - *Quadrature encoder pulse circuit* used to interface with encoder sensors. It is able to decode and count the quadrature encoder pulses from an optical encoder.
    - Interrupt logic and interface with the central processing module. A *special power drive protection interrupt* logic is included for fast shutdown of the PWM and power stage.

The *Event Manager* included in the '24x device was more complex and had additional functions. However, these functions were not used much and were replaced by other implementation alternatives.

8.8.2 SOFTWARE IMPLEMENTATION OF CARRIER-BASED PWM

Chapter 3 presented different carrier-based PWM algorithms. Their implementation is straightforward with a comparison of the reference signal with a triangular high-frequency waveform. The triangular waveform can be generated with a *General-Purpose Timer* programmed with the sampling interval period. Each of the continuous counting up, counting down, or up- and down-counting modes may be used resulting in asymmetrical or symmetrical PWM generation. The comparison with the reference is ensured by the three available compare units, each one already tied within the Event Manager to a pair of PWM outputs. The compare units are updated at each sampling interval interrupt. The time intervals to be uploaded into the compare units of the Event Manager are centered around the constant corresponding to half of the sampling interval.

For instance, in order to implement a sinusoidal PWM with subunitary modulation index m and a sampling interval T_s, the period register of the general-purpose timer should be programmed with the integer N_T_s corresponding to the sampling interval, and each compare unit should be updated with

$$\begin{cases} N\_T_a = \dfrac{1}{2}N\_T_s[1 + m\sin\alpha] \\[2mm] N\_T_b = \dfrac{1}{2}N\_T_s\left[1 + m\sin\left(\alpha + \dfrac{2\pi}{3}\right)\right] \\[2mm] N\_T_C = \dfrac{1}{2}N\_T_s\left[1 + m\sin\left(\alpha + \dfrac{4\pi}{3}\right)\right] \end{cases} \qquad (8.2)$$

where α is calculated in increments depending on the sampling period and the fundamental period.

8.8.3 SOFTWARE IMPLEMENTATION OF SVM

The simplest solution for software-based implementation of the SVM algorithm on the Event Manager of Texas Instruments' DSPs is to use the symmetrical PWM-generation feature previously described in Section 8.8.2. The ON-time intervals for each switch can be calculated based on Equations 5.41 through 5.46 and on the proper definition of the switching reference function. Each sector has the option to generate a switching pattern starting from the zero vector 000 or from the zero vector 111. This can be programmed by selecting the polarity of the PWM output signals. Figure 8.3 is an example of SVM generation of a vector position within the first 60° sector, using active vectors 100 and 110.

The time constants used within the PWM generation on the first 60° sector with compare modules are (Equations 8.3 and 8.4):

When starting from 000:

$$\begin{cases} N\_T_a = \dfrac{T_0}{4} \\ N\_T_b = \dfrac{T_0}{4} + \dfrac{T_1}{2} \\ N\_T_C = \dfrac{T_0}{4} + \dfrac{T_1}{2} + \dfrac{T_2}{2} \end{cases}$$ (8.3)

where T_1 and T_2 are calculated using Equations 5.41 through 5.46 and T_0 using Equation 5.47.

When starting from 111:

$$\begin{cases} N\_T_a = \dfrac{T_0}{4} + \dfrac{T_1}{2} + \dfrac{T_2}{2} \\ N\_T_b = \dfrac{T_0}{4} + \dfrac{T_1}{2} \\ N\_T_C = \dfrac{T_0}{4} \end{cases}$$ (8.4)

where T_1 and T_2 are calculated with Equations 5.41 through 5.46 and T_0 with Equation 5.47.

A different set of equations should be calculated for each 60° sector. The software routine for SVM generation also requires the proper selection and programming of the first zero vector used for each sector. Because this implementation uses both zero vectors on each sampling interval, all the sampling intervals over one fundamental frequency cycle can be programmed to start from the same zero vector, either 000 or 111. No additional programming is required on each sampling interval but for the change of the time constants within the compare units.

Similar implementations based on the reference function can be conceived with a timer that counts up or down.

8.8.4 HARDWARE IMPLEMENTATION OF SVM

The Event Manager provides a hardware solution to implement the SVM. The user has to program the required polarity of the output PWM signals to enable the hardware implementation of the SVM and to start a general-purpose timer in the continuous up- or down-counting modes.

At each sampling interval, the appropriate interrupt routine has to receive or calculate the vector coordinates of the desired position of the voltage vector (V_d, V_q) to define the sector the vector belongs to. Each sector is characterized by two adjacent vectors V_x and V_{x+60} and they are programmed with the calculated vector codes in the Event Manager. Note that *Texas Instruments'* code programming definition of the sectors is the reverse of the convention used in this book. The hardware SVM generator also provides an option to select the vector to be used first in the sampling interval. The Event Manager will next use the switching pattern corresponding to these vectors for specified amounts of time.

The user software has to calculate the time intervals allocated to the first (T_1) and the second (T_2) active vectors on the sector as well as the time interval required for the zero state (T_0). Section 8.5 gives some hints to implement this calculation. The resulting constants are loaded as $0.5 * T_1$ in the first compare unit and $0.5 * (T_1 + T_2)$ in the second compare unit.

At the beginning of each sampling period the outputs are compared to the switching pattern representing the first preprogrammed active vector. We get the first compare match at $0.5 * T_1$ from the beginning of the timer up counting. The outputs are switched to the pattern corresponding to either V_x, if this is selected as the first vector on the sector, or V_{x+60}, if this is selected as the second vector on the sector.

At the second compare match, at $0.5 * (T_1 + T_2)$, the pattern is switched to a zero vector (000 or 111), whichever is closer to the last active vector (or switching pattern). The timer continues to count up to the programmed period, then slopes downwards, counting down. The next match occurs for the compare register loaded with $0.5 * (T_1 + T_2)$. This is used to enable the switching pattern for the "second" active vector. At the second match of the slope counting down, the switching pattern of the "first" active vector is transferred to the outputs.

The resulting waveforms are symmetric with respect to the middle of the sampling period. Figure 8.19 shows the two possible ways to generate a vector within the first sector (active vectors 100 and 110) while using the hardware generator within the Event Manager (Figure 8.20). Switching patterns only for the high-side IGBTs are shown. Low-side IGBTs are controlled in a complementary mode.

Notice that only a zero vector is used on each sampling interval, allowing the use of same zero vector for a whole 60° sector. The first consequence relates to lack of switching on one of the inverter legs for 60° and this reduces switching losses. In other words, the hardware implementation of the SVM method within the Event Manager represents a reduced-loss PWM algorithm, suitable for motor control, in which current usually lags behind the voltage. This method *(Method DD1)* has been already analyzed in Chapter 5, Figure 5.26. At the price of some extra switchings per fundamental period, other reduced-loss SVM algorithms may be implemented on the same hardware.

8.8.5 DEADTIME

A deadtime generator is also included within the Event Manager. A single value should be programmed for all six outputs and if the deadtime generator is enabled, the rise-up (turn-ON) of each of the output signals is delayed by a preprogrammed time interval. The deadtime interval depends on clock period, clock prescaling, and the 4-bits programming constant covering all time intervals required by modern power semiconductors.

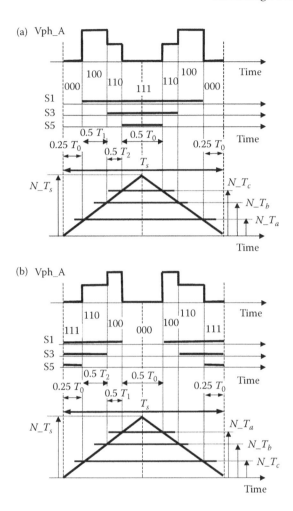

FIGURE 8.19 Example of vector synthesis using the hardware SVM generator. (a) Starting the sampling interval with 100. (b) Starting the sampling interval with 110.

8.8.6 INDIVIDUAL PWM CHANNELS

Texas Instruments' DSPs also include several PWM channels able to work individually directly from a timer. The procedure for programming each individual PWM channel is very easy, starting from setting up a timer as the period counter and loading a compare constant in a special register. When the timer reaches the value preloaded within the compare register, an interrupt can be generated and an output toggles.

8.9   USING FLASH MEMORIES

Any new development and implementation of the PWM generators follow the development in digital hardware technologies. One of the most important achievements

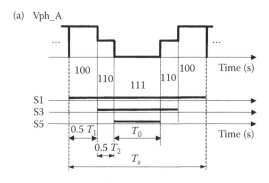

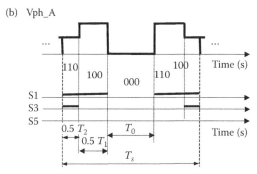

FIGURE 8.20   Example of vector synthesis using the hardware SVM generator.

over the last 5 years was the explosion of the flash memory technology [30,31] and it created a new implementation opportunity for the PWM generators [34].

Flash memory is a nonvolatile memory that can be electrically erased and reprogrammed. It provides a lower cost with more functionality than previous technologies for nonvolatile memory. The first flash memory device was invented in 1981 by Fujio Masuoka at Toshiba Corporation. The technology has been adopted steadily since then, topping $20B in 2006, which meant 34% of the entire market of memory devices. Later on, it went up to 50% of the entire memory market.

The success of the flash memory technology reflected in the control of power converters as well. The 2011 generation of power conversion microcontrollers (formerly known as "*Motor Control DSPs*") are introducing more on-chip flash memory as well as DMA (*Direct Memory Access*) circuitry [32,33]. The principles of using DMA circuitry in power conversion control is known from the early 1980 s [35] when external DMA interface circuits were paired with microcontrollers for better memory access. However, the microcontroller devices did not use this feature too much until 2011. Now, both the TI's *Piccolo* series and the Zilog's *Z16FMC* series are introducing DMA control for both internal and external flash memory [32,33].

Another recent technology development is related to access to external large-size flash memory with improved single-, dual-, or quad-I/O SPI interfaces. The *Serial Peripheral Interface* principles are known for quite a while, and this serial communication interface is mostly used as a simpler alternative to more complex solutions like RS232, RS485, and so on. The baud-rate yields limited by its 1-bit serial transfer of information. The newer versions extend the same principle to communication on 2 or 4 wires along with the appropriate communication management. The speed can be thus increased up to 320 MHz by sending and receiving 2 or 4 bits of data at a time.

The use of flash memory within the digital control platforms benefits from the principles shown previously in Section 3.7 as binary programming PWM. Instead of the conventional comparison of a reference with a carrier signal implemented on different timer structures, the novel approach uses optimized PWM stored in large look-up tables. This allows complex optimization at the pulse level otherwise impossible to be achieved within power converter products.

Numerous algorithms for PWM optimization are reported over the last 30 years. They were not very successful in industrial application due to the extensive digital resources necessary for implementation as well as the somewhat reduced generality. Extensive memory look-up tables can be nowadays implemented within flash memory, independent of the control system. Moreover, synchronized sampling at various frequencies is possible within the same peripheral.

A novel architecture involving flash memories is proposed in [34] and shown in Figure 8.21. Flash memory chips are organized in 2^n blocks, each made up of 2^m sectors, each made up of 2^q pages of memory. The 2-axis output of the control system provides the start address for a memory look-up table containing a table of 128 words of 8-bit, each of them defining the inverter state at a given moment of time. Similar to the binary programmed PWM presented in Section 3.7, this table is read with a constant clock frequency, and the data is transferred to the 6 gate drivers of the three-phase power converter.

The solution shown in Figure 8.21 [34] is based on a 64MB flash memory. As of 2012, external flash memory chips at $2 are available from companies like Numonyx (M29W640), ST-Microelectronics (SST25VFO64C), Spansion (MirrorBit 64). This low-cost compares to that of stand-alone bus compatible PWM interfaces like Siemens SLE4520 and Ixys IXDP610 [21,22], or companion FPGAs [16].

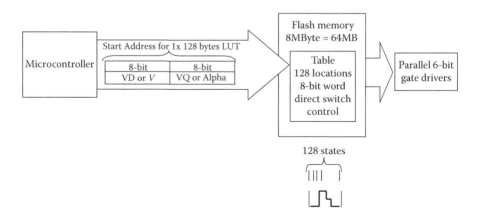

FIGURE 8.21 Hardware architecture with flash memory.

Alternatively, complete PLC and FPGA solutions can be considered to include control logic and memory-based PWM generators within the same device.

The emerging flash memory technology provides an opportunity for the power conversion control designer to reconsider the PWM generation. It is our conclusion that the optimal PWM algorithms previously reported in academic laboratories become now more viable for industrial success with flash memories.

The specific of the new architecture is that either the DMA circuitry or the multi-I/O SPI protocols allow reading a 128-word page of data with a start address only, independent of the microcontroller software run. Contrary to the organization and operation of conventional memory look-up tables where reading each location needs microcontroller intervention, the *(vd,vq)* address is sent to the memory only once as a start address, and the entire 128-words PWM sequence is transferred to the gate drivers without other intervention of the main microcontroller.

This allows a large variety of state sequences over the sampling frequency that corresponds to the 128 clock periods. Within the conventional PWM generators, the sequence of states over a sampling period is the same over the entire fundamental cycle. As shown in Section 4.2, the state sequence is derived from operation of counters as up-counting, down-counting, or up/down-counting. Previous solutions also agree on the use of 4 states over the sampling period like [*zero state -> active vector 1 -> active vector 2 -> zero state*], for the entire fundamental cycle. The novel architecture eliminates such limitation.

Possible criteria to be considered for optimization of the PWM waveforms follow the previous sections and are not detailed herein:

- Optimal sharing of the zero sequence states (as demonstrated in Figures 5.17 and 5.18)
- Variation of the pulse frequency by alternating the use of 3, 4, or 5 identical pulses for each position within a 60° sector, depending on the angular coordinate
- Over-modulation based on optimal pattern

As shown within the previous chapters of the book, these criteria are independent of each other and the same look-up table can be used at different fundamental frequencies, different clock frequencies, or different operation points. The use of these optimization criteria within PWM generators leads to the results incorporated in Figure 8.22.

8.10   ABOUT RESOLUTION AND ACCURACY OF PWM IMPLEMENTATION

Given the large number of digital platforms able to accommodate the implementation of PWM generators, it is worthwhile to master the specifics of accuracy and resolution for PWM algorithms. Let us explain these for the two major types of implementation: counter-based and memory-based. Since we limit our discussion to the core PWM generator, we will consider the counter-based structure as generic for any digital implementation, including DSP and PLC/FPGA solutions.

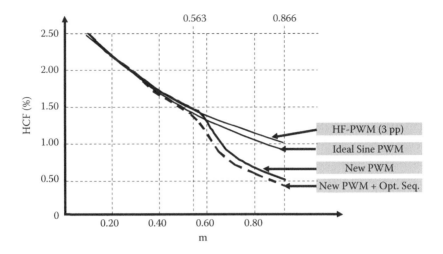

FIGURE 8.22 HCF for 3.6 kHz switching frequency and the optimal PWM.

Most of PLC/FPGA solutions used to implement the PWM algorithms follow the same counter-based structure as the microcontrollers [7–11]. PLC/FPGA(s) offer a higher level of parallelism than microcontrollers [11], and this is used for implementation of faster protection and compensation algorithms like current-based dead-time compensators, or deadbeat current control for bandwidth challenged applications. Such features do not interfere with the core counter-based implementation of the PWM algorithms.

By definition, the resolution of a PWM generator represents the smallest change it can detect in the ideal reference it is provided. Due to the counter-based implementation, modern microcontrollers measure resolution as time increments. High resolution PWM channels go down to 150 ps, and this is most useable for switching frequencies above 200 kHz. Typically, operation within 1–25 kHz does not require high resolution PWM. For instance, a TMS320x280x microcontroller with a system clock of 100 MHz can generate 20 kHz PWM with 12.3 bits resolution on regular PWM, and 18.1 bits on high resolution PWM mode [35].

Limit cycles in voltage-controlled high frequency power supplies may result from the signal quantization and refer to steady-state output oscillations at frequencies lower than the converter switching frequency [36]. This is not the usual case in three-phase converters (open-loop or current controlled) since the current/voltage ripple is way larger than the measurement resolution, and the uncertainty is given by harmonics/ripple rather than by quantization.

The comparison is schematically shown in Figure 8.23.

Figure 8.24 shows that reproducing a sine-wave reference on 12 and respectively 8 bits, in order to generate PWM at 3.6 kHz (72 pulses per 50 Hz fundamental period) produces comparable results in the output voltage due to low-frequency sampling.

By definition, the accuracy of a PWM generator is the degree of closeness of the generated pulse to the actual desired value. The physical size of the memory look-up table limits the accuracy for the memory-based PWM implementation. What is

	Conventional Counter-Based DSP/ FPGA	Flash Memory	Other Comments on Comparison
Micro-controller calculation	Vector calculations on Q30 format. Sine generation by: – Additional memory OR – Taylor exp. from 256 word sine memory, up to Q29	Vector calculations on Q30 format Sine generation by: – Additional memory OR – Taylor exp. from 256 word sine memory, up to Q29	The same properties. – Calculation with high resolution.
Reference to the PWM Generator	Duty cycle [denoted w/R] – maximum 12 bit on regular PWM – maximum 18 bit on high resolution PWM module.	*Vector coordinate* [VD,VQ] or [V,Alpha] – 8 × 8 bit resolution	– Both truncate results.
PWM Generator	PRINCIPLE Timer/Counter	PRINCIPLE Binary program-ming on 7 bit (128 sequences).	*Different principle.* – We need to com-pare accuracy, NOT resolution
Pulse to power stage	Gate driver (IR2110, 2A): – LS/HS matching error (20 ns) – Deadtime resolution (3bit) – Propagation delay (2 × 150 ns) – IGBT Turn-on/off (150 ns) – Finite rise time (dv/dt)	Gate driver (IR2110, 2A): – LS/HS matching error (20 ns) – Deadtime resolu-tion (3 bit) – Propagation delay (2 × 150 ns) – IGBT Turn-on/off (150 ns) – Finite rise time (dv/dt)	The same properties. – No resolution guarantee below 2 × 250 ns. – Pulses with finite dv/dt

FIGURE 8.23 Comparison of two methods.

considered today a feasible solution based on widely available and affordable flash memories of 64 MB allows an accuracy of 0.4%. Larger memory devices with appro-priate faster access will emerge to better carry on the implementation of the mem-ory-based PWM generator architecture with multiple optimization criteria.

8.11   CONCLUSION

This chapter presented the trends in the control of medium and high-power convert-ers. Different theoretical principles for defining a PWM channel by analog or digital means were introduced along with various implementation solutions using modern

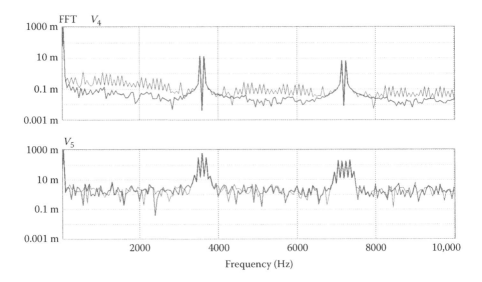

FIGURE 8.24 The effect of quantization resolution on harmonics: top figure = sinusoidal reference quantized on 12, respectively, 8 bits, and sampled at 3.6 kHz by software program; bottom figure = output PWM voltage for a single leg converter, at 50 Hz fundamental, with a reference quantized on 12, respectively, 8 bits, and sampled at 3.6 kHz.

microcontrollers and DSPs. This chapter is rich in presenting novel solutions to a state-of-the-art power converter control structure.

REFERENCES

1. Bob, C. 2003. Getting more out of your motor—Motor driver ICs integrate more functions. Allegro Microsystems, *Application Note.*
2. Neacsu, D.O. 1993. A new digital controller for SVM algorithms. *IEEE SCS*, 1, 101–104.
3. Tzou, Y.Y. and Hsu, H.J. 1998. FPGA realization of space-vector PWM control IC for three-phase PWM inverters. *IEEE Trans. PE*, 12(6), 953–963.
4. Sangchai, W., Wiangtong, T., and Lumyong, P. 2000. FPGA-based IC design for 3-phase PWM inverter with optimized space vector modulation schemes. *IEEE Midwest Symposium on CAS2000*, pp. 106–109.
5. Deng, D., Chen, S., and Joos, G. 2001. FPGA implementation of PWM pattern generators for PWM invertors. *Canadian Conference on Electrical and Computer Engineering*, pp. 225–230.
6. Tonelli, M., Battaiotto, P., and Valla, M.I. 2001. FPGA implementation of a universal space vector modulator. *IEEE IECON*, pp. 1172–1178.
7. Cirstea, M., Aounis, A., McCormick, M., and Urwin, P. 2001. Vector control system design and analysis using VHDL (for induction motors). *IEEE PESC*, pp. 81–84.
8. Iverson, K.E. 1962. *A Programming Language*, John Wiley and Sons, New York.
9. Zadeh, L.A. 1965. Fuzzy sets. *Inf. Control* 8, 338–353.
10. Negoita, C.V. and Ralescu, D.A. 1975. *Applications of Fuzzy Sets to Systems Analysis.* Birkhauser Verlag and New York Halsted Press, Basel.
11. Mamdani, E.H. and Assilian, S. 1974. An experiment in linguistic synthesis with a fuzzy logic controller. *Int. J. Man Mach. Stud.* 7, 1–13.

12. Neacsu, D., Stincescu, R., Raducanu, I., and Donescu, V. 1994. Fuzzy logic control of a PWM V/f inverter-fed drive. *ICEM'94*, Paris, France, vol. III, pp. 12–18.
13. Saetieo, S. and Torrey, D. 1998. Fuzzy logic control of a space vector PWM current regulator for three-phase power converter. *IEEE Trans. PE* 13, 419–426.
14. Anon. 2000. NEC 78098x MCU, NEC, Datasheet Document No. U12804EJ2V1DS00 (2nd edition). Published February 2000.
15. Moynihan, F. 2000. Fundamentals of DSP-based control for AC machines. *Analog Dialogue*, 34(1), 334.
16. Takahashi, T. 2001. FPGA-based high performance AC servo motor drive—Accelerator TM configurable servo drive design platform. *International Rectifier.*
17. Mohan, N., Undeland, T., and Robbins, W. 2002. *Power Electronics*, 3rd edition. John Wiley and Sons, New York.
18. Thorborg, K. 1988. *Power Electronics*. Prentice Hall, London, UK.
19. Holtz, J. 1992. Pulsewidth modulation—A survey. *IEEE Trans. IE* 39, 410–420.
20. Anon. 2003. Mixed signal DSP controller ADSP-21990. Analog devices. Datasheet.
21. Anon. 1998. Inverter Interface and Digital Deadtime Generator for 3-Phase PWM Controls, IXYS Datasheet Documentation, pp. 1–20.
22. Anon. 2001. Bus compatible digital PWM controller. IXYS IXDP610 Documentation.
23. Anon. 2001. *TMS320LF/LC240xA DSP Controllers Reference Guide—System and Peripherals.* TI Literature no. SPRU357B.
24. Yu, Z. 1999. Space Vector PWM with TMS320C24x/F24x using hardware and software determined switching patterns. TI Literature no. SPRA524, March 1999.
25. Doval-Gandoy, J., Iglesias, A., Castro, C., and Penalver, C.M. 1999. Three alternatives for implementing space vector modulation with the DSP TMS320F240. *IEEE PESC.*
26. Trzynadlowski, A.M. 1994. *The Field Orientation Principle in Control of Induction Motors.* Kluwer, Dordrecht.
27. Anon. 2001. DSP56301 User's Manual—24-Bit Digital Signal Processor, Motorola Documentation no. DSP56301UM/AD, Revision 3, March 2001.
28. Anon. 2001. MC68HC908MR32 & MC68HC908MR16 Advance Information—HCMOS Microcontroller Unit, Motorola Documentation no. MC68HC908MR32/D, Rev. December 2001. REV 5.
29. Neacsu, D. 2005. Implementation of PWM algorithms. *IEEE PESC.*
30. Yinug, F. 2007. The rise of the flash memory market: Its impact on firm behavior and global semiconductor trade patterns. US International Trade Commission, *Journal of International Commerce and Economics*, Internet Version, July 2007, pp. 137–185.
31. Zitlaw, C. 2011. The future of NOR flash memory. *EE Times*, May.
32. Anon. 2011. *Piccolo Microcontrollers. Texas Instruments Corp., March.*
33. Anon. 2011. Motor Control MCUs Z16FMC Series Product Specification. Zilog Corporation.
34. Neacsu, D.O. 2012. Novel microcontrollers with direct access to flash memory benefit implementation of multi-optimal space *vector modulation. IEEE Trans. Industrial Informat.* 8(3), 528–535.
35. Rajashekara, K.S., Vithayathil, J. 1982. Microprocessor based sinusoidal PWM inverter by DMA transfer. *IEEE Trans. IE* 29(1), 46–51.
36. Anon. 2011. *TMS320x280x—High Resolution PWM—Reference Guide.* TI literature SPRU924F, Revised, October.
37. Peterchev, A., Sanders, S. 2003. Quantization resolution and limit cycling in digitally controlled PWM converters. *IEEE Trans. PE*, 18(1), 301–308.

9 Practical Aspects in Closed-Loop Control

9.1 ROLE, SCHEMATICS

The performance of power converters can be improved with the use of closed-loop control [1,2]. Because the large majority of power converters start from a voltage source, closed-loop current control is very useful (Figure 9.1). Given the operations at high voltages and with high-frequency switching, the implementation of a current control loop faces a series of specific problems. This chapter discusses these problems and attempts to provide solutions.

9.2 CURRENT MEASUREMENT—SYNCHRONIZATION WITH PWM

The most important module in the current closed-loop control relates to current measurement. The main requirements for the sensor and the acquisition system relate to their capability to detect the presence of electrical noise, temperature, and electromagnetic interference (EMI) radiation in the measurement system. A series of dedicated sensors have been developed to overcome these difficulties.

9.2.1 SHUNT RESISTOR

The older solution for current measurement uses a low-value resistor in the current path and measures the voltage drop across it. The shunt resistor's resistance will likely be in the order of milliohms or microohms, so that only a modest amount of voltage will be dropped at full current. The sensing resistor's value should be very stable with current level and temperature and should have a small equivalent inductance. For instance, a 1 W, 15 A, 0.005 Ω surface-mount resistor can have as much as 5 nH of package inductance.

The low value of the shunt resistor is comparable to wire-connection resistance, which means voltage is measured across the shunt to avoid detecting the voltage drop across the current-carrying wire connections. Shunts are usually equipped with four connection terminals so that the voltmeter measures only the voltage dropped by the shunt resistance itself, without any stray voltages originating from wire or connection resistance. Such a measurement method, able to avoid errors caused by wire resistance, is called the Kelvin or 4-wire method. The measurement connection wires are insulated from the power wires at the hinge point and are in contact only at the tips where they clasp the wire or terminal of the subject being measured. Thus, current passing through the measurement circuit does not go through the power path and will not create any error-inducing voltage drop along its length. In other words,

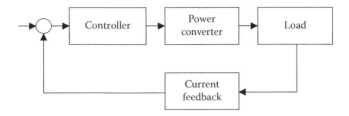

FIGURE 9.1   System diagram for a closed-loop current control.

there is no common path for the measurement and power currents. Shunt resistors with Kelvin contacts have four connections.

Shunt resistors are usually made of a low-temperature-coefficient metal foil on an anodized aluminum substrate and can be packaged in either conventional TO-247 or TO-220 packages or SMT packages for lower power level.

Manganin wire (an alloy of copper, manganese, and nickel) has a low temperature coefficient within 15 ppm/°C from 0° to 80°C. Another commonly used low-temperature-coefficient material is nickel–chromium, or nichrome. This has a resistivity of about 110 mV/cm and requires less wire length than manganin's 44 mV/cm. This helps reduce the inductance for very low-value resistors. Manganin is superior to nichrome in temperature coefficient and long-term stability of resistance value. Another similar alloy is Constantane (Eureka) with a resistivity of 49 mV/cm.

With the advent of Integrated Power Modules during the last decade, several changes occurred with the packaging of lower power level converters as they got more and more on circuit board. This helped integration of thin-film power resistors onto the board layout as individual surface-mount components.

Contemporary designs for low voltage DC/DC converters are using copper traces for building the sensing resistors [3]. The same principles of Kelvin connections are followed and the trace's length and width are appropriately calculated from knowledge about the material properties. As copper has a high temperature coefficient, additional compensation may be required or operation at elevated temperature may be needed during operation. As we are seeing more and more printed-circuit-board integration of IPM devices, we may see similar designs for the high voltage applications, in the horsepower range. However, accurate designs consider for now dedicated shunt resistors made of low-temperature coefficient materials.

One advantage of the shunt resistor is its practically infinite bandwidth. However, isolation is usually required after the shunt resistor.

The signal from a current-sensing resistor is usually processed with an Operational Amplifier with a high common-mode rejection, as the useful signal is usually floating from ground under a large common-mode voltage. Examples in this class of instrumentation amplifiers include Texas Instrument's INA148 or INA117 with +200 V common-mode high input, INA146 (100 V cm), INA 149 (275 V cm) or Analog Devices' AD626. As such devices cannot accommodate a high enough DC common-mode voltage, the sensing resistor should be placed close to ground.

The signal processing usually continues with a low-pass filter implemented around an Operational Amplifier. Among multiple design options, one may

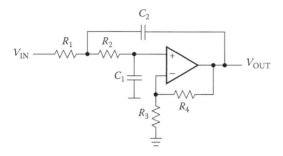

FIGURE 9.2 Sallen-Key topology for low-pass filter on the measurement path.

consider a second order Bessel-type filter able to allow a constant group delay within the pass-band frequencies, thus providing a lesser waveform distortion [4]. However, the drop of the gain characteristic is less sharp than with Butterworth and Tschebyscheff filter designs. If the switching frequency is rather close to the frequencies of interest, these designs may be more effective. Either way, the actual circuit implementation is recommended to follow the Sallen-Key topology (Figure 9.2).

Another solution for signal processing consists of a high-voltage integrated circuit (IC), such as the IR217x [5]. The IR217x is a monolithic current-sensing high-voltage IC designed for servo-drive applications. It senses the current through an external shunt resistor and modulates a fixed frequency train of pulse with the sensing information. These pulses are transferred to the low side. The output format is a discrete pulse width modulation (PWM) that eliminates the need for an A/D input interface and can be directly connected to a timer circuit within any digital signal processor (DSP) or microcontroller.

Selection and rating of the shunt resistor yield from the trade-off between the desire to have a larger voltage drop for easier signal processing and the allowable power dissipation. A larger voltage drop means a larger resistance that would also produce more power loss across the sensing resistor [6,7].

For a motor drive application, where AC currents are quasi-sinusoidal, it is suggested to select the sensing resistor starting from its power rating (P_{max}) and the known maximum load (motor) phase current (I_{rms}). It yields:

- $R = \dfrac{P_{max}}{I_{rms}^2}$ For a sensing resistor connected directly on the output phase

  current line:

- $R = \dfrac{2 \cdot P_{max}}{I_{rms}^2}$ For a sensing resistor placed in the emitter of the low-side

  IGBT:

Similar rating and layout recommendation can be found in [8].

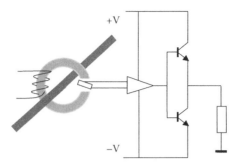

FIGURE 9.3   Open-loop Hall sensor.

9.2.2   HALL EFFECT SENSORS

Shunt resistors are less used today in high-current applications due to the inherent voltage drop. The alternative lies in the use of Hall-effect sensors.

In 1879, Edwin Hall, a graduate student in physics, used a magnetic field to manipulate the charge carriers in a strip of gold foil. He created in the strip a current flowing perpendicular to the field. As the charges that made up the current were moving perpendicular to the field, the magnetic field exerted a force that pushed some of these charges to the top of the strip. Later, scientists discovered the electron and, today, we say that Hall discovered that it was the motion of electrons that caused the current he observed.

An open-loop Hall-effect current sensor is represented in Figure 9.3. It has a block of semiconductor as the sensing element, supplied by a constant current source, and a programmable amplifier to raise the millivolt output to a reasonable value. A current proportional to the measured current is produced in a sensing resistor through the Hall-effect. Older devices used laser-trimmed, thick-film resistors to adjust the programmable amplifier to give a standard output voltage under standard conditions of a magnetic field. Newer devices use a flash memory to hold the amplifier gain setting. A Hall-effect current sensor provides a noise-immune signal and consumes very little power.

Better performance can be achieved with closed-loop current sensors. They represent a different class of Hall-effect current sensors that include an application-specific integrated circuit (ASIC) to provide extremely low offset drift with temperature, resulting in stable, repeatable, accurate measurements.

Hall-effect current sensors are available in hundreds of amperes and provide highly accurate measurement for a large class of power electronic applications. Their bandwidth is usually around 100 kHz, enough for high-power converter applications [9].

9.2.3   CURRENT SENSING TRANSFORMER

For a long time, current-sensing transformers have been considered the best solution for current measurement. The advent of Hall-effect sensing devices, however, reduced the market share of current transformers. They are still used, though, in a limited class of applications, including power converters with high switching frequency. Current-sensing transformers can usually ensure a bandwidth larger than the Hall-effect sensors.

9.2.4 Synchronization with PWM

An analog circuit follows the sensor to adapt the range and bandwidth of the signal to the input of the digital circuit. Given the generic inductive type of load, the current will have a quasi-linear variation during each interval characterized by a pulse of voltage.

The current ripple around an average value is determined by the value of inductance, the switching frequency, and the magnitude of the voltage pulse. Sampling the current at any moment during the switching interval introduces a small amount of ripple in the measurement result, leading to aliasing and offset effects (Figure 9.4).

To alleviate these effects, a synchronized PWM is selected to ensure current acquisition during the zero states, when there is no variation in the current and the value already follows the average value of the current. This approach has been recently adopted in the single-phase and three-phase inverter designs, but it is well known from the control of DC/DC converters, such as the phase-shift, full-bridge, zero-voltage switching (ZVS) converter. It has been previously incorporated in a class of Unitrode circuits.

The current sampling synchronized with the PWM signal is used within the Texas Instruments' family of DSP circuits. This ensures an automatic sampling of the currents or A/D channels at preselected moments when the carrier's triangular signal

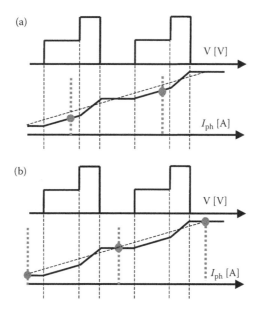

FIGURE 9.4 (a) Current sampling at a random position within the switching interval; (b) synchronized current sampling.

changes slopes. In the language of digital circuits, this is equivalent to sampling the analog inputs when the counter reaches the lowest or largest value.

9.3 CURRENT SAMPLING RATE—OVERSAMPLING

As a large majority of modern converters are controlled by digital structures, the conversion of the analog input representing the current into a digital signal should be done at a given sampling rate. The selection of the sampling rate is the result of a compromise among many factors [1,2,10].

First, the power stage switches states at a rate given by the switching frequency. As the goal of the PWM operation is to produce pulses of voltage following a reference signal, sampling current at a rate higher than that of the switching frequency does not have any meaning given the bandwidth limitation at the power stage. Sampling current at the highest frequency possible, that is, the switching frequency of the power stage, may be limited by the real time required to compute the control algorithm.

It is, however, a good practice to sample the current at the highest possible rate even if the control algorithm computes at a lower rate. In this case, we have more samples available than required and this is called oversampling. Oversampling is able to relax the filter requirements in the initial sampling and convert this high rate signal to the desired sample rate using linear digital filters. We basically use the additional samples to filter the final result.

The lowest sampling frequency is determined by the time constants of the electrical circuit or load that influence the performance of the control system. This constraint can also be described as the tracking effectiveness of the control system. The sampling theorem requests sampling at least twice as fast as the highest frequency contained in the signal. If the closed-loop system is required to track a signal with a given bandwidth, the sampling rate should be at least twice the highest frequency in the closed-loop system bandwidth, which can be different from the highest frequency in the plant model. However, defining the lower sampling frequency from the sampling theorem may not satisfy all requirements of the response time of the closed-loop system.

9.4 CURRENT CONTROL IN (a,b,c) COORDINATES

Both motor control and grid applications use the rotating-reference frame to control currents in the so-called d–q system of reference. The current components become quasi-DC and the control is simplified to a low requirement in bandwidth. For a conventional inductive load, the control system reduces to a simple proportional-integral (PI) controller. Variables in the rotating-reference frame must be restored in the stationary three-phase reference frame using inverse transformation.

However, if the system is single-phase or three-phase without an isolated neutral, the control system should be able to track a sinusoidal or harmonic reference. The harmonic reference occurs when the power converter is used within an active filter structure. In either case, the synchronous coordinate transform cannot be applied.

Consider a power converter and load characterized by a plant model $G_p(s)$. The control system is characterized by a transfer function $G_C(s)$. The open-loop transfer function yields:

$$G_{OL}(s) = G_C(s) \cdot G_p(s) = \frac{A(s)}{B(s)} \tag{9.1}$$

Considering a sinusoidal reference:

$$f(t) = I \cdot \sin \omega t \Rightarrow F(s) = \frac{I \cdot s}{s^2 + \omega_0^2} \tag{9.2}$$

The error of the feedback signal can be calculated as

$$E(s) = \frac{F(s)}{1 + G_0(s)} = \frac{B(s) \cdot F(s)}{B(s) + A(s)} = \frac{B(s)}{B(s) + A(s)} \cdot \frac{I \cdot s}{s^2 + \omega^2} \tag{9.3}$$

Applying the Final Value Theorem defines the constant steady-state value of a time function given its Laplace transform. This uses the partial fraction expansion.

$$E(s) = \frac{a_1}{s + \omega_1} + \cdots + \frac{a_i}{s + \omega_i} + \frac{b_1}{s - j \cdot \omega_0} + \frac{b_2}{s + j \cdot \omega_0} \tag{9.4}$$

If any of the poles ω_i is in the right half of the s-plane, the time-domain signal will increase to an unbounded limit. We will consider these poles with a negative real part. The other pair of imaginary poles derived from the sinusoidal character of the reference would introduce in the time-domain error signal a sinusoidal wave that persists forever and makes impossible the definition of the steady-state error. To avoid this situation, the open-loop transfer function should have the same poles $+/- j\omega_0$, so that these poles disappear from the error-transfer function, guaranteeing the reduction of the steady-state error to zero if the signal frequency is well known.

Several solutions are therein possible.

- Harmonic reference tracking with a P-I-S controller (Figure 9.5) [10]
- Harmonic reference tracking with a feed-forward controller (Figure 9.6) [11]

Both solutions will be presented in more detail in Chapter 11 for the particular case of AC/DC conversion along with another method based on switching directly the converter states (similar to the hysteretic controller) [12].

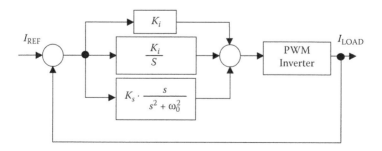

FIGURE 9.5   Current control and harmonic reference tracking with a P-I-S controller.

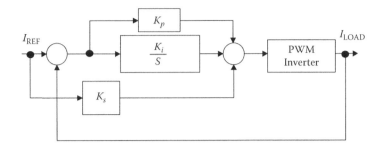

FIGURE 9.6 Current control and harmonic reference tracking with a feed-forward controller.

The stability of the system is, however, dependent on the gain of the s component added to the control system. The transfer function of the open loop exhibits a large phase change around the resonant frequency where the gain is large. The phase margin of the open loop decreases with an increase in the compensation gain. However, a proper selection of the gain can ensure sufficient phase margin.

The problems of tracking a sinusoidal signal can be alleviated with a proper controller, including a term for the effect of the sinusoidal waveform. Despite the success of this solution, current control with reference tracking is more successful in the rotating d–q reference frame. The d and q components are constrained to fix DC values that are easy to control using conventional PI regulators.

Even if the system is either single-phase or three-phase with a connected neutral, the phasor theory can be employed to calculate the d–q components for each phase (independent of the existence of other phases) [13,14].

9.5 CURRENT TRANSFORMS (3->2)—SOFTWARE CALCULATION OF TRANSFORMS

The most common implementation of the current control uses the Park/Clarke set of transforms (Equations 9.5 through 9.7).

$$\begin{bmatrix} I_\alpha \\ I_\beta \\ I_0 \end{bmatrix} = \frac{2}{3} \cdot \begin{bmatrix} 1 & -\dfrac{1}{2} & -\dfrac{1}{2} \\ 0 & \dfrac{\sqrt{3}}{2} & -\dfrac{\sqrt{3}}{2} \\ \dfrac{1}{2} & \dfrac{1}{2} & \dfrac{1}{2} \end{bmatrix} \cdot \begin{bmatrix} i_X \\ i_Y \\ i_Z \end{bmatrix} \tag{9.5}$$

$$\begin{bmatrix} I_d \\ I_q \\ I_0 \end{bmatrix} = \begin{bmatrix} \cos\theta & \sin\theta & 0 \\ -\sin\theta & \cos\theta & 0 \\ 0 & 0 & 1 \end{bmatrix} \cdot \begin{bmatrix} I_\alpha \\ I_\beta \\ I_0 \end{bmatrix} \tag{9.6}$$

The same transforms can be grouped within a single form.

$$\begin{bmatrix} I_d \\ I_q \\ I_0 \end{bmatrix} = \frac{2}{3} \cdot \begin{bmatrix} \cos\theta & \cos\left[\theta - \dfrac{2 \cdot \pi}{3}\right] & \cos\left[\theta - \dfrac{4 \cdot \pi}{3}\right] \\ \sin\theta & \sin\left[\theta - \dfrac{2 \cdot \pi}{3}\right] & \sin\left[\theta - \dfrac{4 \cdot \pi}{3}\right] \\ \dfrac{1}{2} & \dfrac{1}{2} & \dfrac{1}{2} \end{bmatrix} \cdot \begin{bmatrix} i_X \\ i_Y \\ i_Z \end{bmatrix} \qquad (9.7)$$

These equations are similar to (Equations 5.8 through 5.11) and more details are provided in Chapter 5.

What concerns the software calculation of these transforms (Equations 8.5 through 8.7), dedicated routines are part of any motor control or grid control library. A look-up table of a trigonometric function, optimized for a 90° sector, is used.

Using closed-loop control in (d, q) coordinates often requires a careful look into the load-circuit equations. As the load may include a first-order system (inductance or capacitance), the controlled measure appears under a derivative in the load-circuit equation. The three-phase equations converted in the (d, q) components should take into account the derivative term. This produces a phase shift of 90° changing a real component into an imaginary one or an imaginary one into a real one. These terms should be considered within the control system and they are called cross-coupling terms [12].

9.6 CURRENT CONTROL IN (d, q)—MODELS—PI CALIBRATION

The generic control system in (d, q) components is shown in Figure 9.7

The PI control system is described mathematically by

$$D_c(s) = k_p + \frac{k_p}{T_I \cdot s} \qquad (9.8)$$

The time domain equivalent variation results in

$$u(t) = k_p \cdot e(t) + k_I \cdot \int_0^t e(t)dt \qquad (9.9)$$

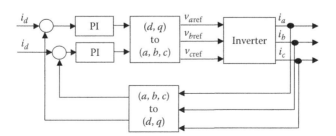

FIGURE 9.7 (d, q) current control of a symmetrical three-phase system.

Considering a linear system, the digital approximation of this equation yields:

$$u[kT_s + T_s] = k_p \cdot e[kT_s + T_s] + k_I \cdot \int_0^{kT_s} e(t)dt + k_I \cdot \int_{kT_s}^{kT_s + T_s} e(t)dt \approx$$

$$\approx k_p \cdot e[kT_s + T_s] + u_I[kT_s] + k_I \cdot \frac{T_s}{2} \cdot \{e[kT_s + T_s] + e[kT_s]\}$$

(9.10)

There are several ways possible for the approximation of the last integral term. Equation 9.10 is considering an approximation using a trapezoidal form with the base T_s. Furthermore, the calculation of the next action term is usually achieved in one of the following ways:

- Accumulator method: A large register uI is used as an accumulator for the integral term and the integral component is continuously added to this register. This is the most used method, but its drawback is in the possible wind-up or overflow of the accumulator.
- Incremental controller: An incremental controller is used to calculate the change in the action.

$$\Delta u = u(kT_s + 1) - u(kT_s) = k_p \cdot \left(e[k \cdot T_s + T_s] - e[k \cdot T_s] \right) + k_I \cdot e[k \cdot T_s + T_s] \quad (9.11)$$

This implementation is faster and uses a shorter code, but covers the information contained within the accumulator.

In order to design the control system and to define the most appropriate gains for the PI-control system, a model of the load is defined in (d, q) components. Multiple methods are available to develop a controller that will meet given requirements for steady-state and transient response. These methods require a precise dynamic model of the process in the form of equations in motion or a detailed frequency response over a certain range of frequencies. Such methods include:

- Design with the root locus method
- Design with the frequency response method
- Design with the state space method

The design requirements are related to

- System stability
- Performance for steady-state operation (steady-state error Err)
- Performance for dynamic operation (transient time t_p, response time t_r, stabilization time t_s, overshoot M_p) (Figure 9.8)

All of these design methods lead actually to a simplified case for the current control of a mostly inductive load.

In practice, the operator will tune the regulator by trial-and-error. Tuning of the proportional-integral-derivative controllers has been the subject of continuing

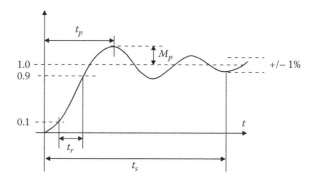

FIGURE 9.8   Parameters of the transient response in time domain.

studies since Callender (1936) [15]. Many of these solutions are based on estimates of the plant model derived from experiment and they can be found in reference text-books such as [1].

Ziegler and Nichols provided [16,17] two experimental methods for tuning the PI controller. The first suggests tuning of the control parameters until a decay ratio of 25% is achieved within the step-response transient. This is equivalent to a decay of the transient response to a quarter of its value after one value of oscillation (over-shoot). The gains of the PI controller yield $k_p = 0.9/RL$ and $TI = L/0.3$, where R represents the slope of the step-up response and L represents the lag time at a step change.

Another approach is called the ultimate sensitivity method [1], as it relies on the estimation of the amplitude and frequency of the system oscillations at the limit of stability. The proportional is first increased until the system becomes marginally stable. This can be seen in the existence of continuous oscillations limited by the saturation of the actuator. The gain k and the period T of these oscillations are called the ultimate gain and period. The PI parameters are then calculated as $k_p = 0.45k$ and $TI = T/1.2$.

9.7   ANTI-WIND-UP PROTECTION—OUTPUT LIMITATION AND RANGE DEFINITION

The real characteristics of the system can cause the actuator to saturate. For instance, a three-phase system has a limited range of the available output voltage, and any requirement from the control system beyond this range would translate in a satu-ration of the output and loss of controllability. If the error signal continues to be applied to the integrator input under these conditions, the accumulator will grow (wind-up) until the sign of the error changes and the integration turns around. The system behaves as an open-loop system and the accumulator becomes a source of instability in it.

The solution is an integrator antiwind-up circuit, which turns-off the integral action when the actuator saturates. To prevent this, an integrator antiwind-up circuit is used, which turns-off the integral action when the actuator saturates. A simple solution is shown in Figure 9.9.

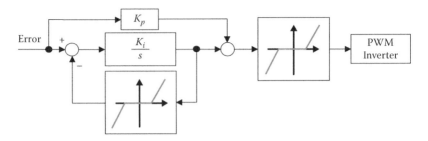

FIGURE 9.9 Anti-windup compensation of a PI controller.

There are many digital control solutions for the implementation of an antiwind-up control system. The system described here shows a linear dependency of the feedback during saturation, which is able to introduce a first-order lag equivalent of an antiwind-up integrator during saturation.

9.8 CONCLUSION

Current control within power converters is subject to noise and distortion. Special precautions need to be taken to filter and measure current in the presence of large ripples. Digital current control is somewhat simple, as a large number of applications use only conventional PI controllers. Several other particular aspects related to implementation are presented in this chapter.

REFERENCES

1. Franklin, G.F., Powell, J.D., and Emami-Naemi, A. 2002. *Feedback Control of Dynamic Systems*, Prentice-Hall, Upper Saddle River, NJ.
2. Ogata, K. 1997. *Discrete Time Control Systems*. Prentice-Hall, Upper Saddle River, NJ.
3. Kmetz, J. 1997. Minimum size copper sense resistors. Micrel Application Hint 25.
4. Mancini, R. 2002. OpAmps for everyone. Application Note SLOD006B, Texas Instruments.
5. Adams, J. 2003. Using the IR217x linear current sensing ICs. Application Note AN-1052, *International Rectifier*.
6. Bille, S.M. 2012. Two or three shunt resistor based current sensing circuit design in 3-phase inverters. ST Microelectronics AN4076. Application Note.
7. Wood, P., Battello, M., Keskar, N., Guerra, A. 2002. IPM application overview integrated power module for appliance motor drives. Application Note AN-1044.
8. Anon. 2012. Current sense resistors. Application Note, TT Electronics.
9. Anon. LEM Sensors. Internet Documentation, www.lemusa.com.
10. Fukuda, S. and Yoda, T. 2001. A novel current-tracking method for active filters based on a sinusoidal internal model. *IEEE Trans. Ind. Appl.* 37, 888–894.
11. Neacsu, D.O. 2010. Analytical investigation of a novel solution to AC waveform tracking control. *IEEE International Symposium in Industrial Electronics*, Bari, Italy, July 2010, pp. 2684–2689.
12. Neacsu, D.O. 2004. Current control with fast transient for three-phase AC/DC boost converters. *IEEE Trans. Ind. Electron.* 51, 1117–1121.
13. Dong, G. and Ojo, O. 2005. Design issues of natural reference frame current regulators with applications to four-leg converters. *IEEE PESC*, Recife, Brasil, pp. 1370–1376.

14. Miranda, U.A., Rolim, L.G.B., and Aredes, M. 2005. A DQ synchronous reference frame current control for single-phase converters. *IEEE PESC*, Recife, Brasil, pp. 1377–1381.
15. Callender, A., Hartree, D.R., Porter, A. 1936. Time lag in a control system. *Philos. Trans. R. Soc. London A*, 235, 415–444.
16. Ziegler, J.G. and Nichols, N.B. 1942. Optimum settings for automatic controllers. *Trans. ASME* 64, 759–768.
17. Ziegler, J.G. and Nichols, N.B. 1943. Process lags in automatic control circuits. *Trans. ASME* 65(5), 433–444.

10 Intelligent Power Modules

10.1 MARKET AND TECHNOLOGY CONSIDERATIONS

10.1.1 HISTORY

Currently, most power converter products are already well defined and accepted by many customers. Hence, the corporate effort is to produce it cheaper and more efficiently by excellence in operations. The process of improving the well-known topologies into fabrication is based on new packaging methods for the hardware as well as novel control architectures and platforms. Moreover, both the fabrication process and the component selection within the design process are now highly optimized.

Different R&D institutions have adopted this philosophy and embarked in product optimization. For instance, the *Office of Naval Research* (ONR)—a top research sponsorship institution of the U.S. Department of Defense—has sponsored and led a series of programs in mid-1990s and early 2000s meant to demonstrate and improve capabilities in power conversion with focus on packaging, power density, and control architecture. The two major programs including both academia and industry were the *Power Electronics Building Block* (PEBB) and later the *Advanced Electrical Power Systems* (AEPS).

Despite targeting mostly naval applications, these programs set-up the basis of standardization and packaging for any medium power application (tens to hundred kW). Later on, *ABB* has extended the PEBB program into the multimegawatt range. More recently, different aviation electronics producers adopted PEBB as a technology of choice for the development of future *more-electric aircraft systems.*

In early and mid-2000, ideas that emerged from the PEBB and AEPS programs were assimilated by large power semiconductor manufacturers and designed-into the development of *Intelligent Power Modules*. The basic description of these IPM devices follows herein, while Chapter 18 a will enter an advanced research topic for developing novel technologies and circuits based on these IPM devices.

The IC technology has experienced an impressive development during the last 30 years. The number of transistors on the same chip has continuously doubled at each other year. This tremendous technology capacity combined with the progress in computer aided-design has allowed emergence of very-large-scale integrated circuits (VLSI) able to achieve high performance in signal processing and size reduction of the electronics equipment.

Similar advancements were possible within the collateral efforts of the power semiconductor manufacturers. Progress has been achieved in packaging of the gate driver, sensing logic, and power semiconductor under the same hybrid IC package. This emerged into a new set of devices—often called *"intelligent power modules"*—that steadily reached a certain standard of usage in low power range (Figures 10.1

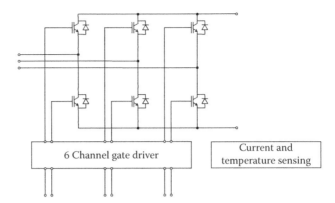

FIGURE 10.1   Typical circuitry within an IPM device.

and 10.2). These IPM devices contain a full gate driver with sensing and protection capabilities [1,2].

10.1.2   ADVANTAGES AND DRAWBACKS

Advantages of the IPM-type devices consist of

*   Improved reliability since the power semiconductor module could contribute better performance for the system's reliability than individual components [19].
*   Improved reliability since a power module provides a better thermal design and layout, both with effects on the system reliability. Using a power module supplied by manufacturer rather than multiple individual components is recommended for the inverter application [19].
*   Two to three times better power cycling capability than using conventional power switches.

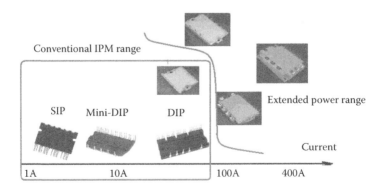

FIGURE 10.2   Examples of packages and power rating for IPM devices.

- Lower parasitic inductance than within the discrete solutions with benefits in voltage spike reduction, and possible operation at higher switching frequency with lower switching loss.
- Simplified power connection (VDC + , VDC−, A,B,C).
- Microcontroller connection through 6 logic-level inputs.
- Propagation delays for all low-side and high-side IGBTs are matched.
- Protection against over-current and over-temperature faults is secured.
- Reduction of system's volume and weight.
- Easier debugging and field repair of the electronics equipment.

The drawbacks of IPM modules are

- Maximum power ratings were reduced up to recent products.
- Switching frequency is limited due to thermal constraints.
- Still require DC bus capacitor and passive filtering since these modules follow the conventional back-to-back topology.

Performance improvement when using IPM devices instead of discrete devices came from a series of technology achievements such as:

- IGBT device technology –> reduction of power loss.
- Packaging materials –> better heat extraction.
- Improved gate driver control.

10.1.3 IGBT Chip

Recent IGBT devices aim to deliver benchmark $V_{CE(ON)}$, with zero temperature coefficient, at lowest conduction loss possible, together with soft switching transients able to reduce EMI. The IGBT technology evolved sub sequentially since early 1990s as it follows:

- Punch-through planar IGBT
  - Suitable for voltages 250 V–1200 V,
  - Cost effective technology, optimized for either speed or short-circuit rating up to 10 μs.
- Nonpunch-through planar IGBT
  - Suitable for voltages 600 V–1200 V,
  - Optimized for ruggedness,
  - Short-circuit tolerance up to 10 μs at higher switching frequency.
- Field-stop trench IGBT
  - Suitable for voltages 350 V–650 V,
  - Optimized for both conduction and switching performance,
  - Rated for 5 μs short-circuit capability,
  - Allows higher current rating in smaller packages.

- Punch-through trench IGBT
  - Suitable for voltages 600 V
  - Cost effective technology at low switching frequencies (below 1–5 kHz),
  - Rated for 3 μs short-circuit capability,
  - Allows higher current rating in smaller packages.
- Field-stop trench IGBT
  - Excellent conduction and switching characteristics,
  - Rated for 10 μs short-circuit capability,
  - Allows higher current rating in smaller packages.

Different manufacturers have marketed technologies similar with each other, with small variations from this historical evolution. The multiple options for technology development allowed the specialization of the IGBT chip depending on application. The IPM devices emerged first for application to appliance and *"white goods"* products. The application is therein a motor drive that does not need a switching frequency higher than 5 kHz, with transitions generally below 5 kV/μs (or 5 V/ns). Hence with the main focus of the design process yields on reduction of the conduction loss.

10.1.4 GATE DRIVER

In order to get the most benefit from these new IGBT chips, *the gate driver* itself is optimized. The gate driver ICs aiming at improved performance are using at least different gate resistors for turn-on and turn-off. The most advanced devices are also considering dynamic control of the transition slope on dependence of the switched current [3–5]. Since the gate driver can change its driving speed based on the switching current as discussed in Chapter 2.5, it is possible to achieve simultaneously low loss and low noise (Figures 10.3 and 10.4).

10.1.5 PACKAGING

The third key technological advent that made IPM success possible relates to package fabrication (Figure 10.2).

The transfer mold technology was first used and it is based on copper lead frames. For larger power levels, the requirements for heat extraction are more demanding and the heat transfer through the copper leads was not enough. A step forward came with integration of cooling structures like mold resin and aluminum heat-sink. Further on, higher power levels require even more heat extraction which is nowadays possible with integrated ceramic substrates.

As opposed to the low-power IC technology, the substrate of a power module must carry higher currents, provide larger voltage insulation, and deal with increased amount of power loss and heat. The role of the substrate in a power module is to provide the circuit connections and to provide cooling since it is common for the substrate to face operation up to 150° or 200°C. The most used substrate technologies are the *Direct Bonded Copper* substrate, and the *Insulated Metal Substrate*.

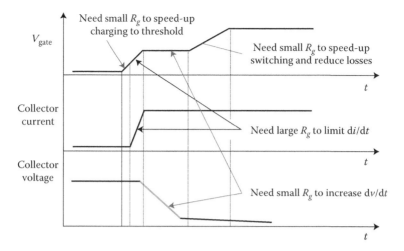

FIGURE 10.3 Nonlinear gate resistance.

Direct Bonded Copper (DBC) substrates are composed of a ceramic tile like alumina and a sheet of copper bonded to one or both sides. The copper and ceramic tile are heated to a controlled temperature, in an atmosphere of nitrogen and oxygen, until a copper-oxygen eutectic forms which bonds successfully both to copper and the oxides used as substrates. The top copper layer can use printed circuit board technology to draw an electrical circuit, while the bottom copper layer is usually kept plain for cooling by attachment to a heat spreader or heat-sink.

The advantages of *Direct Bonded Copper* technology are

- Low coefficient of thermal expansion, which ensures good thermal cycling performances (up to 50,000 cycles)
- Excellent electrical insulation
- Good heat spreading characteristics

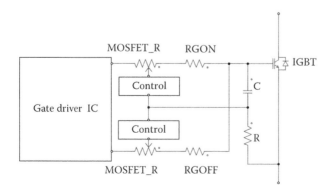

FIGURE 10.4 Possible implementation of the controlled (nonlinear) gate resistance.

The ceramic material used within *Direct Bonded Copper* technology can be:

- *Alumina* (Al_2O_3) = widely used because of its low cost despite the lower thermal performance (24-28 W/mK) and despite being somewhat brittle. Over the years, the thickness of this substrate was reduced from 0.63 to 0.38 mm for most applications in order to reduce the thermal resistance (R_{th}) from the semiconductor chip to the heat sink.
- *Aluminum nitride* (AlN) = offers better thermal performance (> 150 W/mK) at higher cost.
- *Beryllium oxide* (BeO) = despite representing another option for good thermal performance, it is not very much used since it is somewhat toxic.

Insulated metal substrate (IMS) technology starts from a metal base-plate like aluminum, covered with a thin layer of dielectric like an epoxy-based layer, followed by a layer of copper (35 μm to more than 200 μm thick). Due to this structure, the *Insulated metal substrate* (IMS) technology is a single-sided substrate and it can accommodate components on the copper side only.

Power modules can be designed with or without a base-plate. If the base-plate is missing, the *Direct Bonded Copper (DBC)* substrate is directly placed on the heat-sink. Alternative solutions for the copper base-plate include composite materials such as *AlSiC* or *Cu-Mo*, mostly used in traction applications due to their low thermal conductivity and high costs.

The heat removal performance is obviously related to the performance of the materials used for packaging of the power modules. Consequently, the development and improvement of new materials influence size and weight of the power semiconductor modules. Therefore, the contemporary R&D effort is dedicated to justifying the new opportunities within the power topology itself. It is thus interesting that the application circuitry yields to follow timely the advent of novel packaging technologies.

10.1.6 OTHER APPROACHES

A completely different approach considers the integration of a printed-circuit-board within the same package of a conventional base-plate module [6]. Hence, some of the control features are implemented locally within the same power module.

As an extreme case, there are some varieties of IPM devices that contain the microcontroller on the same package with the power semiconductor devices. However, they are not very much used since it was not proven that such approach can pass all the EMI and safety standards. There is also a somewhat limited adaptability to any control platform already developed in-house by the converter system integrator.

10.2 REVIEW OF IPM DEVICES AVAILABLE

Each major power semiconductor manufacturer came with its own solution for power modules since the first IPM products in late 1990s [7]. Table 10.1 provides some examples of conventional IPM devices. The quoted numerical data is just an example able to sustain the theory developed within this chapter and there is

TABLE 10.1
Examples of 3-Phase IPM Technologies in Arbitrary Order [2,8–10]

Family Name	Manufacturer International	Launch Year	Bus Voltage	Rated Current
IRAMS	Rectifier	2004	600 V	3.20 A

Packaging

- Package SIP
- Heat spreader for the power die along with full transfer mold structure
- Max. power dissipation 73 W.
- Low VCE(on) short-circuit rated Punch-Through IGBT
- Standard 3-ph gate driver
- Isolation 2000 V_{rms}

Features

- 5 V Schmitt-triggered input logic
- Under-voltage lockout.
- Internal bootstrap diode
- Interlocking function
- over-current and over-temperature protections
- Built-in temp. monitor;

Family Name	Manufacturer ST	Launch Year	Bus Voltage	Rated Current
SLIMM (STGIP)	Microelectronics	2011	600 V	10.20 A

Packaging

- Package SDIP-25/38 L (ECOPACK)
- Direct bond copper (DBC) substrate leading to low thermal resistance
- Max. power dissipation 42 W.
- Short-circuit rated IGBT with VCE(sat) negative temperature coefficient
- Advanced gate driver
- Isolation 2500 V_{rms}

Features

- 3.3 V, 5 V, 15 V CMOS/TTL hysteresis comparator inputs with pulldown and pullup resistors
- Under-voltage lockout.
- Internal bootstrap diode
- Interlocking function
- Over-current and over-temperature protections
- Op amps for advanced current sensing
- 5 kΩ NTC for temp. control

Family Name	Manufacturer Fairchild	Launch Year	Bus Voltage	Rated Current
Motion SPM	Semiconductor	2005	600 V	3.20 A

Packaging

- Mini-DIP package (SPM27).
- Direct die-bond on the copper lead frame (DBC), w/bare ceramic material attached to the frame, and then molded into epoxy resin.
- Short-circuit rated optimized PT planar IGBT design
- Advanced gate driver
- Isolation 2500 V_{rms}

Features

3.3 V inputs
Under-voltage lockout.
Interlocking function
Over-current and over-temperature protections

(continued)

TABLE 10.1 (continued)

Examples of 3-Phase IPM Technologies in Arbitrary Order [2,8–10]

Family Name	Manufacturer Allegro	Launch Year	Bus Voltage	Rated Current
SCM1100	Microsystems (SANKEN)	2007	600 V	15 A

Packaging

- PowerDIP package (Propr.).
- exposed thermal pad for enhanced power dissipation capacity.
- Max. power dissipation 41.7 W
- Short-circuit rated IGBT
- Advanced gate driver
- Isolation 2000 V_{rms}

Features

- 3.3 V/5 V CMOS inputs
- Under-voltage lockout.
- Interlocking function
- over-current and over-temperature protections
- Internal bootstrap diode

Family Name			Bus Voltage	
L-series		Launch Year	600 V &	Rated Current
L1-series	Manufacturer Mitsubishi	2005/ 2009	1200 V	215 A & 144 A

Packaging

- Small package/medium package/large package
- Max. power dissipation 128 W/390 W/833 W
- Integration of 5th generation trench chip (CSTBTTM) achieves lower saturation voltage
- Isolation 2500 V_{rms}

Features

- Requires individual GD supplies
- Individual GD output for short circuit protection, UV protection, OT protection
- On-chip temperature sensor diode on the IGBT chip

Family Name	Manufacturer Infineon	Launch Year	Bus Voltage	Rated Current
CIPOS	(Under LS Power	2010	600 V	8.22A (Mini)
(Code IGCM)	Semitech Ltd.)			6.20 A (SIL)
(Code IKCS)				

Packaging

- Two versions in Mini package and SIL package.
- Transfer-molded technology.
- Max. power dissipation 29 W/p.u. (mini) & 59 W/p.u. (SIL)
- Infineon reverse conducting IGBTs with monolithic body diode (mini) and TrenchStop® IGBT (SIL)
- isolation 2000 V_{rms}

Features

- 3.3 V LSTTL, CMOS inputs.
- bootstrap circuit
- Temperature sense
- UV lockout
- Over-current protection
- Cross-conduction prevention

no intention to compare performance. Many other manufacturers like *Cyntec* and *Sanyo (now within ON Semiconductor)* may have not been included herein simply for space reasons.

A new generation of high power devices has been launched in late 2011 and they can be seen as an important leap forward for their expanded power in the range of 100 s Amp (Table 10.2). We are seeing this as a very important technological advance since it pushes the envelope of this technology across the entire low-voltage IGBT product range (low voltage in the sense of power systems that is below 1500 V).

TABLE 10.2
Latest High Power IPM Product Announcements [11]

Family Name V series	Manufacturer FUJI	Launch Year Fall 2011	Bus Voltage	Rated Current
			600	20, 30, 50
			1200	10, 15, 25
			600	50, 75
			600	50–200
			1200	25–100
			1200	25, 35, 50
			600	200–400

Packaging

- Ceramic insulated (silicon nitride substrate) package
- Sizes
  49.5 × 70 × 12.5
  49.5 × 70 × 12.5
  50 × 87 × 12
  84 × 128.5 × 14
  84 × 128.5 × 14
  50 × 87 × 12
  110 × 142 × 27

Features

- Latest 6th-generation IGBT chips (trench gate with field stop)
- Most recent drive IC enable the modules to achieve the industry's least amount of power loss (20% lower than Fuji Electric's previous products) and lowest level of noise
- Modules are fitted with four alarm lights Overall size and thinness have been reduced by 80% compared to previous products (N,S,U-series)

10.3 USE OF IPM DEVICES

The advent of IPM devices has opened some new R&D opportunities for the system designer which does not have to worry about the optimization of the low-level design for the gate driver and protection. The power electronics designer can therefore focus on performance achievements at system or power circuitry level. New multiswitch topologies can hence be created and some of them are presented in Chapter 18.

These IPM devices are integrating semiconductor devices along with their cooling structures, with very limited or without passive components. The power electronics designer benefits now from all these topologies that do not contain passive components. It generally does worth trading the integration of multiple semiconductors against the reduction of the passive components count in power electronics equipment. This is the core idea of the advanced research presented in Chapter 18. They all can be subscribed to a design trend for all-semiconductor power conversion solutions able to increase the reliability, and maintainability.

10.3.1 LOCAL POWER SUPPLIES

Designing a power circuit with IPM devices includes setting up the following low power external components [12]:

- Bootstrap power supply should supply the high-side gate circuit without dropping below the UVLO threshold (typically around 11 V);

- Bypassing the local power supply for the HVIC;
- Certain IPM devices come with internal thermistor for temperature monitoring and over-temperature protection;
- Certain IPM devices come with an option for external resistor to sense current, and others have the sensing circuitry already internally.

Among all these tasks, the design decision for the floating (high-side) power supply is the most open to debate. The conventional supply of each channel is done with a topology called bootstrap power supply (Figure 10.5). This is based on charging the supply (bootstrap) capacitor for the high-side circuitry during the conduction time of the low-side IGBT from the low voltage power supply used for the circuitry related to the low-side device. After turning-off the low-side IGBT, the gate driver (protection and control) of the top-side IGBT is supplied from the energy stored within the bootstrap capacitor.

The bootstrap capacitor is usually a ceramic MLCC capacitor able to behave well in high frequency while storing the required energy. While a combination of ceramic and electrolytic capacitors was traditionally of choice, the recent availability of ceramic capacitors with large capacitance value suggests the setup of the bootstrap capacitor with a single ceramic capacitor. This is usually selected with a value of 1 to 47 µF, and rated at 25 V.

The bootstrap diode needs to withstand 1.5 times the bus voltage, and this becomes usually a requirement for a 1000 V diode when used along a conventional 600 V IPM. Moreover, it has to be a fast recovery diode, 50 ns or better, at 1–2 A.

Finally, inserting a damping resistor (R_{boot} = 1 to 5 ohms) in series with the bootstrap diode is an effective way to eliminate some important problems like:

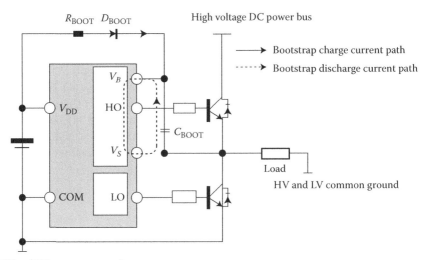

FIGURE 10.5   Principle of bootstrap power supply.

TABLE 10.3
High-Side Power Supply Requirements

Topology	Figure	Comments about t_{on}/t_{off}	Bootstrap Possible?
Simple parallel	Figure 14.1	Normal PWM	Yes
Interleaved	Figure 14.22	Normal PWM	Yes
Parallel assembly diode/CSI	Figure 18.5	1/3 cycle	Maybe w/special design
Multilevel	Figure 18.11 Figure 18.12	Normal PWM	Yes
Conventional matrix w/CSI	Figure 18.15 Figure 18.18	Large t_{off}	No
Direct converter w/VSI	Figure 18.9	Normal PWM	Yes
Novel topology	Figure 18.19	Large t_{off}	No

- EMI noise due to possible reverse recovery charge of the bootstrap diode.
- Overcharge of the bootstrap capacitor by a negative spike on the VS voltage, spike induced by the stray inductance on the negative line.

Some other minor precautions are required at the design of the HVIC and at the start-up of the entire operation.

The issue with this cost-effective bootstrap power supply comes when using the IPM in a nonconventional way. For instance, the special case of using IPMs within the special topologies is proposed later on in Chapter 18. These cannot benefit from the low-side conduction due to the large interval of pause in operation.

Table 10.3 reviews the power supply requirements in this respect.

A possible solution to this matter consists of the independent supply of each gate driver. This is done in industrial systems with special flyback configurations. In the case of topologies built with multiple modules, such option is not feasible and also departs from the goals of modularization and reduction of passive components. The possible elimination of passive components and a design close to the all-semiconductor approach improves reliability and reduces loss [13].

A switch-mode unregulated dc/dc converter can be used instead and this is an ideal circuit substitute for the bootstrap diode (Figure 10.6). For instance, the RECOM dc/dc switch-mode converter comes in a small SIP4 package, works without a requirement for external components and without heat-sink, while offering unregulated 15 V/15 V conversion, 1000 V isolation, 2 W output (133 mA) at an efficiency of 80–85%. It can accommodate a capacitive load of up to 680 µF, more than enough for a large majority of gate driver applications. Multiple other manufacturer options are available.

For a numeric design, consider a maximum gate current request of 4.5 A for 500 ns, the minimum supply capacitor required to minimize the voltage drop at 1 V yields:

$$C_x \geq \frac{(4.5\ \text{A}) \cdot (0.5 \cdot 10^{-6})}{1} = 2.25\ \mu\text{F} \langle\langle\ 680\ \mu\text{F} \tag{10.1}$$

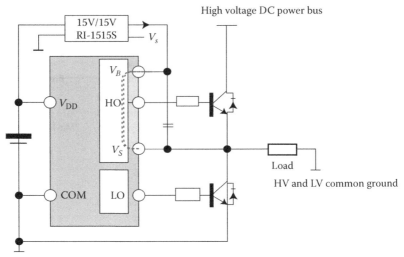

FIGURE 10.6   Using the unregulated 15/15 V power supply, with 1000 V isolation.

A ceramic capacitor MLCC of 10 μF at 25 V, in a surface mount 1206 package can be adopted for any IRF IRAMS. Alternatively 22 or 47 μF nominations are also available in ceramic surface mount technology.

The use of a dc/dc hybrid converter is obviously more expensive, but it solves a series of problems with bootstrap power supply and can provide for a great laboratory setup needed to develop new control algorithms for all the novel topologies proposed in Table 10.3.

10.3.2 CLAMPING THE REGENERATIVE ENERGY

The main use of IPM devices relates to AC loads and the designer should address the operation with an inductive load after the operation of the switches ceases for voluntary or involuntary reasons. This may be the case of a faulty condition when all the switches turn-off for the protection of the drive and the energy from the load needs a path to discharge. Otherwise damage to the power semiconductor devices can occur.

The most used method for de-energizing the load in such situations consists of a current path through diodes. Moreover, the intermediary DC bus capacitor of a back-to-back configuration may need sometimes a quicker discharge or voltage limitation through a brake resistor. Such brake or discharge resistor is connected across the DC capacitor with a 7th switch. While the back-to-back converter topology is inherently using the existing diodes within the two IGBT six-packs, any of the direct (matrix) converters proposed in Chapter 18 needs to use an additional dual diode bridge circuitry from input to output. Such additional clamp circuitry is mostly needed even if it is not shown in the main circuit topology drawing.

On the other hand, the conventional IPM devices are intended for low power applications, with currents below 50 A. In such case, the preferred protection is conventionally based on *varistors* as shown in Section 6.2.4. For all the other operation modes requiring a regenerative discharge of the load, these converters are four-quadrant and they can easily manage the energy flow from the load to the grid.

REFERENCES

1. Motto, E. 1998. Application specific intelligent power modules—A novel approach to system integration in low power drives. *PCIM Conference*.
2. Bhalerao, P. and Wiatr, R. 2008. New intelligent power module series. *PCIM Europe*, 2, pp. 21–22.
3. Motto, E.R. and Donlon, J. 2006. Speed shifting gate drive for intelligent power modules. *IEEE APEC*.
4. Neacsu, D.O., Nguyen, H.H. 2002. US Patent #6,459,324—Gate drive circuit with feedback-controlled active resistance.
5. Neacsu, D.O. 2001. Active gate drivers for motor drive applications. IEEE PESC2001, Vancouver, Canada, June.
6. Giacomini, D., Bianconi, E., Martino, L., Palma, M., 2000. A new fully integrated power module for three-phase servo motor driver applications. *International Rectifier Application Paper*.
7. Tolbert, L. 2012. Smart integrated power module. Project ID: APE046, 2012 U.S. *DOE Hydrogen and Fuel Cells Program and Vehicle Technologies Program Annual Merit Review and Peer Evaluation Meeting*.
8. Anon. 2008. IRAMS10UP60B-iMotion Series. Integrated power hybrid IC for appliance motor applications. International Rectifier's Datasheet.
9. Anon. 2010. STGIPS10K60A—IGBT intelligent power module. ST Microelectronics' Datasheet.
10. Lee, J.B., Lee, J.H., Cho, J., Chung, D.W., and Frank, W. 2010. AN-CIPOS mini-1-Technical Description, Infineon/LS PowerSemitech Application Note.
11. Shimizu, N., Karasawa, T., Takagiwa, K. 1012. New lineup of large-capacity "V-Series" intelligent power modules. *Fuji Electr. Rev.* 58(2) 65–69.
12. Schimel, P. 2011. A few useful tricks for modular inverter design. *IEEE APEC special session SP-2.2.1*.
13. Neacsu, D.O. 2010. Towards an all-semiconductor power converter solution for the appliance market. *IEEE Industrial Electronics Conference IECON*, Glendale, AZ, USA 2010.

Part II

Other Topologies

11 Resonant Three-Phase Converters

11.1 REDUCING SWITCHING LOSSES THROUGH RESONANCE VERSUS ADVANCED PWM DEVICES

It has been shown in the introduction that a switching operation is adopted in high-power converters in order to reduce losses and to improve efficiency. However, the operation of the power semiconductor devices is far from ideal and losses still occur. It has also been shown in Chapter 2 that these losses arise during *ON*-time and during transient events.

Losses during *ON*-time are called conduction losses and they depend entirely on the voltage drop across the switching device. Modern MOSFET devices feature very low $R_{ds(on)}$ [1]. For instance, the CoolMOS devices can switch up to 85 A at 600 V and benefit from an $R_{ds(on)}$ of 35–70 mΩ (see the topmost IXYS IXKK 85N60C), while the newest Q2-Class HiPerFET devices can switch up to 80 A at 500 V and benefit from an $R_{ds(on)}$ of about 60 mΩ. Other conventional high voltage MOSFETs have $R_{ds(on)}$ in the range of 150–200 mΩ. Versions of these technologies of MOSFETs can be switched up to 1200 V. Analogously, new technologies reduce the ON-state voltage drop across insular gate bipolar transistors (IGBTs) to the range of 2–3.5 V. Trench-gate IGBTs are recommended for low conduction loss with their collector–emitter voltage drop of 1.5–2.2 V (for instance, see Powerex CM200DU-12F for 200 A at 600 V) [2]. All these technology advancements are remarkable, and efforts will continue in the coming years on the same lines. Conduction loss, however, is technology-dependent and cannot be minimized by application topology. The only thing the designer can do is to estimate the weigh of the conduction versus switching loss within the application and select the proper power switch. If low conduction loss is more important for the application, devices with lower conduction loss should be selected. If, on the contrary, operation at a high switching frequency is required, devices with short transient times should be selected.

The second major category of losses is due to the transients of the voltage and current at turn-on and turn-off of the power semiconductor devices. When power devices change their conduction state, voltage and current have finite transitional slopes that superimpose for a short time, creating switching loss. The amount of switching loss at turn-on or turn-off of any switching device depends on both the technical characteristics of the power device and the application circuit. Here, there is room for improvement and for energy saving by the designer of the circuit. We dedicate this chapter, therefore, to understanding what energy savings are achievable with help from the circuit design engineer.

A complete analysis of the switching processes within power semiconductor devices has been presented in many books or papers. We limit this presentation to understanding the timing of a switching process. Figure 11.1 shows device behavior at turn-on. These waveforms have also been presented in Chapter 2 along with the definition of the most appropriate gate-driver design.

The semiconductor's power loss is calculated with the product of drain-source (collector–emitter) voltage and source (emitter) current. This product is obviously large during the switching process. At turn-on, the current changes its state before the voltage change and the sequence of transitions is reversed at turn-off. Furthermore, the turn-off of the IGBT or bipolar power transistors is characterized with a tail current that increases the switching loss.

The minimization of the switching loss is possible by reducing the time interval when both voltage and current are not close to zero. Different gate-driver techniques account for a controlled slope of voltage and current transitions. There are however limits to this minimization. The transition of current cannot be too steep as it would produce large voltage spikes in all the circuit parasitic inductances. The voltage slope is generally limited within the semiconductor device technology and the resulting parasitic capacitances have finite values.

Observing Figures 11.1 and 11.2 brings into focus the timing of the current and voltage waveforms as another possible solution for loss minimization [3]. Obviously, the switching loss depends on the amount of time between the voltage and current transitions. If somehow, we could move the voltage transition before the current transition, as shown in Figure 11.1, the switching loss would approach zero. Analogously, moving the transition of current before the transition of the voltage minimizes loss in the turn-off process.

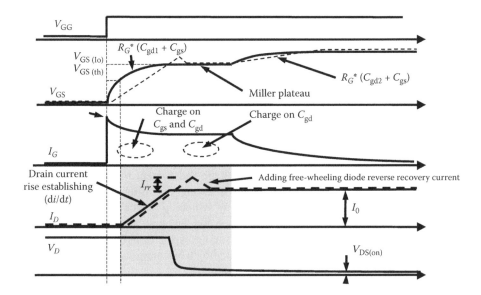

FIGURE 11.1 Generic turn-on waveforms for an IGBT/MOSFET power device.

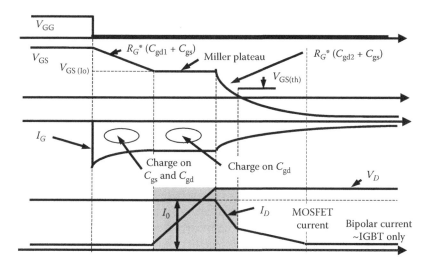

FIGURE 11.2 Generic turn-off waveforms for an IGBT/MOSFET power device.

Chapter 1 explained the role of power semiconductor devices in processing high power and the importance of maintaining simple circuit schematics. Changing the sequence of voltage and current slopes will definitely complicate the circuit schematic and this is the major trade-off a power circuit designer will face when selecting the proper topology for a power-conversion system. To keep the circuit schematics at a reasonable level of complexity, resonant circuits are used to change the sequence of voltage and current slopes. These can freely oscillate or they may require synchronization with the power-switching pattern. Such synchronization usually requires the introduction of more switches, and a pertinent analysis should be done to justify the energy lost in the newly added devices versus the switching-loss savings in the main power stage.

This chapter introduces the Reader to the philosophy of using resonant power converters in high-power conversion systems and provides several examples for circuits. Given the dynamic market for power semiconductor devices, the goal here is not to provide a comprehensive analysis of all possible switching devices, but to present the reasoning that will help a power electronics engineer to make the right topology selection decision.

11.2 DO WE STILL GET ADVANTAGES FROM RESONANT HIGH POWER CONVERTERS?

Let us start with a bit of history. The first widely used power semiconductor device was the silicon-controlled rectifier (SCR), also called thyristor. This device could be turned-on by a control signal and conduct current like a diode before the external circuit conditions turned it off. The use of this device in inverter type of applications required special turn-off circuits. Some of these turn-off circuits were built with resonant L–C networks.

The development of GTO (gate turn-off) devices for high-power conversion circuits simplified the inverter building. The GTO devices had a very large tail current at turn-off and efficiency optimization brought back the need for resonant circuits.

Later on, IGBTs became the main choice for power semiconductor devices. In 1982, RCA and General Electric virtually simultaneously announced the discovery of this device [4]. It seemed to be the perfect device for switching high power and, in the early 1990s, almost all production of power converters in the 10–100 kW range was based on IGBTs.

First generation IGBTs, however, continued to have large switching losses. Reducing switching loss in IGBTs was one of the most important subjects of R&D efforts in the early 1990s. Hundreds of papers or patent applications were written and, probably, every researcher in this field was, in one form or another, involved in researching new, resonant-power converters for high-power applications. These circuits were actually not entirely new, as they could be adapted from resonant converters built much earlier with SCR or GTO devices.

Two important research directions have emerged from these preoccupations [5]. The first addresses the invention and development of new, resonant-converter topologies. A possible classification of these solutions for inverter applications follows and more detail is provided in a further section.

Classification by the position of the resonant circuit:

- Resonant circuit in the DC bus allows a simplified six-switch converter topology
- Resonant circuit on each converter leg (pole voltage)
- Resonant circuit around each semiconductor switch
- Resonant circuit in the output

Classification by the signal used in reducing losses:

- Zero voltage switching: the voltage is kept at zero during the switching process
- Zero current switching the current is null during the change of conduction state

Classification based on the complexity of the resonant circuit:

- Free-running, continuous, resonant operation
- Synchronized resonant swing before the desired switching of the main device

This R&D direction has also addressed the mathematics of calculating the resonant-circuit operation, the resonant-component selection, and the possible variations in the values of the resonant circuit's passive components.

The second research direction addresses the system implications of resonant circuits and efficiency improvements. The following topics have been analyzed:

- Understanding the reduction in switching loss at the power-switch level by comparison with a hard-switched device.
- Evaluating the additional loss introduced by the resonant circuit and the switching circuit managing the release of the resonant swing.
- Understanding the additional stress in the power semiconductor devices and addressing the trade-offs between efficiency improvement and weak switch utilization.
- Implementation of the digital controller with the most appropriate switching pattern timing in the power stage and for the resonant circuit.

The major merit of this second direction in R&D efforts was to acknowledge the potential drawbacks of using resonant converters and to establish a proper comparison of losses at the system level rather than at the switch level. Some researchers defend the use of resonant converters by claiming that the spread of losses over several devices and components would help the cooling system. This is entirely true and moves the use of resonant converters into a much general class of applications.

In the early 1990s, power converters with soft-switching operations were reported to save up to 10% of the switching loss. More recent implementation solutions are claiming a reduction in power loss of between 2% and 5%. Attention should be paid to how these savings are estimated.

Throughout the 1990s, semiconductor technology evolved continuously and the latest IGBT generation features excellent switching times and substantial loss reduction. The IGBTs' relatively long turn-off time—they required as much as 2 ms to turn off—was a major shortcoming of the first generation. Technology improvements made possible turn-off in less than 200 ns. The latest generations of IGBTs especially designed for switched-mode power supplies and UPS applications can turn off in less than 100 ns. This reduces loss and makes IGBTs compatible to or better than MOSFETs in high-voltage applications that operate at more than 100 kHz. The maximum attainable switching frequency has also changed over the years from 10 kHz to more than 100 kHz [6,7].

To better assess the developments in IGBT technology, let us consider data from Tables 11.1 and 11.2. These are datasheet extracts. We acknowledge that

---

TABLE 11.1

Switching Loss Comparison for 100 A IGBT Devices (Energy Per Pulse)

	Energy at Turn-on			Energy at Turn-off		
	40 A (mJ)	60 A (mJ)	80 A (mJ)	40 A (mJ)	60 A (mJ)	80 A (mJ)
Powerex 600 V F-Series	0.90	1.06	1.08	1.80	2.50	3.00
Powerex 600 V H-Series	1.20	1.80	2.10	2.00	2.60	3.20
Powerex 600 V NF-Series	0.70	1.00	1.05	2.00	3.00	3.30
Powerex 600 V U-Series		1.36			0.80	
IRF WARP series (50 A)		0.80			0.50	

TABLE 11.2
Switching loss comparison for 400 A IGBT devices (energy per pulse)

	Energy at Turn-on			Energy at Turn-off		
	100 A (mJ)	200 A (mJ)	300 A (mJ)	100 A (mJ)	200 A (mJ)	300 A (mJ)
Powerex 600 V F-Series	4.00	6.00	10.0	5.00	10.0	20.0
Powerex 600 V H-Series	6.00	11.00	21.0	4.50	10.0	20.5
Powerex 600 V NF-Series	3.50	6.00	10.0	6.00	11.0	20.0
Toshiba IGBT ++ Series				11.00	16.0	21.0

measurement and reporting conditions vary from manufacturer to manufacturer, and we show this data only to assess the level of expected switching loss in hard switched converters. The data are in no way a direct performance comparison. It is important to note that such levels of switching loss make it difficult to further improve resonant circuits.

The advent of technology in the power semiconductor industry and the required complexity in the control circuit of resonant converters minimized the application of this technique within the large-scale production of power converters in the range of 10–100 kW. Almost all manufacturers of power converters took a conservative approach in maintaining production of hard-switched converters and targeting efficiency improvement by reducing parasitics, timely commissioning of new semiconductor devices, and working towards the most optimal gate-driver circuits.

However, there are many applications where resonant circuits are the best choice [5,8,9].

- Many high-voltage and high-power converters continue to use SCRs, GTOs, or older generation IGBTs that may benefit from resonant-circuit techniques.
- Power converters used in very high-temperature environments may benefit from a reduction of losses in the main switching devices or a slight transfer of losses in the resonant circuit.
- The use of very modern or advanced power semiconductors at the edge of their ratings. For instance, building power converters with the new CoolMOS or HyperFET devices allow switching at 600 V and 200 kHz of a current of 20–50 A. If increasing the switching frequency further provides any benefit to the application (size of magnetics, for instance), resonant switching is a good choice.
- Building power converters to fit extremely small spaces may require the use of resonant converters.

Let us consider the switching of power semiconductor devices under zero voltage or zero current and make this analysis independent of the circuit topology.

11.3   ZERO VOLTAGE TRANSITION OF IGBT DEVICES

11.3.1   POWER SEMICONDUCTOR DEVICES UNDER ZERO VOLTAGE SWITCHING

Switching loss can be reduced by bringing the voltage across the semiconductor device at zero before the turn-on process. The resonant circuit should be activated at switching instant only and this suggests the use of the term "quasi-resonant" for this class of converters.

It is important to understand the physics inside the power semiconductor devices switched at zero voltage in order to better assess the energy savings and possible failure associated with this operation mode.

Figures 11.1 and 11.2 emphasize the drain-source (emitter–collector) voltage trip during the Miller plateau of the gate voltage. The gate-drain capacitance provides a feedback path from drain to the gate that increases the equivalent charge needed for switching in the gate circuit. The total dynamic input capacitance results are greater than the sum of the static electrode capacitances. This effect, called the Miller effect, was first studied by John Miller for vacuum tubes. During the first voltage rise of the gate voltage, the gate-to-source capacitance gets charged, and during the flat portion (Miller plateau), the gate-to-drain capacitance gets charged. The total drive charge is typically higher for the Miller capacitance than for the gate-to-source capacitance. The width of the Miller plateau strongly depends on the amount of voltage seen on the drain (collector) of the power semiconductor device. Figure 11.3 shows the dependence of the gate voltage on drain current and voltage at turn-on. At the second voltage slope, both capacitances are charged as required by the switching of both voltage and current in the power stage.

During a zero-voltage transient, the Miller plateau disappears, as there is no voltage difference requiring additional charge into the gate-to-drain capacitance. This is important at the design of the gate circuit, as the overall charge is reduced and the stress in the gate driver is diminished.

Keeping in mind this behavior of the gate circuit, let us consider the selection of the proper power semiconductor device for a zero-voltage transition (ZVT). Usually,

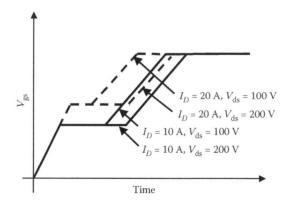

FIGURE 11.3   Gate voltage at turn-on for different current and voltage levels.

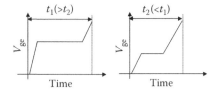

FIGURE 11.4 Different gate characteristics.

the switching performance is analyzed based on a combination of required gate charge and transconductance. Let us compare two devices with the theoretical characteristics shown in Figure 11.4. The first slope of the gate voltage is determined by the gate-source capacitance, which is larger for the second device. The second device has a higher transconductance and therefore requires less voltage on its gate for the given amount of collector current. This results in a faster device and in the interesting conclusion that the device with the smaller gate capacitance is not always the fastest.

Considering the same power devices for a comparison of their operation under zero-voltage transient outlines the advantage of selecting devices with smaller gate capacitances, as the transconductance effect is reduced by the zero voltage present in the drain (collector).

Let us consider Figure 11.2 in relation to the turn-off of a power semiconductor device. The turn-off process can be seen as the reverse of the turn-on, except for the tail current characteristic of IGBT devices and not present in the switching of the power MOSFETs. The existence of the tail current can be explained with the pseudo-Darlington connection of two transistors in the IGBT model (Figure 11.5). The base of the second (the PNP) bipolar transistor is not accessible for additional control and its turn-off is totally dependent on the internal physics of the device. As the MOSFET channel stops conducting, electron current ceases, and the IGBT current drops rapidly to the level of the whole recombination current at the inception of the tail. The lifetime of the minority carriers at this junction, therefore, slows down the overall transient time by introducing a time interval to remove the tail current.

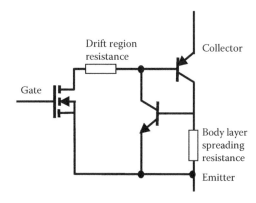

FIGURE 11.5 Equivalent circuit for an IGBT device.

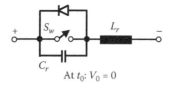

At t_0: $V_0 = 0$

FIGURE 11.6 Quasi-resonant circuit for ZVT, with initial zero voltage on capacitor.

Traditional lifetime-killing techniques and/an n1 buffer layer to collect the minority charges at turn-off are commonly used to speed-up this recombination process. Because these techniques reduce the gain of the PNP transistor, they also increase the voltage drop on the IGBT device. Moreover, this solution of lifetime killing to collect minority charges at turn-off may increase turn-on losses due to a quasi-saturation condition at turn-on. The existence of the tail current limits the use of a zero-voltage transient at the turn-off of IGBT devices [5,10,11].

The operation of the quasi-resonant power converters providing ZVT can be defined on the basis of the initial conditions in the resonant circuits or the circuit topology. A first solution is shown in Figure 11.6, where the resonant cycle starts with zero voltage across the resonant capacitor. After the switch turns-off, a resonant circuit is formed with a resonant capacitor C_r and inductor L_r. If the period of the resonant circuit is chosen such that the voltage across the switch is again zero when turn-on is desired, then switching loss is minimized.

An alternative to this solution is shown in Figure 11.7. The C_r is now connected to an external potential V_{ref} constant during the resonant cycle. The equivalent small-signal models of the circuits shown in Figures 11.6 and 11.7 are identical.

11.3.2 STEP-DOWN CONVERSION

Let us take a very simple example to illustrate this principle [5].

Figure 11.8 shows a single-switch buck converter with commutation at zero voltage. Chapter 3 has already shown the reduction of three-phase converters to simple buck or boost power stages. Understanding this simple resonant circuit helps the development of complex three-phase resonant converters.

The operation of the power switch within this power converter follows the same control characteristics as the conventional buck converter and we can consider the

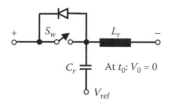

At t_0: $V_0 = 0$

FIGURE 11.7 Quasi-resonant circuit for ZVT, with initial zero voltage on capacitor.

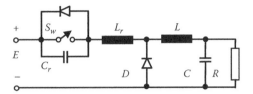

FIGURE 11.8 Quasi-resonant circuit for ZVT, with initial zero voltage on capacitor.

load filter L–C as being ideal (Figure 11.9). Its equivalent effect is a constant DC load current denoted by I_L.

Let us start the analysis with the turn-off process. The voltage across is initially zero. The load current I_L circulates through the resonant inductor L_r and the resonant capacitor C_r. This produces the linear increase of the voltage at the capacitor terminals.

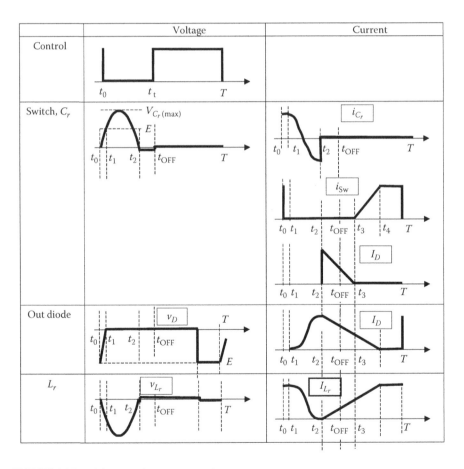

FIGURE 11.9 Voltage and current waveforms.

$$v_{C_r}(t) = \frac{I_L}{C_r} \cdot t \tag{11.1}$$

The voltage across the resonant inductor l_r is maintained null due to the constant load current. This implies a linear variation of the voltage across the buck diode D.

$$v_D(t) = E - \frac{I_L}{C_r} \cdot t \tag{11.2}$$

At moment t_1, the diode D has a positive bias and turns-on.

$$t_1 = \frac{E}{I_L} \cdot C_r \tag{11.3}$$

The resonant circuit can be characterized with a resonant frequency f_r and a characteristic impedance Z_r.

$$f_r = \frac{1}{T_r} = \frac{1}{2 \cdot \pi \cdot \sqrt{L_r \cdot C_r}} \tag{11.4}$$

$$Z_r = \sqrt{\frac{L_r}{C_r}} \tag{11.5}$$

These are also related by the following equations:

$$\omega_r \cdot Z_r = \frac{1}{C_r} \Rightarrow C_r = \frac{1}{\omega_r \cdot Z_r} \tag{11.6}$$

$$L_r = \frac{Z_r}{\omega_r} \tag{11.7}$$

Equation 11.3 can be written in dependence with these variables.

$$t_1 = \frac{E}{I_L} \cdot C_r = \frac{E}{I_L} \cdot \frac{1}{\omega_r \cdot Z_r} \tag{11.8}$$

If the characteristic impedance Z_r is chosen very large, the time interval t_1 can be considered very small or at least much smaller than the switching period of the buck converter.

During the following time interval, the diode D conducts the load current and the switch S_w remains in the *off* state. The input voltage E is seen across the resonant circuit L_r–C_r and the voltage across the resonant capacitor C_r is given by:

$$L_r \cdot C_r \cdot \frac{d^2 v_{C_r}(t)}{dt^2} + v_{C_r}(t) = E \tag{11.9}$$

$$v_{C_r}(t) = E + Z_r \cdot I_L \cdot \sin\left[\omega_r \cdot (t - t_1)\right] \tag{11.10}$$

where the initial value of the current through the resonant inductor has also been considered.

The sinusoidal voltage across the resonant capacitor C_r reaches a peak voltage $v_{C_r(max)}$ and then decreases to zero.

$$v_{C_r(max)} = E + Z_r \cdot I_L \tag{11.11}$$

The moment of time when voltage reaches zero yields:

$$t_2 = t_1 + \frac{1}{\omega_r} \cdot \left[\pi + \arcsin\left(\frac{E}{Z_r \cdot I_L}\right)\right] \tag{11.12}$$

The current through the resonant circuit in this moment is given by:

$$i_{C_r}(t_2) = i_{L_r}(t_2) = -I_L \cdot \sqrt{1 - \left[\frac{E}{Z_r \cdot I_L}\right]^2} \tag{11.13}$$

The current through the output diode at t_2 yields:

$$i_D(t_2) = I_L + I_L \cdot \sqrt{1 - \left[\frac{E}{Z_r \cdot I_L}\right]^2} \tag{11.14}$$

When the resonant voltage prepares for the negative swing, the anti-parallel diode turns-on. This ensures the circulation of the current through the inductance L_r until the energy is discharged. During this interval, the voltage across the L_r is maintained constant. The current passing through L_r and the anti-parallel follows a linear variation:

$$i_{D_{S_w}}(t) = i_{L_r}(t) = \left[-I_L \sqrt{1 - \left[\frac{E}{Z_r \cdot I_L}\right]^2}\right] + \frac{E}{L_r} \cdot (t - t_2) \tag{11.15}$$

The current through the output diode D yields:

$$i_D(t) = I_L \cdot \left[1 + \sqrt{1 - \left[\frac{E}{Z_r \cdot I_L} \right]^2} \right] + \frac{E}{L_r} \cdot (t - t_2) \tag{11.16}$$

The moment of time t_3 when the current through the resonant inductor and the anti-parallel diode vanishes yields from Equation 11.14:

$$t_3 = t_2 + \frac{1}{\omega_r} \cdot \frac{Z_r \cdot I_L}{E} \cdot \sqrt{1 - \left[\frac{E}{Z_r \cdot I_L} \right]^2} \tag{11.17}$$

When the switch S_w turns-on, the current circulates through S_w, the resonant inductor L_r and the output diode. Because the output diode is still in the ON state, the voltage across the resonant inductor equals the input voltage E. The current will continue the linear variation from zero to the load current I_L determining the same variation of the current through D_{sw}, L_r, and output diode D.

$$i_{D_{S_w}}(t) = i_{L_r}(t) = \left[-I_L \sqrt{1 - \left[\frac{E}{Z_r \cdot I_L} \right]^2} \right] + \frac{E}{L_r} \cdot (t - t_2) \tag{11.18}$$

$$i_D(t) = I_L \cdot \left[1 + \sqrt{1 - \left[\frac{E}{Z_r \cdot I_L} \right]^2} \right] + \frac{E}{L_r} \cdot (t - t_2) \tag{11.19}$$

The output diode current will eventually vanish at a moment of time t_4 that is given by:

$$t_4 = t_3 + \frac{1}{\omega_r} \cdot \frac{Z_r \cdot I_L}{E} = t_2 + \frac{1}{\omega_r} \cdot \frac{Z_r \cdot I_L}{E} \cdot \left[1 + \sqrt{1 - \left[\frac{E}{Z_r \cdot I_L} \right]^2} \right] \tag{11.20}$$

where the results from Equations 11.16 and 11.17 are considered.

Finally, after the moment t_4, the switch is the only device carrying current towards the load through the resonant inductor L_r. The state of this switch can be changed at any time and the entire operation cycle is repeated.

It is important to note that the moments t_2 and t_4 are very important, as they limit the possible variation of the ON and OFF time intervals within the controller operation. Their values are given by Equations 11.12 and 11.20 and these are dependent on both the passive components L_r and C_r as well as on the load current and supply voltage. The load current is the only variable parameter during the operation of the power stage.

The proper operation of the resonant circuit should respect the following constraint:

$$Z_r \cdot I_L \geq E \tag{11.21}$$

It can be seen that the ZVT can be obtained for certain current levels and that light loads do not concur to reduction of the switching loss. One can also notice that the switching loss for a light load is anyhow reduced.

This converter can control the output voltage only by changing the period of the entire cycle that modifies the value of T or the moment when the switch is turned off. The duration of the OFF state is dictated by the circuit conditions. The averaged output voltage can be calculated by:

$$V_C = E \cdot \left[1 - \left(\frac{1}{2} + \frac{Z_r \cdot I_L}{2 \cdot \pi \cdot E} \right) \right] \cdot \frac{2 \cdot \pi}{\omega_r \cdot T} \tag{11.22}$$

This, unfortunately, is dependent on both the load current and the input voltage and can be controlled only by the cycle period. However, this analysis illustrates the operation of a simple resonant circuit and does not have as its objective the design of a complete power converter.

The voltage across the power switch is increased to $C_r(\text{max})$ from the input voltage E. Depending on the current level, this can be up to twice as much as the input voltage. This obviously means to overrate the power switch. When using IGBTs, there is not much difference in the conduction losses in devices of 600 and 1200 V. However, using MOSFETs implies a substantial difference between $R_{ds(on)}$ of devices rated at 250, 500, and 600 or 1000 V, as they are the result of different technologies. The ratio of $R_{ds(on)}$ can be twice as much. Moreover, there is more current passing through the switch and the conduction loss is larger. A fair loss comparison should definitely include both the optimization of the switching loss and the change in the conduction loss.

11.3.3 STEP-UP POWER TRANSFER

Many power conversion applications require a power transfer from a low-voltage source to a high-voltage load through a step-up conversion [3]. Figure 11.10 shows a

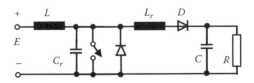

FIGURE 11.10   Step-up power converter.

possible resonant circuit for this condition. The input inductance L can be modeled with a current source.

Let us assume that the general operation of this boost converter is not modified by the presence of the resonant circuit, and let us start the analysis at the moment when the switch is turned-off. At that moment, the current through the L_r and the voltage across the C_r are zero. The output diode is also in the *OFF* state. The quasi-constant input current is linearly charging the C_r.

$$C_r \cdot \frac{dv_{C_r}(t)}{dt} = I_L \tag{11.23}$$

$$v_{C_r}(t) = \frac{I_L}{C_r} \cdot t \tag{11.24}$$

At the time t_1, the voltage across the capacitor C_r reaches the output voltage V_0 and the voltage across the output diode reverses its polarity. The time interval t_1 can be calculated with:

$$t_1 = \frac{V_0}{I_L} \cdot C_r = \frac{1}{\omega_r} \cdot \frac{V_0}{Z_r \cdot I_L} \tag{11.25}$$

where the previous notations for ω_r and Z_r are considered.

At the moment t_1, the diode turns on and the switch S_w maintains its OFF state. The input current is now shared between the resonant capacitor C_r and the output branch of L_r and diode D. The inductance L_r and the capacitor C_r form a resonant circuit supplied by the load equivalent voltage V_0. This resonant circuit starts with the voltage across the capacitor equaling the load voltage.

$$L_r \cdot C_r \cdot \frac{d^2 v_{C_r}(t)}{dt^2} + v_{C_r}(t) = V_0 \tag{11.26}$$

The voltage across the resonant capacitor increases from V_0 to a maximum value and then decreases to zero.

$$v_{C_r}(t) = V_0 + Z_r \cdot I_L \cdot \sin \omega_r (t - t_1) \tag{11.27}$$

At moment t_2, this voltage reaches zero.

$$t_2 = t_1 + \frac{1}{\omega_r} \cdot \left[\pi + \arcsin \left(\frac{V_0}{Z_r \cdot I_L} \right) \right] \tag{11.28}$$

The resonant swing of the voltage reaches zero only if $Z_r{*}I_L > V_0$. The current through the C_r is given by:

$$i_{C_r}(t) = C_r \cdot \frac{dv_{C_r}(t)}{dt} = I_L \cdot \cos \omega_r (t - t_1) \tag{11.29}$$

and it has the final value

$$i_{C_r}(t_2) = -I_L \cdot \sqrt{1 - \left(\frac{E}{Z_r \cdot I_L}\right)^2} \qquad (11.30)$$

Analogously, the resonant current through the L_r and output diode is given by:

$$i_{L_r}(t) = I_L - i_{C_r}(t) = I_L - I_L \cdot \cos\omega_r(t - t_1) \qquad (11.31)$$

After moment t_2, the resonant circuit has the tendency to swing the voltage across capacitor to negative values, but the anti-parallel diode turns-on and clamps this voltage. The L_r has a constant voltage across its terminals and the current through this inductance varies linearly from its initial value at t_2.

$$i_{L_r}(t) = I_L \cdot \left[1 + \sqrt{1 - \left(\frac{V_0}{Z_r \cdot I_L}\right)^2}\right] - \frac{V_0}{L_r} \cdot (t - t_2) \qquad (11.32)$$

The current through the anti-parallel diode is given by:

$$i_{D_{S_W}}(t) = I_L - i_{L_r(t)} = \left[-\sqrt{1 - \left(\frac{V_0}{Z_r \cdot I_L}\right)^2}\right] + \frac{V_0}{L_r} \cdot (t - t_2) \qquad (11.33)$$

The moment t_3 when this current vanishes is determined as:

$$t_3 = t_2 + \frac{1}{\omega_r} \cdot \frac{Z_r \cdot I_L}{V_0} \cdot \sqrt{1 - \left(\frac{V_0}{Z_r \cdot I_L}\right)^2} \qquad (11.34)$$

Unfortunately, this time interval depends on both the input current and the output voltage and it cannot be influenced by control.

The power switch should be controlled for the ON state at any moment during the time interval (t_2, t_3), before the anti-parallel diode would turn-off. In this way, the switch turns-on after the resonant current reverses its direction at t_3. The linear discharge of the resonant current through the output L_r continues until t_4.

$$t_4 = t_2 + \frac{1}{\omega_r} \cdot \frac{Z_r \cdot I_L}{V_0} \cdot \left(1 + \sqrt{1 - \left(\frac{V_0}{Z_r \cdot I_L}\right)^2}\right) \qquad (11.35)$$

The switch stays in the ON state for the rest of the switching cycle and the load diode remains in the OFF state. The entire operation can be understood with the waveforms showed in Figure 11.11.

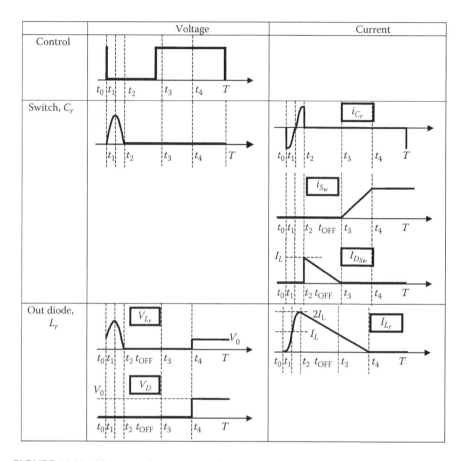

FIGURE 11.11   Voltage and current waveforms.

11.3.4   BI-DIRECTIONAL POWER TRANSFER

Many applications require a bi-directional circulation of the load current and the previous power stage is enhanced by the use of a dual module of MOSFET devices. To derive the resonant operation of such a two-switch converter, first let us note that the switch and the resonant inductor are in series, and their sequence can be changed without altering the operation of the converter (Figure 11.12).

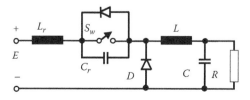

FIGURE 11.12   Another form of circuit from Figure 11.8.

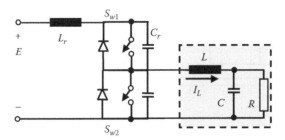

FIGURE 11.13   Power stage with dual module.

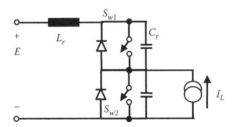

FIGURE 11.14   Equivalent model for the single leg converter of Figure 11.13.

Replacing the buck diode with an assembly of switch, diode, and resonant capacitors leads to the sequence shown in Figure 11.13. The output filter does not have any effect on the resonant operation in the power stage and only a constraint of a constant load has been considered (Figure 11.14). The same operation will occur in a power stage without filter but with a constant load.

Understanding the resonant operation of this converter implies analyzing separately the step-up and step-down power transfer from the "load" to the "input". This can be achieved by considering the S_{w2} as the power-commuting device and the diode D_1 as the freewheeling diode for the step-up conversion from load, and S_{w1} as the power commuting device and the diode D_2 as the freewheeling diode for the step-down conversion. The same intermediate states occur for a negative load current.

Let us consider now the switch S_{w1} in conduction and the current getting out of converter and circulating towards the load. When S_{w1} is turned-off, both switches are in the OFF state and the parallel capacitors are charging from the load current. The high side capacitor that was initially discharged increases its voltage. The low-side resonant capacitor charged at the bus voltage, is now discharged with the load current. The pole voltage decreases according to the resonant circuit. The resonant swing of the voltage reaches zero before commanding the turn-on of the switch S_{w2} if the initial value of the load current is large enough. If there is not enough energy in the load current, the switch S_{w2} will be commanded when there is still voltage across it and it will turn on with switching loss (Figure 11.15).

Of most interest to us is the extension to the three-phase conversion systems. Figure 11.16 builds upon the theory developed in Chapter 3 and illustrates the

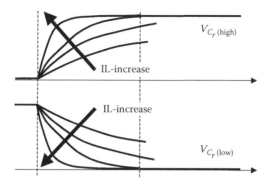

FIGURE 11.15 Resonant capacitors voltage.

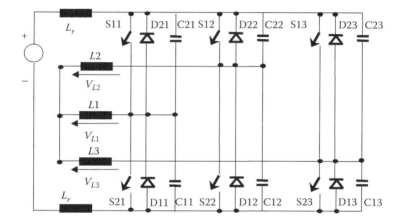

FIGURE 11.16 Three-phase system with resonant circuits.

principle of resonant circuitry applied to a three-phase conversion system. It is interesting to note that the resonant circuits shown in figure somehow occur naturally from the building of the power stage. The output capacitances (C_{oss}) of the power MOSFETs or IGBTs and the bus-bar parasitic inductances represent a good starting point for constructing this resonant circuit.

One can notice the equivalence between this presentation of the Zero Voltage Transition (ZVT) methods and the resonant snubbers shown in Chapter 6, Section 6.2.5.5.

11.4 ZERO CURRENT TRANSITION OF IGBT DEVICES

11.4.1 Power Semiconductor Devices under Zero Current Switching

Let us re-analyze the switching of the power devices shown in Figures 11.1 and 11.2 with the condition of zero current in the drain circuit [1,12]. The first slope of the gate voltage ends faster due to the reduced Miller threshold at zero current.

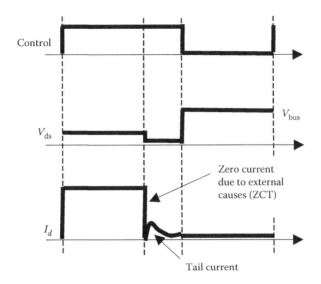

FIGURE 11.17   Turn-off characteristics with Zero Current Transition (ZCT).

This is also explained in Figure 11.3. However, the Miller plateau is still dependent on the voltage from the drain-source (collector–emitter) circuit. After the voltage swings towards zero within a turn-on process, the current may be allowed to increase slowly to the actual value of the load current through a series inductance (Figure 11.17).

The zero current switching at turn-off is ensured with an external circuit that cancels any drain current before the actual turn-off command. Excess charges thus get trapped within the power semiconductor device and they start decaying through internal recombination. The drain-source voltage has a fast rise after the turn-off command. This voltage is supported within the reverse-biased p-base drift region junction (Figure 11.5).

After the supply voltage has reached the drain-source (collector–emitter) circuit, the remaining process is a recombination that characterizes the tail current. This results in a turn-off current bump in the drain (collector) circuit. The excess carrier in the drift region can be swept away into the MOSFET channel in parallel with the recombination process if the gate voltage is still applied to maintain the inversion layer and the MOSFET channel. It is important to maintain the control voltage in the gate circuit, before all carriers are swept out through the MOSFET channel, in order to reduce the current tail and the current bump in the power circuit. This reduces switching loss accordingly. This brief analysis outlines the benefits of zero current transition (ZCT) in turn-off switching.

The ZCT class of resonant power converters is characterized by placing an inductor in series with the power switch and counting on a zero current at turnoff. Figure 11.18 shows a simplified circuit based on an additional voltage potential V_{ref} and Figure 11.19 presents a solution without any other voltage source. This resonant circuit reduces losses at the turn-off of the power switch. Achieving zero current

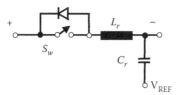

FIGURE 11.18   Quasi-resonant circuit for ZCT.

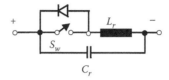

FIGURE 11.19   Quasi-resonant circuit for ZCT.

through the switch at turn-off is possible with a sinusoidal variation of the current through the resonant inductor L_r. Unfortunately, this swing of the current reaches a peak value of twice the load current and this large current passes through the power switch during the *ON* state. The rating of the switch should be increased and the conduction losses will increase accordingly.

This resonant circuit reduces losses at the turn-off of the power switch. Achieving zero current through the switch at turn-off is possible with a sinusoidal variation of the current through the resonant inductor L_r. Unfortunately, this swing of the current reaches a peak value of two times the load current and this large current passes through the power switch during the on state. The rating of the switch should be increased and the conduction losses will increase accordingly.

11.4.2   STEP-DOWN CONVERSION

Let us explain the operation of a resonant circuit for ZCT within a buck converter. The simplified circuit diagram is shown in Figure 11.20.

The operation of the buck converter should not change from the conventional power converter and the load current is assumed constant during the resonant cycle in Figure 11.21. Let us start this analysis with a zero current through the L_r and a zero voltage across the C_r, the load current being sustained by the output diode D.

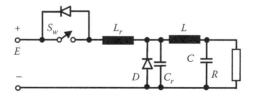

FIGURE 11.20   Quasi-resonant circuit for ZCT.

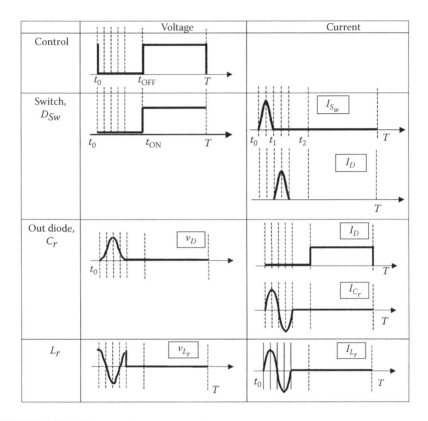

FIGURE 11.21   Voltage and current waveforms.

At t_0, the switch is turned on and current starts to circulate through the L_r and the power switch with a linear variation.

$$L_r \cdot \frac{di_{L_r}(t)}{dt} = E \tag{11.36}$$

$$i_{L_r}(t) = \frac{E}{L_r} \cdot t \tag{11.37}$$

The current through the output diode D is given by the difference between the load current and the current through the resonant inductor:

$$i_D(t) = I_L - \frac{E}{L_r} \cdot t \tag{11.38}$$

At the moment of time t_1, the current through the output diode reaches zero and the diode turns-off. Using Equation 11.7 yields:

$$t_1 = \frac{L_r \cdot I_L}{E} = \frac{1}{\omega_r} \cdot \frac{Z_r \cdot I_L}{E} \tag{11.39}$$

After t_1, the load current passes through the power switch and the L_r. The C_r is no longer clamped at zero voltage and it can produce an additional current through the switch and the L_r. This capacitor starts to charge from zero voltage. The current through the L_r is now calculated for the resonant circuit L_r–C_r with initial zero voltage on the capacitor:

$$L_r \cdot C_r \cdot \frac{d^2 i_{L_r}(t)}{dt^2} + i_{L_r}(t) = I_0 \tag{11.40}$$

$$i_{L_r}(t) = I_L + \frac{E}{Z_r} \cdot \sin \omega_r (t - t_1) \tag{11.41}$$

This current has a sinusoidal variation from the initial value equaling the load current to a maximum value and it is decreasing later to zero. The moment when the current vanishes yields:

$$t_2 = t_1 + \frac{1}{\omega_r} \cdot \left[\pi + \arcsin \frac{Z_r \cdot I_L}{E} \right] \tag{11.42}$$

The voltage across the capacitor has a harmonic variation during the resonant cycle.

$$v_{C_r}(t) = E \cdot \left[1 - \cos \omega_r \cdot (t - t_1) \right] \tag{11.43}$$

The maximum voltage across the C_r is calculated with:

$$V_{Cr}(t) = E \cdot \left[1 + \sqrt{1 - \left(\frac{Z_r \cdot I_L}{E} \right)^2} \right] \tag{11.44}$$

It is important to calculate the amount of energy required by the resonant circuit in order to extend its swing to zero. This happens when the current in Equation 11.41 can reach zero and it yields:

$$Z_r \cdot I_L \leq E \tag{11.45}$$

It can be seen that this solution does not work for large load currents. The maximum current through the power switch and resonant inductor is calculated as:

$$i_{Lr(max)} = I_L + \frac{E}{Z_r} \tag{11.46}$$

This current can be more than two times the load current. The conduction loss within the switch is definitely increased by this additional current circulation.

After t_2, the switch turns-off at zero current and both the power switch and the output diode are in the off state. The C_r takes over the entire load current and this produces the capacitor discharge.

$$C_r \cdot \frac{dv_{C_r}(t)}{dt} = -I_L \qquad (11.47)$$

The linear variation of the voltage can be expressed as:

$$v_{C_r}(t) = E \cdot \left[1 + \sqrt{1 - \left(\frac{Z_r \cdot I_L}{E} \right)^2} \right] - \frac{I_L}{C_r} \cdot (t - t_2) \qquad (11.48)$$

The output diode turns-on at t_3 when this voltage equals zero.

$$t_3 = t_2 + \frac{1}{\omega_r} \cdot \frac{E}{Z_r \cdot I_L} \cdot \left[1 + \sqrt{1 - \left(\frac{Z_r \cdot I_L}{E} \right)^2} \right] \qquad (11.49)$$

After the output diode turns-on, the load current is circulated through this diode while there is no current circulation through the power switch and the L_r. After this moment t_3, the power switch can be turned on at any time and the entire cycle is repeated.

Because the main power semiconductor switch turns-off according to the resonant cycle, the output voltage does not depend on the duty cycle of the control signal but on the resonant period.

$$\frac{V_0}{E} = \frac{T_r}{T} \qquad (11.50)$$

The current through the power switch is several times larger than the load current and this should be rated for larger currents. Further, the conduction loss is increased with this current circulation and this trade-off should be considered at the design of the power stage.

11.4.3 STEP-UP CONVERSION

The principle of ZCT can also be applied to step-up converters. It is assumed that the operation of the resonant circuit does not affect the main function and operation of the original step-up converter. The only modifications are the presence of the series' resonant inductor L_r and the parallel resonant capacitor C_r.

Let us start the analysis with the turn-on event (Figure 11.22). In the beginning, both the power switch and the output diode are in conduction and the voltage across the C_r is kept equal to the load voltage (Figure 11.23). The input current is split between the L_r and the load current. The current through the L_r inductance has a linear variation:

$$L_r \cdot \frac{di_{L_r}(t)}{dt} = V_0 \tag{11.51}$$

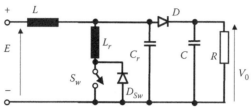

FIGURE 11.22   Step-up converter with ZCT.

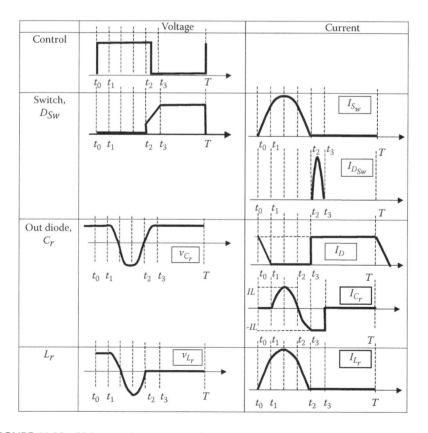

FIGURE 11.23   Voltage and current waveforms.

$$i_{L_r}(t) = \frac{V_0}{L_r} \cdot t \tag{11.52}$$

The load current equals:

$$i_D(t) = I_L - i_{L_r}(t) = I_L - \frac{V_0}{L_r} \cdot t \tag{11.53}$$

and reaches zero at t_1.

$$t_1 = \frac{L_r \cdot I_L}{V_0} \tag{11.54}$$

After t_1, the parallel resonant circuit L_r–C_r remains in the circuit and the L_r current is the result of the differential equation:

$$L_r \cdot C_r \cdot \frac{d^2 i_{L_r}(t)}{dt^2} + i_{L_r}(t) = I_L \tag{11.55}$$

with the following solution:

$$i_{L_r}(t) = I_L + \frac{V_0}{Z_r} \cdot \sin \omega_r \cdot (t - t_1) \tag{11.56}$$

The voltage across the parallel resonant circuit varies based on:

$$v_{L_r}(t) = v_{C_r}(t) = L_r \cdot \frac{d i_{L_r}(t)}{dt} = V_0 \cdot \cos \omega_r (t - t_1) \tag{11.57}$$

The current through the L_r increases from the initial value to a maximum value and decreases to zero at the moment of time t_2.

$$t_2 = t_1 + \frac{1}{\omega_r} \cdot \left[\pi + \arcsin \frac{Z_r \cdot I_L}{V_0} \right] \tag{11.58}$$

This current also turns-off the switch under a ZCT. The resonant voltage at this moment reaches:

$$v_{L_r}(t) = v_{C_r}(t) = -V_0 \cdot \sqrt{1 - \left(\frac{Z_r \cdot I_L}{V_0} \right)^2} \tag{11.59}$$

After t_2, the switch is in the OFF state and the resonant current continues through the anti-parallel diode. After the entire negative cycle of the current, the anti-parallel diode turns-off when current reaches zero again.

$$t_3 = t_1 + \frac{1}{\omega_r} \cdot \left[2 \cdot \pi - \arcsin \frac{Z_r \cdot I_L}{V_0} \right] \tag{11.60}$$

The voltage across the capacitor C_r and inductance L_r moves to a level given by the following:

$$v_{C_r}(t_3) = v_{L_r}(t_3) = -V_0 \cdot \sqrt{1 - \left(\frac{Z_r \cdot I_L}{V_0} \right)^2} \tag{11.61}$$

The next switching of the main power device should occur after this time moment in order to be ready for another conduction interval.

The operation of this very simple resonant circuit determines an output voltage depending on the input voltage through a relationship with the resonant period as parameter. More complex circuits should be developed to ensure full control of the duty cycle and output voltage.

11.5 POSSIBLE TOPOLOGIES OF QUASI-RESONANT CONVERTERS

All the circuits presented and analyzed in the previous sections allow for low switching loss due to the zero voltage turn-on. Unfortunately, they represent a very simple solution without too many control features. Constraints in reduction of the switching harmonics push for precise ON and OFF time control. This implies use of more evolved circuit configurations. Several examples are shown in this section without a comprehensive analysis.

11.5.1 Pole Voltage

We can define a group of resonant three-phase power converters based on resonant circuits on each leg of the three-phase power converter. The operation of these resonant circuits is based on inverter pole measures. Several examples are shown in Figures 11.24 and 11.25. The first one represents a derivative of the circuit shown in Figure 11.13 and it is suitable for a zero voltage transient. The next evolutionary step takes place in the Auxiliary Resonant Commutated Pole Inverter (ARCPI, Figure 11.25). This has the ability to stop and release the resonant process at precise moments, controlling it by the use of the bi-directional switch. Pulse width modulation algorithms can therefore be improved and the harmonics in the load optimized. Different versions of these circuits have been proposed and their optimization focuses on the reduction of the ratings for the power semiconductors and passive components used in the filter.

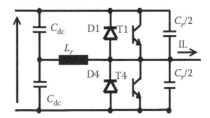

FIGURE 11.24   Conventional resonant pole inverter.

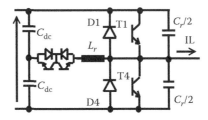

FIGURE 11.25   Auxiliary resonant commutated pole inverter.

11.5.2   RESONANT DC BUS

Another class of converter moves the resonant circuit on the DC side in an effort to leave the main power stage unchanged from the hard-switched operation [10–16]. Advantages in packaging at module- or converter- level are therefore achieved. The resonant circuit on the DC bus produces a resonant swing of the voltage that is able to reduce the entire bus to zero. All power semiconductor devices in the main converter can, therefore, change their conduction state during this zero voltage interval.

The immediate drawback is the operation of the resonant cycle with pulse width modulations (PWMs), as there are limitations in the resolution of the PWM algorithm. The main six-switch power stage disperses the resonant train of pulses from the DC bus to the three phases according to a sinusoidal law. All power semiconductor devices should be rated for double the DC bus voltage in the circuit shown in Figure 11.26. However, this is the simplest solution for a resonant DC bus.

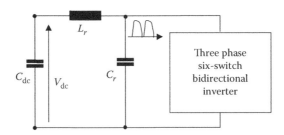

FIGURE 11.26   Simple resonant DC link three-phase inverter.

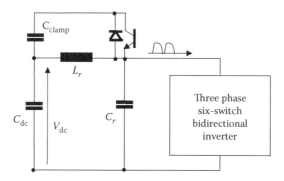

FIGURE 11.27 Resonant DC link three-phase inverter with a clamp circuit.

As this rating constraint is not easily achieved in a three-phase converter, a clamping circuit is introduced (Figure 11.27) to limit the voltage trip to high peaks by paralleling another capacitor that changes the period and impedance of the resonant circuit. The peak voltage is now limited to

$$K = 1 + \sqrt{\frac{C_r}{C_r + C_c}} \qquad (11.62)$$

However, this solution does not allow any control of the pulse width and all pulses should be constructed with multiples of the resonant period. An alternative solution is shown in Figure 11.28 and consists in breaking the resonant cycle during the desired pulse width of the output voltage.

11.6 SPECIAL PWM FOR THREE-PHASE RESONANT CONVERTERS

The major issue with PWM control of resonant PWM three-phase converters consists in the minimum pulse-width required for allowing the resonant swing of either voltage or current waveforms. Several excellent PWM methods suitable for PWM

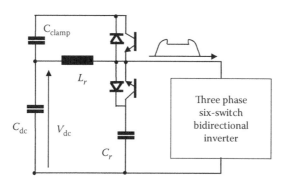

FIGURE 11.28 Fully controlled quasi-resonant DC link inverter.

control of resonant converters have been presented in Chapter 4, Section 5. Both the staircase PWM and the third harmonic injection PWM are good for the control of resonant power converters.

A special class of three-phase resonant power converters is based on a continuous oscillation of the resonant L_r-C_r circuit. The PWM algorithm should therefore consider pulses with their width as multiples of the resonant period. Special optimization routines can be used to improve the PWM pattern. A good example in this direction has been already presented in Chapter 3, Section 7.3 as a binary programmed PWM algorithm.

PROBLEMS

P11.1   Determine the requirements of both the power switch S_w and the diode D for the converter shown in Figure 11.8 and operated according to Figure 11.9. Use all equations from 11.1 to 11.20 for this calculation.

P11.2   Determine the requirements of both the power switch S_w and the diode D for the converter shown in Figure 11.10 and operated according to Figure 11.11. Use all equations from 11.23 to 11.35 for this calculation.

P11.3   Determine the requirements of both the power switch S_w and the diode D for the converter shown in Figure 11.20 and operated according to Figure 11.21. Use all equations from 11.36 to 11.50 for this calculation.

P11.4   Determine the requirements of both the power switch S_w and the diode D for the converter shown in Figure 11.22 and operated according to Figure 11.23. Use all equations from 11.51 to 11.60 for this calculation.

P11.5   Demonstrate Equation 11.62.

REFERENCES

1. Anon. 2002. What is the benefit of CoolMOS in phase shifted ZVS bridge topology? *Infineon Application Note*, January.
2. Anon. 2000. Using IGBT modules. *Powerex Application Notes*.
3. Mohan, N., Undeland, T., and Robbins, W. 2002. *Power Electronics*. 3rd edition. John Wiley and Sons, New York.
4. Wheatley, C.F., Jr. and Becke, H. 1982. U.S. Letters Patent No. 4,364,073: "Power MOSFET with an Anode Region", RCA, December 14.
5. Bose, B.K. 1992. Power electronics—A technology review. *Proc. IEEE*, 80, 1303–1334.
6. Travis, B. 2000. IGBTs come of age in switchers—New high-speed IGBTs can beat MOSFETs in conversion efficiency and silicon area in switching supplies operating at 100 kHz and faster. *EDN* 27, 42–46.
7. Ambarian, C. 1997. WARP SpeedTM IGBTs—Fast enough to replace power MOSFETs in switching power supplies at over 100 kHz. *IRF Application Note*, IRF Technical Paper, 1–6.
8. Huth, S. and Winterheimer, S. 1993. The switching behavior of an IGBT in zero current switch mode. *Fifth European Conference on Power Electronics and Applications*, vol. 2, pp. 312–316, 13–16 September.
9. Vlatkovic, V., Borojevic, D., and Lee, F.C. 1994. Soft-transition three-phase PWM conversion technology. *IEEE PESC Conference Record 1*, Taipei, pp. 79–84.

10. Divan, D.M. 1986. The resonant DC link converter—A new concept in static power conversion. *IEEE-IAS Annual Meeting Conference Record*, pp. 648–656.
11. Divan, D.M. and Skibinski, S. 1987. Zero switching loss inverters for high power applications. *IEEE-IAS Annual Meeting Conference Record*, pp. 627–634.
12. Imbertson, P. and Mohan, N. 1993. Asymmetrical duty cycle permits zero switching loss in pwm circuits with no conduction penalty. *IEEE Trans. Ind. Appl.* 29, 212–125.
13. Divan, D.M., Venkataramakan, G., and De'Doncker, R.W.A.A. 1993. Design methodologies for soft switched inverters. *IEEE Trans. Ind. Appl.* 29, 126–135.
14. Dehmlow, K., Heumann, R., and Sommer, R. 1992. Resonant inverter systems for drive applications. *EPE J* 2, 225–232.
15. Alexa, D. 1995. Resonant circuit with constant voltage applied on the clamp capacitor for zero voltage switching at the power converters. *Elec. Eng.* 78, 169–174.
16. Trivedy, M., Shenai, K., and Larson, E. 1997. Critical evaluation of IGBT performance in zero current switching environment. *IEEE Conference Record*, pp. 989–993.

12 Component-Minimized Three-Phase Power Converters

12.1 SOLUTIONS FOR REDUCTION OF NUMBER OF COMPONENTS

Previous chapters have extensively analyzed the three-phase converter based on six power semiconductor switches. One of the major preoccupations for the use of this conventional topology within industrial products is related to cost reduction. It has been shown that the largest cost share corresponds to building the power stage.

Accordingly, one approach to cost reduction would rely on seeking new topologies with a reduced number of components. This is especially important for applications in the horsepower range up to several tens of kilowatts.

Different solutions have been reported in the literature during the last twenty years or so. An important research direction has been dedicated to new grid interfaces with power factor correction. Some of them, including the single-switch, three-phase AC/DC converter, were analyzed in Chapter 9.

Constraints of variable frequency and magnitude for AC motor-drive applications have limited the efforts for new power converters used for simplified three phase AC sources. All reported solutions combine advantages and disadvantages versus the conventional six-switch inverter and none of them have really captured the market. These new solutions different from the conventional six-switch converter must be understood and studied for their merits and for the opportunity they provide to open up new directions for further research. They can be grouped in two categories:

- New inverter topologies: with reduced component count.
- Direct converters: to employ a single-power stage without intermediate DC link capacitor to perform direct AC/AC conversion.

12.1.1 New Inverter Topologies

The most well-known topology with reduced component count is shown in Figure 12.1 [1–12]. It is based on two inverter legs while the third phase is taken from the DC capacitor mid-point. This topology has proven merit to be considered for industrial implementation. For this reason, a special part of this chapter is dedicated to its full analysis. Similar application to AC/DC conversion stages has been considered.

An interesting combination of this topology, with a single-phase front-end converter, enabled use of a six-pack module for implementation of the whole

341

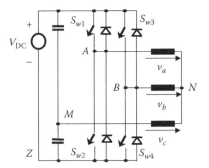

FIGURE 12.1   Topology for a B4 inverter.

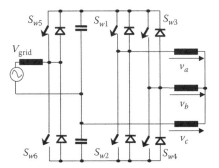

FIGURE 12.2   Reduced component count AC/DC/AC converter.

AC/DC/AC conversion (Figure 12.2) [13]. The drawback is in the larger value of the DC capacitor bank. The two conversion stages (AC/DC and DC/AC) are controlled independently of each other and the additional control module manages the DC voltage within two threshold levels. A quick comparison with conventional topologies outlines the larger DC voltage on the bus. This aspect will be detailed in Sections 12.2 and 12.3.

   A completely different approach to reducing the number of components in the AC/DC/AC power electronic conversion supports a unidirectional load current (Figure 12.3a) [14–17]. If the load is a three-phase AC machine, it can be proven that this DC component does not affect the torque production but only increases the machine losses. Each leg is reduced to a DC/DC converter with a variable reference, as shown in Figure 12.3b.

   These waveforms can be characterized mathematically by

$$i_{1d} = I_m \cdot \left[u(\omega t) - u(\omega t - 120) \right] \cdot \sin \omega t + I_m \cdot$$
$$\left[u(\omega t - 120) - u(\omega t - 240) \right] \cdot \sin(\omega t - 60) \qquad (12.1)$$

$$i_{2d} = i_{1d}(\omega t - 120) \qquad (12.2)$$

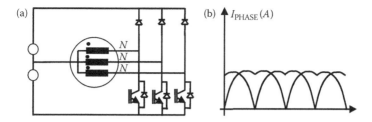

FIGURE 12.3 New DC/AC topology with unidirectional currents.

$$i_{3d} = i_{1d}(\omega t - 240) \tag{12.3}$$

One can verify that the difference between two reference currents is a purely sinusoidal wave. For instance: $i_{1-2d} = I_m \sin\varphi t$. It is very important to note that this set of phase currents produces a rotating magnetic field in the machine. In order to demonstrate this, let us first apply the vectorial transform (5.1) to the current waveforms shown in Figure 12.3b. This yields

$$\bar{I} = \frac{2}{3} \cdot \left[i_{1d} + \bar{a} \cdot i_{2d} + \bar{a}^2 \cdot i_{3d} \right] \tag{12.4}$$

where

$$a = e^{j\frac{2\pi}{3}} \tag{12.5}$$

After some calculation:

$$\bar{I} = \frac{1}{\sqrt{3}} \cdot I_m \cdot e^{j\left[\omega t - \frac{2\pi}{3} \right]} \tag{12.6}$$

The set of currents from Figure 12.3b generates a rotating magnetic field with constant angular speed and magnitude. A set of currents will be induced in the short circuited rotor and this will generate another magnetic field. The interaction between the stator and rotor magnetic fields produces a constant electromagnetic torque. The special shape of currents operates the induction machine under unstable conditions and the whole theory of induction machine dynamic model is not applicable here. Analysis should be performed on each interval separately. Stator and mutual equivalent inductances yield:

$$L_{sX} = L_{sA} + \frac{1}{2} \cdot M_{s(A,C)} = L_s + \frac{1}{2} \cdot M_s$$

$$L_{sY} = L_{sC} - \frac{1}{2} \cdot M_{s(A,C)} = L_s - \frac{1}{2} \cdot M_s \tag{12.7}$$

$$M_{s(X,Y)} = \frac{\sqrt{3}}{2} \cdot M_{s(A,C)} = \frac{\sqrt{3}}{2} \cdot M_s$$

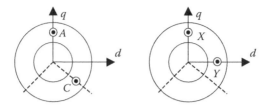

FIGURE 12.4 Transformation of (A,C) to (X,Y) system to prove the operation of the IM drive with the proposed currents.

where $L_{sA} = L_{sB} = L_{sC} = L_s$ and $M_{s(A,C)} = M_s$. The resulting voltage equations are similar to traditional dynamic equations for machine modeling except for the values of these inductances. The waveforms of the rotor currents are identical with those carried out for the same drive fed by sinusoidal currents with $I_m/\sqrt{3}$.

A physical explanation of the machine operation is shown in Figure 12.4. Let us consider the time interval with a current passing through the wirings A and C of the stator while the current through the second phase B is zero. An equivalent bi-phase system can be derived for the first $2\pi/3$ rad only if the flux in each wiring produced by the (X,Y) bi-phase system is the same as that produced by the (A,C) system.

Despite the interesting mathematical demonstration of the torque production, this three-phase power converter cannot be practically implemented in the form shown in Figure 10.3a. This is because the currents through the DC mid-point and DC capacitors flow in the same direction and have a tendency to discharge C1 and overcharge C2. In order to prevent this, a special DC/DC converter is proposed in [1,2] to regulate the voltage of the capacitors. This complicates the power stage design.

An alternative solution is proposed in Figure 12.5 [17]. One of the phases is built with a reversed direction and a different number of turns. This solves the problems in the DC bus if that phase current is double the value of each of the other two currents.

Unfortunately, this implies a specially built electrical machine, as shown in Figure 12.5.

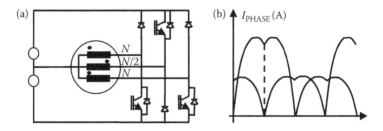

FIGURE 12.5 Implementation of the idea from Figure 12. 3.

Using the proposed converter and waveform solution increases losses in the induction machine. A simplified loss estimation is shown here.

- Stator copper losses

$$P_{Cus} = R_s \cdot \left[\frac{1}{2 \cdot \pi} \cdot \int_0^{2\pi} i_s^2 d\omega t \right] \Rightarrow$$

$$P_{Cus} = 0.40 \cdot R_s \cdot I_m^2 = 1.2067 \cdot R_s \cdot \left[i_{ds}^2 + i_{qs}^2 \right]$$

(12.8)

When using the same induction machine (IM) as in a conventional case, an increase of 20.67% stator copper loss occurs.

- Rotor copper losses

    When supplied with the proposed current waveforms, the rotor current's sinusoidal waves and the rotor copper losses can be approximated as identical with the conventional case.

- Iron losses

    A resistor equivalent to the stator iron losses can be included in the simplified IM equivalent model. This will be passed by a current equal with the difference between the stator current and the current through the magnetizing inductance. For a given torque, the stator losses can be somewhat reduced by a proper adjustment of the magnetic flux.

Even if these converter solutions are not very practical, they represent good conceptual advances in power converter technologies. Taking into account advanced mathematical theory may lead to new topologies in the future and understanding these advanced converters is a great start in researching emerging conversion approaches.

12.1.2 Direct Converters

Given the cost and size of the passive filter components on the intermediary DC bus, solutions for direct conversion have been sought. First, a very simple but practical solution is presented in [18] and it represents the IGBT equivalent (Figure 12.6b) of a conventional SCR-based AC controller (Figure 12.6a) able to generate a three-phase system of voltages with a constant frequency and variable magnitude.

When high-side IGBTs are controlled, they connect the load to the grid. In contrast, when the low side IGBTs are turned-on, they short the load and separate it from the grid. A train of pulses is created and the root mean square value can be regulated appropriately.

These two topologies are recently revisited by employing Reverse Conducting IGBT devices (IGBT-RC), able to optimize the construction of the power stage.

In many motor-drive applications, both frequency and voltage need to be controlled and this can be achieved with different topologies of matrix converters. Given the importance of matrix converters within the power electronics landscape, Chapter

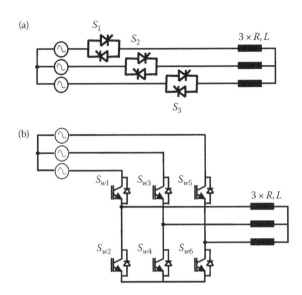

FIGURE 12.6 IGBT-based AC controller.

16 will present the conventional 9-switch matrix converters, and Chapter 18 will introduce special matrix converter topologies built of Voltage Source or Current Source 6-switch converters.

12.2   B4 INVERTER

12.2.1   VECTORIAL ANALYSIS OF THE B4 INVERTER

A new inverter topology with reduced count of components is built with two legs of the conventional inverter while the third phase is collected from the mid-point of the DC capacitor bank (Figure 12.1).

Let us first understand how this power converter operates. The third phase is taken from the mid-point of the capacitor bank and its pole voltage (M-Z) is always fixed at $V_{DC}/2$. The phase voltages are constructed through a control on the other two inverter legs. The first consequence is the lack of zero states: there is no way the three pole voltages can be at the same potential during operation. This means that pulse width modulation (PWM) should be developed based on the remaining active vectors. A rule of operating a voltage source three-phase inverter says that we should always have one switch ON across each inverter leg.

This limits the number of the control states to four as shown in Table 12.1. A set of phase voltages can be generated by a combination of these states.

Phase voltages can be calculated from the pole voltages, as shown in Chapter 3. The appropriate equations are just adapted here to our inverter topology.

$$v_{NZ} = \frac{\left[v_{AZ} + v_{BZ} + v_{MZ}\right]}{3} \tag{12.9}$$

TABLE 12.1

Possible Inverter States and Their Appropriate Voltages

S_{w1}/S_{w2}	S_{w3}/S_{w4}	V_{AZ}	V_{BZ}	V_{AN}	V_{BN}	V_{MN}	V_d (Re)	V_q (Im)
1/0	1/0	V_{DC}	V_{DC}	$V_{DC}/6$	$V_{DC}/6$	$-V_{DC}/3$	$\dfrac{V_{DC}}{6}$	$\dfrac{V_{DC}}{2\cdot\sqrt{3}}$
1/0	0/1	V_{DC}	0	$V_{DC}/2$	$-V_{DC}/2$	0	$\dfrac{V_{DC}}{2}$	$-\dfrac{V_{DC}}{2\cdot\sqrt{3}}$
0/1	1/0	0	V_{DC}	$-V_{DC}/2$	$V_{DC}/2$	0	$-\dfrac{V_{DC}}{2}$	$-\dfrac{V_{DC}}{2\cdot\sqrt{3}}$
0/1	0/1	0	0	$-V_{DC}/6$	$-V_{DC}/6$	$V_{DC}/3$	$-\dfrac{V_{DC}}{6}$	$-\dfrac{V_{DC}}{2\cdot\sqrt{3}}$

$$\begin{cases} v_a = v_{AZ} - \dfrac{\left[v_{AZ} + v_{BZ} + v_{MZ}\right]}{3} \\[2mm] v_b = v_{BZ} - \dfrac{\left[v_{AZ} + v_{BZ} + v_{MZ}\right]}{3} \\[2mm] v_c = v_{MZ} - \dfrac{\left[v_{AZ} + v_{BZ} + v_{MZ}\right]}{3} \end{cases} \qquad (12.10)$$

The first attempts of using this topology have generated PWM with a conventional reference-triangle carrier based method. The reference waveforms have been considered sinusoidal with a 60° phase shift between each other. The pole voltages for the inverter legs can be written as:

$$\begin{cases} v_{AZ} = V \cdot \cos(\omega t) + \dfrac{V_{DC}}{2} + f_a(t) \\[2mm] v_{BZ} = V \cdot \cos\left(\omega t - \dfrac{\pi}{3}\right) + \dfrac{V_{DC}}{2} + f_b(t) \end{cases} \qquad (12.11)$$

where f_a and f_b account for the high-frequency components due to switchings. If the load is heavily inductive or a motor drive, it is a good approximation to neglect these harmonics and to consider the effect of the waveforms in fundamental frequency only. It yields:

$$v_{NZ} = \frac{V_{DC}}{2} + \frac{1}{3} \cdot \left[V \cdot \cos[\omega t] + V \cdot \cos\left(\omega t - \frac{\pi}{3}\right)\right] =$$

$$\frac{V_{DC}}{2} + \frac{1}{3} \cdot V \cdot \left[2 \cdot \cos\left[\omega t - \frac{\pi}{6}\right] \cdot \cos\left(\frac{\pi}{6}\right)\right] \qquad (12.12)$$

$$v_{NZ} = \frac{V_{DC}}{2} + \frac{1}{\sqrt{3}} \cdot V \cdot \cos\left[\omega t - \frac{\pi}{6}\right] \tag{12.13}$$

and

$$\begin{cases} v_a = \left[\frac{V_{DC}}{2} + V \cdot \cos(\omega t)\right] - \left[\frac{V_{DC}}{2} + \frac{1}{\sqrt{3}} \cdot V \cdot \cos\left[\omega t - \frac{\pi}{6}\right]\right] \\ v_b = \left[\frac{V_{DC}}{2} + V \cdot \cos\left(\omega t - \frac{\pi}{3}\right)\right] - \left[\frac{V_{DC}}{2} + \frac{1}{\sqrt{3}} \cdot V \cdot \cos\left[\omega t - \frac{\pi}{6}\right]\right] \\ v_c = \left[\frac{V_{DC}}{2}\right] - \left[\frac{V_{DC}}{2} + \frac{1}{\sqrt{3}} \cdot V \cdot \cos\left[\omega t - \frac{\pi}{6}\right]\right] \end{cases} \tag{12.14}$$

$$\begin{cases} v_a = V \cdot \left[\cos(\omega t) - \frac{1}{\sqrt{3}} \cdot \cos(\omega t) \cdot \frac{\sqrt{3}}{2} - \frac{1}{\sqrt{3}}\sin(\omega t) \cdot \frac{1}{2}\right] \\ v_b = V \cdot \left[\cos\left(\omega t - \frac{\pi}{3}\right) - \frac{1}{\sqrt{3}} \cdot \cos\left[\omega t - \frac{\pi}{6}\right]\right] \\ v_c = -\frac{1}{\sqrt{3}} \cdot V \cdot \cos\left[\omega t - \frac{\pi}{6}\right] \end{cases} \tag{12.15}$$

$$\begin{cases} v_a = \frac{1}{\sqrt{3}} \cdot V \cdot \left[\frac{\sqrt{3}}{2} \cdot \cos(\omega t) - \sin(\omega t) \cdot \frac{1}{2}\right] = \frac{1}{\sqrt{3}} \cdot V \cdot \cos\left(\omega t + \frac{\pi}{6}\right) = \\ \quad \frac{1}{\sqrt{3}} \cdot V \cdot \sin\left(\omega t + \frac{2\pi}{3}\right) \\ v_b = V \cdot \left[\cos(\omega t) \cdot \frac{1}{2} + \sin(\omega t) \cdot \frac{\sqrt{3}}{2} - \frac{1}{\sqrt{3}} \cdot \cos(\omega t) \cdot \frac{\sqrt{3}}{2} - \frac{1}{\sqrt{3}} \cdot \sin(\omega t) \cdot \frac{1}{2}\right] = \\ \quad \frac{1}{\sqrt{3}} \cdot V \cdot \sin(\omega t) \\ v_c = -\frac{1}{\sqrt{3}} \cdot V \cdot \cos\left[\omega t - \frac{\pi}{6}\right] = \frac{1}{\sqrt{3}} \cdot V \cdot \cos\left[\pi - \omega t - \frac{\pi}{6}\right] = \frac{1}{\sqrt{3}} \cdot V \cdot \sin\left[\omega t - \frac{2\pi}{3}\right] \end{cases}$$

$$\tag{12.16}$$

In conclusion, this topology can be controlled by two sinusoidal references with 60° out of phase from each other in order to produce a symmetrical three-phase

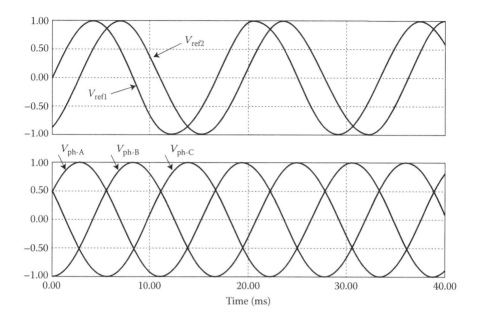

FIGURE 12.7 Phase information for the reference functions and desired fundamental phase voltages on the load (see Equations 12.11 and 12.12).

system. Moreover, these two references should be phase shifted with 30° from the desired first phase voltage (Figure 12.7).

A similar control can be achieved by vectorial analysis. Let us apply transform relationship (5.1) to the phase voltages shown in Table 12.1. A vector in the complex plane will correspond to each operation mode. It is important to note that a different notation of the phase voltages in Figure 12.1 will produce other positions in the complex plane for the active vectors. The vectors shown in the complex plane of Figure 12.8 have magnitudes of $V_{DC}/\sqrt{3}$ and, respectively, $V_{DC}/3$.

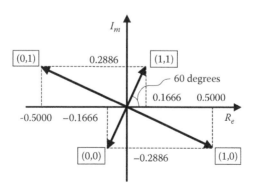

FIGURE 12.8 Vectorial representation of the active vectors corresponding to the B4 inverter.

The generation of a symmetrical three-phase system assumes displacement of the tip of the voltage vector on a circular trajectory. The maximum radius of this circular locus is achieved when the trajectory is tangent to the polygon. It yields:

$$V_{max} = \frac{V_{DC}}{2 \cdot \sqrt{3}} \Rightarrow m = \frac{V}{\frac{V_{DC}}{2 \cdot \sqrt{3}}} \qquad (12.17)$$

Let us remember that the maximum voltage obtainable from a conventional six switches inverter was $1/\sqrt{3} \cdot V_{DC}$ (see Equation 5.31). The B4 inverter needs a double DC voltage in order to produce the same output phase voltage and this is a serious drawback of this topology. It produces increased voltage stress on the power semiconductor devices and electrical machine. The absolute value of the peak-to-peak ripple on the DC bus capacitor voltage is also increased. Finally, the third phase current circulates through the DC bank capacitor and this produces large variations of the voltage between the two capacitors. This can be corrected with large capacitors.

The two-leg inverter produces asymmetrical phase voltages as shown in Figure 12.9.

Since one leg circulates currents through the DC capacitor bank, asymmetries of the operation may introduce a third harmonic on the line-to-line voltages. The

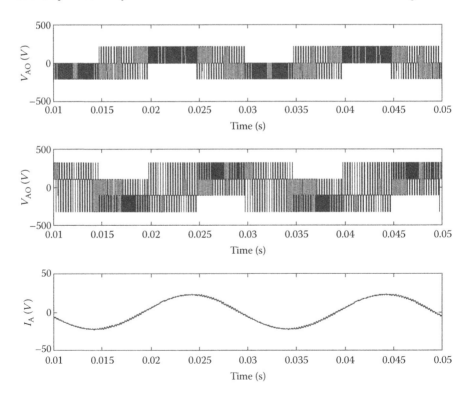

FIGURE 12.9 Typical phase voltages for the converter of Figure 12.1.

content of the third harmonic for the leg connected to the capacitor bank is twice as large as the third harmonic in the other two legs as they add up into the first one.

PWM generation takes into account the following constraints due to the asymmetries in operation:

- Decomposition on two adjacent vectors provides the average relationship.
- Because there is no true zero vector, two vectors with opposite directions are considered for equal time intervals to synthesize a zero-voltage state.
- There are always two possibilities for zero vector generation. It is preferred to use the "short" vectors for this since the "long" vectors produce larger voltage drop on the inductive load and larger ripple.
- In consequence, each vector generation is made with minimum three vectors: two short and one long.
- Several state sequences are possible.

The reduction of the number of switches determines cost reduction as well as conduction loss reduction in the power stage.

12.2.2 Definition of PWM Algorithms for the B4 Inverter

Two Vector PWM methods are herein investigated.

Method 1

The idea of vectorial decomposition on two neighboring vectors is considered for the two-leg inverter. If we consider the desired position of the voltage vector in between V_2 and V_3, the time intervals associated with each state are given by:

$$V_s \cdot \cos\phi = \frac{t_2}{T_s} \cdot \lfloor V_2 \rfloor$$
$$V_s \cdot \sin\phi = \frac{t_3^1}{T_s} \cdot \lfloor V_3 \rfloor$$

(12.18)

$$
\begin{cases}
t_2 = \dfrac{V_s}{\lfloor V_2 \rfloor} \cdot T_s \cdot \cos\phi \\[2mm]
t_3^1 = \dfrac{V_s}{\lfloor V_3 \rfloor} \cdot T_s \cdot \sin\phi \\[2mm]
t_0 = 2 \cdot t_3^1 = 2 \cdot t_1 = T_s - t_2 - t_3^1
\end{cases}
\Rightarrow
$$

$$
\begin{cases}
t_2 = \dfrac{V_s}{\lfloor V_2 \rfloor} \cdot T_s \cdot \cos\phi \\[3mm]
t_1 = \dfrac{1}{2} \cdot \left[T_s - \dfrac{V_s}{\lfloor V_2 \rfloor} \cdot T_s \cdot \cos\phi - \dfrac{V_s}{\lfloor V_3 \rfloor} \cdot T_s \cdot \sin\phi \right] \\[3mm]
t_3 = \dfrac{V_s}{\lfloor V_3 \rfloor} \cdot T_s \cdot \sin\phi + \dfrac{1}{2} \cdot \left[T_s - \dfrac{V_s}{\lfloor V_2 \rfloor} \cdot T_s \cdot \cos\phi - \dfrac{V_s}{\lfloor V_3 \rfloor} \cdot T_s \cdot \sin\phi \right]
\end{cases}
$$

(12.19)

$$\begin{cases} t_2 = \dfrac{V_s}{\left[V_2\right]} \cdot T_s \cdot \cos\phi \\[3mm] t_1 = \dfrac{1}{2} \cdot \left[T_s - \dfrac{V_s}{\left[V_2\right]} \cdot T_s \cdot \cos\phi - \dfrac{V_s}{\left[V_3\right]} \cdot T_s \cdot \sin\phi \right] \\[3mm] t_3 = \dfrac{1}{2} \cdot \left[T_s - \dfrac{V_s}{\left[V_2\right]} \cdot T_s \cdot \cos\phi + \dfrac{V_s}{\left[V_3\right]} \cdot T_s \cdot \sin\phi \right] \end{cases} \tag{12.20}$$

When vectors V_1–V_3–V_4 are used, these equations are the same with V_4 instead of V_2. Moreover, there are several possibilities for the state sequence:

$$00\text{-}10\text{-}11\text{-}11\text{-}10\text{-}00 \quad \text{or} \quad 11\text{-}10\text{-}00\text{-}00\text{-}10\text{-}11$$

As a final verification, the *ON*-time of the upper switch on the first leg can be calculated as the sum of the time spent on V_2 and V_3:

$$t_a = \dfrac{1}{2} \cdot \left[T_s + \dfrac{V_s}{\left[V_2\right]} \cdot T_s \cdot \cos\phi + \dfrac{V_s}{\left[V_3\right]} \cdot T_s \cdot \sin\phi \right] =$$

$$\dfrac{1}{2} \cdot T_s \cdot \left[1 + \dfrac{1}{2} \cdot m \cdot \cos\phi + \dfrac{\sqrt{3}}{2} \cdot m \cdot \sin\phi \right] \tag{12.21}$$

$$t_a = \dfrac{1}{2} \cdot T_s \cdot \left[1 + m \cdot \sin\left(\phi + \dfrac{\pi}{6} \right) \right] \tag{12.22}$$

The ON-time of the high side switch of the second leg is t_3:

$$t_b = \dfrac{1}{2} \cdot \left[T_s - \dfrac{V_s}{\left[V_2\right]} \cdot T_s \cdot \cos\phi + \dfrac{V_s}{\left[V_3\right]} \cdot T_s \cdot \sin\phi \right] =$$

$$\dfrac{1}{2} \cdot T_s \cdot \left[1 - \dfrac{1}{2} \cdot m \cdot \cos\phi + \dfrac{\sqrt{3}}{2} \cdot m \cdot \sin\phi \right] \tag{12.23}$$

$$t_a = \dfrac{1}{2} \cdot T_s \cdot \left[1 + m \cdot \sin\left(\phi - \dfrac{\pi}{6} \right) \right] \tag{12.24}$$

These results are similar with the previous carrier based PWM generation but with a different reference for angular coordinate measurement.

Method 2

The ON-time calculation is achieved in the same way. The difference is in the PWM sequence. The bisectrix of the angles between active vectors split the complex plane in four sectors. A different state sequence is considered for each such sector:

$285 < \phi < 15°$:	11-10-00-10-11
$15 < \phi < 105°$:	01-11-10-11-01
$105 < \phi < 195°$:	00-01-11-01-00
$195 < \phi < 285°$:	10-00-01-00-10

Comparative results

Comparative results between the two methods are shown in Table 12.2. It is important to extend the comparison to each phase or line-to-line voltages since they are different even when controlling the same power converter (Figure 12.9).

12.2.3 INFLUENCE OF DC VOLTAGE VARIATIONS AND METHOD FOR THEIR COMPENSATION

The DC link voltage is subject to larger ripple than in a conventional six-switch inverter [19,20]. Over and above the expected ripple produced by the rectifier power

TABLE 12.2
Adaptation from [8]

		$m = 0.2$	$m = 0.4$	$m = 0.6$	$m = 0.8$	$m = 1.0$
The content on the low harmonics—3rd harmonic; VB-C and VA-C are half						
Method 1	VA-B	0.1734	0.0191	0.0059	0.0022	0.0015
Method 2	VA-B	0.1752	0.0198	0.0064	0.0025	0.0016
The content on the low harmonics—5th harmonic						
Method 1	VB-C	0.0542	0.0115	0.0026	0.0023	0.0019
	VA-B	0.0519	0.0109	0.0054	0.0020	0.0015
Method 2	VB-C	0.0535	0.0110	0.0045	0.0018	0.0015
	VA-B	0.0504	0.0102	0.0050	0.0024	0.0015
The content on the low harmonics—7th harmonic						
Method 1	VB-C	0.0472	0.0111	0.0029	0.0020	0.0016
	VA-B	0.0481	0.0114	0.0032	0.0019	0.0014
Method 2	VB-C	0.0445	0.0096	0.0042	0.0012	0.0010
	VA-B	0.0516	0.0122	0.0038	0.0014	0.0009
The global content in harmonics—HLF calculated with first 500 harmonics						
Method 1	VB-C	9.196	4.207	2.504	1.248	0.955
	VA-B	4.669	2.282	1.526	1.287	1.048
Method 2	VB-C	12.132	4.601	2.680	1.323	1.089
	VA-B	6.613	3.130	1.963	1.063	0.931

Source: Blaabjerg F., Kragh, H., Neacsu, D.O., and Pedersen, J.K. 1997. Comparison of modulation strategies for B4-inverter. *EPE '97*, vol. 2, pp. 378–385.

TABLE 12.3

Considering Individual Voltages V_{D1}, V_{D2} on Capacitors

S_{w1}/S_{w2}	S_{w3}/S_{w4}	V_{AZ}	V_{BZ}	V_{AN}	V_{BN}	V_{MN}
1/0	1/0	$V_{D1} + V_{D2}$	$V_{D1} + V_{D2}$	$V_{D1}/3$	$V_{D1}/3$	$-2V_{D1}/3$
1/0	0/1	$V_{D1} + V_{D2}$	0	$(V_{D1} + V_{D2})/2$	$-(V_{D1} + V_{D2})/2$	$(V_{D2} - V_{D1})/3$
0/1	1/0	0	$V_{D1} + V_{D2}$	$-(V_{D1} + V_{D2})/2$	$(V_{D1} + V_{D2})/2$	$(V_{D2} - V_{D1})/3$
0/1	0/1	0	0	$-V_{D2}/3$	$-V_{D2}/3$	$2V_{D2}/3$

supply, there is another set of opposite components in each of the DC voltages caused by the phase current circulating through the capacitor bank. These components are stronger depending on the load current level. Moreover, variations of the DC bus voltage have a different influence on each phase voltage in comparison with the six-switch inverter where all phase voltages have been influenced in the same manner.

Chapter 5 has shown that variations of the DC bus voltage can be corrected by a proper adjustment of the SVM algorithm through a feed-forward compensation. The same general concept can be applied to the B4 inverter as well. In this respect, Table 12.1 is rewritten in Table 12.3, based on individual voltages on DC capacitors.

Now, we can rewrite Equation 12.20 by taking into account these values.

$$\left\{ \begin{array}{l} t_2 = \dfrac{3 \cdot V_s}{V_{D1} + V_{D2}} \cdot T_s \cdot \cos\phi - \dfrac{V_{D2} - V_{D1}}{V_{D1} + V_{D2}} \cdot T_s \\[3mm] t_1 = \dfrac{1}{2} \cdot \left[T_s - \dfrac{3 \cdot V_s}{V_{D1} + V_{D2}} \cdot T_s \cdot \cos\phi - \dfrac{\sqrt{3} \cdot V_s}{V_{D1} + V_{D2}} \cdot T_s \cdot \sin\phi \right] \\[3mm] t_3 = \dfrac{1}{2} \cdot \left[T_s - \dfrac{3 \cdot V_s}{V_{D1} + V_{D2}} \cdot T_s \cdot \cos\phi + \dfrac{\sqrt{3} \cdot V_s}{V_{D1} + V_{D2}} \cdot T_s \cdot \sin\phi \right] \end{array} \right. \qquad (12.25)$$

The effectiveness of compensation can be understood by observing Figures 12.10 and 12.11.

This example of feed-forward compensation works well before the modulator saturates or tends to go in overmodulation. In other words, the time constants calculated with Equation 12.20 should remain positive at any operation point. Let us note ΔV as the absolute value of the ripple within any of the V_1 or V_2 and RV the normalized value of this ripple:

$$RV = \frac{\Delta V}{\dfrac{V_{DC}}{2}} \qquad (12.26)$$

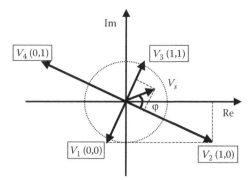

FIGURE 12.10 Vector decomposition.

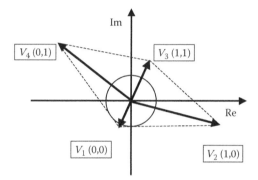

FIGURE 12.11 Deformation of the active vectors due to ripple in the DC bus voltages.

After some calculation, these limits can be computed as

$$\begin{cases} RV \le \dfrac{\sqrt{6}}{4} \cdot m \\[2mm] RV \le 1 - m \end{cases} \tag{12.27}$$

Graphical representation of these constraints defines the maximum operational range of this feed-forward method (Figure 12.12).

12.3 TWO-LEG CONVERTER USED IN FEEDING A TWO-PHASE IM

Two-phase IMs (induction machines) are used in several low-power applications around or below 1 kW, especially in automation equipment. They have a reduced starting torque, when compared to the DC machines, linear control characteristics, and can be controlled through advanced methods such as field-orientation. The major drawbacks are the reduced efficiency and power factors.

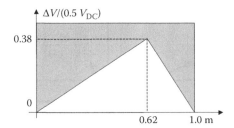

FIGURE 12.12   Maximum correctable ripple by the proposed method.

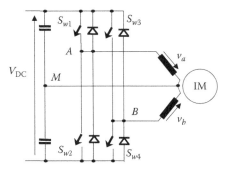

FIGURE 12.13   Two-phase motor drive.

A two-phase inverter with the same power stage as the two-leg inverter (B4) is used for feeding this motor drive (Figure 12.13). This inverter can be supplied from a single-phase diode rectifier suitable to this power level. PWM operation of the IGBT or MOSFET devices building the two-phase inverter ensures variation of both frequency and magnitude of the output voltages. These voltages should be 90° out of phase for a proper torque production within the induction motor. Each inverter leg is PWM-operated between voltage levels of $(-V_{DC}/2)$ and $(V_{DC}/2)$ as shown in Figure 12.14.

12.4   Z-SOURCE INVERTER

The advent of hybrid and electric vehicles opened the door to an impressive development of power electronics solutions for the automotive market. The typical motor drive connection to a high voltage battery pack is shown in Figure 12.15. It consists of a boost (step-up) converter able to increase and regulate the battery voltage to a certain established value. The second stage is a single-phase or a three-phase inverter able to generate a controlled waveform with a high content in fundamental.

The efficiency and cost are driving automotive applications and this supported the appearance of a new inverter topology able to minimize the component count. Both the boost and inverter functions are achieved within the same power stage

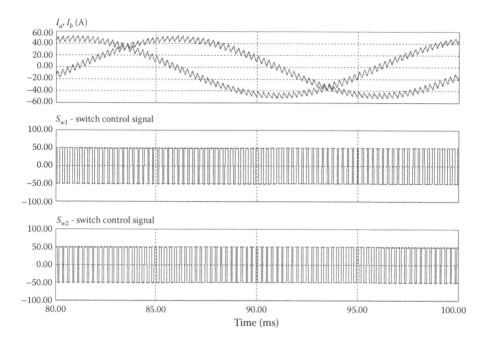

FIGURE 12.14 Significant waveforms at control of a two phase motor drive with 4 kHz switching frequency. Waveforms shown for $m = 0.5$ and 50 Hz fundamental.

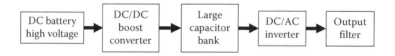

FIGURE 12.15 Typical dual-stage automotive motor drive.

that is usually called *Z*-source converter [21,22] (Figure 12.16). This new converter introduces additional zero vector states. During the zero-states from the conventional inverter operation, the DC link is short-circuited like within the operation of a current source converter. Hence the "*Z-source*" name. When all IGBTs are short-circuited across the DC link (shoot-through), the entire converter works like a boost converter, allowing an increase of the current through the inductors. During the conventional active states of the inverter operation, the DC side voltage equals $2*V_{cap} - V_{batt}$, and that is supposed to be regulated by the boost interval. The voltage across capacitors is increased by the current from the two boost inductors even if the load current also acts against the charging of the capacitors.

Waveforms of Figure 12.17 show the quite large second harmonics present in the capacitor voltage, inherent for the operation of a single-phase inverter. A detail of the control signals is shown in Figure 12.18.

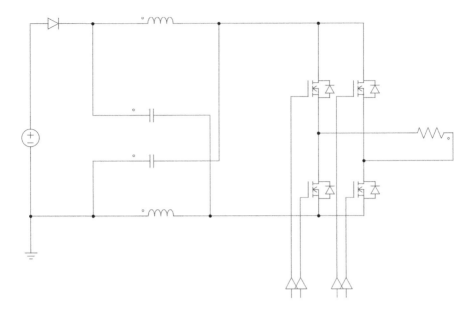

FIGURE 12.16   The Z-source inverter (single-phase version).

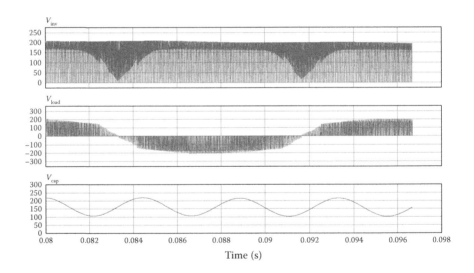

FIGURE 12.17   Waveforms for the generation of 120 V_{ac} from a 120 V_{dc} battery. The output passive LPF filter is not shown ($C = 1$ mF, $L = 0.5$ mH, $f_{sw} = 10$ kHz, sinusoidal modulation).

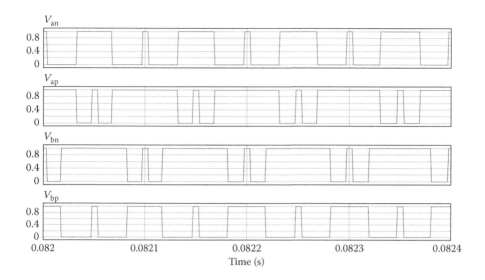

FIGURE 12.18 Zoom on the control signals for the IGBT devices.

The design reduces to the proper sizing of the short-circuit interval along with the conventional sinusoidal modulation control. The peak value of the fundamental of the output voltage yields:

$$V_{out, Peak} = M \cdot V_{inv} \tag{12.28}$$

where V_{inv} is the voltage at the input of the power converter stage. This is like the output of a boost converter and can be calculated from:

$$V_{inv} = B \cdot \frac{V_o}{2} = \frac{V_o}{2} \cdot \frac{1}{1 - 2 \cdot \dfrac{T_o}{T}} \tag{12.29}$$

where T_o is the shoot-through time interval over a switching cycle T, or $D = T_o/T$ is the shoot-through duty ratio. Hence, one can define a global voltage gain as:

$$\frac{V_{out, Peak}}{V_o} = M \cdot B \cdot \frac{1}{2} = M \cdot \frac{1}{2} \cdot \frac{1}{1 - 2 \cdot \dfrac{T_o}{T}} \tag{12.30}$$

The boost factor B is limited by the modulation index M, since a large modulation index M will decrease the available zero-state interval and allow an even shorter shoot-through interval (Figure 12.19).

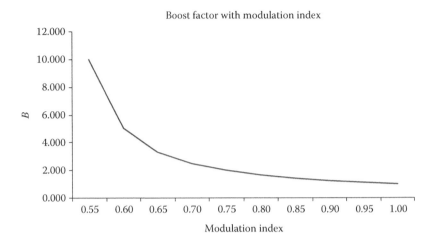

FIGURE 12.19 Maximum ideal boost factor (B_{max}) at different modulation index (M).

For this reason, improved control solutions have been proposed over the years to increase the available boost factor [22]. An alternative design can consider generation of the desired output voltage from a larger V_{inv} voltage, that is a lower modulation index M and a larger boost factor B. However, the product $B*M$ yields is also limited.

12.5 CONCLUSION

During the last 20 years of emerging development in power converters, engineers have tried to explore new three-phase solutions for energy conversion. Several topologies are presented in this chapter and they are intended to open new ideas in design of power converters. Each of these solutions has advantages and disadvantages that make them usable in special applications only.

REFERENCES

1. Covic, G.A., Peters, G.L., and Boys, J.T. 1995. An improved single phase to three phase converter for low cost AC motor drives. *Proceedings of PEDS'95*, Singapore, vol. 1, pp. 549–554.
2. Eastham, J.F., Daniels, A.R., and Lipcynski, R.T. 1980. A novel power inverter configuration. *IEEE IAS* 2, 748–751.
3. van Der Broeck, H.W. and Van Wyk, J.D. 1984. A comparative investigation of a three-phase induction machine drive with a component minimized voltage-fed inverter under different control options. *IEEE Trans. IA* 20, 309–320.
4. van Der Broeck, H.W. and Skudelny, H.C. 1988. Analytical analysis of the harmonic effects of a PWM AC/AC drive. *IEEE Trans. PE* 3(2), 216–223.
5. Blaabjerg, F., Freysson, S., Hansen, H.H., and Hansen, S. 1995. A new optimized space vector modulation strategy for a component minimized voltage source inverter. *IEEE APEC* 2, 577–585.

6. Blaabjerg, F., Freysson, S., Hansen, H.H., and Hansen, S. 1995. Comparison of a space-vector modulation strategy for a three-phase standard and a component minimized voltage source inverter. *Proceedings of EPE'95*, vol. 1, 806–813.

7. Blaabjerg, F., Neacsu, D.O., and Pedersen, J.K. 1997. Adaptive SVM to compensate DC Link voltage ripple for component minimized voltage source inverters. *IEEE PESC* 1, 580–589.

8. Blaabjerg F., Kragh, H., Neacsu, D.O., and Pedersen, J.K. 1997. Comparison of modulation strategies for B4-inverter. *EPE'97*, vol. 2, pp. 378–385.

9. Jacobina, C.B., da Silva E.R.C., Lima, A.M.N., and Ribeiro, R.L.A. 1995. Vector and scalar control of a four switch three phase inverter. *IEEE IAS* 3, 2422–2429.

10. Kim, G.T. and Lipo, T.A. 1995. VSI-PWM inverter/rectifier system with a reduced switch count. *IEEE IAS Thirteenth Annual Meeting*, Orlando, FL, USA, vol. 3, pp. 2327–2332, 8–12 October.

11. Ribeiro, R.L.A., Jacobina, C.B., da Silva, E.R.C., and Lima, A.M.N. 1996. AC/AC converter with four switch three-phase structures. *IEEE PESC* 1, 134–139.

12. Alexa, D. 1995. Static frequency converter with double-branch inverter for supplying three phase asynchronous motors. *EPE J.* 5(1), 23–26.

13. Enjeti, P. and Rahman, A. 1993. A new single-phase to three-phase converter with active input current sharing for low cost AC motor drives. *IEEE Trans. Industry Appl.* 29(4), 806–813.

14. Pan, C.T., Chen, T.C., Hong, Y.H., and Hung, C.M. 1995. A new DC link converter for induction motor drives. *IEEE Trans. EC* 10(1), 71–77.

15. Neacsu, D.O., Gatlan, C., and Gatlan, L. 1999. A three-phase sinusoidal current rectifier with three unidirectional switches and input transformer. *EPE'99*, September 2–4.

16. Neacsu, D.O. 1999. Current-controlled AC/AC voltage source matrix converter for open winding induction machine drives. *Proceedings of the 25th Annual Conference of the IEEE Industrial Electronics Society*, San Jose, CA, USA, vol. 2, pp. 921–926, 29 November– 3 December, 1999.

17. Welchko, B. and Lipo, T.A. 2001. A novel variable frequency three-phase induction motor drive system using only three-switches. *IEEE Trans. IA.* 37(6), 1739–1745.

18. Ziogas, P.D., Vicenti, D., and Joos, G. 1992. A practical PWM AC controller topology. *Conference Record of the 1992 IEEE IAS Tenth Annual Meeting, IEEE-IAS*, Houston, TX, USA, pp. 880–887, 3–5 October.

19. Lee, J.Y. and Sun, Y.Y. 1986. Adaptive harmonic control in PWM inverters with fluctuating input voltage. *IEEE Trans. IE* 33(1), 92–98.

20. Enjeti, P. and Shireen, W. 1992. A new technique to reject DC link voltage ripple for inverters operating on programmed PWM waveforms. *IEEE Trans. PE* 7(1), 171–179.

21. Peng, F.Z., 2003. The Z-source inverter. *IEEE Trans. Industry Appl.* 39(2), 504–510.

22. Peng, F.Z., Shen, M., and Qian, Z. 2005. Maximum boost control of the Z-source inverter. *IEEE Trans. Power Electron.* 20(4), 833–838.

13 AC/DC Grid Interface Based on the Three-Phase Voltage Source Converter

13.1 PARTICULARITIES—CONTROL OBJECTIVES—ACTIVE POWER CONTROL

As energy is mostly transported on AC lines, electronic circuits able to convert AC to DC voltages have been the first application ever for power semiconductor devices [1–6]. The diode rectifier is the simplest power converter grid interface (Figure 13.1). At larger power levels, energy is transported and distributed within three-phase systems, and conversion from three-phase AC to DC voltage is used. All design aspects of diode rectifiers are thoroughly presented in university textbooks for power electronics and will not be reproduced here. However, Table 13.1 reviews the possible diode rectifier solutions and the appropriate factors for the waveform quality.

It is important to note that the high harmonic content of the grid currents may not be in accordance with the modern standards for power quality for certain power levels. Chapter 1 has shown some of the main requirements expected from power converters in different countries. It is easy to observe that above a certain level of the grid current, this class of topologies does not satisfy power quality requirements.

The second obvious disadvantage of this topology is the lack of control for the output voltage. Historically, this drawback was first tackled with thyristor (SCRs)-based converters (Figure 13.2). Their operation assumes a phase control and an output voltage lower than the diode rectifier's output voltage. The waveforms corresponding to the operation of this power stage outline the low power factor and large reactive power circulated in the system.

The advent of power semiconductor devices with turn-off capability has improved the power quality factors for the grid currents. The first use of gate turn-off thyristor (GTO) devices within the diode or SCR-rectifier topologies has opened a new class of power converters. Examples of pioneering solutions are shown in Figure 13.3.

The output voltage results in a train of pulses characterized by a DC component and an HF-switching component. The grid current equals the algebraic sum of the currents through two devices and follows the pulse control strategy. This allows total harmonic distortion (THD) or, generally, the harmonic content of the grid current to

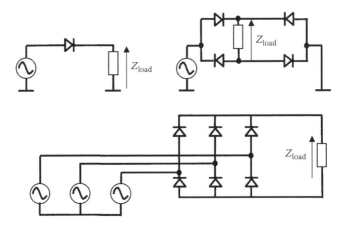

FIGURE 13.1   Different topologies of diode rectifiers.

improve. Given the large inductance in the load, a pulse is seen in the line current during each interval when a voltage pulse is generated on the load (Figure 13.4).

If the number of pulses is small, a control strategy able to eliminate low harmonics (5th, 7th, 11th, 13th, and so on) of the current waveform is employed. Details of designing a pulse width modulation (PWM) controller suitable for such an operation are presented in Chapter 4. If the power semiconductor devices allow a higher switching frequency, a sinusoidal PWM algorithm may be considered. Two switches are controlled at a given moment and the pulse width is derived with a sinusoidal law.

All these circuits switch current through the grid lines. They introduce a line inductance in the path of the switch current, creating overvoltage at switching. To overcome this, snubber capacitors are needed across the power semiconductor devices (Chapters 2 and 3).

All the previously introduced power converters ensure energy conversion towards a load with a large inductance. This is the case when a DC machine drive or a DC magnet supply is used. However, a large group of applications require supply of the load with a constant DC voltage. This includes also the case of a grid interface for AC machine drives in which the DC circuit is dominated by a very large capacitor. Details about rating the DC bus and the importance of the proper value for a specific AC machine drive application are discussed in Chapter 4. We present here the details of the grid interface.

The simplest circuit able to create voltage on the DC intermediary circuit is based on a diode rectifier. As both the grid and the DC bus are voltage sources, an inductance should be used to take over instantaneously the differences between these two voltage sources. Different solutions use the inductance on the DC side, on the grid side, or on both sides. When the inductance is on the DC side, the circuit operation is identical to the previous diode rectifier with a large inductive load. The rectifier bridge produces an output voltage following the envelope of the grid voltages, and the inductance acts as a filter to produce the constant DC bus voltage (Figure 13.5).

When the inductance is on the AC side, the diodes have a shorter conduction angle depending on the actual voltage of the DC bus capacitor. Figure 13.6 shows the

TABLE 13.1

Performance of Different Diode Rectification Schemes without Output Capacitor Filter

Topology	# Diodes	Semiconductor Ratings [V/I]	Output Average Voltage
		Single Phase	
	1	V_{in}, I_{pk}	$V_{d0} = V_{ph} \cdot \dfrac{\sqrt{2}}{\pi}$
	2	V_{in}, I_{pk}	$V_{d0} = V_{ph} \cdot \dfrac{2 \cdot \sqrt{2}}{\pi}$
	4	$V_{in}/2, I_{pk}$	$V_{d0} = V_{ph} \cdot \dfrac{2 \cdot \sqrt{2}}{\pi}$
		Three Phase	
	3	V_{LL}, I_{pk}	$V_{d0} = \sqrt{2} \cdot V_{ph} \cdot \dfrac{3 \cdot \sqrt{3}}{2 \cdot \pi}$
	6	V_{LL}, I_{pk}	$V_{d0} = 2 \cdot \sqrt{2} \cdot V_{2} \cdot \dfrac{3 \cdot \sqrt{3}}{2 \cdot \pi}$

topology and the main waveforms for this AC/DC converter. When the difference between two line voltages is larger than the DC bus voltage, a pair of diodes turns on and the capacitor voltage follows the grid line-to-line voltage (VLL).

This interval corresponds to charging energy within the DC bus capacitor. When the DC bus voltage reaches the level of VLL, the diodes turn-off and the load is supplied solely from the capacitor bus. During the short conduction time interval of the diodes, the charging current is quite high, as it is produced across a small grid inductance from a large voltage difference. Moreover, the charging current at the start-up is very high until the bus voltage reaches the envelope of the grid voltages.

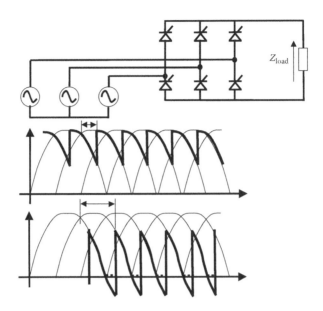

FIGURE 13.2 SCR based bridge converter and output voltage waveforms for different phase angle control.

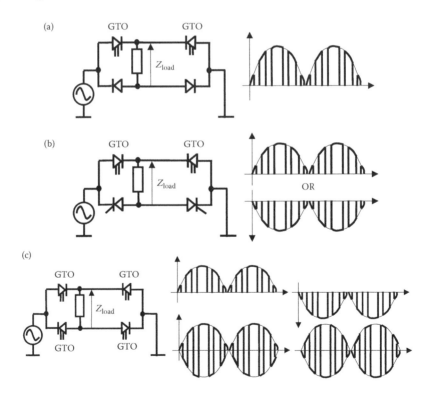

FIGURE 13.3 Use of GTO devices within the single-phase grid interfaces.

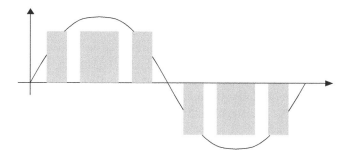

FIGURE 13.4   Pulses within the grid current.

The output of the rectifier bridge represents a quasi-constant DC voltage with a level dictated by the grid peak voltage. The level of the DC bus voltage depends slightly on the load. If there is no load, the DC voltage follows exactly the peak of V_{LL}. The higher the load current, the larger the ripple of the voltage across the capacitor produced by subsequent charge–discharge events. The maximum ripple corresponds to the diode rectification waveform of V_{LL}.

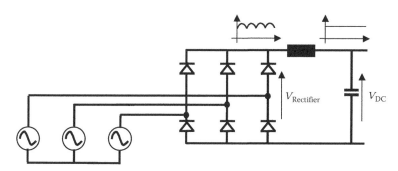

FIGURE 13.5   Diode rectifier with an inductive filter on the DC side.

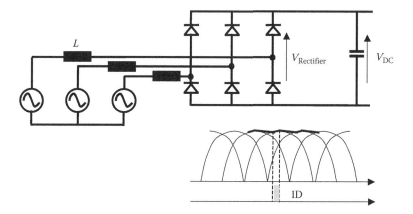

FIGURE 13.6   Diode rectifier with an inductive filter on the DC side.

13.2 MEANING OF PWM IN THE CONTROL SYSTEM

13.2.1 SINGLE-SWITCH APPLICATIONS

As previously shown, there is no control of the level of the DC bus voltage, and the grid currents are formed of pulses of current. A simple solution would consist of a buck or boost converter following up the DC capacitor (Figure 13.7).

There is a double DC stage within these converter topologies and this implies large filter DC capacitors. A closer analysis shows that the output voltage can be achieved after a multiplication of the switching functions corresponding to the two power-converter stages. This is equivalent to considering a single-capacitor filter at the output of the second power converter. In other words, it is possible to use the rectified voltage directly at the input of the buck or boost converter.

Moreover, it is desirable that the grid current be as close as possible to a sinusoidal waveform with reduced harmonics. The operation of the boost converter ensures this behavior of a continuous input current, with a waveform close to a given reference. This is because the inductor appears on the grid side and the inductor current is not chopped during operation. The resulting possible topologies are shown in Figure 13.8. Among these, the circuit with grid-side inductance is preferred because of the AC character of the current through inductors. The second major advantage of this topology consists of the inherent high power factor of the grid current.

Rectifying the AC input using a diode rectifier and chopping it at a high frequency to achieve voltage control has the advantages of simplicity, performance, and reliability. A diode rectifier cascaded with a PWM boost chopper is analyzed in [7–13]. In the first solution [7], the boost inductance is present in the grid side as

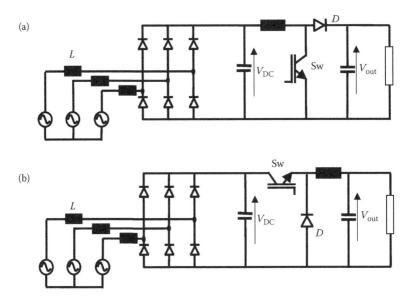

FIGURE 13.7 DC/DC converters following up a diode rectifier stage.

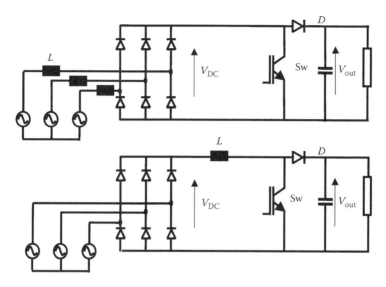

FIGURE 13.8   Three-phase single-switch boost converter.

phase inductance (Figure 13.8a). The analysis and design of this converter is further described in the work of Kolar et al. [8,9].

The step-up (boost) converter operates at constant frequency in the discontinuous mode. This mode presents some disadvantages:

- Higher voltage/current stress
- EMI propagation

Advantages consist of

- High input power factor
- Reduced power loss through zero-current switching
- The absence of reverse recovery problem in the diodes
- The possibility of single control loop to control output power and voltage

Using this approach in high-power applications operated with low switching frequency and producing a low output voltage leads to a lower grid power factor and a current THD greater than 5%. One solution consists of injecting a harmonic content within the control of the switch in order to compensate for the main current harmonics.

The conventional control of the boost converter grid interface is simple, but the advantages of PWM are not fully utilized yet. It has been shown in the work of Weng and Yuvarajan [12,13] by extensive computer analysis that the input current distortion for a single-phase converter is reduced when employing a PWM with a second harmonic signal injected within the reference. The reference signal for the PWM generation is usually a DC value able to define the output voltage level. By injecting

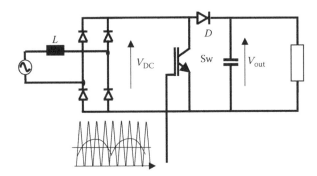

FIGURE 13.9 Injection of the second harmonic within the reference signal of a single-phase boost converter.

a second-order harmonic synchronized with the grid, the inherent variations of the output voltage and input current with the phase of the grid are reduced (Figure 13.9).

The circuit contains a low-pass grid filter that allows the input voltage to be seen directly at the converter input. The second inductance of the filter (L_b) also acts as a boost inductance for the converter. The switch used in this boost converter can be power MOSFET or insulated gate bipolar transistor (IGBT), depending on the application.

When the IGBT "Sw" is ON, the boost inductances are supplied by phase voltages and energy is stored in their magnetic fields. Depending on the phase of the input voltages, three diodes are in conduction at any time. This ON-time interval usually has a constant width and it can modify the output voltage on the basis of the duty ratio. When IGBT is turned-off, the energy stored within the boost inductance is transferred to the output capacitor. As with any boost converter in discontinuous conduction mode, the transfer to the output capacitor is ended when the current vanishes. The duration of this operation (t_{off1}) until the first phase current reaches zero depends on the stored energy and, therefore, on the value of the lowest phase voltage. Observing the phase voltages within a three-phase system defines 12 intervals for analysis, each interval having a different phase voltage with the lowest value. The following discharge interval (t_{off2}) is characterized by the conduction of two diodes only.

Figure 13.10 shows PSPICE simulation results, for example, for the current waveforms given for the case $v_R > 0$, $v_s < 0$, $v_T < 0$. The first current to reach zero is on phase T. Maximum values of the current are denoted by $I_{m,r}(s,t)$ and current levels in the first two phases when the third-phase current vanishes are denoted by $I_{0,r}(s)$. The second waveform always represents voltage shape at the diode rectifier input on the corresponding phase.

From this generic operation of the power stage, different PWM control solutions are considered in the work of Kolar et al. [9].

- Constant ON time and constant switching frequency.
- Variable switching frequency depending on IGBT's turn-on immediately after the end of the inductance's discharge interval, with operation at the border of discontinuous conduction mode.

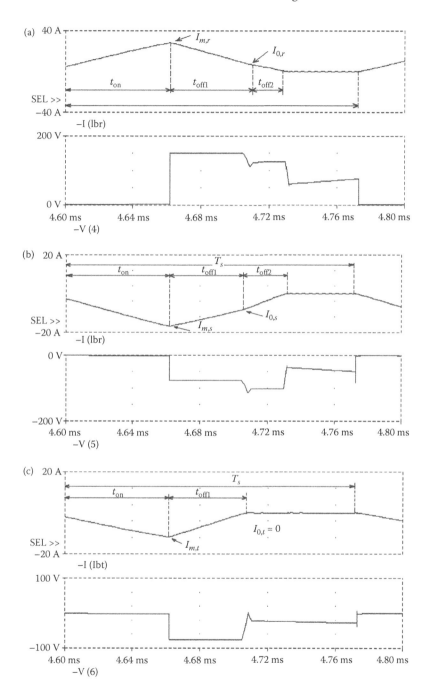

FIGURE 13.10 Shapes of the converter input currents and voltages (a) Phase *R*, (b) Phase *S*, (c) Phase *T*. (From Neacsu, D.O., Yao, Z., and Rajagopalan, V. 1986. *IEEE PESC Conference, June 1996*, Baveno, Italy, 24–27, pp. 727–732, IEEE Paper 0-7803-3500-7/96. © (1986) Printed with permission from IEEE.)

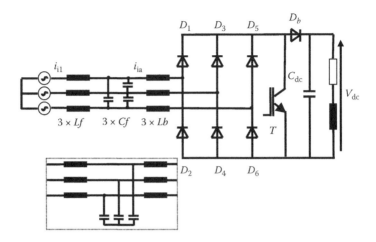

FIGURE 13.11   Basic Single-Switch Three-Phase Boost Converter.

- Maintaining a constant DC power flow on the DC side by a highly dynamic output voltage control. This solution requires a very large bandwidth of the system.

Understanding all aspects of the operation of the power converter shown in Figure 13.11 and the principles of the harmonic injection, as explained in Figure 13.9, for the single-phase converter allows the development of a special PWM algorithm with performance improvements. The duration of the first time interval after the IGBT's turn-off depends on the instantaneous value of the lowest phase voltage, and this can be compensated for by an appropriate modulation of the reference signal for the PWM control. If the constant duty cycle of the conventional operation is denoted by D_0, the new reference signal for the PWM control is given by:

$$D(t) = D_0 \cdot (1 + f(t)) \tag{13.1}$$

The injected harmonics are included in the $f(t)$ function. The ON time of the switch is defined as:

$$t_{on} = (D_0 + f(t)) \cdot T_s = D \cdot T_s \tag{13.2}$$

A qualitative analysis of the converter operation reveals the important six order harmonic as a component of the $f(t)$ function. There are different methods to define the exact or mathematical form of this function. Many solutions are mainly based on repetitive simulation or experimental analysis. In what follows, an analytic solution for a problem of optimality imposed to the theoretically derived expressions of phase currents is provided and implemented based on a computer program.

However, this optimal PWM improves the harmonic performance at the input of the power converter stage and the actual filter also influences the grid harmonic performance.

Let us derive the mathematical expressions for currents through the boost inductance [13] assuming:

- Symmetrical three-phase system with no neutral components.
- Neglect of losses in the converter.
- Same value of the boost inductors in all three phases.
- All currents are positive when entering the power stage and negative when they flow into the grid.
- High switching frequency allowing an approximation of constant grid voltages over the sampling period and a linear variation of the inductor current over t_{on} and t_{off1}.
- The input filter effect to evaluate grid currents by averaging the converter input currents in each sampling interval.
- The symmetry of a three-phase system to reduce the analysis for a 30° interval (symmetrical evolution on the next 30° is achieved and evolution on the other 60° sectors can be defined by changing the phase sequence).
- The interval shown in Figure 13.10 is considered.

A mathematical form for the phase voltages is shown as:

$$v_R = V \cdot \cos(\omega t)$$
$$v_S = V \cdot \cos\left(\omega t - \frac{2 \cdot \pi}{3}\right)$$
$$v_T = V \cdot \cos\left(\omega t - \frac{4 \cdot \pi}{3}\right)$$

(13.3)

We develop the analysis for the 30° interval before the peak of the grid voltage on phase R.

The average relationship for the input current on phase R yields as (Figure 13.10):

$$\frac{t_{on} \cdot I_{m,r}}{2} + \frac{t_{off1} \cdot (I_{m,r} + I_{0,r})}{2} + \frac{t_{off2} \cdot I_{0,r}}{2} = I_{R,av} \cdot T_s$$

(13.4)

Now, let us define mathematically each time interval. The ON time has already been defined by (13.2) and the first discharge interval depends on the instantaneous value of the voltage on the last phase.

$$t_{off1} = \frac{t_{on} \cdot V_1}{V_2 - V_1} \Rightarrow t_{off1} = \frac{v_T}{-\dfrac{1}{3} \cdot V_{dc} - v_T} \cdot t_{on} \Rightarrow t_{off1} = \frac{-v_T}{\dfrac{1}{3} \cdot V_{dc} + v_T} \cdot t_{on}$$

(13.5)

The bend-point value of the current on phase R when the current on phase T vanishes yields as:

$$I_{0,r} = I_{m,r} - \Delta I = \frac{t_{on} \cdot v_R}{L} - \frac{t_{off1} \cdot \left[\frac{2}{3} \cdot V_{dc} - v_R\right]}{L} \tag{13.6}$$

$$I_{0,r} = \frac{t_{on} \cdot v_R}{L} - \left[\frac{-v_T}{V_{dc} + 3 \cdot v_T} \cdot t_{on}\right] \cdot \frac{[2 \cdot V_{dc} - 3 \cdot v_R]}{L} \tag{13.7}$$

$$I_{0,r} = \frac{t_{on}}{L} \cdot V_{dc} \cdot \frac{v_R + 2 \cdot v_T}{V_{dc} + 3 \cdot v_T} \tag{13.8}$$

During the first discharge interval, D_1, D_4, D_6 were in conduction. The second discharge interval for phase R (t_{off2} in Figure 13.11) is characterized by conduction of D_1 and D_4 only.

$$I_{0,r} = \frac{t_{off2}}{L} \cdot \frac{1}{2} \cdot [V_{dc} - v_R + v_S] \tag{13.9}$$

It yields:

$$t_{off2} = 2 \cdot V_{dc} \cdot t_{on} \cdot \frac{v_R + 2 \cdot v_T}{V_{dc} + 3 \cdot v_T} \cdot \frac{1}{V_{dc} - v_R + v_S} \tag{13.10}$$

Taking account of all these equations in the current average relationship yields:

$$I_{R,av} = \frac{D^2 \cdot T_s}{2 \cdot L} \cdot \left\{ \begin{array}{l} v_R - \dfrac{3 \cdot v_T}{[V_{dc} + 3 \cdot v_T]^2} \cdot [v_R \cdot (V_{dc} + 3 \cdot v_T) + V_{dc} \cdot (v_R + 2 \cdot v_T)] \\[2ex] + \left[V_{dc} \cdot \dfrac{v_R + 2 \cdot v_T}{V_{dc} + 3 \cdot v_T}\right]^2 \cdot \dfrac{2}{V_{dc} + v_S - v_R} \end{array} \right\} \tag{13.11}$$

Currents on the other two phases are herein given without demonstration, but they can be calculated as an exercise.

$$I_{T,av} = \frac{D^2 \cdot T_s}{2 \cdot L} \cdot \frac{v_T}{V_{dc} + 3 \cdot v_T} \cdot V_{dc} \tag{13.12}$$

$$I_{S,av} = \frac{D^2 \cdot T_s}{2 \cdot L} \cdot \left\{ \left[v_S - \frac{3 \cdot v_T}{\left[V_{dc} + 3 \cdot v_T \right]^2} \cdot \left[v_S \cdot (V_{dc} + 3 \cdot v_T) - V_{dc} \cdot (v_R + 2 \cdot v_T) \right] \right] - \left[V_{dc} \cdot \frac{v_S - v_T}{V_{dc} + 3 \cdot v_T} \right]^2 \cdot \frac{2}{V_{dc} + v_S - v_R} \right\}$$

(13.13)

These expressions are analogous with the results presented in the work of Kolar et al. [8,9], but with an explicit dependence on the circuit voltages and are very suitable for computer implementation and solving.

The goal for computer optimization is to determine the modulation function $f(t)$, which reduces the input current harmonics. In this respect, the optimization condition is written as the cumulative error from a set of input currents with sinusoidal waveforms and synchronized with the grid voltages. In a real implementation, the root mean square value of these reference currents is calculated from the power transfer condition through the power converter.

$$\min \left[\left(I_{R,av} - i_R \right)^2 + \left(I_{S,av} - i_S \right)^2 + \left(I_{T,av} - i_T \right)^2 \right]$$

(13.14)

The first possibility to solve this optimal problem consists of using a mathematical program like *MathCad*. One should solve the constraint (13.14) for each phase angle of the input voltage within a sector of 30° before the peak of the voltage on phase R. Results

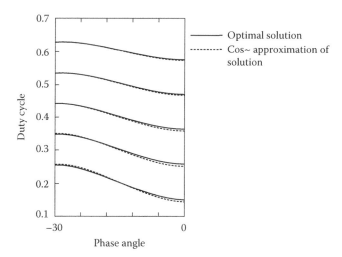

FIGURE 13.12 Results for the duty cycle modulation for the same phase as in Equation 11.3. (From Neacsu, D.O., Yao, Z., and Rajagopalan, V. *IEEE PESC Conference, June 1996*, Baveno, Italy, 24–27, pp. 727–732, IEEE Paper 0-7803-3500-7/96. © (1986) Printed with permission from IEEE.)

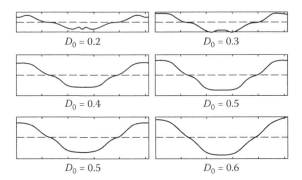

FIGURE 13.13 Input current theoretical waveform: solid line, without modulation; dashed line, with modulation. (From Neacsu, D.O., Yao, Z., and Rajagopalan, V. *IEEE PESC Conference, June 1996*, Baveno, Italy, 24–27, pp. 727–732, IEEE Paper 0-7803-3500-7/96. © (1986) Printed with permission from IEEE.)

for different duty cycles are shown on Figure 13.12. As the preliminary understanding of the problem indicated that a sixth order harmonic may be the optimal injected signal, the appropriate modulation signal for a cosine approximation is also shown. This approximation is very important for the actual implementation of the controller.

Notice a linear dependence of the amplitude of the cosine modulation waveform on the switch duty cycle. The amplitude is given by this linear dependence, whereas the phase variation is achieved with a phase locked-loop (PLL)-based circuit for synchronization with input line voltages. The modulation wave thus obtained is used with a classical PWM integrated circuit (IC) to control the switch.

Taking into account the power conservation principle can help demonstrate that each $2n$th harmonic in the power converter's output voltage is related to the input current harmonics of orders $2n + 1$ and $2n - 1$ (output constant load, input symmetrical system). For instance, an action against the 6th harmonic in the output would also minimize the effect of both 5th and 7th harmonics in the input current.

Injection of a harmonic signal within the reference for the PWM control is able to correct the ideal low-frequency shape of the input current. Figure 13.13 presents theoretical low-frequency components for the cases with and without harmonic injection.

The actual input current harmonics are also influenced by the input filter. One of the negative effects of the input filter that adds up to delays in the control system and the effect of the sampling interval is the phase shift between the input voltage and current. To compensate for this phase delay, a small phase shift should be considered within the harmonic injection reference.

Figures 13.14 and 13.15 show harmonic results when this solution is considered within a power converter-based grid interface. The THD of the input current is still above 5%, but Figure 13.14 does not include the effect of the input filter. The remaining harmonics are at higher frequencies, and they can be removed with a conventional low-pass filter. As the optimal PWM reduces the content in the low frequencies, 5th and 7th harmonics of the input current for the case without harmonic injection and for the case with harmonic injection are shown in Figure 13.15. It can be seen that

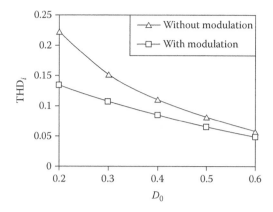

FIGURE 13.14   THD for input current over the switch duty cycle computed with the first 500 harmonics. (From Neacsu, D.O., Yao, Z., and Rajagopalan, V. *IEEE PESC Conference, June 1996*, Baveno, Italy, 24–27, pp. 727–732, IEEE Paper 0-7803-3500-7/96. © (1986) Printed with permission from IEEE.)

the cumulative effect of both 5th and 7th harmonics is reduced by this approach: $(\min(V_5^2 + V_7^2))$. For a switching frequency of 5 kHz, the content in low harmonics for the case without optimal modulation is I5 = 6.77% and I7 = 1.01%, whereas the case with optimal PWM leads to I5 = 3.84% and I7 = 3.83%. However, both harmonics are inside the limits imposed by IEC 555-2 ($I(5)$ = 1.14A and $I(7)$ = 0.77A) for the output powers in the kW region. For $D > 0.5$, the content in both harmonics is less than 4% as required by IEEE 519-1992.

The harmonic injection in the PWM reference signal of a single-switch three phase PWM boost converter has the following advantages:

- Reduced instability risk due to discontinuous conduction mode;
- Applicable to the single-switch three-phase boost converter topologies (including the modern resonant ones) with minimal and low-cost modifications of the command circuit;
- Implementation with any conventional PWM IC owing to the operation with constant switching frequency;
- Reduced requirements for the mains filter and enlarged domain of having input THD current less than 5% and input power factor greater than 95%.

13.2.2   SIX-SWITCH CONVERTERS

Many applications require a bidirectional power transfer to the grid, and this implies moving the boost power converter stage closer to the grid without the intermediate stage of diode rectification. Figure 13.16 shows the three-phase six-switch power converter. Chapter 3 introduced this power stage with applications to DC/AC conversion and AC motor drives. The same power converter with a different control is the most important three-phase electronic grid interface [14–16].

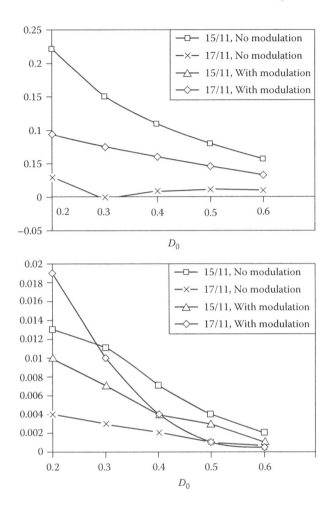

FIGURE 13.15 Low harmonics content of the input currents. (From Neacsu, D.O., Yao, Z., and Rajagopalan, V. *IEEE PESC Conference, June 1996*, Baveno, Italy, 24–27, pp. 727–732, IEEE Paper 0-7803-3500-7/96. © (1986) Printed with permission from IEEE.)

This topology allows a bidirectional power flow with the control of both the DC bus voltage and the input power factor, although the input currents have low harmonics [17–23]. This ensures a very high performance interface to the grid for different industrial equipment, such as electrical drives or DC load supply. Electrical drives with AC/DC boost bridge converters also benefit from energy savings during braking or acceleration during drive dynamics. Controlling the DC bus voltage allows capacitor discharge into the grid instead of on a dummy brake resistor (Figure 13.16).

Let us develop a mathematical model for the analysis of this power converter starting from the grid electrical circuit. The grid voltages are defined as:

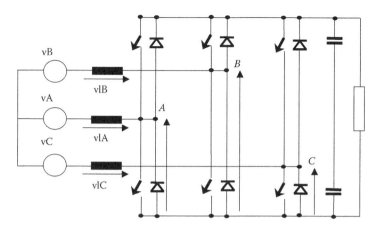

FIGURE 13.16   Three-phase grid interface with six switches.

$$e_{a0} = E \cdot \sin(\omega t)$$

$$e_{b0} = E \cdot \sin\left(\omega t - \frac{2\pi}{3}\right)$$

$$e_{c0} = E \cdot \sin\left(\omega t - \frac{4\pi}{3}\right)$$

(13.15)

The voltage equations for each phase can be expressed in dependence with the each pole voltage from the power converter.

$$e_{a0} = R_r \cdot i_{1a} + L_r \cdot \frac{di_{1a}}{dt} + v_{a0}$$

$$e_{b0} = R_r \cdot i_{1b} + L_r \cdot \frac{di_{1b}}{dt} + v_{b0}$$

$$e_{c0} = R_r \cdot i_{1c} + L_r \cdot \frac{di_{1c}}{dt} + v_{c0}$$

(13.16)

The first approach to the analysis of this converter was developed with a scalar control of the DC voltage through the modulation index of the PWM references synchronized with the AC grid voltage [24]. The difference between the fundamental of the voltage generated by PWM and the grid voltage is applied to the boost inductance, defining the input currents. Given the inherent phase shift produced by an inductor, the control references should be appropriately shifted and synchronized.

A second approach consisted of the so-called delta control of the power converters [25], which is an equivalent of the hysteresis control method.

Later on, vectorial methods [9–13] were preferred for control of this power converter [13,26,27]. The vectorial methods for three-phase systems have been presented in Chapter 5 for DC/AC converters and they can be extended to AC/DC conversion. They provide high performance current control in the d–q frame and have the particular advantage of separating the active and reactive power components. As one of the most important requirements for the grid interfaces is the unity or controllable power factor, this features become very important.

The three-phase system of equations from Equation 13.16 can be transformed in an equivalent two-phase system expressed by:

$$e_{x0} = R_r \cdot i_{1x} + L_r \cdot \frac{di_{1x}}{dt} + v_{x0}$$
$$e_{y0} = R_r \cdot i_{1y} + L_r \cdot \frac{di_{1y}}{dt} + v_{y0}$$

(13.17)

The transformation of these equations into the synchronous reference frame

$$v_d = v_{x0} \cdot \cos\theta + v_{y0} \cdot \sin\theta$$
$$v_q = -v_{x0} \cdot \sin\theta + v_{y0} \cdot \cos\theta$$

(13.18)

yields the input voltage equations in the synchronous d–q reference frame

$$e_d = R_r \cdot i_d + v_d + L_r \cdot \frac{di_d}{dt} - \omega \cdot L_r \cdot i_q$$
$$e_q = R_r \cdot i_q + v_q + L_r \cdot \frac{di_q}{dt} + \omega \cdot L_r \cdot i_d$$

(13.19)

The grid voltages are expressed by two constant components:

$$e_d = E$$
$$e_q = 0$$

(13.20)

Equations 13.19 and 13.20 demonstrate that the grid current can be decomposed into two components:

- i_q allows the control of the input reactive power.
- i_d allows the control of the input active power.

Working with unity power factor means to keep $i_q = 0$. The i_d current reference is many times provided by the output of a proportional-integral (PI) controller for the DC bus voltage control (Figure 13.17), as this component has the role of active power transfer to the DC bus.

The module *Compensation, Voltage limit, and Transform* will be explained later in this chapter. The PWM module has to transform the reference voltage orthogonal coordinates into six gating signals for the power devices. A previous chapter for PWM algorithms showed several solutions for the DC/AC conversion. All of them can be redesigned for the grid interface application as they can be reduced to a set of

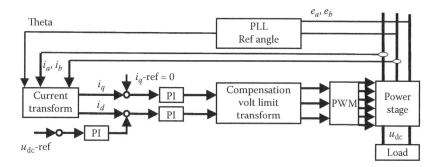

FIGURE 13.17   Base control structure.

reference voltages to be transformed into gate pulses. A group of methods are based on a conversion from the orthogonal system in phase reference voltages followed up by a comparison with a carrier waveform.

The space vector PWM concept allows better utilization of the DC bus voltage compared to carrier-based PWM methods and it represents, hence, a most convenient selection. The maximum voltage achievable at the power converter input is:

$$V_{1a}^{max} = 0.61 \cdot V_{dc} \tag{13.21}$$

It is important to be able to have a large input voltage for the dynamic range of the system. The more the voltage is available to be applied to the boost inductors, the faster the transients can be achieved. Full analysis of the closed loop options is later included.

According to this SVM method, any set of instantaneous three-phase voltages can be obtained by switching the inverter between six active states defining the six-step operation and the null states in order to approximate a uniform rotation of the space vector corresponding to the three-phase input voltage system. Any desired position of the space vector in the complex plane is calculated by weighed addition of the neighboring switching vectors.

After the time intervals have been defined, there are several possibilities to distribute the active states over the sampling period. An interesting alternative does not switch each device for 60° on each fundamental cycle. This has also been discussed in Chapter 4 for generic PWM algorithms. As the input-phase current should be in phase with the input-phase voltage, the 60° no-switching interval should be chosen around the inverter-switching vectors (Figure 13.18). The main waveforms corresponding to this operation are presented in Figure 13.19.

13.2.3 TOPOLOGIES WITH CURRENT INJECTION DEVICES

Power electronics technologies are nowadays used for higher and higher power level applications. Higher power delivery is achieved on three-phase systems rather than single-phase systems. The design engineer faces thus an important trade-off between

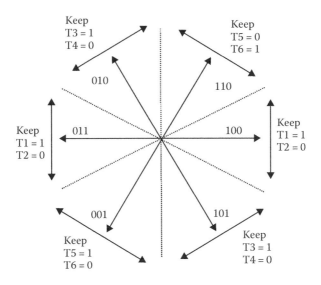

FIGURE 13.18 Switching vectors and definition of the no-switching intervals.

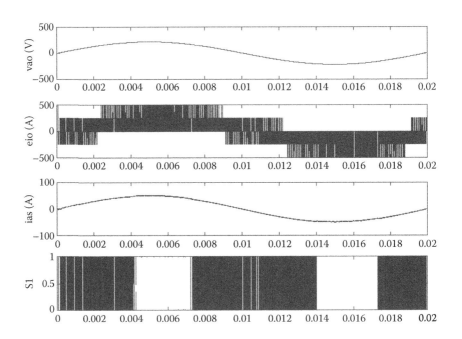

FIGURE 13.19 Waveforms characterizing the modulation with no-switching for 60° over a fundamental cycle.

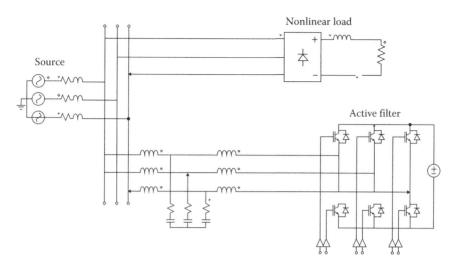

FIGURE 13.20   Principle of an active filter.

the standard requirements for harmonic reduction in three-phase systems and the cost reduction with keeping low the power installed in electronics. A compromise solution relies in using a combination of a high power diode rectifier and low-power compensation equipment [28]. The simplified idea of an active filter is shown in Figure 13.20.

The idea of adding up a compensation current to the diode rectifier's input current has been further exploited with the proposal of simplified passive injection circuits. It is worthwhile to mention here the work of Mohan et al. in early 1990s [29,30]. The core idea was to detect a 3rd harmonic current from the load side and to inject this current into all the input currents as a zero sequence component (Figure 13.21). They used a specially constructed magnetic device able to create the three identical third harmonic currents to be added to the input currents. Other similar principle solutions were reported in [31] and they consisted of a pure R-L-C network tuned on the third harmonic and able to depict this harmonic component for injection into the rectifier's input currents. The injection network sent a third harmonic current of 0.50*I into each phase, that circulated through the conducting diodes as a 0.75*I current, to add up from both $DC+$ and $DC-$ into a 1.50*I current at node N. The phase without conducting rectifier diode had the 0.50*I current projected directly into the grid, while the phases with conducting diodes had seen the load current plus the 0.25*I. Improvements to these waveforms became possible with optimized control.

Later on, the advances in power semiconductor technology and the control (Figures 13.22 and 13.23) circuitry allowed the efficiency improvement with the use of active injection circuits [32] (Figure 13.24). Their operation is similar to the original principle of active filtering (Figure 13.25).

There is a large variety of other passive and active injection circuits, and the above circuits are shown for illustration only.

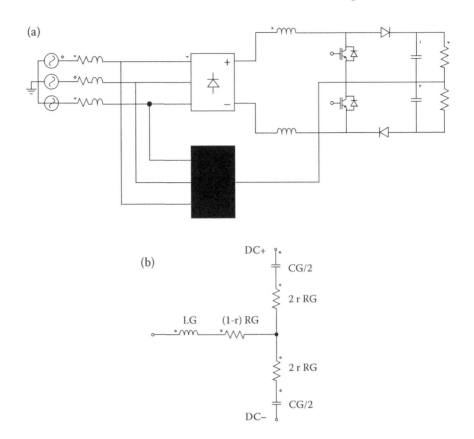

(a)

(b)

DC+

CG/2

2 r RG

LG   (1-r) RG

2 r RG

CG/2

DC−

FIGURE 13.21   Principle of current injection.

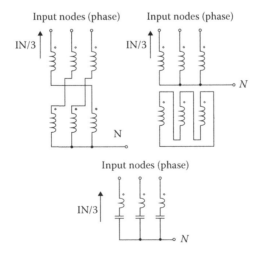

FIGURE 13.22   Third harmonic current injection with passive circuits.

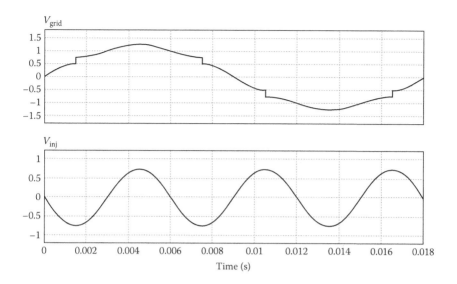

FIGURE 13.23   Injection current and grid current for circuit of Figure 13.22.

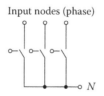

FIGURE 13.24   Active current injection circuits.

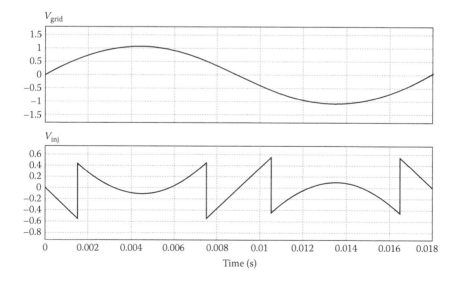

FIGURE 13.25   Injection current and grid current for circuit of Figure 13.24.

13.3 CLOSED-LOOP CURRENT CONTROL METHODS

13.3.1 INTRODUCTION

As shown in Figure 13.17, the most important module of the control system consists of the closed-loop current controllers. The model of the system is expressed by Equation 13.19 and PI controllers are considered on each axis. As Equation 13.19 is different for each axis, different approaches to compensate the cross-coupling terms are considered. Such terms are considered within the following module at the output of the PI current controller.

13.3.2 PI CURRENT LOOP

The PI current loop should be mainly designed to compensate the load voltage produced by the boost inductance term of Equation 13.19. A schematic representation is given in Figure 13.26.

There are many methods to define the gains for the PI control and they are reviewed in Chapter 9. A simplified approximation is given here based on the boost circuitry parameters. The condition of unity gain of the closed-loop transfer function leads, after some calculation, to the values of the PI gains:

- Gain of the integrative term

$$k_I = \frac{T_{RL}}{2 \cdot K_{RL} \cdot T_\Sigma} = \frac{T_{RL}}{2 \cdot K_{RL} \cdot \left(T_s + \dfrac{T_s}{2}\right)} \tag{13.22}$$

- Gain of the proportional term

$$k_P = \frac{T_s}{2 \cdot K_{RL} \cdot T_\Sigma} = \frac{T_s}{2 \cdot K_{RL} \cdot \left(T_s + \dfrac{T_s}{2}\right)} \tag{13.23}$$

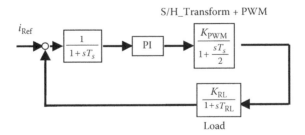

FIGURE 13.26 Schematics of the PI control system.

The concept of the vectorial control of three-phase power converters has already been discussed in Chapter 5. The application to grid interface systems encounters some limitations in the transient response performance due to limited voltage available across the boost inductance. A derivative of a direct application of the PI control without any special compensation is that the transient response time for input current step-up is shorter than the transient response time for the current step-down and the latter can be bothersome in many cases.

13.3.3 TRANSIENT RESPONSE TIMES

On the basis of PI gains, one can calculate the minimum response time for step-up or step-down transitions. As the resistive component is very small, the term $R_r i$ can be even neglected. Owing to the largely inductive character of the load, a fast transient of the active current can be achieved with a large voltage applied across the boost inductance.

As the first equation in Equation 13.19 contains the E term, there are two different cases for analysis depending on the sign of the current change:

- Step-up: In order to increase the active current (i_d), it is necessary to provide the minimum available voltage across L_r (Figure 13.27). The response time will be small and it yields:

$$\Delta T_{up} \approx L_r \cdot \frac{|\Delta i|}{E + |V_{max}|} \tag{13.24}$$

- Step-down: In order to decrease the active current (i_d), it is necessary to provide the maximum available across L_r. The response time yields:

$$\Delta T_{down} \approx L_r \cdot \frac{|\Delta i|}{|V_{max}| - E} \tag{13.25}$$

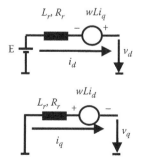

FIGURE 13.27   Equivalent circuits for d and q axis.

13.3.4 Limitation of the (v_D, v_Q) Voltages

It is easy to notice that different step-up and step-down time intervals are achieved with a smaller value for the step-up transient time. The step reference to the PI controller will have the tendency to force the output of the controller at a higher than possible value. Usually, a software limitation block is included as presented in Figure 13.28. This figure also illustrates the PI implementation in the context of Equation 13.19. The cross-coupling terms and the E term are also included. The cross-coupling terms are taken from the current feedback, but an alternative would be the reference currents.

It is important to note that many implementation solutions of the PI control of the grid interface converter use control systems similar to the motor control ones, without the term E. The problem with such an approach is that the dynamic range at the output of the PI controller is limited further by working with the voltage vd component always close to the E value. The digital range for voltage may also be close to the value of E and the output of the PI controller may saturate digitally by reaching the end of its range. Figure 13.29 shows results for the control structure defined with Figure 13.28.

A multitude of solutions is reported in the technical literature to overcome this problem by artificially augmenting the voltage available to be applied across the input boost inductor [33,34]. Observing Figure 13.27 and Equation 13.19 defines two easy approaches to increase the maximum available voltage at the converter input:

- Using the cross-coupling terms.
- Overlooking the PI outputs and applying all the maximum available voltage during transients.

13.3.5 Minimum Time Current Control

The *minimum time current controller* leads to excellent performance by finding the optimal control voltage to track the reference current with minimum time based on

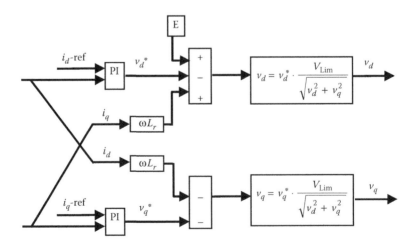

FIGURE 13.28   Schematic of the usual control with limitation.

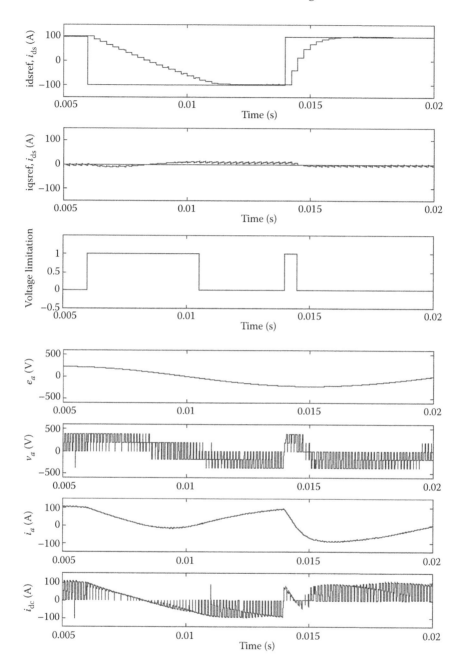

FIGURE 13.29   Simple current controller with voltage limitation for $L_r = 2.5$ mH, $R_r = 0.1$ V, $V_{dc} = 570$ V, $E = 220$ V, $f = 50$ Hz. (From Neacsu, D.O. *IEEE International Symposium on Industrial Electronics*, Bled, Slovenia, July 1996, 12–14, pp. 527–553. IEEE Paper 0-7803-5662-4/99. © (1996) Printed with permission from IEEE.)

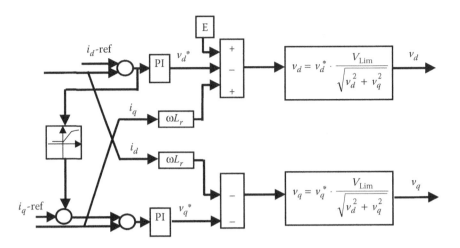

FIGURE 13.30 Schematic of the control system using cross-coupling terms.

the optimal control theory. However, the applicability of this method is jeopardized by the large computational burden. The minimum time current controller is very similar to the following solution that will be discussed in more detail.

13.3.6 CROSS-COUPLING TERMS

The cross-coupling term $(\omega \cdot L_r \cdot i_q)$ [34] of Equation 11.19 is used to increase the voltage capability by increasing the i_q-axis current during a step in the active current i_d. In this case, the i_q current is used to maximize the applied voltage on the inductance L_r. Pushing a high i_q current while respecting $i_q^2 + i_d^2 \leq i_{d_{max}}$ can increase the term $(\omega \cdot L_r \cdot i_q)$ and improve the voltage capability on the active current (i_d) axis. The unity power factor is given up during transients.

The increase in voltage can be expressed as:

$$\frac{\Delta v}{V_{\lim}} = \frac{\omega \cdot L_r \cdot i_q}{0.61 \cdot V_{dc}} \tag{13.26}$$

Results for step-up and step-down transients of the active current while applying the method shown in Figure 13.30 are shown in Figure 13.31.

Changing the i_q current reference or working with current on the q axis requires voltage to be applied on this axis (v_q). As the maximum available voltage is at any moment limited by the maximum radius of the space vector trajectory in the complex plane, a voltage applied on the q-axis may reduce the maximum voltage available on the d-axis, and this jeopardizes somewhat the merits of this method. For this reason, an alternative is next analyzed by applying all voltage on the d-axis and suspending temporarily the effect of the PI control at large current errors.

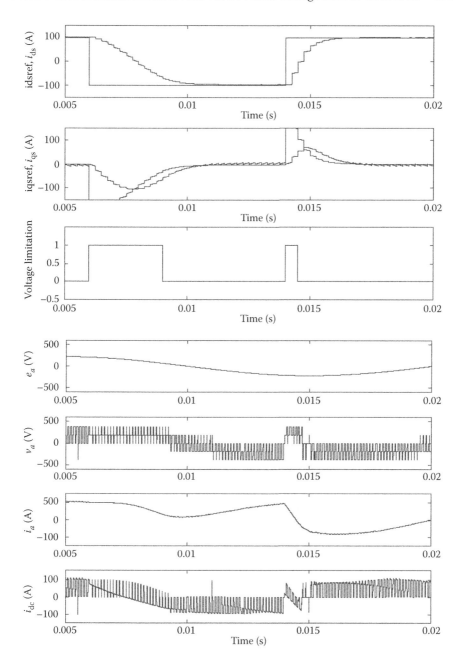

FIGURE 13.31 Current controller with d–q axis cross-coupling for the case $L_r = 2.5$ mH, $R_r = 0.1$ V, $V_{dc} = 570$ V, $E = 220$ V, $f = 50$ Hz. (From Neacsu, D.O. *IEEE International Symposium on Industrial Electronics*, Bled, Slovenia, July 1996, 12–14, pp. 527–553. IEEE Paper 0-7803-5662-4/99. © (1996) Printed with permission from IEEE.)

13.3.7 Application of the Whole Available Voltage on the d-Axis

The limit of the maximum available voltage is generally expressed as

$$v_d^2 + v_q^2 = V_{\lim}^2 = (0.61 \cdot U_{dc})^2 \tag{13.27}$$

When applying all the voltage on the d-axis, $v_d = V_{\lim}$, $v_q = 0$ and the q-axis current is negative from Equation 13.19, will further help increasing the d-axis voltage.

Re-writing Equation 13.19 for the case presented in Figure 13.32 and neglecting the voltage drop across the resistance in the boost circuitry yields:

$$E - V_{\lim} = L_r \cdot \frac{di_d}{dt} - \omega \cdot L_r \cdot i_q$$
$$0 = L_r \cdot \frac{di_q}{dt} + \omega \cdot L_r \cdot i_d \tag{13.28}$$

Solving this system yields the solutions:

$$i_d(t) = I_d \cdot \cos\omega t - \frac{V_{\lim} - E}{\omega \cdot L_r} \cdot \sin\omega t$$
$$i_q(t) = -I_d \cdot \sin\omega t + \frac{V_{\lim} - E}{\omega \cdot L_r} \cdot (1 - \cos\omega t) \tag{13.29}$$

The response time for a step-down transient from I_d to $2I_d$ can be calculated by imposing the condition $i_d(\Delta T) = -I_d$. Figure 11.27 presents a theoretical response without taking into consideration the sampling effect.

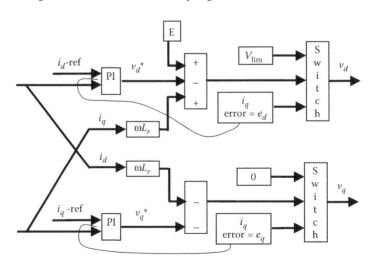

FIGURE 13.32 Control schematic for a system with application of the whole available voltage on d-axis during transients.

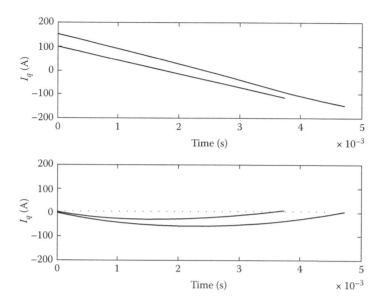

FIGURE 13.33   Theoretical response at i_d current step-down (100 A to −100 A and 150 A to −150 A) for $L_r = 2.5$ mH, $R_r = 0$, $V_{dc} = 570$ V, $V_{lim} = 0.61 * V_{dc}$, $E = 220$ V. (From Neacsu, D.O. *IEEE International Symposium on Industrial Electronics*, Bled, Slovenia, July 1996, 12–14, pp. 527–553. IEEE Paper 0-7803-5662-4/99. © (1996) Printed with permission from IEEE.)

It is very interesting to note that a step-down current modification from I_d to $2I_d$ brings i_q back to zero, and this will not jeopardize the behavior of the i_q current controller when this is again engaged (Figure 13.33) [27].

Figure 13.34 shows results for the application of the entire voltage on d-axis during a transient. A comparison can be made with Figures 13.29 and 13.31. At larger values of the i_d current, applying the whole available voltage on the active current axis leads to better results. As discussed before, the solution involving the cross-coupling terms requires some voltage on the q-axis to form the current i_q and the remaining voltage on the d-axis is reduced. Furthermore, asking for step modifications in both axes can produce instabilities during and following a transient.

13.3.8   SWITCH TABLE AND HYSTERESIS CONTROL

Fast current transients can be achieved with hysteresis control instead of PI control. Accounting for the symmetries within the three-phase systems, hysteresis control is considered for the two orthogonal (x, y) components of the grid current.

A novel hysteresis method in the synchronous reference (d, q) is introduced in Figure 13.35 and Table 13.2. The major advantage of this method, in comparison with the conventional stationary reference for the hysteresis control, consists in its simplified implementation in vector-control systems. Current double-level hysteresis control in (d, q) axes produces logic signals (y_d, y_q) that can uniquely be converted in (v_d, v_q) components of the voltage at the converter input. The sampling frequency can also be limited. The switch table is employed to select the switching vector closest to

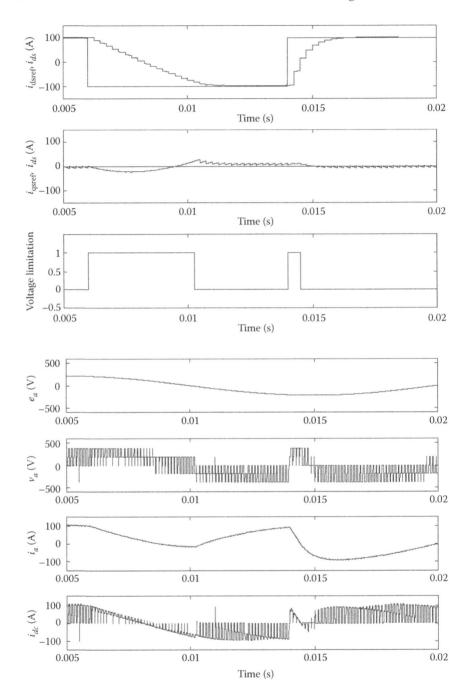

FIGURE 13.34 Whole voltage on d-axis for the case $L_r = 2.5$ mH, $R_r = 0.1$ V, $V_{dc} = 570$ V, $E = 220$ V, $f = 50$ Hz. (From Neacsu, D.O. *IEEE International Symposium on Industrial Electronics*, Bled, Slovenia, July 1996, 12–14, pp. 527–553. IEEE Paper 0-7803-5662-4/99. © (1996) Printed with permission from IEEE.)

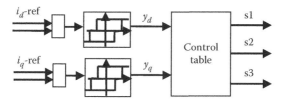

FIGURE 13.35   Schematics of the hysteresis control in synchronous frame.

the coordinates (v_d, v_q) resulting from this algorithm (Figures 13.36 and 13.37). As an alternative, this method can be applied with or without employing the cross-coupling terms. In a sense, this method is the equivalent of direct torque control used for the AC machine drives.

13.3.9   Phase Current Tracking Methods

13.3.9.1   P-I-S controller

Single- and three-phase grid connected power converters can also be controlled in a stationary reference frame directly employing phase or sinusoidal currents. The current control system reduces to a PI controller with a sinusoidal reference (Figure 13.38). Unfortunately, this type of system has been proven to introduce a nonzero steady-state error. To mitigate this issue, a solution based on an internal model principle [35] is discussed.

Let us consider a sinusoidal reference

$$i_R = I_R \cdot \cos(\omega t) \tag{13.30}$$

with the *Laplace* transform

$$I_R(s) = \frac{I_R \cdot s}{s^2 + \omega_0{}^2} \tag{13.31}$$

TABLE 13.2
Switching Table (θ Represents the Angle of the Grid Voltages)

y_d	y_q	v_d	v_q	Choose Closest Vector to the Angle =
−1	−1	+	+	$\theta + \pi/4$
−1	0	+	0	θ
−1	1	+	−	$\theta - \pi/4$
0	−1	0	+	$\theta + \pi/2$
0	0	0	0	Zero vector
0	1	0	−	$\theta - \pi/2$
1	−1	−	+	$\theta + 3\pi/4$
1	0	−	0	$\theta - \pi$

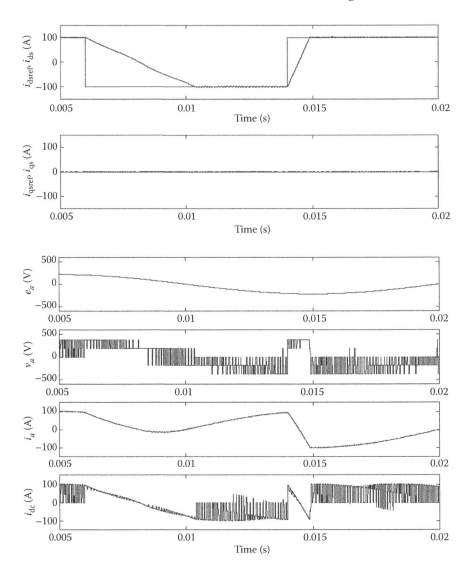

FIGURE 13.36 Switch table and hysteresis control for the case $L_r = 2.5$ mH, $R_r = 0.1$ V, $V_{dc} = 570$ V, $E = 220$ V, $f = 50$ Hz. (From Neacsu, D.O. *IEEE International Symposium on Industrial Electronics*, Bled, Slovenia, July 1996, 12–14, pp. 527–553. IEEE Paper 0-7803-5662-4/99. © (1996) Printed with permission from IEEE.)

The *Laplace* transform for the error signal can be expressed as:

$$E(s) = \frac{I_R(s)}{1 + G_0(s)} = \frac{I_R \cdot s}{s^2 + \omega_0^2} \cdot \frac{1}{1 + G_0(s)} \tag{13.32}$$

where $G_0(s)$ represents the open-loop transfer function of the control system.

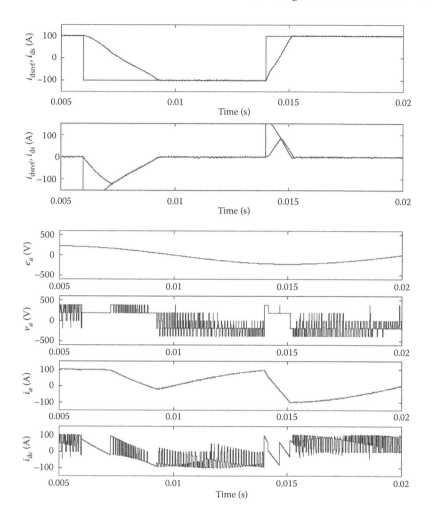

FIGURE 13.37   Switch table and hysteresis control using cross-coupling terms for the case $L_r = 2.5$ mH, $R_r = 0.1$ V, $V_{dc} = 570$ V, $E = 220$ V, $f = 50$ Hz. (From Neacsu, D.O. *IEEE International Symposium on Industrial Electronics*, Bled, Slovenia, July 1996, 12–14, pp. 527–553. IEEE Paper 0-7803-5662-4/99. © (1996) Printed with permission from IEEE.)

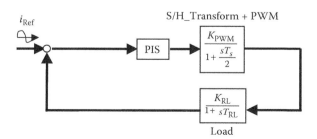

FIGURE 13.38   Schematics of the PI control system.

We would like to find the optimal form of the open-loop transfer function able to cancel the effect of the two imaginary poles $(+/-j\omega_0)$ introduced by the sinusoidal reference term. In this respect, the open-loop transfer function is written as a ratio of polynomials with real number coefficients.

$$G_0(s) = \frac{N(s)}{D(s)} \Rightarrow \frac{1}{1 + G_0(s)} = \frac{D(s)}{D(s) + N(s)} \tag{13.33}$$

$D(s) = 0$ should have the same solutions $(+/-j\omega_0)$ in order to cancel the effect of the sinusoidal reference. Moreover $(+/-j\omega_0)$ should not be at the same time solutions of $N(s) = 0$. This guarantees the elimination of the complex number solutions and the reduction of the steady-state error to zero.

The implementation of this control method is shown in Figure 13.39. The conventional PI regulator is enhanced by a term corresponding to the two imaginary solutions $(+/-j\omega_0)$.

The equivalent transfer function of the regulator becomes:

$$G_{\text{PIS}}(s) = K_p + \frac{K_i}{s} + \frac{K_s \cdot s}{s^2 + \omega_0^2} = \frac{K_p \cdot s \cdot \left[s^2 + \omega_0^2\right] + K_i \cdot \left[s^2 + \omega_0^2\right] + K_s \cdot s^2}{s \cdot \left[s^2 + \omega_0^2\right]} \tag{13.34}$$

The open-loop transfer function is calculated as:

$$G_0(s) = G_{\text{PIS}}(s) \cdot G_{\text{INV}}(s) \cdot G_{\text{LOAD}}(s) \tag{13.35}$$

$$G_0(s) = \frac{F_1(s)}{\left[s^2 + \omega_0^2\right] \cdot F_2(s)} \tag{13.36}$$

and

$$E(s) = \frac{I_R \cdot s}{s^2 + \omega_0^2} \cdot \frac{1}{1 + G_0(s)} = \frac{I_R \cdot s}{\left[s^2 + \omega_0^2\right] \cdot F_2(s) + F_1(s)} \tag{13.37}$$

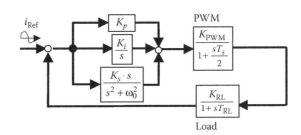

FIGURE 13.39 Block diagram of the compensated controller.

$E(s)$ has no imaginary poles introduced by the sinusoidal reference. It can be observed that $(+/-j\omega_0)$ are imaginary poles of $G_0(s)$ as proposed by the new control algorithm.

13.3.9.2   Feed-Forward Controller

The control method previously named as *Proportional-Integrative-Sinusoidal* is currently very much used in controlling currents of both grid interactive and active filtering applications. Improved forms are proposed in [36,37], as well as a $H\infty$ application. We can also mention a variation of these controllers used for higher harmonic cancelation [38]. The main drawback of this method consists in its dependency on the accurate knowledge of the grid frequency as this is part of the control law. If the grid frequency slightly changes, it yields in a steady-state error of the tracking system (Figure 13.40).

Observing that the current controller acts upon an inverter leg that is composed of two IGBT devices, one can separate the operation on the positive and negative polarities of the current with independent control of each power device. This yields into the control system shown schematically in Figure 13.41.

The control problem changes from tracking a sinusoidal waveform (harmonic, periodic) waveform, into tracking [occasionally] a varying waveform [39]. The intervals with a null reference would reset the steady-state error if any. We can thus consider the reference as approximating a piece-wise waveform, and following up a control system design similar to the case of a ramp input (also called "*velocity*" input) [40].

The reference input is equivalent to

$$i_R = \omega_0 \cdot I_R \cdot t \cdot 1(t) \tag{13.38}$$

with the *Laplace* transform

$$I_R(s) = \frac{\omega_0 \cdot I_R}{s^2} \tag{13.39}$$

In order to achieve one additional degree of freedom on the *zero* assignment within the control law, a feed-forward component is added to compensate nonideal behavior of the control loop under varying input waveform (Figure 13.42).

The controller output yields:

$$C_{\text{PIFF}}(s) = \left(K_p + \frac{K_i}{s} \right) \cdot \left(I_R(s) - I_m(s) \right) + K_n \cdot I_R(s) \tag{13.40}$$

$$I_m(s) = C_{\text{PIFF}}(s) \cdot G_{\text{INV}}(s) \cdot G_{\text{LOAD}}(s)$$

It yields:

$$I_m(s) = \frac{\left(K_p + \dfrac{K_i}{s} + K_n \right)}{\left(\dfrac{1}{G_{\text{INV}}(s) \cdot G_{\text{LOAD}}(s)} + K_p + \dfrac{K_i}{s} \right)} \cdot \left(I_R(s) \right) \tag{13.41}$$

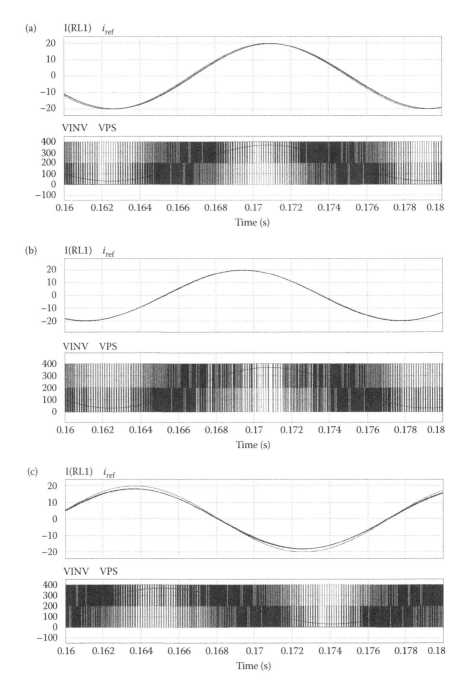

FIGURE 13.40 Results for different control methods: (a) Conventional PI controller, (b) PI + Resonant, (c) PI + Resonant out of synchronism (grid at 56 Hz instead of 60 Hz). Top waveform = Phase current reference and measurement. Bottom waveform = Inverter pole voltage and grid voltage.

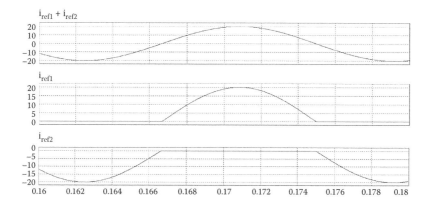

FIGURE 13.41 Principle of the control system.

And finally

$$E(s) = I_R(s) - I_m(s) = \left[\frac{1 - K_n \cdot G_{INV}(s) \cdot G_{LOAD}(s)}{1 + \left(K_p + \dfrac{K_i}{s} \right) \cdot G_{INV}(s) \cdot G_{LOAD}(s)} \right] \cdot I_R(s) \quad (13.42)$$

The *Final Value Theorem* is applicable since there is no right-half pole and the steady-state error yields:

$$e_{ss} = \lim_{t \to \infty} e(t) = \lim_{s \to 0} s \cdot E(s) =$$

$$= \lim_{s \to 0} \left\{ s \cdot \left[\frac{1 - K_n \cdot G_{INV}(s) \cdot G_{LOAD}(s)}{1 + \left(K_p + \dfrac{K_i}{s} \right) \cdot G_{INV}(s) \cdot G_{LOAD}(s)} \right] \cdot \frac{I_R}{s^2} \right\} \quad (13.43)$$

$$= \lim_{s \to 0} \left\{ \left[1 - K_n \cdot \frac{K_1}{s + K_2} \right] \cdot \left[\frac{s}{s + GK_0} \right] \cdot \frac{I_R}{s} \right\} = 0$$

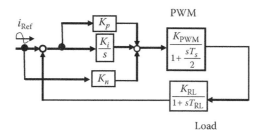

FIGURE 13.42 Feed-forward controller.

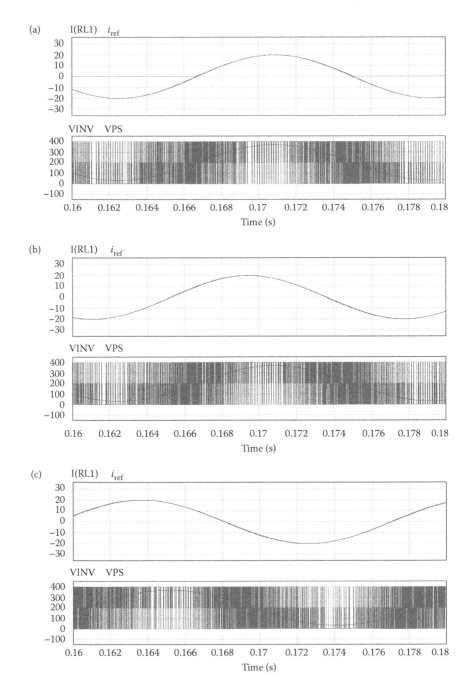

FIGURE 13.43 Results for novel concept. (a) The use of two conventional PI controllers, (b) the use of the feedforward component, (c) same as b, with grid at 56 Hz instead of 60 Hz. Top waveform = Phase current reference and measurement. Bottom waveform = Inverter pole voltage and grid voltage.

Sample results are shown in Figure 13.43. It shows the improvement from Figure 13.40 when the grid frequency varies. It is also important to note the role of proper synchronization of the operation with the grid.

More experimental results are shown in [40]. The performance of the feed-forward controller is comparable with the performance of the *proportional-integral-sinusoidal* controller (also called *resonant controller*). In certain implementations like the analog power management ICs, the system is simpler than the resonant controller. This is especially evident for compensation of currents with multiple harmonic components.

Advantages in improved power efficiency and faster transient response are achieved with the split of the control action on the two IGBT composing an inverter leg and the operation of the power switches for 180° only. This means reduction of the power loss, especially in the control and driver circuits.

13.4   GRID SYNCHRONIZATION

Grid-connected power converters are seen as a current source by the grid while the input voltages are defined by the grid. This implies a synchronization of the PWM pattern with the phase of the grid in order to control the power factor. There are different methods based on a PLL circuit that are used to achieve grid synchronization.

The simplest solution used to synchronize the control of a single-phase grid connected power converter consists in a zero-crossing detector followed by a counter circuit (Figure 13.44).

The grid voltage is sensed before any filter associated with the power converter and reduced to a lower level with a signal transformer or resistive divider. The resulting signal is compared against a threshold device, such as the base-emitter junction of a bipolar transistor or an IC comparator. This provides a logic signal corresponding to the alternating of the grid voltage. There are several sources of errors in this detection, including the noise in the line voltage due to unfiltered switching, dispersion and temperature variations of the threshold level, or delay in detecting the exact zero-crossing moment. However, these errors are generally smaller than the quantization step of the digital system.

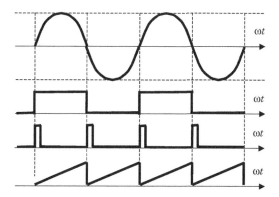

FIGURE 13.44 Principle of a simple synchronization circuit.

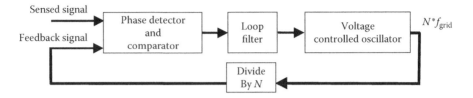

FIGURE 13.45   Principle of a PLL circuit.

The edges of the detected logic signal reset a digital counter that accounts for the phase coordinate of the PWM pattern. Improved systems include a phase-locked loop configuration able to filter unwanted small variations in the detecting process. This helps eliminating the jitter around the zero-crossing moments.

Let us consider a synchronization scheme based on a PLL circuit. A generic PLL circuit schematic is shown in Figure 13.45 and it is composed of a voltage-controlled oscillator (VCO) working at a frequency multiple of N times the detected frequency. In power converters, it makes sense to select the VCO frequency equal to the sampling frequency of the PWM pattern (Figure 13.46). The resulting train of pulses is divided by N resulting in a logic signal with a frequency comparable with the sensed voltage. A phase detector compares the phase of the sensed and feedback voltages and it increases or decreases the control voltage accordingly. This control voltage is used as a reference for the VCO that modifies its frequency based on the phase difference. The goal of this closed-loop approach is to align the sensed signal with the generated one. If the frequency or phase of the sensed signal varies for any reason, this control loop acts as a filter and the feedback signal does not jitter. The feedback signal can further be used as a reference for the generation of the PWM pattern. Usually, the same counter is used both for closing the PLL loop and for counting the angular coordinate of the PWM pattern.

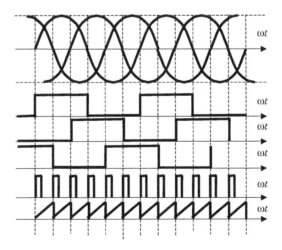

FIGURE 13.46   Principle of a simple synchronization circuit.

The major problem people face when implementing this PLL control in single phase systems relates to the limited number of PWM cycles within one grid period. The discrete nature of the PWM-generation process limits the possibility of adjusting properly the controller phase. For instance, a converter with a 9 kHz switching frequency allows for 150 steps in the phase coordinate. The decision of the feedback loop can only be to adjust the phase by ±1 PWM pulse at a time. The effect is a possible step in the filtered voltage at each zero crossing when the phase is abruptly reset to zero. An alternative consists in using asynchronous PWM generators, but they do not guarantee the best harmonic content of the grid current. However, evolved systems can afford to work with asynchronous PWM that allows a noninteger ratio of switching and grid frequencies.

The PLL function can be implemented with dedicated analog or digital ICs or a proper software routine in the microcontroller or digital signal processor (DSP) control system. The frequency of the main signal is 50 or 60 Hz, which is very low compared to the performance of modern digital controllers. Therefore, a microcontroller implementation requires a zero crossing detector with direct interface to the digital hardware, a timer/counter counting with the clock frequency of the PWM sampling frequency, and a software routine running at the same sampling frequency.

Grid synchronization for three-phase converters benefit from multiple zero crossings and the symmetry of a grid generated three-phase system. The detection module includes zero-crossing detection for all three phases, and this results in a signal with a frequency six times larger than the grid frequency. The PLL circuit can therefore provide more resolution with a much finer tuning of the detected frequency.

Again the implementation can be achieved with analog or digital PLL ICs or with a software routine in the digital controller system.

PROBLEMS

P13.1 Calculate the average and RMS values for the output voltage of rectifier systems shown in Table 13.1.

P13.2 What is the effect of a line or grid inductance in the commutation process of the converter shown in Figure 13.2.

P13.3 Draw the switching pattern of the GTO devices of Figure 13.3c to obtain the output voltage waveforms shown for both cases. How would you define the semiconductor loss for both control methods?

P13.4 Consider the two topologies shown in Figure 13.7. What happens if the intermediate DC capacitor is missing? What should be the minimum value of this capacitor to maintain the operation as desired?

P13.5 Compare the rating of the rectifier diodes in both solutions of Figure 13.8.

P13.6 Compare the input filter-capacitor rating for the two solutions shown in Figure 13.10.

P13.7 Following the example provided by Equations 13.3 through 13.13, calculate the shapes of the current in the much simpler case of a single phase converter and justify the second harmonic injection.

P13.8 Compare the ratings of the IGBT devices used in building the power converters shown in Figure 13.11 and Figure 13.16, respectively, for the same grid-and-load conditions (say, 2 kW with a 120 V/60 Hz grid).

P13.9 What are the drawbacks of neglecting the term E in the control system shown in Figure 13.22? Many people adapt their DSP control software from a motor drive application to a grid-connected converter without adding up this term. The system apparently works, but at what risk? Can you complete this response with an imaginary example?

P13.10 Determine the reasoning behind the switching sequence shown in Table 13.2 by using a vector diagram for both current and voltage.

P13.11 Usually, practical systems of single-phase tracking control do not compensate for the sinusoidal reference. The system apparently works, but at what risk? Can you complete this response with an imaginary example?

P13.12 Consider a single-phase grid-connected converter switched at 3.6 kHz, that is, a frequency ratio of 30. If the PWM generator maintains the synchronization between the generated PWM pattern and zero crossing, how big can be the steps in filtered voltage (or in the voltage reference for the PWM) at each zero crossing reset?

REFERENCES

1. Anon. 1998. Ecostar—Power converter anti-islanding capability. *Application Notes*, Ballard Corporation.
2. Anon. Ecostar—Power converter designed for distributed generation. *Application Notes*, Ballard Corporation.
3. Anon. 1998. Ecostar—Multi-unit capability for grid and standalone operation. *Application Notes*, Ballard Corporation.
4. Chiang, H.D., Wang, J.C., Tong, J., and Darling, G. 1995. Optimal capacitor placement, replacement and control in large scale unbalanced distribution systems: System modeling and a new formulation. *IEEE Trans. Power Syst.* 10(1), 363–369.
5. Anon. 2001. PV system installation and grid interconnection guidelines in selected IEA countries, Fronius.
6. Anon. 2003. IEEE standard. IEEE P1547 std draft 01 standard for distributed resources interconnected with electric power systems. June 2003.
7. Prasad, A.R., Ziogas, P.D., and Manias, S. 1989. An active power factor correction technique for three-phase diode rectifiers. *IEEE PESC Conference Record*, pp. 58–66.
8. Kolar, J.W., Ertl, H., and Zach, F.C. 1993. A comprehensive design approach for a three-phase high-frequency single-switch discontinuous-mode boost power factor corrector based on analytically derived normalized converter component ratings. *IEEE IAS Conference Record*, Part II, pp. 931–938.
9. Kolar, J.W., Ertl, H., and Zach, F.C. 1993. Space vector-based analytical analysis of the input current distortion of a three-phase discontinuous-mode boost rectifier system. *IEEE PESC Conference Record*, pp. 696–703.
10. Tou, M., Al-Haddad, K., Olivier, G., and Rajagopalan, V. 1995. Analysis and design of single controlled switch three-phase rectifier with unity power factor and sinusoidal input current. *IEEE APEC Conference Record*, vol. II, pp. 856–862.

11. Ismail, E., Olivieira, C., and Erikson, R. 1995. A low-distortion three-phase multi-resonant boost rectifier with zero current switching. *IEEE APEC Conference Record*, vol. II, pp. 849–855.
12. Weng, D. and Yuvarajan, S. 1995. Constant-switching-frequency AC–DC converter using second-harmonic-injected PWM. *IEEE APEC Conference Record*, vol. II, pp. 642–646.
13. Weng, D. and Yuvarajan, S. 1995. AC–DC converter using second-harmonic-injected PWM. *IEEE PESC Conference Record*, vol. II, pp. 1001–1006.
14. Malesani, L., Rossetto, L., Tenti, P., and Tomasin, P. 1993. AC/DC/AC PWM converter with minimum storage energy in the DC link. *IEEE APEC Conference Record*, pp. 306–313.
15. Verdelho, P. and Soares, V. 1997. A unity power factor PWM voltage rectifier based on the instantaneous active and reactive current id-iq method. *IEEE International Symposium in Industrial Electronics*, July 7–11, pp. 411–416.
16. Habetler, T.G. 1993. A space vector based rectifier regulator for AC/DC/AC converters. *IEEE Trans. Power Electron.* 8(1), pp. 30–36.
17. Blasko, V. and Kaura, V. 1997. A new mathematical model and control of a three-phase AC/DC voltage source converter. *IEEE Trans. Power Electron.* 12(1), 116–123.
18. Verdelho, P. and Soares, V. 1997. A unity power factor PWM voltage rectifier based on the instantaneous active and reactive current id–iq method. *IEEE International Symposium in Industrial Electronics*, July 7–11, pp. 411–416.
19. Trznadlowski, A.M. and Legowski, S. 1994. Minimum-loss vector PWM strategy for three-phase inverters. *IEEE Trans. PE* 9(1), 26–34.
20. Enjeti, P. and Xie, B. 1990. A new real time space vector PWM strategy for high performance converters. *IEEE IAS*.
21. Chung, D.W. and Sul, S.L. 1997. Minimum-loss PWM strategy for a 3-phase PWM rectifier. *IEEE PESC*, 2, 1020–1026.
22. Habetler, T.G. 1993. A space vector based rectifier regulator for AC/DC/AC converters. *IEEE Trans. Power Electron.* 8, 30–36.
23. Kazmierkowski, M., Dzieniakowski, M.A., and Sulkowski, W. 1991. Novel space vector based current controller for PWM inverters. *IEEE Trans. Power Electron.* 6, 158–166.
24. Henze, C.P. and Mohan, N. A digitally controlled AC/DC power conditioner draws sinusoidal input current. *IEEE PESC*, pp. 531–540.
25. Malesani, L., Rossetto, L., Tenti, P., and Tomasin, P. 1993. AC/DC/AC PWM converter with minimum storage energy in the DC link. *IEEE APEC Conference Record*, pp. 306–311.
26. Neacsu, D.O., Yao, Z., and Rajagopalan, V. 1996. Optimal PWM control of a single-switch three-phase AC–DC boost converter. *IEEE PESC*, pp. 521–526.
27. Neacsu, D.O. 1999. Vectorial current control techniques for three-phase AC/DC boost converters. *IEEE ISIE*, pp. 527–532.
28. Kanaan, H.Y. and Al-Haddad, K. 2012. Three-phase current injection rectifiers. *IEEE Industrial Electron. Magazine* 6(3), 24–40. May/June 2012.
29. Naik, R., Rastogi, M., and Mohan, N. 1992. Third-harmonic modulated power electronics interface with three-phase utility to provide a regulated DC output and to minimize line current harmonics. *Proceedings of the IEEE IAS Annual Meeting Conference*, pp. 689–694.
30. Mohan, N., Rastogi, M., and Naik, R. 1993. Analysis of a new power electronics interface with three-phase utility currents and a regulated DC output. *IEEE trans. Power Delivery* 8(2), 540–546.
31. Pejovic, P. and janda, Z. 1999. An analysis of three-phase low harmonic rectifiers applying the third harmonic current injection. *IEEE Trans. Power Electron.* 14(3), 397–407.

32. Vasquez, N., Rodriguez, H., Hernandez, C., Rodriguez, E. and Arau, J. 2009. Three-phase rectifier with active current injection and high efficiency. *IEEE Trans. Industrial Electron.* 56(1), 110–119.
33. Choi, J.W. and Sul, S.K. 1998. Fast current controller in three-phase AC/DC boost converter using d–q axis crosscoupling. *IEEE Trans. Power Electron.* 13(1), 179–185.
34. Choi, J.W. and Sul, S.K. 1997. New current control concept—Minimum time current control in 3-phase PWM converter. *IEEE Trans. Power Electron.* 12(1), 124–131.
35. Fukuda, S. and Yoda, T. 2001. A novel current-tracking method for active filters based on a sinusoidal internal model. *IEEE Trans. Industry Appl.* 37, 888–895.
36. Timbus, A., Liserre, M., Teodorescu, R., Rodriguez, P., and Blaabjerg, F. 2009. Evaluation of current controllers for distributed power generation systems. *IEEE Trans. Power Electron.* 24(3), 654–664.
37. Lisserre, M., Teodorescu, R., and Blaabjerg, F. 2006. Multiple harmonics control for three-phase grid converter systems with the use of PI-RES current controller in a rotating fame. *IEEE Trans. Power Electron.* 21(3), 836–841.
38. Bojoi, R.I., Griva, G., Bostan, V., Guerrero, M., Farina, F., and Profumo, F. Current control strategy for power conditioners using sinusoidal signal integrators in synchronous reference frame. *IEEE Trans. Power Electron.* 20(6), 1402–1412.
39. Franklin, G., Powell, D., Emami-Naeimi, A. 2002. *Feedback Control of Dynamic Systems*. Prentice Hall, Upper Saddle River, NJ.
40. Neacsu, D.O. 2010. Analytical investigation of a novel solution to AC waveform tracking control. *IEEE International Symposium in Industrial Electronics*, Bari, Italy, July, pp. 2684–2689.
41. Neacsu, D.O, Yao, Z., and Rajagopalan, V. 1996. Optimal PWM control of a single-switch three-phase AC-DC boost converter, *IEEE PESC Conference 1996*, Baveno, Italy, June 24–27, pp. 727–732.

14 Parallel and Interleaved Power Converters

14.1 COMPARISON BETWEEN CONVERTERS BUILT OF HIGH-POWER DEVICES AND SOLUTIONS BASED ON MULTIPLE PARALLEL LOWER-POWER DEVICES

Chapter 1 showed the current interest in parallel or interleaved converters used for grid interfaces or motor drives. This interest has risen due to the continuous development in power semiconductor technologies leading to high-power single-die insulated gate bipolar transistors (IGBTs).

The first step toward parallel converters was hybrid (custom) IGBTs. These modules parallel IGBTs at silicon level and have many advantages. First, these chips and gate resistors are placed very close together, reducing parasitic inductance between chips. This ensures uniform temperature distribution across the chips inside the package.

A layer of thick copper directly bonded on a ceramic substrate forms the conduction path for the emitter current and the thermal spreading layer for the heat from the IGBT chips. The IGBT chips are mounted via solder on this copper layer. The ceramic substrate can be alumina, aluminum nitrite, or beryllium oxide, materials with good thermal conductivity and voltage isolation up to 6000 V. Both thermal conductivity and voltage isolation are better than the thermal pads used for isolation in parallel discrete devices.

During the design of a new power stage, a practicing engineer has to decide whether to use hybrid (custom) IGBT modules or equivalent discrete devices. Here are some points to consider:

- Hybrid (custom) IGBT modules are advantageous when more than five chips are considered. Their advantages lie in size, electrical isolation, thermal management, cost of the whole system, and the lack of noise owing to switching. The last advantage saves the need for extra bus capacitance.
- Discrete devices are advantageous when two or three chips in parallel are required to develop a design. They ensure that the smallest footprint with both electrical isolation and parasitics is manageable.

The next possibility consists of paralleling IGBTs for high-current applications, as shown in Figure 14.1. There is an obvious engineering question whether it is better to realize a high-current switch by paralleling discrete devices or by a single higher rated device.

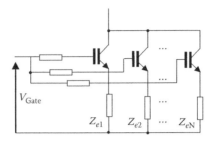

FIGURE 14.1 Parallel IGBTs.

Which solution is better to carry out a 1200 A, 1200 V switch: A high-power 1200 A IGBT or four low-power, parallel-connected 300 A IGBTs? To respond, we use a numerical example and datasheet comparison, using datasheet information from *Powerex* IGBTs. Further, ripple differences and different snubber requirements for both devices will be ignored.

Thermal features:

- IGBT/Diode thermal resistances of 0.13/0.18 [C/W] (300 A) versus 0.022/0.050 [C/W] (1200 A).
- Heatsink touch areas of 163.67 sq. in (300 A) versus 133.01 sq. in (1200 A).

Diode recovery loss:

- 300 A device: $T_{rr} = 250$ ns, $Q_{rr} = 17.6$ μF
- 1200 A device: $T_{rr} = 300$ ns, $Q_{rr} = 9.0$ μF

Switching loss and conduction loss: From the datasheet information regarding switching and conduction losses, one can conclude that paralleling power semiconductor switches represents a competitive solution for building three phase power converters (Figures 14.2 and 14.3).

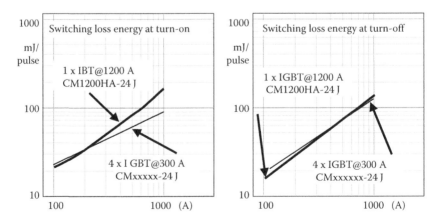

FIGURE 14.2 Switching loss at turn-on and turn-off.

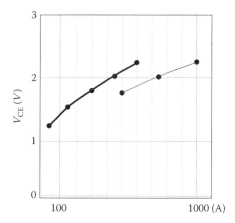

FIGURE 14.3 Conduction loss for both solutions: (left) 1200 A device, (right) 4 × 300 A device.

14.2 HARDWARE CONSTRAINTS IN PARALLELING IGBTs

Having demonstrated the suitability of parallel connection of IGBTs, ask for understanding the limits of the practical implementation. The idea of paralleling IGBT devices is based on using low-power devices that share currents equally. When the sharing is not equal, one of the currents may go beyond the rated value for that particular IGBT. Special care should be taken to avoid current mis-sharing so that the IGBT can be protected.

Differences in the current level can occur in steady state or in dynamic operation. The steady-state current imbalance is produced by different $V_{ce}(I_c)$ characteristics as well as by the differences in the circuit parasitic resistances. Differences in the turn-on and turn-off imbalances are produced by the distribution of the module transconductance characteristics as well as by the differences in parasitic inductances. The operation is influenced by temperature imbalance and thermal instability. Each of these sources of instability is discussed next.

Two IGBTs cannot have identical characteristics due to technological dispersion. The ON-time V_{ce} (I_c) characteristics are shown in Figure 14.4 and they can be approximated for each of the two parallel IGBTs by:

$$V_{ce} = V_{01} + r_1 I_{C1}$$

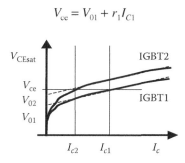

FIGURE 14.4 Sketch of the $V_{ce}(I_c)$ characteristics for two parallel IGBTs.

$$V_{ce} = V_{02} + r_2 I_{C2} \tag{14.1}$$

Reporting each individual current to the total current of the parallel IGBTs yields

$$I_{C1} = \frac{V_{02} - V_{01} + r_2 I_{Ctotal}}{r_1 + r_2} \tag{14.2}$$

$$I_{C2} = \frac{V_{01} - V_{02} + r_1 I_{Ctotal}}{r_1 + r_2}$$

Observing these two equations shows that the device with the lower V_{ce} carries most of the current. Moreover, the total current is lower than $N_p * I_r$ due to the characteristic dispersion of a number of N_p IGBT devices. Higher temperature allows better current sharing. For instance, imbalance in static characteristic represents 15% at 25°C and 5% at 125°C.

To measure and match sets of IGBTs with close V_{ce} (I_c) characteristics, semiconductor manufacturers like Powerex came up with a special marking for different levels of the saturation voltages.

Example: ... C: 1.70–1.95 V D: 1.90–2.15 V

After the design has been matched to IGBTs within the same ratings, a derating coefficient should be used to select design and device. The value of this coefficient depends on the semiconductor family as shown in Table 14.1.

A static derating factor can be defined as

$$\delta = 1 - \frac{I_{Total}}{N_p I_M} \tag{14.3}$$

and it can be estimated with an empirical relationship:

$$\delta = 1 - \frac{((N_p - 1)(1 - x)/(1 + x)) + 1}{N_p} \tag{14.4}$$

where N_p is the number of parallel devices and x is the mis-sharing factor, which is equal to 0.10 for 250 and 600 V devices, 0.15 for 1200 V devices, and 0.20 for 1700 V devices (Figure 14.5).

TABLE 14.1

Derating of Different IGBT Families

250 V Trench Gate	10%
600 V	10%
1200 V and 1400 V	15%
1700 V	20%

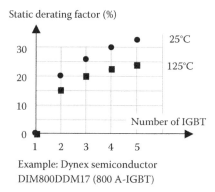

Example: Dynex semiconductor
DIM800DDM17 (800 A-IGBT)

FIGURE 14.5 Dependence of the static derating factor on the number of IGBTs.

Parasitic resistances within the circuit are the second factor that can influence current sharing during IGBT's ON-time. Mounting, busbar, and terminal connection of IGBT devices introduce parasitic resistances in both emitter and collector. The emitter resistance has a greater effect on current sharing as it influences the gate circuitry too. Depending on the emitter resistance, there will be a voltage drop on the emitter that decreases the actual gate control voltage ($V_{GE} = V_{control} - V_{RE}$). This changes the IGBT output characteristics and modifies the appropriate collector current. Different IGBT characteristics define different currents for the same collector–emitter voltage. To reduce this imbalance, it is necessary to make the wiring on the emitter as short and uniform as possible. This reduces parasitic resistances.

It is also important to analyze current sharing during turn-on or turn-off transients. The output IGBT characteristic (V_{ce}-I_c) does not have any direct influence on the transient turn-off voltage imbalance. In contrast, the transconductance characteristic (Figure 14.6) determines the current sharing during transient.

If the same gate conditions are ensured for both IGBTs, the device with the largest value of transconductance (steeper characteristic) carries a larger current and incurs the highest switching loss. A *dynamic derating factor* analogous to a *static derating factor* can be defined, and the comparison is shown in Figure 14.7.

Dynamic current sharing is better than static current sharing. However, dynamic sharing is more sensitive to external circuit factors, such as the stray inductance in

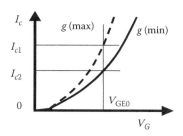

FIGURE 14.6 Transconductance characteristics.

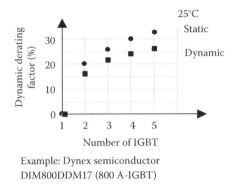

FIGURE 14.7 Dynamic derating factor in comparison with the static derating factor.

the gate-emitter circuit loop. The transient current passes through parasitic inductances in the emitter and decreases the effective gate voltage. Different values of the emitter inductance lead to different voltage drops in the gate circuitry, producing different lengths of delay of the transition and different shapes of the current. To reduce this imbalance, it is necessary to make the wiring on the emitter as short and uniform as possible (Figure 14.8).

Another improvement is related to the way the gate driver should be connected. It is better to use the same gate driver with separate gate resistors to eliminate the risk of parasitic oscillations. Further, the gate resistors should be tied closely through the gate control terminal with a separate emitter pin for gate control.

Finally, there is the effect of temperature imbalance. All IGBT characteristics depend on temperature and the current sharing between two or more parallel IGBTs is influenced by their individual temperature at the junction. Examples of temperature dependence of each characteristic can be seen in any IGBT datasheet.

Temperature has some influence on current sharing in steady-state conditions. An immediate solution is to use a common heatsink for all parallel IGBTs to keep

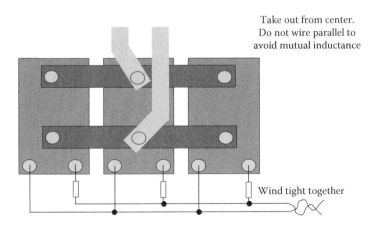

FIGURE 14.8 Emitter connection for reduced parasitics.

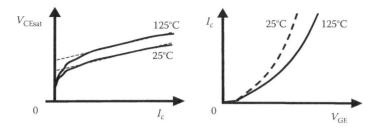

FIGURE 14.9 IGBT characteristics.

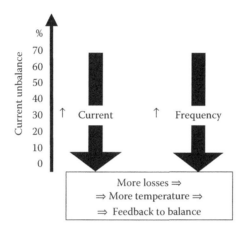

FIGURE 14.10 Current unbalance dependence on operation current and frequency.

temperature spread within 10°C. The common heatsink introduces a thermal feedback between junctions. The IGBT with a higher power dissipation has a higher temperature and influences the temperature of the other through the common heatsink. The operation point of the second IGBT changes and current sharing is modified. Even if the temperature coefficient is negative for all devices, the IGBT with the lower voltage drop has a lower temperature coefficient. Lower voltage drop also implies larger current share. Both current and temperature are increasing and this is producing the decrease of the voltage drop and current compensation (Figure 14.9).

Figure 14.10 presents this idea schematically.

14.3 GATE CONTROL DESIGNS FOR EQUAL CURRENT SHARING

Modern solutions using active gate control for static or dynamic current balancing are shown in Figure 14.11. They require high bandwidth analog processing of the sense current feedback. The levels of the gate voltage and current are controlled during the ON-time of each individual IGBT, whereas the delay of the transient is controlled in a dynamic manner.

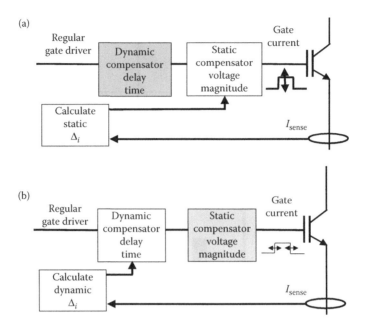

FIGURE 14.11 Gate driver control for current sharing. (a) Static current balancing. (b) Dynamic current balancing by delaying turn-on and turn-off moments.

14.4 ADVANTAGES AND DISADVANTAGES OF PARALLELING INVERTER LEGS WITH RESPECT TO USING PARALLEL DEVICES

The manufacturing volume in high-power applications is not very large and many projects end up with a unique prototype or a very short product series. As high-power devices require a special or individual design, many engineers are tempted by parallel low-power hardware. Paralleling IGBTs or MOSFETs has already been discussed earlier in this chapter, but this is not always a convenient option as the power stage needs a new design and package anyway. However, paralleling power stages is a very easy and cheap approach. The effort lies in adapting the existing control system to paralleling [3]. This is mainly done through software and the physical effort is thus limited.

Whilst there are many engineering teams which have tried parallel operation in their R&D laboratories, only a few manufacturers tried to have a systematical analysis of this new solution and to review its use in larger scale volume products. The idea is to use power converter modules from an existing series production and to add especially manufactured hardware that can support parallel connection of these modules, while using revised or additional software.

There are two possible paralleling approaches: through galvanic isolation and by direct connection on the DC side. All solutions with isolation use separate power supplies for each power converter and transformers for adding up the paralleling effects. Such a system becomes expensive and bulky, with a decreased efficiency and seriously limiting the number of parallel devices.

The second approach is based on direct connection on the DC side with inter-phase reactors on the load side. The control system must ensure equal current sharing [4,5,11]. The instantaneous output voltages are not equal even if their average value is. This implies different voltage levels on each side of the inter-phase reactors. This voltage drop produces circulation currents between modules that increase the required rating for the power semiconductor devices.

14.4.1 INTER-PHASE REACTORS

What are the major challenges to direct connection of power converters? Design of inter-phase reactors is limited by constraints of steady-state or transient operation. Steady-state ripples require a large inductance to limit them and reduce the IGBT conduction losses and filter losses. A smaller value of the inductance is preferred to provide a large current slew rate. The selected inductance should support the whole DC bus voltage over the sampling interval while producing a maximum ripple variation of less than a specified amount. Moreover, this inductance should limit the circulation currents on fundamental frequency owing to difference in the reference waveform of the paralleled/interleaved power converters [6,9,10,12]. Finally, a some-what smaller value is usually selected from these constraints to favor dynamic performance and to allow a reasonable circulating current.

There are different possible connections between power stages. The way the inter-phase reactor becomes part of the circuit is shown in Figure 14.12.

Separate inductors or mid-point inductors are often used. New solutions employ coupled inductors with direct or inverse coupling. Analysis of directly coupled or inversely coupled inductors reveals the value of the equivalent inductance at its terminals.

$Leq < L$ for direct coupling $\Rightarrow$ Increases steady $-$ state ripple

$Leq > L$ if $(M/L) < (d1/d2)$ for inverse coupling $\Rightarrow$ Decreases steady $-$ state ripple

$$(14.5)$$

Another alternative is to spread the reactor inductance between the AC and DC sides to improve the harmonic filtering on the AC side. This idea is adapted from the conventional diode rectifiers where power structures above 15 kW have inductors on both AC and DC sides.

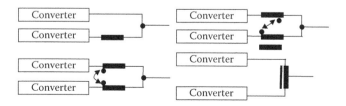

FIGURE 14.12 Different connections of the inter-phase reactor.

14.4.2 CONTROL SYSTEM

The major tasks of the control system are [4,5,11]:

- To equalize current sharing
- To limit circulating currents

As always, design of the control system is based on the load character. An inductive load (Figure 14.13a) suggests an open-loop control based on the switching-pattern selection, as the voltage on the load-side node is not well defined and it forms a *section of inductances*. In contrast, capacitive or strong voltage source character of the load (Figure 14.13b) implies a firm closed-loop control as the voltage across the load is well defined.

14.4.3 CONVERTER CONTROL SOLUTIONS

The selection of the sequence within the pulse width modulation (PWM) of all parallel converters is important. First, a single PWM circuit can be built and the same switching sequence can be used to control all parallel converters. The drawback of this solution is the difficulty of defining separate protection circuits for each power stage. Moreover, any difference in the inter-phase inductors can create zero-sequence circulating currents.

A second solution uses different control hardware on each power converter leading to almost identical PWM patterns while retaining the individual protection circuitry of each power converter. The drawback consists of not having too much improvement in the ripple during operation.

The third solution implies the control of the parallel structure by considering all possible switching states. Such an approach is limited to a small number of parallel converters to retain the number of states to a minimum. Let us herein consider the case of only two power converters. Different switching patterns in the two converters produce different circulation currents through inductors. Each difference in the PWM pattern results in a voltage drop across the inter-phase inductances and a change in current owing to the inductive nature of the load. In contrast, identical patterns in both converters maintain the same current error. These circulation currents can be produced by the difference in the reference waveforms due to the control loops, difference in the active vectors used at a given moment, or differences between the zero states used by each individual converter. Importantly, the effect of the difference in the active vectors used at a given moment can be corrected by conventional (d, q) control.

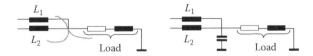

FIGURE 14.13 Different load types.

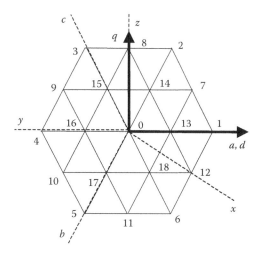

FIGURE 14.14 Switching states.

The voltage vectors derived from all switching states can be seen in Figure 14.14. Generation of any PWM uses a combination of these vectors very similar to the Space Vector Modulation (SVM) presented with the conventional three-phase six-switch converter. Each state of the system exposes different inverter legs to circulation currents. The rate of change of these circulation currents leads to changes in the zero sequence current. The rate of change of the zero sequence current is presented in Table 14.2. Table 14.3 shows changes in the cross-currents.

Once this converter analysis has been accomplished, goal is to suppress current harmonics while providing small $d\Delta i/dt$ variation. The switching pattern is selected to minimize both cross-currents and zero-sequence currents with the data from previous tables. First, the cross-current vector is defined in one of the seven regions formed as complex plane sectors and a small region near zero while the zero sequence current is compared with a small window comparator to detect values from Table 14.2. The currents should remain within some tolerance margins throughout the operation (Table 14.4).

TABLE 14.2
Switching Table

	000	001	010	011	100	101	110	111
000	0	17	15	16	13	18	14	0
001	12	5	16	10	18	11	0	17
010	15	16	3	9	14	0	8	15
011	16	10	9	4	0	17	15	16
100	13	18	8	15	1	12	7	13
101	18	11	0	17	12	6	13	18
110	14	0	8	15	7	13	2	14
111	0	17	15	16	13	18	14	0

TABLE 14.3
Zero Sequence Currents Function of the Switching Table

	000	001	010	011	100	101	110	111
000	0	−1	−1	−2	−1	−2	−2	−3
001	1	−1	0	−1	0	0	−1	−2
010	1	0	0	−1	0	−1	−1	−2
011	2	1	1	0	1	0	0	−1
100	1	0	0	−1	0	−1	−1	−2
101	2	1	1	0	1	0	0	−1
110	2	1	1	0	1	0	0	−1
111	3	2	2	1	2	1	1	0

14.4.4 CURRENT CONTROL

Each inductor current on each power converter can be controlled individually or on the three-phase concept of $(d, q, 0)$ control (Figure 14.15). Generally, a system of N identical subsystems can be transformed into a set of state equations described by a vector that is the average of the N subsystems. This creates a common-mode system that describes the dynamics of the average vector and it is used to define the control of the output voltage. Deviations from the average vector of all subsystems form another set of $N − 1$ state vectors adding up to zero. The system dynamics are therefore described by the deviation from the average vector through this set of $N − 1$ state vectors.

The proportional-integral (PI) controller can have any internal structure as discussed in the relevant chapter. The gains and limits of each current controller can be defined with the equivalent circuits on (d, q) coordinates. Understanding these gains and limits is very important as the paralleling modifies the structure of the plant model and the closed-loop transfer function. Depending on the direction of the power transfer, these equivalent circuits correspond to buck or boost DC/DC converters.

TABLE 14.4
Cross-Currents $i_1 − i_2$

	000	001	010	011	100	101	110	111
000	0	−ic	−ib	+ia	−ia	+ib	+ic	0
001	+ic	0	+$\sqrt{3}ix$	−ib	−$\sqrt{3}ix$	−ia	−$2c$	+ic
010	+ib	−$\sqrt{3}iz$	0	−ic	+$\sqrt{3}iy$	+$2ib$	−ia	+ib
011	−ia	+ib	+ic	0	−$2ia$	+$\sqrt{3}iy$	+$\sqrt{3}ix$	−ia
100	+ia	+$\sqrt{3}ix$	+$\sqrt{3}iy$	+$2ia$	0	−ic	−ib	+ia
101	−ib	+ia	−$2ib$	+$\sqrt{3}iy$	+ic	0	+$\sqrt{3}iz$	−ib
110	−ic	+$\sqrt{3}ix$	+ia	+$\sqrt{3}ix$	+ib	−$\sqrt{3}iz$	0	−ic
111	0	−ic	−ib	+ia	−ia	+ib	+ic	0

"+" = Increase on axis; "−" = Decrease on axis.

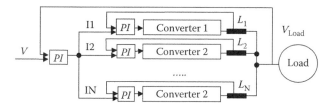

FIGURE 14.15 Structure of the control system.

14.4.5 SMALL-SIGNAL MODELING FOR (D, Q) CONTROL IN A PARALLEL CONVERTER SYSTEM

Separate equivalent circuits in quasi-DC variables are considered on the (d, q)-axes (Figure 14.16). They help in deriving a small-signal model with a control variable D *(duty cycle)* on each axis (Figure 14.17). D_d and D_q are duty cycles on (d, q)-axes and m is the converter modulation index so that $D_d^2 + D_q^2 = m^2$. The small-signal model is derived from the state variable equations that have all inductor currents and the capacitive output voltage as state variables (all measured in the power transfer direction).

$$x = \begin{bmatrix} i_L \\ v_C \end{bmatrix} \quad u = \begin{bmatrix} v_{IN} \\ D \end{bmatrix} \quad y = \begin{bmatrix} v_{OUT} \\ i_L \end{bmatrix} \tag{14.6}$$

Simplified control functions are presented next. A more detailed analysis including all the instabilities due to nonlinearity and nonminimum phase system

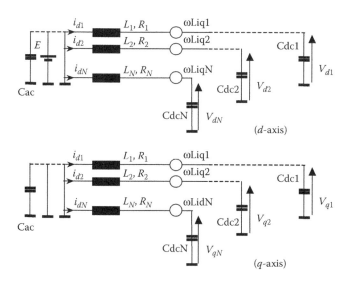

FIGURE 14.16 Boost DC/DC converter. (From Neacsu, D.O., Wagner, E., Borowy, B. 2002. *37th IAS Annual Meeting. Conference Record of the Volume 3*, 13–18 October, pp. 1958–1965. With permission.)

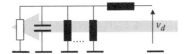

FIGURE 14.17 Equivalent circuits on (d, q)-axis.

characteristics are provided on [1,2]. The big difference between current control of a single converter and current control of parallel-connected converters is seen in a different model of the plant (Figures 14.18 and 14.19).

Let us first analyze each (d, q) equivalent circuit when it behaves as a DC/DC buck converter and the small signal model of the equivalent load that results from Figure 14.20. Transfer function from *duty cycle to the inductor current*:

$$H(s) = \frac{k}{sL} \frac{\left(s^2 + (1 / C_{AC} R_{AC})\right)s + (N - 1) / L_1 C_{AC}}{\left(s^2 + (1 / C_{AC} R_{AC})\right)s + N / L_1 C_{AC}} \tag{14.7}$$

The inductances of the other converters introduce a dumping effect. The *modulation index to inductor current* transfer function is different for each interleaved stage. The huge difference from the single-stage case is the integrative character in low frequency [7,8].

The (d, q) modeling for the power transfer through a boost DC/DC converter is shown in Figure 14.19 and the small signal model is accordingly derived. Transfer function from *duty cycle to the inductor current* is also shown in the figure.

Figure 14.19 shows that a change in the first converter duty cycle determines a change in the inductor L_1 current, followed by a change of the load voltage across the capacitor C_L. The voltage on the load changes currents through the other inductors and a feedback change in the output capacitor voltage occurs. The duty cycle

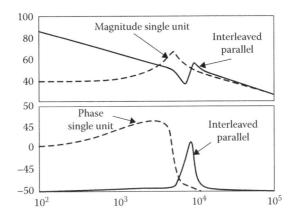

FIGURE 14.18 Bode plots for the buck converter model.

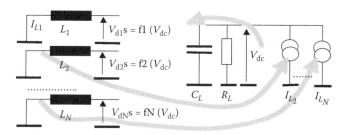

FIGURE 14.19 Equivalent circuit for parallel connection of boost converters.

to inductor current transfer function can be calculated based on the small signal equivalent circuit:

$$H(s) = \frac{k_2}{sL} \frac{\left(s^2 + (1/C_L)\right)s[(1/R_L) + ((1 - D)/V_L))I_L] + (N - 1)(1 - D)^2/L_1C_L}{\left(s^2 + (1/C_LR_L)\right)s + N(1 - D)^2/L_1C_L}$$

(14.8)

In short, inductances on the other converters have a dumping effect on the converter transfer function and the *modulation index to inductor current* transfer function depends on the number of interleaved stages [6,9,10,12]. The biggest difference in small-signal model from the case of a single converter consists in the integrative character at low frequency.

14.4.6 (*D, Q*) VERSUS (*D, Q, 0*) CONTROL

Many versions of vector control implemented by the industry within standard three-phase power converters have considered only the (*d, q*) control, without any

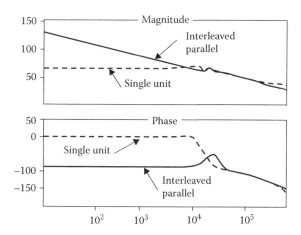

FIGURE 14.20 Bode plots for the boost converter case.

controller for the zero sequence. Using parallel hardware in interleaved power converter applications has raised the question whether conventional (d, q) control is enough or should it be improved with a zero-sequence controller.

Circulation currents between power converters can appear if the switching pattern in one or more converter legs is different from one power stage to another. Assuming that only one phase shows a circulation current, the power converter's three currents can be written as

$$I_a = x + \sin(\omega t), \; I_b = \sin(\omega t + 120), \; I_c = \sin(\omega t + 240)$$

After (a, b, c) transforms to (d, q), $(0.66x)$ adds up to (d, q) components and $(0.33x)$ adds up to the zero-sequence component.

It is important to note the integrative action of the zero sequence components. When a voltage is applied across the inter-phase inductors, owing to the different switching patterns in different power stages, the zero-sequence current increases or decreases depending on the polarity of the voltage. When a zero voltage is applied, the zero-sequence current is kept constant. Repetitive voltage across the inter-phase inductor can increase the zero-sequence component at values dangerous for the power stage.

In current control, the use of PI control on each of the (d, q) axes ideally withdraws the additional component $(0.66x)$ on these axes and does not affect the component on the zero-sequence axis. All three phase currents are subjected to an equal amount of error. If this is too large, we need a zero-sequence controller.

The zero-sequence current is not seen in the common-mode model, as it is a part of the differential mode model. This means that a power system with N parallel converters needs only $N - 1$ controllers, as there are $N - 1$ independent currents. Each zero-sequence current controller is designed from the inter-phase inductances that appear on the zero sequence equivalent circuitry.

14.5 INTERLEAVED OPERATION OF POWER CONVERTERS

The same parallel-connected hardware can be operated in multiple ways leading to reduced ripple in the aggregate input and output waveforms (Figure 14.17). The simplest solution is to use the same PWM clock for all converters. The switching pattern results are identical and the whole power stage behaves like a single high-power converter.

Different power converters can use different clock signals producing IN/OUT current waveforms with a ripple reduced by $\sqrt{N}$ owing to passive (stochastic) ripple cancelation. This is definitely not a practical solution. The existing hardware can be controlled with equal phase displacement $(2\pi/N)$ producing the so-called interleaving (Figure 14.21). The ripple magnitude is reduced by N times and the ripple frequency is increased by N times, simplifying filtering. This solution was first used for DC/DC converters and it is increasingly used in AC applications. Buck or boost topologies have been interleaved for a long time to improve harmonics and to share power between several power devices. Advanced methods send the current sharing information through the harmonic content of these converters. This is mainly possible due to the large ratio between the switching frequency and the control bandwidth.

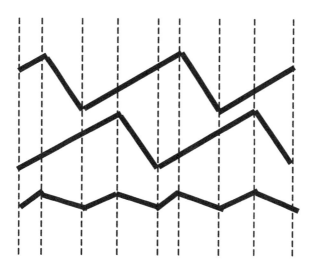

FIGURE 14.21 Ripple composition by interleaving.

In contrast, AC applications use lower switching frequency and the control bandwidth is limited.

Interleaving has been used for other classes of three-phase converters such as the buck-derived three-phase converters (in reference 7 of Chapter 3) (Figure 14.22a) or the B4 inverter modules (in reference 8 of Chapter 3) (Figure 14.22b). All these have given an extensive analysis of interleaved six-switch three-phase converters [6,9,10,12].

There are numerous advantages in using interleaving control of parallel hardware. The most important advantage is in using prior knowledge to build, control and protect the single-unit three-phase inverters and converters. Other advantages are discussed next (Figure 14.23).

The ripple of the sinusoidal phase current is maximum when the area under the curve of inductor voltage (volt seconds) or the phase voltage is maximum. Accordingly, the ripple of the current that results from interleaving is reduced. Both conduction losses and filter losses are minimized by interleaved operation of parallel hardware. Reliability is improved by parallel hardware and a good level of redundancy is ensured. Finally, parallel hardware ensures electronic gearing.

14.6 CIRCULATING CURRENTS

It has been shown that the circulating currents are the result of differences between the switching patterns within different power stages. The amount of the circulating current is influenced by the selection of the PWM algorithm in multiple modes.

First, the circulation currents depend on the difference between the reference waveforms (or ON-times) used on each power converter, as shown in Figure 14.24. In short, the sinusoidal PWM leads to the smaller difference between references and the smaller circulating current between power stages.

Another type of circulating current results from different active vectors used on each power stage. As a result of the use of different active vectors at the same time,

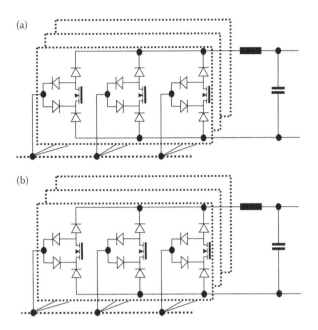

FIGURE 14.22 Different topologies used within interleaved configurations: (a) buck-derived three-phase cenverters. (b) B4-Inverters modules.

one leg of the inverter is connected to the +DC bus, while the same leg on another inverter is connected at –DC bus. Such circulation current gets decomposed on $(d, q, 0)$ components and can be partly corrected through the (d, q) current controllers.

Finally, pure zero-sequence currents are produced when all three switches connected to one terminal of the DC bus are ON, while the three inverter poles on the other inverters are connected at the other terminal of the DC bus. This happens

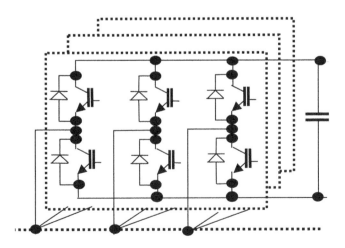

FIGURE 14.23 Circulating currents from differences in the PWM references.

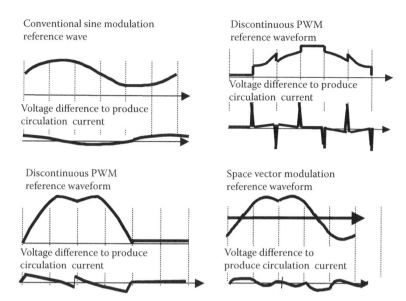

FIGURE 14.24 Different interleaving possibilities for an even number of converters. (From Neacsu, D.O., Wagner, E., Borowy, B. 2002. *37th IAS Annual Meeting. Conference Record of the Volume 3*, 13–18 October, pp. 1958–1965. With permission.)

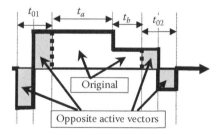

FIGURE 14.25 Generation of the PWM sequence without any zero vector.

when one converter uses a zero vector and another converter uses the other possible zero vector. Special PWM methods can be employed to avoid using any zero vector (Figure 14.25).

An alternative solution is to select the same zero state in all power converters whenever a zero state is necessary. The selected common zero state can be changed in a few seconds to avoid temperature run-up. Using always the same zero vector, no matter what active vectors are employed during the PWM sampling interval eliminates the possibility of circulating currents associated to identical zero states.

14.7 SELECTION OF THE PWM ALGORITHM [1,2]

Previous chapters have presented different PWM algorithms to control a three-phase inverter. Interleaved control of parallel hardware can be based on any of these algorithms. We limit our analysis to carrier-based PWM (asymmetrical or symmetrical), SVM, discontinuous PWM implemented on the SVM support to minimize loss, or special PWM with the same zero state. Any of these methods can be generated by symmetrical (center-aligned) PWM or asymmetrical PWM depending on the state sequence at the switching interval. The type of PWM employed influences the harmonic content and waveform shapes of the phase currents, circulation currents, neutral voltage, and DC side currents.

When using symmetrical PWM on an even number of interleaved converters, there are two possibilities for selecting the most suitable switching sequence, and these are shown in Figure 14.26.

Extensive analysis has established comparative results that can be grouped into the following conclusions:

- General Remarks
  - The first high-frequency components of the phase currents are always at Nf_{sw}, where N is the number of interleaved power stages.
  - The first high-frequency component of the zero-sequence current is always at f_{sw}. The DC bus and the neutral voltages show the first component at Nf_{sw} frequency, except for possible components owing to system asymmetries.
  - Using PWM with an optimal sequence to reduce the load current harmonics produces more harmonics on the individual inverter currents.
  - When circulating currents are reduced, the *peak-to-peak* neutral voltage bounces across a larger interval.
- Carrier PWM
  - Symmetrical PWM produces less harmonics in the inverter currents than does asymmetrical PWM.
  - Interleaving of symmetrical PWM can be implemented *by pulse* or *by period* (Figure 14.26).
  - The zero-sequence currents in symmetrical methods are larger due to greater superposition of different states.
  - Peak-to-peak neutral voltages are also lower for the symmetrical PWM methods.

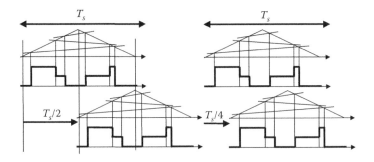

FIGURE 14.26 Different interleaving possibilities for an even number of converters.

- Reduced Loss PWM
  - Reduced-loss PWM shows a very large zero-sequence current at a low frequency if no zero-sequence current controller is used.
  - Zero-sequence controller becomes mandatory with SVM or reduced loss PWM algorithms.

The most important conclusion of this study is that using sinusoidal PWM algorithms in interleaved applications is more advantageous than SVM or Reduced-Loss PWM algorithms.

14.8 SYSTEM CONTROLLER

Each individual three-phase inverter is controlled by the same controller as in a single-unit application. PWM, power stage protection, and current control are implemented locally within each inverter's control module (Figure 14.27).

References for these controllers are provided from a system controller. The main tasks for the system controller are the proper current distribution among inverters, tight output regulation, hierarchical protection, and user interface or control through a reference.

The most important task is to distribute current equally between power converters. This can be achieved with *master-slave control, central-limit control,* or *circular-chain control* [4,5,11] (Figures 14.28 and 14.29).

Ultimately, each of these methods has to have an effective communication protocol or interface. As local control is ensured with hardware controllers located at each power converter, each system controller must be equipped with at least two communication ports. The first communication port is needed to communicate between power stages to ensure current sharing and fault management [13]. The information sent over this communication channel depends on the selection of the system controller structure such as the master-slave control, central-limit control, and the circular-chain control.

The second channel is used for PC interfacing for online monitoring and development and it can be considered local.

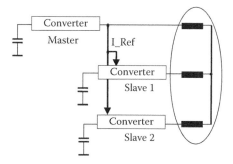

FIGURE 14.27 Parallel connection of six-switch three-phase converters.

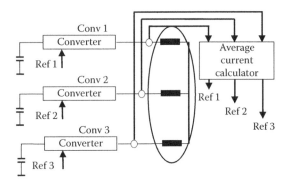

FIGURE 14.28 Master-slave structure.

Both required communication channels are bidirectional and a serial support is preferred owing to system complexity. Possible communication links for multiconverter applications are CAN, RS232, or RS485. The converter control S/H is strongly affected by the communication *baudrate* or bandwidth. To achieve a control S/H at each PWM cycle, the communication link should ensure a full exchange of information at the PWM frequency rate. This is a very serious constraint and limitation.

The last issue to be solved by the system controller is related to protection at a higher hierarchical level. It has been already shown that the local controller should ensure a fast protection of the power semiconductor devices, and shutdown quickly in case of overcurrent, IGBT desaturation, overvoltage, or overtemperature. These are faults that can occur quickly and require immediate attention. After a shutdown decision has been taken, the shutdown event information is processed by each converter controller and the software enters a special routine for shutdown and fault debouncing. The inter-converter network baudrate is limited and instant global decision is not possible. The solution considered often is to distribute faults depending on the required response time. Some faults are sent over the regular communication link and some are transmitted over an emergency hardware wire for a quick shutdown of all power converters. System monitoring is ensured through periodical status reports for the faults that do not require quick response action. A special software routine appraises the system and corrects existing faults on the basis of these

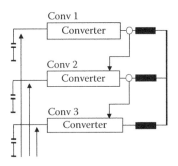

FIGURE 14.29 Circular chain control.

reports. Definition of this software routine for start-up and/or coming back from the shutdown state is very important.

Another fault that could happen during operation is the loss of the communication link even temporarily. The most common solution uses a special communication protocol that has parity checking and security features for surveillance of the communication channel. This further reduces the communication baudrate. The special communication protocol sets a flag for the appropriate fault and the software algorithm initiates a shutdown procedure, otherwise uncontrollable circulating currents can occur within each converter. Local software can detect these faults and shutdown independently. Simplified solutions are often used instead of the special communication channel, by settling a fault bit within the regular fault message sent through the communication channel.

14.9 CONCLUSION

This chapter presents the details of the parallel and interleaved operation of parallel three-phase hardware. It is, therefore, demonstrated that interleaved operation of three-phase systems is a viable solution for medium and high power converters, especially in applications with low voltage and high current [14,15]. Design aspects and selection of the inter-phase reactors or PWM algorithms are also shown.

PROBLEMS

P14.1 A power electronic converter needs to parallel three F-series 600 V Powerex IGBTs to constitute each switch. What is the static derating factor that we should consider?

P14.2 For the same power converter, if the IGBTs are 200 A devices, what is the maximum current this structure will allow through all three IGBTs?

P14.3 Consider an interleaved power converter with two inverter power stages, each controlled with sinusoidal PWM at 10 kHz and operated at grid frequency of 60 Hz and delivering 120 V RMS in open loop. Calculate the low-frequency component applied permanently on the inter-phase inductors. What is the low-frequency circulation current if each inductor has 100 mH.

P14.4 Consider the time constant equations of the SVM algorithm as given by Equation 5.30. Define the new time intervals corresponding to Figure 14.27. Draw the switching pattern within one sampling interval to generate a vector in the first sector. Is there any way by which this can be implemented in a center-aligned PWM-support hardware?

REFERENCES

1. Neacsu, D.O., Borowy, B., and Bonnice, W. 2001. Limiting inter-converter zero sequence currents within three-phase multi-converter power systems—Review and ultimate solution. *IEEE IECON*, Denver, CO, USA, pp. 1255–1261.

2. Neacsu, D.O., Wagner, E., and Borowy, B.N. 2002. Selection of the PWM algorithm for 3-phase interleaved converters. *IEEE IAS*, Pittsburgh, PA, USA, pp. 13–19.

3. Garg, A., Perreault, D.J., and Verghese, G.C. 1999. Feedback control of paralleled symmetric systems with applications to nonlinear dynamics of paralleled power converters. *Proceedings of the 1999 IEEE International Symposium on Circuits and Systems, ISCAS '99*, vol. 5, pp. 192–197, 30 May–2 June.

4. Zhou, X., Xu, P., and Lee, F.C. 2000. A novel current-sharing control technique for low-voltage high-current voltage regulator module applications. *IEEE Trans. PE* 15(6), 1153–1162.

5. Wu, T.F., Chen, Y.K., and Huang, Y.H. 2000. 3C strategy for inverters in parallel operation achieving an equal current distribution. *IEEE Trans. IE* 47(2), 273–281.

6. Perreault, D. and Kassakian, J. 1997. Distributed interleaving of paralleled power converters. *IEEE Trans. CAS* 44(8), 728–734.

7. Mazumder, S., Nayfeh, A., and Borojevic, D. 2002. Comparison of nonlinear and linear control schemes for independent stabilization of parallel multi-phase converters. *IEEE IAS*.

8. Mazumder, S.K., Nayfeh, A.H., and Borojevic, D. 2001. A novel approach to the stability analysis of boost power-factor-correction circuits. *IEEE 32nd Annual Power Electronics Specialists Conference, PESC 2001*, Vancouver, Canada, vol. 3, pp. 1719–1724, June.

9. Kelkar, S. and Henze, C.P. 2001. High performance three-phase unity power factor rectifier using interleaved buck-derived topology for high power battery charging applications. *IEEE 32nd Annual Power Electronics Specialists Conference, PESC 2001*, Vancouver, Canada, vol. 2, pp. 1013–1018, June.

10. Singh, B.N., Joos, G., and Jain, P. 2000. Interleaved 3-phase PWM AC/DC converters based on a 4-switch topology. *IEEE 31st Annual Power Electronics Specialists Conference, 2000. PESC 2000*, Galway, Ireland, vol. 2, pp. 1005–1011, 18–23 June.

11. Wu, T.F., Chen, Y.K., and Huang, Y.H. 2000. 3C strategy for inverters in parallel operation achieving an equal current distribution. *IEEE Trans. IE* 47(2), 273–281.

12. Chang, C. and Knights, M.A. 1995. Interleaving technique in distributed power conversion systems. *IEEE Trans. CAS* 42(5), 245–251.

13. Donescu, V. 2001. Fault detection and management system broadcasts motor drive faults. *PCIM*, June, 38–42.

14. Anon. 2000. FM2000—Modular drive design up to 2000HP. TECO-Westinghouse internet documentation.

15. Anon. 2000. Drive modules offer system integrator flexibility. *ABBACS600 in Engineering Talk*.

16. Neacsu, D.O., Wagner, E., Borowy, B., 2002. *37th IAS Annual Meeting. Conference Record of the Volume 3*, 13–18 Oct. 2002, pp. 1958–1965.

15 AC/DC and DC/AC Current Source Converters

15.1 INTRODUCTION

Most of the other Chapters of the book dealt with the *Voltage Source Converter*, a power converter structure built of 6 bi-directional switches and supplied from a voltage source. An alternative topology consists of a *Current Source Converter* [1–3] (Figure 15.1). Such converter structure is built of 6 unidirectional switches and it is supplied from a current source. This means the converter must always ensure a current circulation and a path for this current. Furthermore, the current circulation is unidirectional from the DC current source to the load (for a Current Source Inverter), or from the AC grid to the DC current load (for an AC/DC Current Source Converter). The possible states are shown in Figure 15.2.

The oldest form of a *Current Source Converter* was the AC/DC SCR-Based Rectifier without the DC-side capacitive filter, where thyristors are controlled to adjust the load current that is flowing from grid to the load. The advent of IGBT devices shifted the focus more and more on voltage source type topologies since the current source topologies required additional blocking diodes to transform the IGBT-diode copack into a true unidirectional device. The recent appearance of IGBT-RB (reverse blocking) devices encourages a new look into the *Current Source Converter* topologies.

The DC/AC Current Source topology is mostly used for motor control and these motors are three-phase. Hence, only the three-phase converters will be analyzed in this chapter.

The advantages of the *Current Source Converter/Inverter* when compared to the counterpart Voltage Source structure are:

- Operation with reduced power loss [4,5].
- Operation with reduced noise.
- The lack of free-wheeling diodes improves the size and weight of the converter.
- The reliability yields improved due to less count of components.

The disadvantages of the Current Source Converter topologies are:

- Slower transient response of the current (considering the topology of converter + filter).

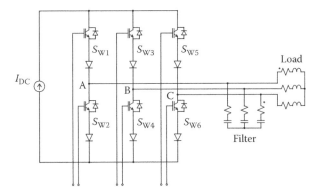

FIGURE 15.1   Three-phase current source inverter.

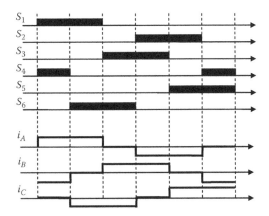

FIGURE 15.2   Switching states and currents for the un-modulated current source converter.

The sampling/switching frequency is usually chosen within the range 1–10 kHz in order to reduce the reactive energy generated in the AC-side leakage reactance during commutation and hence the devices' stress. Other contemporary concerns refer to mitigation of the common-mode voltage [6].

The current control is achieved with an appropriate PWM algorithm able to adjust the current level through the modulation index m. Most of the PWM algorithms are similar to those for the Voltage Source Converters in terms of generating pulses by a comparison between a reference waveform and a carrier. Usually the reference waveform is trapezoidal [10,11] or sinusoidal [7].

15.2   CURRENT COMMUTATION

The control of the power stage needs to ensure the conduction of only two devices at any moment. The transition between the different conduction states needs to be done seamless, without loss of current circulation. This is achieved with a mechanism similar yet opposite to the dead-time generator: a small interval is introduced to superimpose the two adjacent states by allowing conduction of switches defining both the old state and the new state. This concept is true for any PWM algorithm and for either AC/DC or DC/AC conversion.

For instance, at the state change from conduction of S_{w1} and S_{w6} to the conduction of S_{w3} and S_{w6}, a circulation path should be maintained for the current entering the converter on the DC+ conductor (see Figure 15.2). Usually this is ensured with an interval when both S_{w1} and S_{w3} are controlled to be ON (Figure 15.3) [8]. This interval is called *overlap* and it must be long enough to turn ON the OFF switch before turning OFF the conducting switch. This overlap also provides a minimum ON time for the switches.

The actual current commutation between switches S_{w1} and S_{w3} depends on the polarity of the line voltage v_{AB} (Figure 15.1). Both switches have gate signals for the conduction state and the polarity of the voltage drop across them dictates the conduction state. The exact commutation moment occurs either at the beginning or at the end of the *overlap* interval. This ambiguity in the exact commutation moment implies a loss of control and an uncertainty in the switch current, and further on the output line current during the overlap periods. Considering both edges of each PWM pulse of the AC-side current denotes three cases (similar to the analysis of effects of the dead-time interval):

- Pulse is shifted.
- Pulse is losing some current.
- Pulse is increased with certain length.

The overall effect of this change in the pulses' shape yields in a slight distortion of the AC-side sinusoidal current waveform [9]. The waveforms are influenced as shown in Figure 15.4. Using fast switching devices allows operation with a short overlap time that will not distort or otherwise influence the current waveform.

Finally, it is important to note that the gate control cannot be achieved with conventional dual-channel (converter leg) or six-channel (inverter) gate drivers since

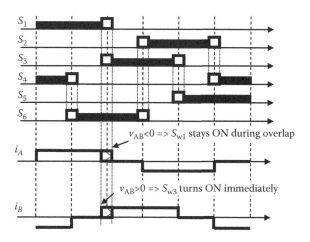

FIGURE 15.3 Control for current commutation.

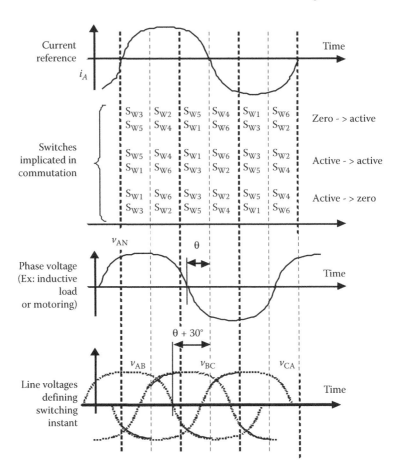

FIGURE 15.4   Influence of the overlap on the switching states.

these have the shoot-through protection implemented. Individual gate drivers would allow the conduction of both switches on the same inverter leg.

15.3   USING SWITCHING FUNCTIONS TO DEFINE OPERATION

We have used switching functions to characterize the operation of a three-phase voltage source converter in Chapter 3, Section 3.8. The same theory is used herein for the Current Source Converter. Switching Functions are defined as periodical signals able to characterize analytically the change of states without entering the details of transitions between states. Let us consider the switching functions as being identical with the signals used for control of switches from Figure 15.1.

$$i_A = i_{DC} \cdot (S_1 - S_4) \tag{15.1}$$

$$i_B = i_{DC} \cdot (S_3 - S_6) \tag{15.2}$$

$$i_C = i_{DC} \cdot (S_5 - S_2)$$ (15.3)

The DC-side voltage can be computed from instantaneous output voltages as:

$$v_i = v_{AN} \cdot (S_1 - S_4) + v_{BN} \cdot (S_3 - S_6) + v_{CN} \cdot (S_5 - S_2)$$ (15.4)

These will further be used for modeling of the power converter when simulated in MATLAB-SIMULINK or PSPICE, and also for development of the PWM algorithm.

System level waveforms are shown in the following figures for a generic PWM algorithm without any compensation for overlap time or canceling of the possible filter resonance. Different improved options for the generation of the gate control pulses will be shown in detail within Section 15.4. These waveforms are shown herein for illustration of purposes at different switching frequencies. It can be seen how important is the selection of the filter components and the avoidance of possible oscillations within the filter structure (Figures 15.5 through 15.8).

The *AC/DC Current Source Converter* (also known as *Switched-Mode Rectifier*) is represented in Figure 15.9. This time the PWM pulses are synchronized with the grid with a PLL-type circuitry. The same switching functions theory is employed for modeling of the operation.

The operation is very similar to the Cur*re*nt Source Inverter, but the energy flows now from the grid to the DC-side load. The proper operation is secured by the

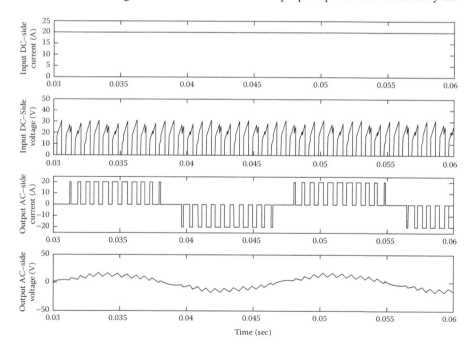

FIGURE 15.5 System-level waveforms for the current source inverter ($f_{PWM}/f_{out} = 24$, $L = 1$ mH, $C = 470$ μF).

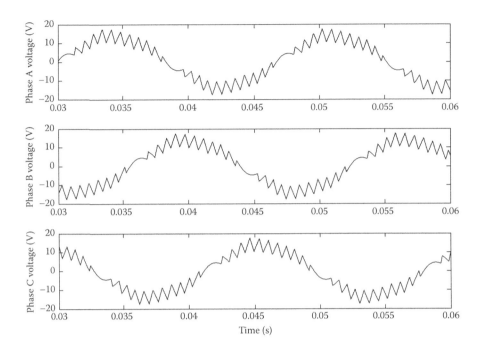

FIGURE 15.6   Output phase voltages after L-C filter ($f_{\mathrm{PWM}}/f_{\mathrm{out}} = 24$, $L = 1$ mH, $C = 470$ µF).

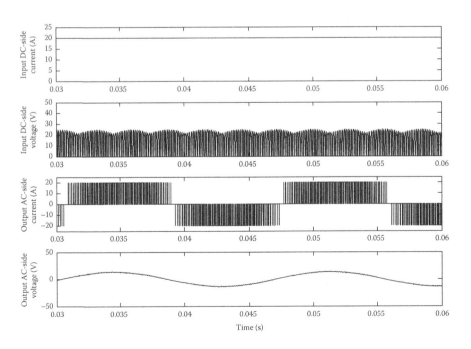

FIGURE 15.7   System-level waveforms for the current source inverter ($f_{\mathrm{PWM}}/f_{\mathrm{out}} = 120$, $L = 1$ mH, $C = 470$ µF).

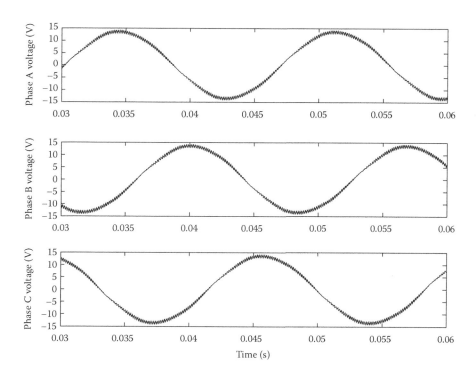

FIGURE 15.8   Output phase voltages after L-C filter ($f_{PWM}/f_{out} = 120, L = 1$ mH, $C = 470$ μF).

presence of an input L-C filter, where the inductance L makes up for the grid inductance and a possibly small added filter inductance. The switching functions are the
same with Equations 15.1 through 15.4.

Simulation results are shown within the following figures. They are given for
illustration purposes as the L-C filter components and the PWM control could be
optimized further. It is important to mention the role of accurate synchronization
between control and grid phase information (Figures 15.10 and 15.11).

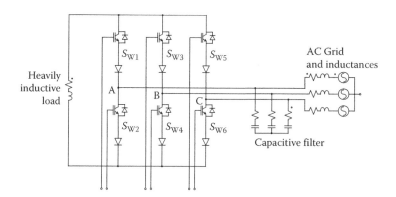

FIGURE 15.9   Switched mode PWM rectifier.

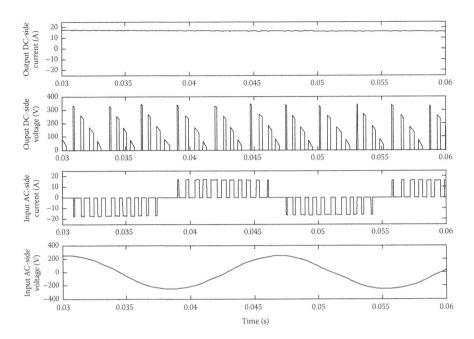

FIGURE 15.10 System-level waveforms for the grid-connected current source converter ($f_{PWM}/f_{out} = 24$, $L = 2$ mH, $C = 470$ μF).

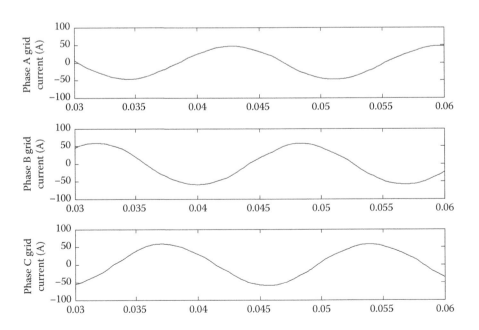

FIGURE 15.11 Grid currents ($f_{PWM}/f_{out} = 24$, $L = 2$ mH, $C = 470$ μF).

The operation described within Figure 15.2 and Equations 15.1 through 15.4 cannot allow the adjustment of the current in the circuit. This is achieved further with various PWM control algorithms. Moreover, optimization within these PWM algorithms is also discussed.

15.4 PWM CONTROL

15.4.1 TRAPEZOIDAL MODULATION

The biggest difference between the PWM algorithms for the CSI and VSI structures consists of operation of only two switches at any given moment within CSI instead of three for the VSI. The two conducting switches belong to two different legs of the inverter and current flows in two phases only. In certain situations the two switches on the same leg are conducting, creating a short-circuit path able to circulate the DC-side current without stressing the AC-side filter capacitors.

The simplest PWM algorithm is following the principles of the carrier-based modulators for Voltage Source Converters (Figure 15.12). The reference signal is a trapezoidal waveform with unity magnitude, and the PWM pulses are produced by comparison during the slopes of the reference waveform. This method is called upon as *Trapezoidal PWM* [10,11]. The middle segment of the waveform is flat for 60°. Similar waveforms are used for the other two phases.

Figure 15.13 shows an example of the phase currents derived after the AC filter when using such a PWM method, when the PWM has a low ratio between the pulse frequency (switching frequency) and the fundamental (reference) waveform's frequency.

15.4.2 HARMONIC ELIMINATION PROGRAMMED MODULATION

Observing the waveforms of Figure 15.12 suggests the PWM generation based on optimal algorithms like those described in Chapter 3, Section 3.7.1. Actually the mathematical results reported there can also be used for control of the *Current Source Inverter* when elimination of certain harmonics is imposed as design requirement. After the angular coordinates for the switching instants are determined by

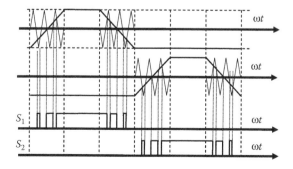

FIGURE 15.12   Carrier-based PWM method.

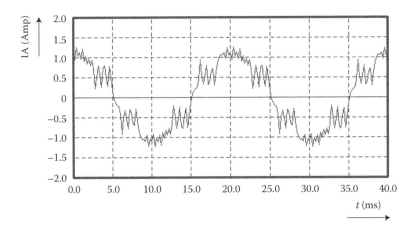

FIGURE 15.13   Example of phase current after the $L\text{-}C$ filter, for a low frequency ratio.

calculus, the actual PWM sequence will be determined respecting the control logic for a CSI converter.

It is worthwhile to mention that any PWM control algorithm for a three-phase voltage source converter can be used for control of a current source type structure when the difference between two control signals on different phases (like a line-to-line signal) is used for control of a single switch from the CSI structure. This is also apparent from comparison of the line-to-line voltage of a VSI converter that has the same shape with a phase current from the CSI operation.

Mathematical details and examples for this method are not repeated herein as they stand true from the previous explanation given in Chapter 3, Section 3.7.1.

15.4.3   SINUSOIDAL MODULATION

Another PWM controller for the *Current Source Inverter/Converter* can be defined starting from a Sinusoidal reference like in the case of a controller for a Voltage Source Converter (Figure 15.14) [12]. First the signals pertaining to the converter states for the Current Source Converter are calculated with simple logic operators (Figure 15.15). This guarantees the generation of the 6 active states. Next the zero states derived by control of Figure 15.14 need to be identified as a safety measure. During such intervals, ON states need to be inserted in order to maintain the circulation of the current. Various logic circuitry can ensure this.

This logic is unfortunately not enough as it is necessary to also be able to generate the zero states and to select the proper sequence of switches. This can be carried out with a combinatorial logic as shown in [15].

A different approach more closely related to microcontroller implementation works with the converter state coded as a switching state. The conventional PWM controller calculates first the time intervals to be spent in different switching states. Then, the actual states are identified and coded. Based on the switching state code, the actual switch is selected with a memory table (Figure 15.16). The memory look-up table is shown as a "*multiplexer + combinatorial logic*" module in the simulation

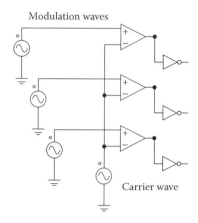

FIGURE 15.14 PWM generator for a voltage source inverter.

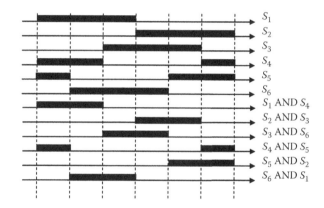

FIGURE 15.15 Determine the control signals from a conventional PWM generator for a VSI.

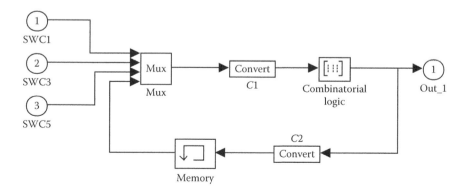

FIGURE 15.16 Selection of the switching state with a memory look-up table, as depicted from a conventional MATLAB-SIMULINK simulation file.

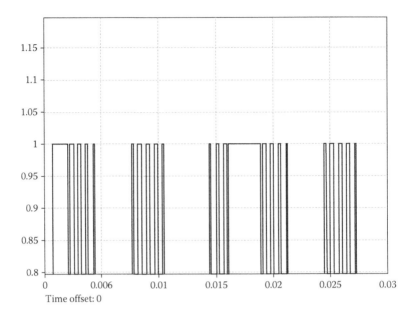

FIGURE 15.17   Control signals for the logic circuitry shown in Figure 15.16.

file, which actually can be another way to implementation. The fourth input to this table secures the generation of the zero state by continuation of the switch state. The gate control signals are shown in Figure 15.17.

15.4.4 SPACE VECTOR MODULATION

The success of *Space Vector Modulation* algorithms for control of the voltage source converters has recommended this method in controlling current source converters.

In this respect, the states of the converter are analyzed and the appropriate current vectors are set on the complex plane [13,14]. One and only one switch in the upper side and lower side must be turned-ON at a time. Each pairing of two switches determines a specific state of the converter. There are nine possible combinations for the ON-state converter switches. For each state, a space vector can be associated to represent the system of AC-side currents. The same theory applies to both the AC/DC and DC/AC converters.

For the ideal case when sinusoidal AC-side currents are expected, the curve following the tip of the space vector I has to be a circle in the complex plane and its track speed a constant. For the conventional phase controlled converters only seven distinct positions of the space vector in the complex plane can be obtained $I_1 - I_6$ and zero I_0 (Figure 15.18, Table 15.1).

The high switching capability of the new devices appropriate for PWM control allows the synthesis of some different positions of the space vector I on a circular polygonal locus with an adequate switching of the converter. The possibility of turning-ON the switches of the same leg allows generation of zero states and so there is always a current path for the converter's AC side inductive current.

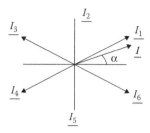

FIGURE 15.18 Possible positions of the current space vector in the complex plane.

The desired position of the space vector I is placed between two space vectors I_a and I_b, $a,b = 1 - 6$ (Figure 15.18) which represent the two states involved in the switching process. Writing the appropriate average relation provides

$$\vec{I}_a \cdot t_a + \vec{I}_b \cdot t_b + \vec{I}_0 \cdot t_0 = \vec{I} \cdot T \tag{15.5}$$

where:

T = sampling period
t_a = time assigned for the state I_a
t_b = the time assigned for the state I_b
t_0 = the time assigned for the state I_0

For the peculiar case of AC/DC conversion, the proposed space vector PWM strategy takes into account both the prescription of the desired load current I and the input AC source unbalance or distortions through the instantaneous grid voltages. If we denote with E the measured AC-side converter input voltage, the relations for the time durations become:

$$\begin{cases} t_a = T \cdot k_v \cdot \sin(60^0 - \alpha) \\ t_b = T \cdot k_v \cdot \sin(\alpha) \\ t_0 = T - t_a - t_b \end{cases} \tag{15.6}$$

TABLE 15.1
Converter States

Vectors	Switches
I_1	$Sw_1 + Sw_6$
I_2	$Sw_3 + Sw_6$
I_3	$Sw_3 + Sw_2$
I_4	$Sw_5 + Sw_2$
I_5	$Sw_5 + Sw_4$
I_6	$Sw_1 + Sw_4$
I_0 [1]	$Sw_1 + Sw_2$
I_0 [2]	$Sw_3 + Sw_4$
I_0 [3]	$Sw_5 + Sw_6$

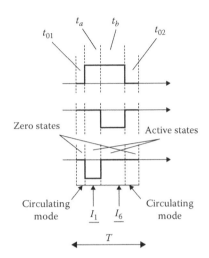

FIGURE 15.19 Pulse generation within the space vector modulation.

where

$$k_v = \frac{R \cdot I_{ref}}{E} \tag{15.7}$$

While these equations are presented herein for the general theory development, they will be again exploited in Chapter 18 within a real practical application.

The duration t_0 must be split into two zero states so that the transition from zero states to active states always involve only one switching. For the sake of simplicity and some generic optimization expectation, $t_0^+ = t_{0-}$. Figure 15.19 shows as an example the synthesis of a current pulse whose associated space vector is placed between I_1 and I_6.

15.5 OPTIMIZATION OF PWM ALGORITHMS

Once the basic details for the Space Vector Modulation have been introduced, the algorithm can be improved further by association with certain optimization criteria [13–16]. Generally, the problems with PWM algorithms for the *Current Source Inverter/Converter* are related to the limited value of the switching frequency.

Since the AC-side of the Current Source Inverter/Converter is usually followed up with a LC filter, it is worth considering this filtering effect with a time-integral function of the inverter current. This function has the meaning of a voltage with gain depending to the values of capacitor C and grid line inductance (AC/DC case) or motor leakage inductance L (DC/AC case).

The representation of this time integral in the complex plane represents a polygonal locus with a shape close to an exact circle. The difference between the polygonal locus and the ideal circular locus represents the error and causes torque fluctuation in the motor drive case.

Several criteria can be considered to reduce the amount of harmonics generated by the noncircular locus of the time integral of switching currents. This is especially useful in the case of limited switching frequency. To help the understanding of these optimal methods, we will show herein results for the motor drive case. They can easily be expanded to the grid interface application.

15.5.1 Minimum Squared Error

The idea is to generate a regular polygonal locus that is to generate the space vector positions by a constant increment of the angular coordinate. Results from this method can only be improved further with a proper positioning of the active states over the sampling interval. However, the additional improvements are minimal.

15.5.2 Circular Corona

Another optimization principle works close to the hysteresis control concept already known from switched mode converters. Maintaining the polygonal locus inside a circular corona is equivalent to reducing the output voltage ripple with effects in increasing the output power efficiency. Since the definition of this strategy is based on a geometrical analysis of the aforementioned constraint, it is worthwhile to use this criterion only for defining a polygonal locus with few sides, in the case of low carrier frequency. The resulting PWM pulse pattern is analogous to the case of delta or hysteresis modulation.

15.5.3 Reducing the Low Harmonics from the Geometrical Locus

Since the AC-side low-pass filter is reducing the higher frequency harmonics of the current, it worth considering the possibility of choosing a locus able to eliminate the low harmonics. The degree of freedom comes from the relationship between the numbers of the edges from the polygonal locus that corresponds to the number of the low harmonics eliminated.

Previous research on the vectorial representation in the complex plane demonstrated that the rotational speed of the vector can be regulated by introducing operation within the zero-vector states. Observing these results outlines the difference between the previous optimal methods (circular corona and optimized geometrical locus) and the conventional Space Vector Modulation algorithms. A compromise solution can be developed in order to obtain the desired trajectories in the complex plane with a frequency modulation or a zero-vector split modulation. Results in the next section will limit to the zero-vector split modulation.

15.5.4 Comparative Results

A comparison in between the harmonic performance for the methods described above is considered based on ideal PWM generation, without considering details of resonance between the filter components or gain loss within the inverter/converter.

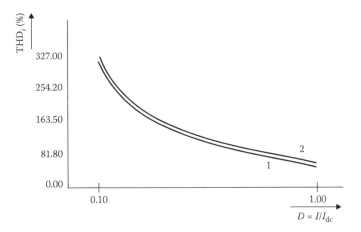

FIGURE 15.20 THD of the AC-side current for (1) space vector modulation, (2) trapezoidal PWM.

Similar to Chapter 3, Section 3.5.3, the Total Harmonic Distortion (THD) content of inverter output current defined as

$$\text{THD}_i = \sqrt{\frac{1}{I_1^2(d)} \cdot \sum_{k=2}^{\infty} I_k^2(d)} \tag{15.8}$$

Figure 15.20 presents a direct comparison between the *Space Vector Modulation* and the trapezoidal modulation for a modulation index of 0.7, an output frequency of 20 Hz, and 24 pulses. It is demonstrated that the harmonics of the output current are lower than the classical trapezoidal modulation waveform and hence the efficiency of converter becomes better since the output current and voltage are nearly sinusoidal waves.

These results set the basis for understanding the benefits of the optimal methods mentioned above. To expand further the comparison while following the theory presented in Chapter 3, Section 3.5.3, a new harmonic performance coefficient is defined herein. The high order harmonics are attenuated by the AC-side filter and a more meaningful comparison is made by taking into account the harmonic order. The distortion coefficient is therefore defined by multiplying each frequency component with $1/(k\omega)$.

$$\text{HCF}_i[\%] = \frac{100}{I_i} \cdot \sqrt{\sum_{n=5}^{\infty} \left[\frac{I_n}{n}\right]^2} \tag{15.9}$$

Other similar distortion factors can be defined [15] to illustrate the influence of the filter:

$$\text{DF}_i[\%] = \frac{100}{I_i} \cdot \sqrt{\sum_{n=2}^{\infty} \left[\frac{I_n}{n^2}\right]^2} \tag{15.10}$$

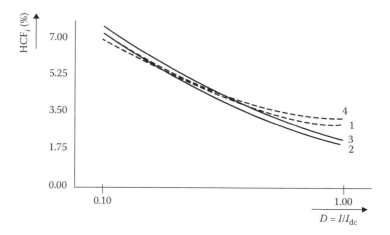

FIGURE 15.21 HCF for (1) = circular corona PWM with 24 edges polygonal locus; (2) = circular corona PWM with 36 edges polygonal locus; (3) = optimized geometrical locus with 36 edges polygonal locus; (4) = minimum square error method.

The following results are based on HCF coefficients only. If all symmetries in a three-phase system are considered, the harmonics used in the calculation of the HCF or DF coefficients can start from the 5th since there are no 2x or 3x harmonics. Otherwise, for a real case when symmetries cannot be considered, all harmonics, starting from the 2nd should be used in calculation.

The optimal methods can be compared with this HCF harmonic coefficient as well as content in the 5th and 7th harmonics of the output current (Figures 15.21 through 15.24).

15.6   RESONANCE IN THE AC-SIDE OF THE CSI CONVERTER–FILTER ASSEMBLY

Since the waveforms of the AC-side currents are made of pulses of current, the *Current Source Converter* is usually followed up with an AC filter, able to retain the higher harmonics of the current and to deliver a sinusoidal current into the AC circuitry. Since either the grid connection or the motor drive connected to the AC-side of the *Current Source Converter* is characterized with an inductance, the filter's capacitance and such inductance constitute a resonant circuit [17,18] (Figure 15.25).

There are possible both hardware and control software solutions to prevent or damp resonance. An additional trap filter can be added to the system to a specified frequency. Or, the PWM and the filter components can be optimally selected to avoid having the resonance frequency near frequencies in the spectrum of the current coming into the filter from the *Current Source Inverter*. For instance, the 5th and 7th harmonics can be cancelled by proper PWM selection, while the filter ($Lf + Cf$) resonance can be selected at around 4.6 times the fundamental frequency [5].

The most academically challenging solutions are coming from proper design of the control loops. Usually such solutions revolve around the idea of introducing by

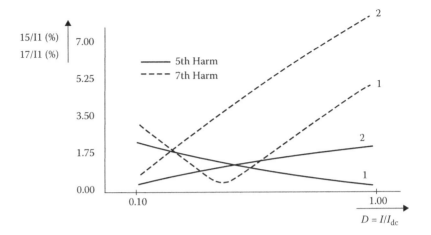

FIGURE 15.22 The 5th and 7th harmonics: (1) = circular corona PWM with 24 edges polygonal locus; (2) = minimum square error method with 24 edges polygonal locus.

control a virtual resistor along the hardware of the resonant filter so that the resonance yields damped [17]. A simple control structure is shown in Figure 15.26 for the case without any compensation. Since the converter's output is a current, the controlled measure is a voltage across the load circuitry, measured across the capacitor Cf.

If the AC-side is at the output of the converter and the load is a motor drive, the circuit equation for the load should contain the back-EMF voltage (e). Also the motor drive operation at different speed and load torque levels would imply change in the resonance frequency. This recommends further the use of a virtual resistance for damping of the resonance against other control methods.

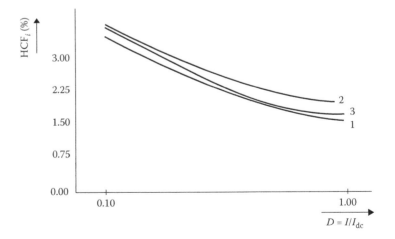

FIGURE 15.23 HCF for: (1) = circular corona with 48 edges polygonal locus; (2) = optimized geometrical locus with 36 edges polygonal locus; (3) = conventional SVM with 48 pulses.

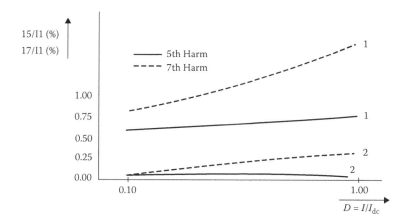

FIGURE 15.24 The 5th and 7th harmonics: (1) = circular corona PWM with 48 edges polygonal locus; (2) = minimum square error method with 48 edges polygonal locus.

The virtual resistor is set-up in software based on the feedback voltage V_c (Figure 15.27). A *High-Pass Filter* is applied to this voltage to avoid interference with the fundamental waveforms, and the current through the virtual resistance is calculated and accounted for within the current reference I_{ref}. The HPF will degrade the transient damping performance for the virtual harmonic damper, especially when there is a disturbance from the filter capacitor feedback voltage. This is caused by the tradeoff between a smaller cutoff frequency in an HPF giving a slow response, and a larger cutoff frequency leading to high frequency signal distortions. A multitude of versions to this approach are reported in literature.

A very similar approach to resonance damping consists in generating a virtual negative inductance along the virtual resistance (Figure 15.28) [19]. The advantage is reduced loss in the *Current Source Converter.*

Alternative methods to eliminate the *LC* resonance include feed-forward control compensation using the *LC* filter model and control shaping using harmonic compensators. Such a compensator can be designed around the *Posicast* controller [17]. This is splitting any step input command into two intermediate steps. By considering the pause interval between the two steps as a parameter, the system response produced by the second step can cancel the resonant response excited by the first step, resulting in an oscillation-free step.

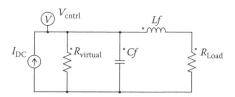

FIGURE 15.25 Circuitry illustrating the resonance with CSI-filter assembly.

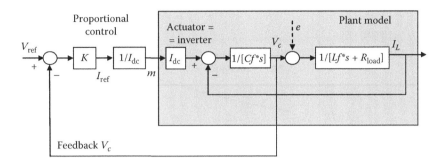

FIGURE 15.26 Simple model of the control system for the capacitor voltage.

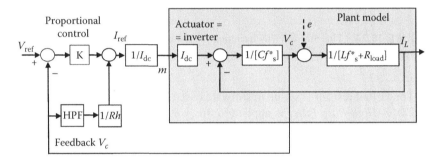

FIGURE 15.27 Control with compensation.

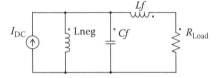

FIGURE 15.28 Circuitry illustrating the compensation with a virtual negative inductance.

15.7 CONCLUSIONS

Merits and drawbacks of the Current Source Converters have been explained and the basics of operation outlined. The current commutation between different switches needs attention for maintaining the current circulation. A large variety of PWM algorithms is available, many of them being derived from the implementation of PWM for voltage source inverters. Both passive and active approaches to damping the filter resonance are illustrated and their need shades the advantages of this class of converters.

More recent efforts consider multilevel current source converters where multiple stages are able to add the load current building up a current waveform with multiple steps. The optimization of the multilevel current waveforms can be achieved very similar to the optimization of voltage source converters at waveform level.

REFERENCES

1. Trzynadlowski, A., 2010. *Introduction to Modern Power Electronics*. John Wiley and Sons, New York.

2. Holmes, D.G. and Lipo, T.A. 2003. *Pulse Width Modulation for Power Converters— Principles and Practice*. John Wiley and Sons, New York.

3. Wu, B., Pontt, J., Rodriguez, J., Bernet, S., and Kouro, S. 2008. Current-source converter and cycloconverter topologies for industrial medium-voltage drives. *IEEE Trans. Industrial Electron.* 55(7), 2786–2797.

4. Rey, J., Doval-Gandoy, J., Penalver, C.M., Lopez, O., Nogueiras, A., and Lago, A. 2005. Evaluation system for current source converter modulation techniques. *31st Annual Conference of IEEE Industrial Electronics Society*. IECON 2005, pp. 5, 2005.

5. Suh, S., Steinke, J., and Steimer, P. 2006. A study on efficiency of voltage source and current source converter systems for large motor drives. *37th IEEE Power Electronics Specialists Conference*. PESC'06, pp. 1–7.

6. Zhu, N., Xu, D., Wu, B., Zargari, N.R., Kazerani, M., and Liu, F. 2013. Common-mode voltage reduction methods for current-source converters in medium-voltage drives. *IEEE Trans. Power Electron.* 28(2), 995–1006.

7. Fukuda, S. and Hasegawa, H. 1988. Current source rectifier/inverter system with sinusoidal currents. *IEEE IAS Annual Meeting*, pp. 909–914.

8. Abu-Khaizaran, M. and Palmer, P. 2007. Commutation in a high power IGBT based current source inverter. *IEEE PESC'07*, pp. 2209–2215.

9. Halkosaari, T., Kuusel, K., and Tuusa, H. 2001. Effect of nonidealities on the performance of the 3-phase current source PWM converter. *Power Electronics Specialists Conference, 2001. PESC. 2001 IEEE 32nd Annual*, vol. 2, pp. 654–659.

10. Nonaka, S. and Neba, Y. 1991. Quick regulation of sinusoidal output current in PWM converter-inverter system. *IEEE Trans. Industry Appl.* 27(6), November/December, pp. 1055–1062.

11. Wu, B., Dewan, S.B., and Slemon, G.R. 1992. PWM CSI inverter for induction motor drives. *IEEE Trans. Industry Appl.* 29(1), January/February, pp. 64–71.

12. To, H.-P., Rahman, M.F., and Grantham, C. 2006. Modulation of current-source converters from gating signals for voltage-source converters. *37th IEEE Power Electronics Specialists Conference*, pp. 1–6, 18–22.

13. Neacsu, D.O., Pastravanu, A., and Lucanu, M. 1993. Space vector PWM AC/DC converter. *Proceedings of the Int'l Symposium SCS'97*, Iasi, Romania, pp. 252–255.

14. Neacsu, D.O. and Donescu, V. Performance characterization of space vector PWM current source inverter. *The 4th Int'l Conference OPTIM'94*, Brasov, Romania, vol. 4, section F, pp. 109–114.

15. Espinoza, J.R. and Joos, G. 1997. Current-source converter on-line pattern generator switching frequency minimization. *IEEE Trans. Industrial Electron.* 44(2), 198–206.

16. Halkosaari, T. and Tuusa, H. 2000. Optimal vector modulation of a PWM current source converter according to minimal switching losses. *IEEE 31st Annual Power Electronics Specialists Conference, PESC 2000*, vol. 1, pp. 127–132.

17. Li, Y.W. 2009. Control and resonance damping of voltage-source and current-source converters with LC filters. *IEEE Trans. Industrial Electron.* 56(5), 1511–1521.

18. Neba, Y. 2005. A simple method for suppression of resonance oscillation in PWM current source converter. *IEEE Trans. Power Electron.* 20(1), 132–139.

19. Morsy, A.S., Ahmed, S., Enjeti, P., and Massoud, A. 2010. An active damping technique for a current source inverter employing a virtual negative inductance. *IEEE PESC*, pp. 63–67.

16 AC/AC Matrix Converters as a 9-Switch Topology

16.1 BACKGROUND

The three-phase AC/AC matrix converters have attracted lots of interest in the 1990s [1–7]. Methods for both the realization of bidirectional switch and the PWM control have been investigated and most of the initial limitations of the matrix converter are nowadays overcome. This continuous effort has been rewarded with system-level products like *Yaskawa Matrix Converter* or power modules from manufacturers like *Infineon*, *Semelab*, *Dynex* or *IXYS*.

The *Yaskawa's AC7 Matrix Converter* [8] represents the world's first low voltage (230 V/460 V) matrix converter to directly convert input AC voltage to output AC voltage without the need for a DC Bus. Using this configuration helps improve energy efficiency and it also overcomes many problems typically associated with conventional drives such as harmonic current distortion, line regeneration, installation space, or motor drive issues with very long cable lengths, bearing currents, and common mode currents. This product line is rated up to 7.5 HP, 15 HP, 30/40 HP, with possible extension to 125HP, and it is recommended for applications like centrifuges, elevators, escalators or various test stands. A second product line is offered by *Yaskawa Corporation* to the European market as *Varispeed F7*.

Despite the academic success concerning the implementation of the matrix converters [9–11], it is worth noting that the *Yaskawa Corporation* changed the name of all its products in 2007 from names ending in the generation number like ..."*G7*," and so on, into names with 1000s like *A2000*. They changed all these names except for the *Varispeed AC Drive* based on matrix converters. These are still shown in the product line-up but they do not show in the annual financial report for 2011–2012 as a separate sales line as it was in 2004–2006. It does not seem that this is a representative or sellable product anymore. Moreover, the Medium Voltage (1000s of Volts) version of the matrix converter made by *Yaskawa Corporation* was only a one-time success for a skin mill corporation [12].

Another Japanese manufacturer, *Fuji* brought up to production a matrix converter series called *Frenic-MX* in 2006. Most of the other manufacturers have designed matrix converters for demonstration purposes, without finding enough benefits and customer appreciation to really replace (or complete) conventional back-to-back topologies with matrix converters within their product line.

Merits are proven usually in space and volume challenged applications like aviation systems [13–15]. Hence, a good opportunity for project development came along the *All Electric Aircraft* concept.

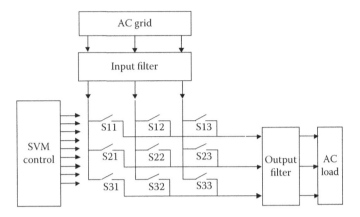

FIGURE 16.1   Basic matrix converter topology.

The academic interest for matrix converter topologies has fuelled industrial research, and power semiconductor manufacturers came up with integrated modules for implementation of the power stage suitable to a matrix converter.

The conventional *AC/AC matrix converter* is a matrix of switches that allows an independent control of frequency, amplitude, and phase of the outputs while the operation is synchronized with the input AC waveforms for operation with controllable displacement power factor (Figure 16.1).

Since three-phase systems are most commonly used, the matrix dimension is herein limited to three. Given all connection possibilities, the three-phase AC–AC matrix converter consists of nine bidirectional switches which allow an output terminal to be connected to any input terminal. Any symmetrical three-phase output voltages can be obtained from a set of input voltages by suitable switching of the matrix. The PWM operation of the switches requires that three voltage sources (or capacitors) are present on only one side to provide bypass paths for the inductive currents. Since the input AC source is conventionally an industrial or residential power grid, with a strong voltage source character, the output side will produce a voltage source character towards the load.

The load current is reflected towards the grid through switches into a pulsed current. When the pulsations of this current poses a problem to the grid, an input LC filter is required. The matrix converter will act as a current-type load, extracting current from the filter's capacitor. In most cases, this capacitor will have its voltage uncontrolled by a feedback loop, and this should be considered in the protection circuitry.

The three-phase AC load can be a motor drive or a power supply load. In either case, the load should present an inductive character towards the power converter. For instance, a three-phase AC power supply will have a three-phase LC filter on the output side to provide the required inductive load character to the converter. Moreover, only one of the switches linked to the same output terminal can be in conduction at a given time. This means there are 27 possibilities of input-to-output connection (Figure 16.2):

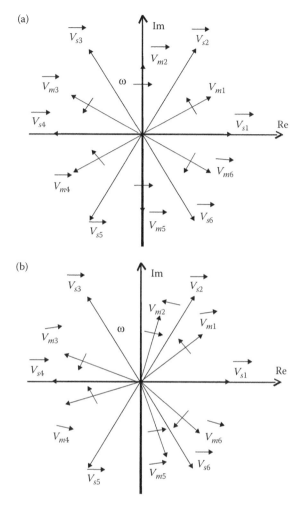

FIGURE 16.2 Possible space vector positions within the complex plane on the load-side reference frame when considering both stationary and rotating vectors: different moments captured in (a) and (b).

- Six possibilities to connect each output terminal to a different input terminal; these are equally $2\pi/3$ out of phase, and each can be represented by a space vector rotating in the complex plane of the load reference frame (let us call them as "*rotating vectors*").
- Eighteen possibilities to connect two output nodes to the same input terminal while the third output is linked to a different input. The matrix converter could be seen as a functional superposition of three three-phase inverters supplied by rectified line voltages. The appropriate switching vectors are fixed in six directions equally $\pi/3$ out of phase over the complex plane of the load reference frame and their magnitudes are variables as the rectified voltages evolve (let us call these vectors "*stationary vectors*").

- Three null-states when all three output nodes are connected to the same input terminal.

Designing a PWM algorithm means averaging in between the available states in order to emulate a rotating vector in the load reference frame. Various methods are hence available.

16.2 IMPLEMENTATION OF THE POWER SWITCH

The most important hardware challenge relates to the implementation of the bi-directional power switch. The bi-directional switch must be able to conduct currents of both directions, and to block voltages of both polarities. Conventional solutions are presented in Figure 16.3, and they are carried out with bipolar transistors, MOSFETs, or IGBTs. The left-side assembly in (a) prevents a large voltage drop across the emitter–collector circuitry, and therefore diodes need to provide good reverse recovery characteristics. Alternatively, diodes can be made on the *SiC* substrate.

The circuit shown in Figure 16.3a is offered by certain power semiconductor manufacturers as copackaged in a single unit. For instance, *Dynex Semiconductor* offers hybrid bi-directional switches up to 400 A, 1700 V, mounted on a 140 × 73 mm metal baseplate (DIM400PBM17-A000, [16]). Alternatively, a group of three bidirectional switches are offered in the same package as an output leg (phase) configuration. Such configuration helps reducing the parasitic inductance between devices and improves the switching process when transferring the load current from one device

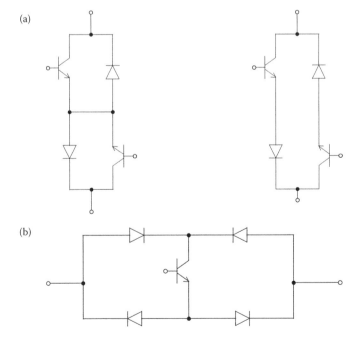

FIGURE 16.3 Conventional implementation of the bi-directional power switch.

to another. Products from *Dynex Semiconductor* and *Semelab* can be mentioned in this category. For instance, SML300MAT06 by *Semelab* is rated at 300 A, 600 V and intended for a matrix converter leg. Further on, more integration is proposed by *Infineon/Eupec* with a full matrix converter integrated on the same module rated at 35 A, 1200 V (EUPEC FM35R12KE3, [17]).

The advent of SiC based devices has provided a chance for improvement of these configurations by replacement of diodes with SiC diodes or even prototypes for replacement of the power switches with SiC JFET devices [18]. This approach enabled power density levels in forced air cooled systems to 20 kW/dm^3.

Given the success of the matrix converter topology as well as of other converter topologies with bi-directional switches, the *RB-IGBT* (reverse blocking IGBT) device has been developed recently [19,20]. This new device has advantages in reduction of the voltage drop during the conduction state since series diodes are no longer necessary for reverse blocking capability. An RB-IGBT has the same fundamental structure as a conventional IGBT. Therefore, the trade-off between switching loss and on-state voltage in a RB-IGBT does not differ from that of a conventional IGBT, except for the need for diodes in the conventional bi-directional switch. When reverse biased, the recovery characteristics are the same as for a conventional diode. Products are already available on the market for applications with currents less than 100 A, in the low voltage range [20].

It is worth noting here that the conventional matrix converter equipped with the conventional implementation of the power switch shown in Figure 16.3a allows a reduction of losses in the entire system by 1/3 when compared to the back-to-back converter made of identical IGBT devices [21], or an efficiency change from 94% to 96% [11,21]. The use of RB-IGBT reduces further the system loss by 40% [21]. The power density is also improved by roughly 1/3 from back-to-back converter to the matrix converter [11,21]. Such ballpark numbers worth be remembered even if numerous other loss or efficiency measurements are accurately reported in literature for a multitude of PWM arrangements.

16.3   CURRENT COMMUTATION

The PWM control of the switches within the matrix converter topology must ensure always a path for the circulation of the load current (usually of inductive nature). On the other hand, two input nodes should never be short-circuited.

The current commutation within bidirectional switches built as shown in Figure 16. 3a has been analyzed in [2,22,23]. Because of the unpredictable time delays at the switch commutation, open- or short-circuit hazards could happen. A multistepped switching procedure is considered in [22,23], which requires independent control of the current flow on the two possible directions of the bidirectional switch.

A possible modeling approach is based on identification of all possible states at current commutation and a local state machine able to secure the multistep procedure. In order to understand this approach, let us herein consider the simplified case of commutation of the load current from an input phase to another (Figure 16.4). Let us assume that we will start from the state with current circulating through leg A, when both switches A12 and A11 have gate signals for turning-on. Conversely, switches B11

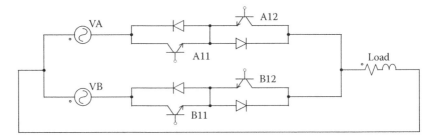

FIGURE 16.4 Sketch for understanding current commutation.

and B12 have the gate signals corresponding to the turn-off state. When the current commutation sequence starts and the voltage in the second leg is larger than the grid phase voltage in the first leg, we can use the current direction information and change the gate control for turning-off the switch (A11 or A12) which is anyway off (states 2 and 4 on the second column of Figure 16.5). Next, we can apply the gate signal for turning-on the switch which does not correspond to the current circulation among B11 and B12. Finally, the other switch among B11 and B12 can be turned-on to carry the load current and to turn-off the conducting switch on the first leg.

This way, based on grid voltage relationship and load current direction, we can judge all possible state sequences. This yields in the diagram shown in Figure 16.5, which includes all possible cases and codes the migration from a state to another from the input voltage relationship and load current direction. For completeness of information, the case with zero current is also included. Since there are four switches

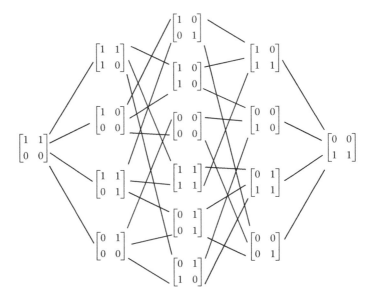

FIGURE 16.5 Diagram of the state transition for proper current commutation.

involved (two would have to turn-off and the other two to turn-on), four columns are shown as four intermediary stages of the current commutation process.

While Figure 16.5 shows the mathematical modeling of states, different practical solutions for physical implementation have been proposed over time. They all count on either sensing both the output current and input voltage, or at least one among the output current and input voltage. In certain situations, soft turn-on and soft turn-off instances can be achieved due to zero current switching.

16.4 CLAMPING THE REACTIVE ENERGY

Along the conventional IGBT protection circuitry, the matrix converter requires special care in handling the recovery of the reactive energy from the load circuitry since there is no natural free-wheeling path. The common solution is illustrated in Figure 16.6. The grid-side diode rectifier establishes the voltage to clamp to. The capacitor is typically very small and its value depends on nature of load. For instance, a 3 kW Matrix Converter Drive for an Aircraft Application [14,15] at a maximum output current of 30 A and a machine inductance of 1.15 mH requires a capacitor of 2 μF. The capacitor is discharged across an additional resistor or used to power up the other auxiliary power supplies (sensing and measurement, microcontroller, gate drivers, and so on).

An alternative solution for lower power levels consists in the use of varistor devices for clamping voltages across the input and output lines [10].

16.5 PWM ALGORITHMS

16.5.1 SINUSOIDAL CARRIER–BASED PWM

Similar to the conventional PWM converter, a modulation function is attributed to each switch and a constant time interval is selected to correspond to the desired carrier frequency. The block diagram is shown in Figure 16.7. The implementation consists of comparing each modulation function with the same carrier triangular signal as in the case of PWM for conventional six-switch converter. Since two input lines should never be short-circuited, the switches on the same row in Figure 16.1 should

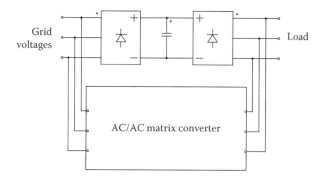

FIGURE 16.6 Clamping circuit for protection.

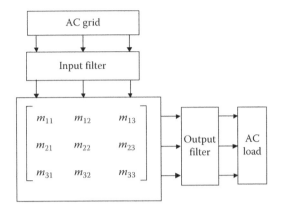

FIGURE 16.7   Introducing the modulation functions.

never conduct at the same moment. Moreover, each output should be connected to an input for the current continuation. This yields:

$$\begin{cases} m_{11} + m_{12} + m_{13} = 1 \\ m_{21} + m_{22} + m_{23} = 1 \\ m_{31} + m_{32} + m_{33} = 1 \end{cases}$$

The output–input relationship is given by [24]

$$[e] = [M] \cdot [v_2] \Rightarrow [v_2] = [M]^T \cdot [e] \tag{16.1}$$

It has been demonstrated that $[M]$ can be decomposed into two matrices that are basically separating *the frequency changer* $[M_f]$ and the *static VAR compensator* $[M_\phi]$ components:

$$[M] = [M_f(t)] + [M_\phi(t)] \tag{16.2}$$

Moreover,

$$[M_f(t)] = \begin{bmatrix} C_{12}(0) & C_{12}\left(-\dfrac{2\pi}{3}\right) & C_{12}\left(\dfrac{2\pi}{3}\right) \\[2ex] C_{12}\left(-\dfrac{2\pi}{3}\right) & C_{12}\left(\dfrac{2\pi}{3}\right) & C_{12}(0) \\[2ex] C_{12}\left(\dfrac{2\pi}{3}\right) & C_{12}(0) & C_{12}\left(-\dfrac{2\pi}{3}\right) \end{bmatrix} \tag{16.3}$$

The $[M_f]$ matrix is derived using the d–q to a–b–c transform and considers the power transfer through a fictitious intermediary DC link when the energy transfer through zero-sequence components is not considered. All possible control cases share the same general expression for the direct transfer function $[M_f]$,

$$[M_f(t)] = [C(\omega_1)]_{3\times2} \cdot [P]_{2\times2} \cdot [C(\omega_2)]_{3\times2}^T \tag{16.4}$$

Where

$$[C(\omega_i)] = [b_1(\omega_i) \quad b_2(\omega_i) \quad b_3(\omega_i)] \tag{16.5}$$

represents a matrix composed of *ortho-normal base vectors* of the input or output three-phase systems and $[P]$ represents a matrix of constant weights p_{ij}, with the meaning of transfer power.

Let us first consider three different cases:

(A) Choosing

$$[P]_{2\times2} = P_f \cdot \begin{bmatrix} 1 & 0 \\ 0 & -1 \end{bmatrix} \tag{16.6}$$

defines conjugated forms

$$C_{12}^A(x) = \frac{2}{3} \cdot P_f \cdot \cos[(\omega_1 + \omega_2)t + x] \tag{16.7}$$

(B) Choosing

$$[P]_{2\times2} = P_f \cdot \begin{bmatrix} 1 & 0 \\ 0 & 1 \end{bmatrix} \tag{16.8}$$

defines nonconjugated forms

$$C_{12}^B(x) = \frac{2}{3} \cdot P_f \cdot \cos[(\omega_1 - \omega_2)t + x] \tag{16.9}$$

(C) Considering both the conjugated and nonconjugated forms leads to

$$C_{12}^C(x) = P_f \cdot \left\{ \frac{1}{3} \cdot \cos[(\omega_1 - \omega_2)t + x] + \frac{1}{3} \cdot \cos[(\omega_1 + \omega_2)t + x] \right\} \tag{16.10}$$

In all of these forms, P_f plays the same the role as the turns ratio of the primary to secondary windings of a magnetic transformer and $x = 0, \dfrac{2\pi}{3}, -\dfrac{2\pi}{3}$.

The second term ($[M_\phi]$) is considered when a transfer through the zero sequence is possible to lead to a DC component on the output side. This can be used for instance to compensate the neutral node voltage. The generation of a DC component on the load side is generally not the case of a conventional matrix converter. However, the mathematical modeling is herein included for completeness of information.

$$[M_\phi(t)] = \begin{bmatrix} C_{1\phi}(0) & C_{1\phi}(0) & C_{1\phi}(0) \\ C_{1\phi}\left(-\dfrac{2\pi}{3}\right) & C_{1\phi}\left(-\dfrac{2\pi}{3}\right) & C_{1\phi}(-\dfrac{2\pi}{3}) \\ C_{1\phi}\left(\dfrac{2\pi}{3}\right) & C_{1\phi}\left(\dfrac{2\pi}{3}\right) & C_{1\phi}\left(\dfrac{2\pi}{3}\right) \end{bmatrix} \tag{16.11}$$

where

$$C_{1\phi}(x) = \frac{\sqrt{2}}{3} \cdot Pm \cdot \cos\left[\omega_1 \cdot t + \phi + x\right] \tag{16.12}$$

and $\quad x = 0, \dfrac{2\pi}{3}, -\dfrac{2\pi}{3}$

Through $[M_\phi]$, the zero sequence voltage V_{20} can be produced and regulated on output side of the converter which projects an AC voltage at angular frequency ω_1 on the input side whose magnitude and phase angle are controllable by P_m and ϕ, respectively.

Any of the above algorithms for $[M_f]$ (the conjugated or nonconjugated forms) produces an output voltage with a limited maximum voltage. A general solution applied in PWM three-phase bridge converters consists of injecting the third harmonic in the modulating waveform. Similar approach is considered for a matrix converter in [25,26] where the output voltage is considered as

$$[V_0] = V_0 \cdot \begin{bmatrix} \cos(\omega_2 t) \\ \cos(\omega_2 t - \dfrac{2\cdot\pi}{3}) \\ \cos(\omega_2 t - \dfrac{4\pi}{3}) \end{bmatrix} + \frac{V_i}{4} \cdot \begin{bmatrix} \cos(3\omega_1 t) \\ \cos(3\omega_1 t) \\ \cos(3\omega_1 t) \end{bmatrix} - \frac{V_0}{6} \cdot \begin{bmatrix} \cos(3\cdot\omega_2 t) \\ \cos(3\cdot\omega_2 t) \\ \cos(3\cdot\omega_2 t) \end{bmatrix} =$$

$$= V_0 \cdot \begin{bmatrix} \cos(\omega_2 t) \\ \cos(\omega_2 t - \dfrac{2\cdot\pi}{3}) \\ \cos(\omega_2 t - \dfrac{4\pi}{3}) \end{bmatrix} + \frac{V_0}{2\cdot\sqrt{3}} \cdot \begin{bmatrix} \cos(3\omega_1 t) \\ \cos(3\omega_1 t) \\ \cos(3\omega_1 t) \end{bmatrix} - \frac{V_0}{6} \cdot \begin{bmatrix} \cos(3\cdot\omega_2 t) \\ \cos(3\cdot\omega_2 t) \\ \cos(3\cdot\omega_2 t) \end{bmatrix} =$$

$$= m \cdot V_{\max} \cdot \left\{ \begin{bmatrix} \cos(\omega_2 t) \\ \cos(\omega_2 t - \dfrac{2 \cdot \pi}{3}) \\ \cos(\omega_2 t - \dfrac{4\pi}{3}) \end{bmatrix} + \dfrac{1}{2 \cdot \sqrt{3}} \cdot \begin{bmatrix} \cos(3\omega_1 t) \\ \cos(3\omega_1 t) \\ \cos(3\omega_1 t) \end{bmatrix} - \dfrac{1}{6} \cdot \begin{bmatrix} \cos(3 \cdot \omega_2 t) \\ \cos(3 \cdot \omega_2 t) \\ \cos(3 \cdot \omega_2 t) \end{bmatrix} \right\}$$

$$(16.13)$$

This leads to the following m_{ij} form of the modulating waveform to produce unity input power factor.

$$
\begin{aligned}
m_{ji}(t) = {} & \frac{1}{3} \cdot \left[1 + P_f \cdot \cos\left[(\omega_1 + \omega_2) \cdot t - (2 \cdot (i+j) - 4) \cdot \frac{\pi}{3} \right] \right. \\
& \left. + P_f \cdot \cos\left[(\omega_2 - \omega_1) \cdot t - 2 \cdot (i - j) \cdot \frac{\pi}{3} \right] \right] \\
& - \frac{1}{18} \cdot \left[P_f \cdot \cos\left[(\omega_1 + 3 \cdot \omega_2) \cdot t - 2 \cdot (j - 1) \cdot \frac{\pi}{3} \right] \right. \\
& \left. + P_f \cdot \cos\left[(\omega_2 - \omega_1) \cdot t - 2 \cdot (1 - j) \cdot \frac{\pi}{3} \right] \right] \\
& - \frac{1}{18 \cdot \sqrt{3}} \cdot P_f \cdot \cos\left[4 \cdot \omega_1 \cdot t - 2 \cdot (j - 1) \cdot \frac{\pi}{3} \right] \\
& + \frac{7}{18 \cdot \sqrt{3}} P_f \cdot \cos\left[2 \cdot \omega_1 \cdot t - 2 \cdot (1 - j) \cdot \frac{\pi}{3} \right]
\end{aligned}
$$

$$(16.14)$$

Using these modulation signals improves the maximum available output voltage from 0.500 to 0.866 of the peak of the input voltage (Figure 16.8). As will be shown later on, similar results can be achieved with Space Vector Modulation [27].

The generation of the gate control pulses is achieved with a comparison between these reference signals and a triangular carrier waveform. The solution proposed in [25] considers a single-ended carrier (Figure 16.9). It can also be improved for reduction of the switching processes with a double-sided symmetrical pulse generation. Either case, the sequence is selected so that the input voltage with the largest instantaneous value corresponds to the state placed in the middle (like phase "2" in the figure).

16.5.2 SPACE VECTOR MODULATION CONSIDERING ALL POSSIBLE SWITCHING VECTORS

Almost all the previously reported *Space Vector PWM* algorithms are based on the stationary vectors and a clear and convincing, comparative analysis of the use of all the switching combinations has not been reported yet. Ref. [3] develops the

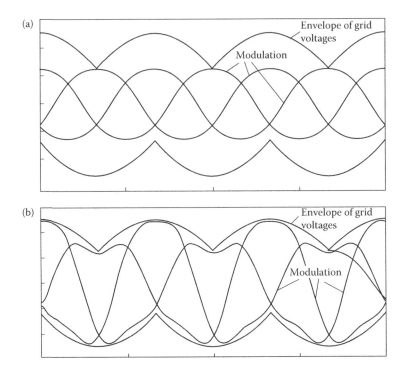

FIGURE 16.8 Reference modulation waveforms for (a) conventional, (b) extended voltage range.

mathematical model and the appropriate control algorithm for the use of all the available switching combinations including the calculus of the time portions, simulation results, and the steps for the implementation of an algorithm based on all the switching vectors.

Intuitively, the generation of the desired space vector by averaging in between two adjacent vectors closer to each other than in the case of a conventional six-switch inverter (that is 60°) would improve the quality of the output voltage waveform. This

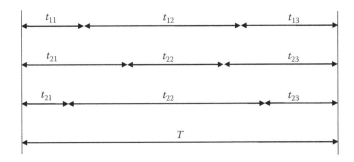

FIGURE 16.9 Example of state sequence.

principle was used successfully at multilevel power inverters. The difference here is the rotation of one of the vectors that suggests the use of a high enough switching frequency. Moreover, there is no obvious advantage in the input (grid side) currents as the PWM cannot easily be optimized for the input when both stationary and rotating vectors are used.

Applying the voltage space vector to a three-phase R–L load leads to a phase shift in between the voltage and current space vectors in the complex plane of the load reference frame. The input currents can be derived by reflecting these output currents through the matrix converter. They can be determined from the output current space vector by analyzing the system of all switching vectors in the grid source reference frame. The input current space vectors that correspond to the main voltage switching vectors are fixed while those corresponding to the "*stationary*" vectors are rotating at a relative speed determined by the mains and output frequencies. It can be inferred that the phase angle in between the voltage and current space vectors at the fundamental frequency is preserved.

16.5.2.1 Selection of the Closest Rotating and Stationary Vectors

The desired vector position is defined with polar coordinates (V,α). The angular resolution is ensured similarly at both small and large magnitudes of desired vector. The selection of the closest *Rotating* and *Stationary* vectors can be simplified in a digital implementation by defining all the angular coordinates as binary words with the three most significant bits set up for sector number, as a divide-by-six counter.

On the load reference frame, the rotating vector V_{m1} is superimposed on the real axis when the v_{in1} voltage is maximum. In order to avoid the dependence of the angular coordinate of V_{m1} on the supply frequency variations, synchronization is achieved by a PLL circuit from the $v_{in2,3}$ input grid voltage. The absolute value of angular coordinate of the rotating vector V_{m1} can be read from a counter placed in the loop of the PLL circuit. A fast selection of the line voltage that defines the *Stationary* vector magnitude is obtained with the three most significant bits of this counter.

Finally, the sequence of states can be selected as

Zero vector → *stationary vector* → *rotating vector* → *zero vector*

and this is applied on each sampling interval. Analogous to the harmonic reference generation from Section 5.7.1, a double-sided PWM can be generated.

16.5.2.2 Definition of Time Intervals

Any instantaneous position of the desired load-side voltage vector is always placed in between a rotating and a stationary vector. A generalized sector bounded by a *rotating* vector (denoted with index "*m*") and a *stationary* vector (denoted with index "*s*") is dynamically defined analogous with [6,7] (Figure 16.10). This simplifies the calculation of the time constants. Definition of the time portions allocated to the switching vectors yields from the averaging relationship:

$$\overrightarrow{V_m} \cdot t_m + \overrightarrow{V_s} \cdot t_s + \overrightarrow{V_0} \cdot t_0 = \overrightarrow{V} \cdot T \qquad (16.15)$$

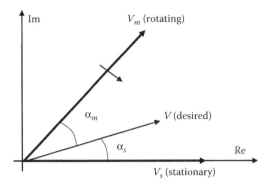

FIGURE 16.10 Calculation of time intervals allocated to each state.

where $t_m + t_s + t_0 = T$

The calculation is performed with the projection of both desired and switching vectors on *Real* and *Imaginary* axis, one of the stationary switching vectors having the same direction as the hypothetical *Real* axis of the load reference frame (Figure 16.2). This yields:

$$\text{Re: } T \cdot V \cdot \cos\alpha = \left[\frac{1}{T} \cdot \int_0^T (\text{Re } V_{s})dt \right] \cdot t_s + \left[\frac{1}{T} \cdot \int_0^T (\text{Re } V_{m})dt \right] \cdot t_m$$

$$\text{Im: } T \cdot V \cdot \sin\alpha = \left[\frac{1}{T} \cdot \int_0^T (\text{Im } V_{s})dt \right] \cdot t_s + \left[\frac{1}{T} \cdot \int_0^T (\text{Im } V_{m})dt \right] \cdot t_m \qquad (16.16)$$

Solving the system for the situation of a high pulse rate—when the effects of the rotation of the reference system can be overlooked—and taking into consideration the dependency of the secondary vector magnitude with α_s and α_m yields the solution:

$$t_m = \frac{V}{V_m} \cdot T \cdot \frac{\sin\alpha_s}{\sin(\alpha_s + \alpha_m)} = \frac{V}{\sqrt{2} \cdot E} \cdot T \cdot \frac{\sin\alpha_s}{\sin(\alpha_s + \alpha_m)}$$

$$t_s = \frac{V}{V_s} \cdot T \cdot \frac{\sin\alpha_m}{\sin(\alpha_s + \alpha_m)} = \frac{V}{\sqrt{2} \cdot E} \cdot T \cdot \frac{\sin\alpha_m}{\sin(\alpha_s + \alpha_m)} \cdot \frac{\cos\dfrac{\pi}{3}}{\cos\left(\dfrac{\pi}{6} - (\alpha_s + \alpha_m)\right)}$$

$$(16.17)$$

These time intervals are the same for both possible successions of the *rotating* and *stationary* vectors. The portions of time allocated to the null-states over the sampling period can be considered equal with each other and are calculated with

$$t_{01} = t_{02} = \frac{1}{2} \cdot T \cdot \left[1 - \frac{V}{V_i} \cdot \frac{\cos(\frac{\pi}{6} - \alpha_s)}{\cos(\frac{\pi}{6} - \alpha_s - \alpha_m)} \right] =$$

$$\frac{1}{2} \cdot T \cdot \left[1 - \frac{V}{\sqrt{2} \cdot E} \cdot \frac{\cos(\frac{\pi}{6} - \alpha_s)}{\cos(\frac{\pi}{6} - \alpha_s - \alpha_m)} \right]$$

(16.18)

The maximum attainable modulation index is defined as V_{rms}/V_{in} and it is calculated to be 0.866. This value equals the maximum achievable with the conventional back-to-back converter as well as with the improved carrier-based PWM for matrix converter [2].

An example of waveforms defining operation of the matrix converter with a PWM algorithm derived from vectorial analysis of the power stage with consideration of both the stationary and rotating vectors is shown in Figure 16.11.

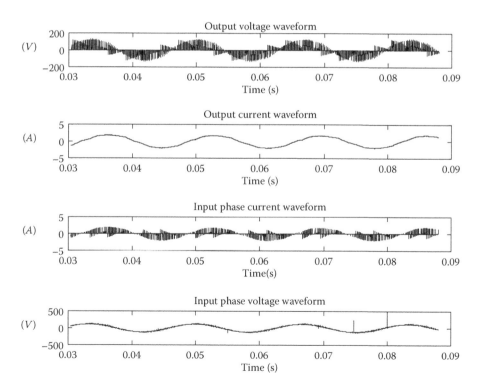

FIGURE 16.11 Waveforms at the matrix converter terminals for $N = 72$, $f_1 = 30$ Hz, $T = 0.001388$ s, $m = 0.7$, load composed of $R = 5$ ohm, $L = 50$ mH.

16.5.3 Space Vector Modulation Considering Stationary Vectors Only

Since the previous PWM algorithm is complex, very dependent on synchronization with the phase of the input grid voltages, and without proven major benefits, a *Space Vector Modulation* based on only the stationary vectors is generally considered [28]. All switching combinations are shown in Table 16.1 along with the values for the generated output voltage. The vectorial representation is shown in Figure 16.12 and we can see that each vectorial position for the output voltage space vector can be achieved with multiple combinations of switches.

The PWM algorithm is very intuitive as it can be seen as a superposition of a rectifier-inverter back-to-back system without an intermediary DC link.

The same Equation 5.30 as in the case of a three-phase inverter can be used here:

$$
\begin{cases}
t_a = \dfrac{3V_s}{2V_{rec}} \cdot T_S \cdot \left(\cos\alpha - \dfrac{1}{\sqrt{3}}\sin\alpha\right) = \dfrac{\sqrt{3} \cdot V_s}{V_{rec}} \cdot T_S \cdot \sin\left[\dfrac{\pi}{3} - \alpha\right] \\[4mm]
t_b = \dfrac{\sqrt{3} \cdot V_s}{V_{rec}} \cdot T_S \cdot \sin\alpha \\[4mm]
t_0 = T_S - t_a - t_b
\end{cases}
\tag{16.19}
$$

where the DC voltage has been replaced with a variable signal V_{rec} corresponding to the instantaneous average of the three-phase PWM rectified voltage.

A simplified approach considers the "rectifier" as a three-phase diode rectifier, operating as a selector of the highest line-to-line voltage in the input, voltage that is applied to the "inverter" side [29,30] (consider Figure 1.2 without the intermediary DC link module). The "rectifier" side is therefore not modulated, with pairs of switches being commutated at each 60°. The DC link voltage V_{rec} follows the line-to-line voltage, and so does its moving average, calculated at the sampling frequency. Compensation of this fluctuation can be achieved with a simple feed-forward use of the rectified voltage into Equation 16.19 (Figure 16.13). Sample results are shown in Figure 16.14. The 120° conduction intervals for each input current can be seen.

If this solution had its obvious merits when implemented with a real rectifier-inverter structure without a DC link capacitor, it does not make use of all resources when implemented within a 9-switch matrix converter.

If the input phase currents are required to be controlled (displacement power factor), a PWM operation of the fictitious "rectifier" should be considered. The fictitious "rectifier" operates now as a *Current Source Converter* with two switches conducting the DC-side current at any moment. The intermediary DC link voltage and current are now shown in Figure 16.15. The moving average calculated at each interval equal to the PWM sampling interval is also shown in the figure, and it looks like a reversed rectified voltage [11] (Figure 16.16).

The calculation of the time intervals allocated to each state can be expressed in dependence to the phase of the input voltage and current using the information about the averaged fictitious DC link voltage as depicted from Figure 16.16. In the most

TABLE 16.1

Switching Sequences and Voltages, Considering Input Nodes as R,S,T and Output Nodes as a,b,c

Vector	Switches			Output Phase Voltages			Output Voltages		Line-Line	Input Currents		
+1	S11	S22	S32	R	S	S	R-S	0	S-R	I_a	I_b+I_c	0
−1	S12	S21	S31	S	R	R	S-R	0	R-S	I_b+I_c	I_a	0
+2	S12	S23	S33	S	T	S	S-T	0	T-S	0	I_a	I_b+I_c
−2	S13	S22	S32	T	S	S	T-S	0	S-T	0	I_b+I_c	I_a
+3	S13	S21	S31	T	R	R	T-R	0	R-T	I_b+I_c	0	I_a
−3	S11	S23	S33	R	T	R	R-T	0	T-R	I_a	0	I_b+I_c
+4	S12	S21	S32	R	S	S	S-R	R-S	0	I_b	I_a+I_c	0
−4	S11	S22	S31	R	R	S	R-S	S-R	0	I_a+I_c	I_b	0
+5	S13	S22	S33	S	S	T	T-S	S-T	0	0	I_b	I_a+I_c
−5	S12	S23	S32	S	T	T	S-T	T-S	0	0	I_a+I_c	I_b
+6	S11	S23	S31	R	R	S	R-T	T-R	0	I_a+I_c	0	I_b
−6	S13	S21	S33	T	R	T	T-R	R-T	0	I_b	0	I_a+I_c
+7	S12	S22	S31	R	S	R	0	S-R	R-S	I_c	I_a+I_b	0
−7	S11	S21	S32	R	R	S	0	R-S	S-R	I_a+I_b	I_c	0
+8	S13	S23	S32	T	T	S	0	T-S	S-T	0	I_c	I_a+I_b
−8	S12	S22	S33	S	S	T	0	S-T	T-S	0	I_a+I_b	I_c
+9	S11	S21	S31	R	R	T	0	R-T	T-R	I_a+I_b	I_c	I_b
−9	S13	S23	S31	T	T	R	0	T-R	R-T	I_c	0	I_a+I_b

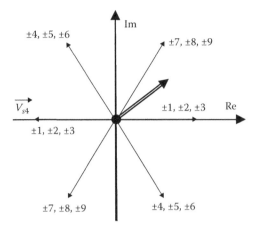

FIGURE 16.12   Possible switching vectors derived from stationary vectors shown in Table 16.1.

general case, when the phase of the input current is required to be controllable, these
equations yield:

$$
\begin{cases}
t_a = \dfrac{2}{\sqrt{3}} \cdot \dfrac{V_s}{V_1 \cdot \cos(\Phi)} \cdot \cos\left[\varphi_i - \dfrac{\pi}{6}\right] \cdot T_S \cdot \sin\left[\dfrac{\pi}{3} - \alpha\right] \\[3mm]
t_b = \dfrac{2}{\sqrt{3}} \cdot \dfrac{V_s}{V_1 \cdot \cos(\Phi)} \cdot \cos\left[\varphi_i - \dfrac{\pi}{6}\right] \cdot T_S \cdot \sin\alpha \\[3mm]
t_0 = T_S - t_a - t_b
\end{cases}
\qquad (16.20)
$$

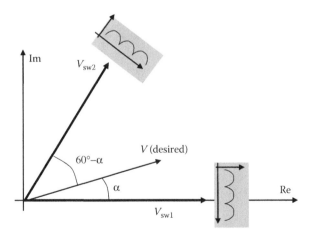

FIGURE 16.13   Calculation of time intervals allocated to each state for the "inverter" operation when considering the DC link fluctuations in the case of fictitious rectifier-inverter modeling, without any modulation on the grid-side.

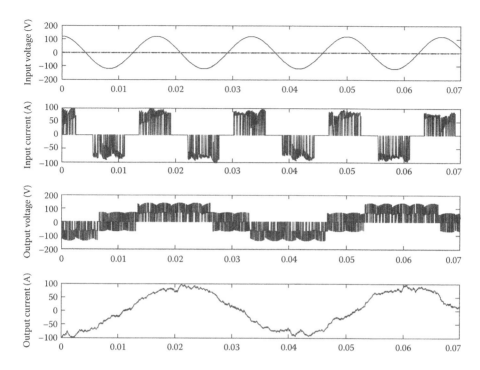

FIGURE 16.14 Characteristic waveforms for operation without modulation of the fictitious rectifier. Operation with modulation index $m = 0.7$, output frequency $f_o = 12.5$ Hz, switching frequency 5.4 kHz, with grid at 120 V and 60 Hz, and $R-L$ load with $L = 1$ mH, $R = 1$ Ohm.

where ϕ_i represents the instantaneous angular coordinate of the input current in the complex plane with input-side reference frame, bounded within a 60° sector in between adjacent current vectors; and $\Phi = \phi_u - \phi_i$ represents the voltage-current phase shift for the input. For the operation with unity power factor, $\cos \Phi = 1$.

The theory presented so far does not guarantee the control of the displacement factor in the input. Since there is no rule for the selection of the state corresponding to a desired active vector, we can use this degree of freedom for selecting the state which allows a control of the displacement factor. The idea is to generate each switching state with two vectors, or to use four active states over a sampling interval. For instance, the V_{sw1} position can be achieved with any of the states ±1, ±2, ±3, and the V_{sw2} position can be achieved with any of the states ±7, ±8, ±9 (Figure 16.12). For displacement factor control, Figure 16.17 shows the input current vectors corresponding to each state.

The same principle for generation of the desired vector position is used as in the case of a *Current Source Converter*. The free-wheeling states (zero vector) are reserved now for the "inverter" stage, and the PWM for the current vector is reduced

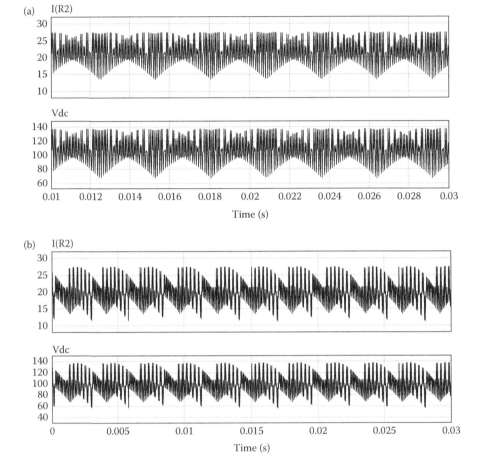

FIGURE 16.15 Generic DC link voltage and current for unity power factor (a) or 45° phase shift (b) on the grid side, and without any capacitor in the DC link (f_{sw} = 10 kHz).

at splitting each of the states t_a and t_b from the "inverter" operation into two states for modulation of the input current. It yields:

$$\begin{cases} t_{a1} = t_a \cdot \dfrac{\sin\left[\dfrac{\pi}{3} - \varphi_i\right]}{\cos\left(\varphi_i - \dfrac{\pi}{6}\right)} \\[4mm] t_{a2} = t_a \cdot \dfrac{\sin\left[\varphi_i\right]}{\cos\left(\varphi_i - \dfrac{\pi}{6}\right)} \end{cases} \quad \begin{cases} t_{b1} = t_b \cdot \dfrac{\sin\left[\dfrac{\pi}{3} - \varphi_i\right]}{\cos\left(\varphi_i - \dfrac{\pi}{6}\right)} \\[4mm] t_{b2} = t_b \cdot \dfrac{\sin\left[\varphi_i\right]}{\cos\left(\varphi_i - \dfrac{\pi}{6}\right)} \end{cases} \qquad (16.21)$$

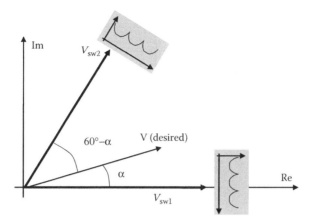

FIGURE 16.16 Calculation of time intervals allocated to each state for the "inverter" operation when considering the DC link fluctuations produced by the PWM operation of the rectifier side within the fictitious rectifier-inverter modeling.

As mentioned, ϕ_i represents the instantaneous angular coordinate of the input current in the complex plane with input-side reference frame, measured from an active current vector into a 60° sector. This is equivalent with

$$
\begin{cases}
t_{a1} = \dfrac{2}{\sqrt{3}} \cdot \dfrac{V_s}{V_1 \cdot \cos(\Phi)} \cdot T_S \cdot \sin\left[\dfrac{\pi}{3} - \alpha\right] \cdot \sin\left[\dfrac{\pi}{3} - \varphi_i\right] \\[2ex]
t_{a2} = \dfrac{2}{\sqrt{3}} \cdot \dfrac{V_s}{V_1 \cdot \cos(\Phi)} \cdot T_S \cdot \sin\left[\dfrac{\pi}{3} - \alpha\right] \cdot \sin[\varphi_i] \\[2ex]
t_{b1} = \dfrac{2}{\sqrt{3}} \cdot \dfrac{V_s}{V_1 \cdot \cos(\Phi)} \cdot T_S \cdot \sin\alpha \cdot \sin\left[\dfrac{\pi}{3} - \varphi_i\right] \\[2ex]
t_{b2} = \dfrac{2}{\sqrt{3}} \cdot \dfrac{V_s}{V_1 \cdot \cos(\Phi)} \cdot T_S \cdot \sin\alpha \cdot \sin[\varphi_i]
\end{cases}
\tag{16.22}
$$

This helps producing a set of sinusoidal input currents with any desired phase shift. As the desired input current vector rotates within this complex plane representation, the most suitable states are selected for PWM generation. Starting from this principle, multiple PWM switching sequences are yet possible and they were reported in literature. The most known method uses four active vector states over each sampling period (Figure 16.18) [1], and this is considered herein for the waveform examples.

Results for this method are shown in the following figures:

- High modulation index $m = 0.7$, output frequency $f_o = 12.5$ Hz (Figure 16.19);
- Low modulation index $m = 0.15$, output frequency $f_o = 12.5$ Hz (Figure 16.20);
- High modulation index $m = 0.7$, output frequency $f_o = 52.0$ Hz (Figure 16.21);
- Low modulation index $m = 0.15$, output frequency $f_o = 52.0$ Hz (Figure 16.22).

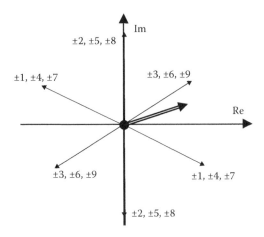

FIGURE 16.17 Vectorial positions for the input current.

The use of stationary vectors means that two input nodes (phases) are connected to the converter at a given moment, and also that two output nodes are connected to the same input node, while the third output node (phase) is connected to another input node (phase).

This form of the *Space Vector Modulation* can be found similar to the results from [9,11] for the *Indirect Matrix Converter*, also presented later on in this Chapter.

16.5.4 INDIRECT MATRIX CONVERTER (SPARSE CONVERTER)

As suggested by the matrix form equations described in Section 16.5.1, the operation of the conventional matrix converter can be understood as a sequence of a rectifier operation followed with an inverter without any intermediary DC link capacitor bank.

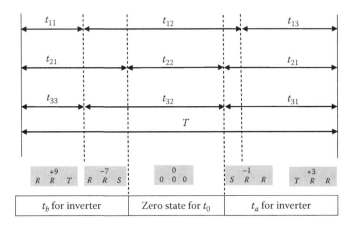

FIGURE 16.18 Example of pulse generation for Direct Space Vector Modulation with four active states.

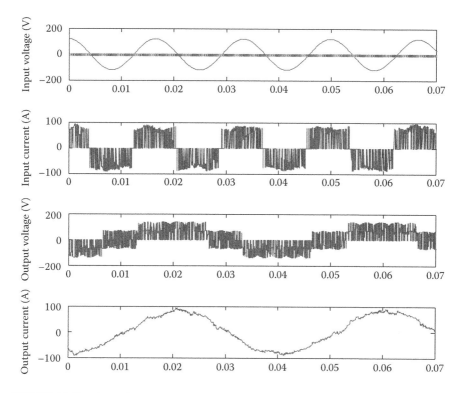

FIGURE 16.19 Operation waveforms for high modulation index $m = 0.70$, output frequency $f_0 = 12.5$ Hz (40 ms), switching frequency 5.4 kHz, with grid at 120 V and 60 Hz, and $R–L$ load with $L = 1$ mH, $R = 1$ Ohm.

This means the three-phase AC input voltages are rectified into a 6-pulse waveform before being applied to a conventional 6-switch three-phase inverter (Figure 16.23).

This PWM principle is also called *Indirect PWM* and it can also allow us to develop a new converter structure that actually does operate with the waveforms considered fictitiously for the description above. This is not identical with a back-to-back converter since it requires bi-directional switches for the implementation of the front-end converter. It comes from the requirement for a full recovery of the reactive energy and a four-quadrant operation of the converter. The circuit is shown in Figure 16.24 and it is referred to as *Indirect Matrix Converter (Sparse Converter)*.

16.5.5 IMPLEMENTATION OF PWM CONTROL

There are a multitude of possibilities for implementation of any of the methods suggested above. The most important aspect relates to the necessity of a dedicated *"glue logic"* circuitry after the conventional counter/timer circuits. This logic circuit is used for selection of the proper switching sequence for the nine bidirectional switches as well as for implementing the logic for the current commutation.

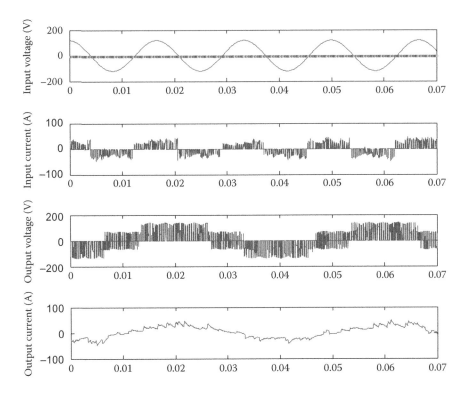

FIGURE 16.20 Operation waveforms for low modulation index $m = 0.15$, output frequency $f_0 = 12.5$ Hz (40 ms), switching frequency 5.4 kHz, with grid at 120 V and 60 Hz, and R–L load with $L = 1$ mH, $R = 1$ Ohm.

A possible architecture is herein described and it is useable for either direct or indirect PWM. The microcontroller software is supposed to manage the characteristics of the desired output waveform depending on application requirements. A PLL type module is required to acquire synchronization with the grid and to sense the phase of the grid power system. The PWM software routine has as inputs:

- Required output phase and magnitude (modulation index);
- Grid instantaneous phase information.

This information is coded in two operation indices: K = number corresponding to the sector where the desired output vector belongs to in the complex plane of output reference frame; J = number corresponding to the sector where the input current vector belong to in the complex plane of input reference frame and which is determined by the external PLL information.

The software routine calculates the time intervals allocated to each state. This may be achieved with SVM type of formula, including compensation for fluctuation within the fictitious DC link voltage. The numerical values for the time intervals can be arranged as shown in Figure 16.25. A single counter with multiple compare features, or 5 separate counters can be used. Counter overflow signals are used along with two

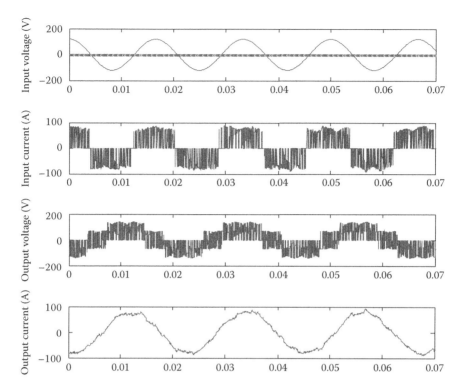

FIGURE 16.21   Operation waveforms for high modulation index $m = 0.70$, output frequency $f_0 = 52.0$ Hz (22 ms), switching frequency at 5.4 kHz, with grid at 120 V and 60 Hz, and $R–L$ load with $L = 1$ mH, $R = 1$ Ohm.

operation indices to generate the actual gate control. A memory look-up table, an IF/THEN software sequence, or a logic circuit can be used to synthesize the gate control of the 9 switches from these 7 signals (5 counter overflow and 2 operation indices).

Designing a PWM algorithm means averaging in between these available states in order to emulate a rotating vector in the load reference frame. Various methods are hence available (Figure 16.26).

16.6   CONCLUSION

Matrix converters are a very attractive subject for academic research and development as it offers work with advanced concepts of PWM and filtering, or voltage and current source operation. The success of the conventional matrix converters is limited when trying to replace the conventional back-to-back converter solutions. Alternatively, other direct converter topologies seem more attractive for future exploration as they overcome issues with both the conventional matrix converter and the back-to-back converter. Such issues are:

- Limited output voltage ($m = 0.866$);
- Complex commutation process;

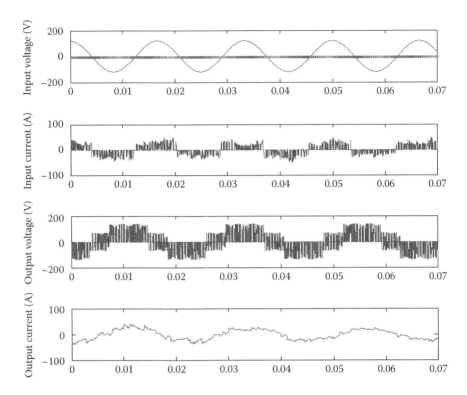

FIGURE 16.22 Operation waveforms for low modulation index $m = 0.15$, output frequency $f_0 = 52.0$ Hz (22 ms), switching frequency 5.4 kHz, with grid at 120 V and 60 Hz, and $R-L$ load with $L = 1$ mH, $R = 1$ Ohm.

- Irregular topology for packaging and volume production;
- Restricted reactive power compensation capability;
- Limited operation at unbalanced input voltages.

These are overcome by solutions further described in Chapter 18.

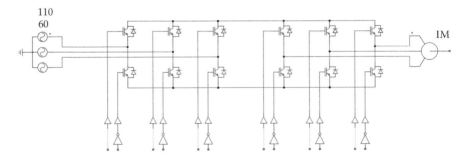

FIGURE 16.23 Conceptual back-to-back converter used for definition of a PWM algorithm.

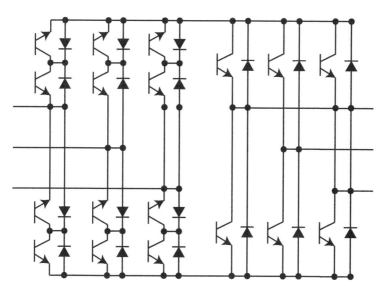

FIGURE 16.24 Indirect Matrix Converter.

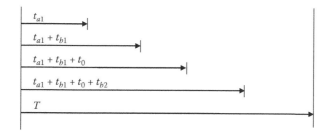

FIGURE 16.25 Time constants and counter arrangement.

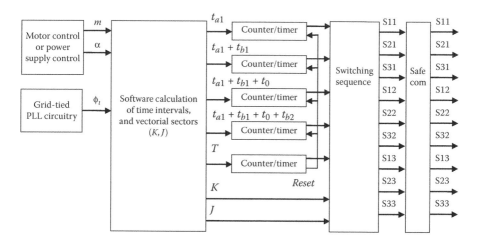

FIGURE 16.26 Digital architecture for implementation of the PWM algorithms.

REFERENCES

1. Borojevic, D. 1991. Space vector modulation in matrix converters. *VPEC Virginia Power Electron. Center* 5(1,2).
2. Huber, L. and Borojevic, D. 1991. Space vector modulation with unity input power factor for forced commutated cycloconverters. *Proceedings of PESC*, pp.1032–1041.
3. Neacsu, D.O. 1995. Theory and design of a space vector modulator for AC/AC matrix converter. *European Trans. Electr. Power Eng.* 4, 285–292.
4. Neft, C.L. and Shauder, C.D. 1988. Theory and design of a 30-HP matrix converter. *Proceedings of IEEE IAS'88*, pp. 248–253.
5. Nielsen, P., Blaabjerg, F., and Pedersen, J.K. 1996. SVM matrix converter with minimized number of switchings and a feed forward compensation of input voltage unbalance. *PEDES'96*, New Delhi, January 8–11.
6. Halasz, S., Schmidt, I., and Molnar, T. 1995. Matrix converters for induction motor drive. *EPE'95*, Sevilla, Spain.
7. Bernet, S., Matsuo, T., and Lipo, T.A. 1996. A matrix converter using reverse blocking NPT-IGBT's and optimized pulse patterns. *IEEE PESC'96*, Baveno, Italy, vol. 1, pp. 107–113.
8. Anon. 2008. AC7 matrix converter—Application Note, Document AN.AC7.01. Yaskawa Electric America, April.
9. Friedli, T. and Kolar, J.W. 2012. Milestones in matrix converter research. *IEEJ J. Industry Appl.* 1(1), 2–14.
10. Matteini, M. 2001. Control techniques for matrix converter adjustable speed drives. University of Bologna, Italy, PhD Thesis.
11. Kolar, J.W. and Friedli, T. 2010. Comprehensive evaluation of three-phase AC-AC PWM converter systems. *Tutorial at IEEE IECON*.
12. Yamamoto, E., Hara. H., Uchino, T., Kume, T.J., Kang, J.K., and Krug, H.P. 2007. Development of matrix converter for industrial applications. Yaskawa White Paper.
13. Lai, R., Pei, Y., Wang, F., Burgos, R., Boroyevich, D., Lipo, T.A., Immanuel, V., and Karimi, K. 2007. A systematic evaluation of AC-fed converter topologies for light weight motor drive applications using SiC semiconductor devices. *Proceedings IEEE Int'l Electric Machines and Drives IEMDC*, vol. 2, pp. 1300–1305.
14. Wheeler, P. and Clare, J. 2005. Matrix converter technology. *Tutorial IEEE IECON*.
15. Wheeler, P., Rodriguez, J., Clare, J.C., Empringham, L., and Weinstein, A. 2002. Matrix converters: A technology review. *IEEE Trans. Industrial Electron.* 49(2), 276–288.
16. Anon. 2010. DIM400PBM17-A000 IGBT bi-directional switch module. DS5524-3 November 2010 (LN27710).
17. Anon. 2001. Datasheet of FM35R12KE3. *Infineon Technologies*, 2001-08-16.
18. Schulz, M., De Lillo, L., Empringham, L., and Wheeler, P. 2011. Pushing power density limits using SiC-Jfet-based matrix converter. In: *PCIM Europe Conference*, VDE VERLAG GMBH, Berlin, Offenbach, vol. 1, 17–19 May 2011, pp. 464–470.
19. Motto, E.R., Donlon, J.F., Tabata, M., Takahashi, M., Yu, Y., and Majumdar, G. 2004. Application characteristics of an experimental RB-IGBT. *IEEE/IAS Ann. Meeting Conf. Rec.*, vol. 3, pp. 1540–1544, October.
20. Takei, M., Naito, T., and Ueno, K. 2010. *The Reverse Blocking IGBT for Matrix Converter with Ultra-Thin Wafer Technology*, Fuji.
21. Itoh, J., Odaka, A., and Sato, I. 2004. High efficiency power conversion using a matrix converter. *Fuji Electr. Rev.* 50(3), 94–98.
22. Burany, N. 1989. Safe control of four-quadrant switches. *IEEE IAS Conference Record 1989*, part I, pp. 1190–1194.
23. Oyama, J., Higuchi, T., Yamada, E., Koga, Y., and Lipo, T. 1989. New control strategy for matrix converter. *IEEE PESC'89 Conference Record*, pp. 360–367.

24. Ooi, B.T. and Kazerani, M. 1994. Application of dyadic matrix converter theory in conceptual design of dual field vector and displacement factor controls. *IAS Annual Meeting 1994*, pp. 903–910.

25. Venturini, M. and Alesina, A. 1980. The generalized transformer: A new bidirectional sinusoidal waveform frequency converter with continuously adjustable input power factor. *IEEE Power Electronics Specialists Conference Record*, pp. 242–252.

26. Watthanasarn, C., Zhang, L., and Liang, D.T.W. 1996. Analysis and DSP-based implementation of modulation algorithms for AC-AC matrix converters. *IEEE-PESC 1996*, pp. 1053–1058.

27. Klumpner, C., Blaabjerg, F., Boldea, I., and Nielsen, P. 2006. New modulation method for matrix converters. *IEEE Trans. Industry Appl.* 42(3), 797–803.

28. Huber, L. and Borojevic, D. 1995. Space vector modulated three-phase to three-phase matrix converter with input power factor correction. *IEEE Trans. Industry Appl.* 31(6), 1234–1246.

29. Ziogas, P.D., Kang, Y., and Stefanovic, V.R. 1986. Rectifier-inverter frequency changers with suppressed DC link components. *IEEE Trans. Industry Appl.* IA-22(6), 1027–1036.

30. Kim, S., Sul, S.K., and Lipo, T.A. 2000. AC/AC power conversion based on matrix converter topology with unidirectional switches. *IEEE Trans. Industry Appl.* 36(1), 139–145.

17 Multilevel Converters

17.1 PRINCIPLE AND HARDWARE TOPOLOGIES

We have seen in Chapter 3, Section 3.6.1, that we can approximate a sinusoidal waveform with a staircase waveform as shown in Figure 3.21. The angular intervals for changing the waveform from a step to another and the levels for each step can be optimized by harmonic constraints. However, such dual optimization may be difficult to implement in hardware. Alternatively, levels of equal height may be easier to construct and implement within a modular approach.

17.1.1 H-BRIDGE MODULES

Let us first imagine that each level is created with a single-phase H-bridge converter, working completely independent of the others with the only purpose of creating stair-like levels in accordance to the control strategy from Figure 17.1 (that is an extension of Figure 3.21). The base topology is shown in Figure 17.2. The operation is very intuitive as each H-bridge inverter can output one polarity or the other of the DC-side voltage by operating IGBTs on the diagonal of the H-bridge [1]. The zero voltage drop can be achieved with a zero-state in control of each module. With the decomposition shown in Figure 3.21, the conduction intervals for each IGBT can be easily depicted.

Obviously, the number of levels can be higher than three. The operation of each converter needs also to secure a constant voltage source for supply. This is assumed here as being carried out with another converter or battery not shown for simplicity.

An alternative for the converter shown in Figure 17.2 is shown in Figure 17.3, where the conventional single-phase H-bridge has been split into two inverter legs, each one contributing eventually to a different alternance of the output voltage (all three inverter legs on the left side of the figure are contributing to the positive alternance, and the three inverter legs on the right side form the negative alternance of the pole voltage). This concept is used by Siemens Corporation within High Voltage DC (HVDC) transmission lines, with multilevel converters rated up to 400 MW, and built with 200 Power Modules per Converter Arm [2].

Additional to the waveform optimization shown in Figure 17.1, another advantage of using this topology in high voltage applications is related to the ratings of the power devices. The converter leg shown in Figure 17.2 features IGBT rated for the individual DC sources, and not for the entire staircase voltage applied to the load. For instance, 1200 V IGBTs can be used in 2400 V RMS load voltage applications with the 3-level converter. The same principle applies for higher level of voltage buses. A 30 kV transmission line can thus have active filtering or voltage control within an 18-level converter, while using medium voltage IGBTs rated at 3000 V.

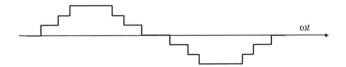

FIGURE 17.1   Multilevel waveform.

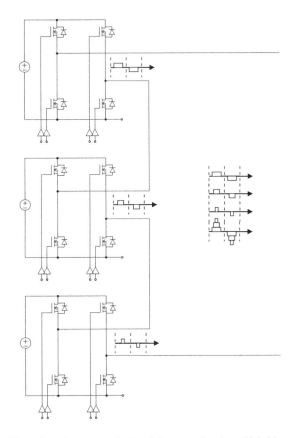

FIGURE 17.2   Three-Level converter built of three single-phase H-bridge modules.

Finally, let us note another advantage of using multilevel converters. As it will be shown later on this chapter, the PWM applied to the control of multilevel converters operates with two adjacent levels, which means that the slopes of the output voltage contain steps of smaller height. Having smaller changes in the output voltages means reduced EMI radiation produced by (dv/dt).

The modular structure of the H-Bridge multilevel converters is advantageous for manufacturing, service, and maintenance. Moreover, the high power applications allow a distributed power supply through multiple lower power DC energy sources, each being usually implemented with another power converter from a high voltage, high power AC line.

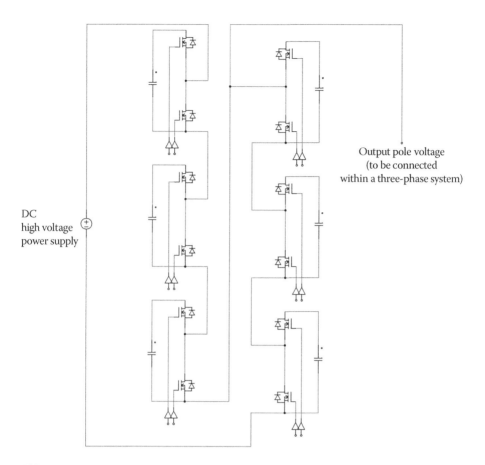

Output pole voltage
(to be connected
within a three-phase system)

DC
high voltage
power supply

FIGURE 17.3   Another 3-level converter topology built of modules.

This approach of building multilevel converters with H-bridge modules is not very economic for multilevel converters when the power rating is relatively low. In this respect, other topologies with reduced number of components are sometimes considered.

17.1.2 FLYING CAPACITOR MULTILEVEL CONVERTER

The principle underlining this topology has been introduced in [3], and it is based on a layering structure of capacitors (Figure 17.4). Three identical legs can constitute a 3-Phase Multilevel Converter with a load connected in star. The voltage across each capacitor needs to be maintained at equal values and a capacitor voltage coincides with the voltage ratings for the power devices in converter. Moreover, the m-level flying-capacitor multilevel inverter will require m capacitors for building the m-levels in the output voltage and $(m − 1) \times (m − 2)/2$ auxiliary capacitors per phase when the voltage rating of the capacitors is identical to that of the main switches. For instance, the particular case shown in Figure 17.4 for the 4-level

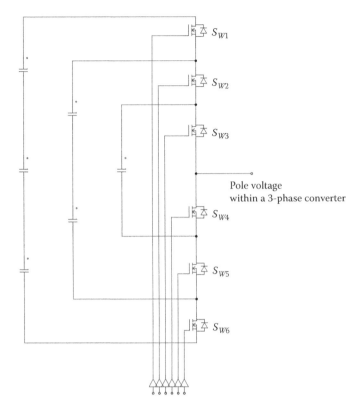

FIGURE 17.4 Flying capacitor multilevel (4-level) converter.

converter will require 3 level-building capacitors and $(4 - 1)*(4 - 2)/2 = 3$ auxiliary capacitors.

An important advantage of the Flying Capacitor Topology consists in its inherent redundancy. Same levels of voltage can be made up with different capacitor connection. This is illustrated within Table 17.1 true for the 4-level converter shown in Figure

TABLE 17.1

Possible Generation of Voltage Levels in a 4-Level Flying Capacitor Converter

	Sw_1	Sw_2	Sw_3	Sw_4	Sw_5	Sw_6
$3*V_{dc}$	X	X	X			
$2*V_{dc} = 3*V_{dc} - VDC$	X	X		X		
$2*_{dc}C$		X	X			X
$V_{dc} = 3*V_{dc} - 2*VDC$	X			X	X	
$V_{dc} = 2*V_{dc} - V_{dc}$		X		X		X
V_{dc}			X		X	X

17.4. This degree of freedom represents an important advantage as the voltage across the flying capacitors can be controlled by the proper selection of the switching states.

Let us also note that there are always 3 devices turned-ON for each inverter leg (phase). This can be generalized: an m-level multilevel converter has $m - 1$ devices (usually IGBTs) turned-ON at a given moment, on each converter leg.

The large number of capacitors amount to a large enough capacitance able to ride through short duration outages and deep voltage sags.

On the downside, the control algorithm is fairly complicated. The importance of capacitors in the proper operation of the converter is jeopardizing performance with ageing and the large capacitor value variation with temperature and ambient.

17.1.3 DIODE-CLAMPED MULTILEVEL CONVERTER

The most known and most used multilevel converter structure is shown in Figure 17.5 [4,5].

The intermediary voltage levels to be applied to the load are set-up on a capacitor divider across the high voltage DC supply voltage. The pole voltage is clamped to these intermediary voltage levels with diodes. For the case shown in Figure 17.5, the number of intermediary levels is one, and we need two capacitors to achieve this.

Similar to the previous topologies, each active switching device is required to block only a voltage level of V_{dc}. The multilevel converters of higher order require the clamping diodes to be chosen for different ratings, for reverse voltage blocking depending on what capacitor divider point are they connected to. Alternatively, the

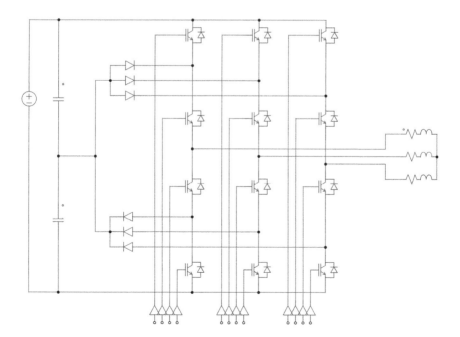

FIGURE 17.5   Diode-clamped 3-level converter.

inverter can be designed such that each blocking diode has the same voltage rating as the active switches with n diodes in series for the diodes that need to block $n \times V_{dc.}$ The total number of diodes required for each pole voltage under such design yields $(m - 1) \times (m - 2)$. For instance, a 3-level converter requires 2 diodes for each pole voltage (phase).

The required amount of capacitance is lower than that of the previous topology.

The IGBTs that are turned-ON on the same inverter leg, at a given moment, are connected in series and located adjacent to each other. Moreover, the number of devices turned-ON is related to the number of levels within the inverter's structure. For instance, a 3-level IGBT has two devices ON, a 4-level inverter has always three devices ON, and a 5-level inverter has four devices on the same leg turned ON.

The drawbacks relate to the large number of required diodes, and the somewhat difficult control algorithm for maintaining the voltage constant on the capacitors. For this reason, the use of this topology is limited to 3-level, 4-level, or 5-level converters, in a range of limited power. They were very successful in medium voltage motor drives.

A topology similar to the diode-clamped multilevel converter was presented in [6], with benefits in increasing the apparent switching frequency.

Finally, let us note that the diode-clamped inverter has redundancy in line-line voltage generation and no redundancy in phase voltage. Denoting as above the number of levels with "n," and the instantaneous voltages on the three output waveforms as (i, j, k), the number of redundancies at a given moment yields

$$N = n - 1 - \left[\left(\max(i,j,k) \right) - \min\left(i,j,k\right) \right] \tag{17.1}$$

As discussed for the H-bridge multilevel converters, these redundancies can be used for adjusting the voltages on the DC side capacitors, adjusting the neutral voltage in the output, or adjusting the sharing of the switching events between switches on the same leg.

17.1.4 COMBINATION CONVERTERS

Another design direction has been around the idea of series-connection of various converter topologies and their operation so that multilevel voltages yield on the load. This principle ranges from simple connection of two conventional bridge converters with phase shift control, to combination of any multilevel converters in order to obtain even more levels of voltages. The roots of this idea reside in multiphase transformers used with controlled rectifiers in early times of semiconductor based processing of electrical energy. We will refer to this class of converters as *Combination Converters* as they can emerge from a combination of any two previously shown topologies. Several examples are shown in Figure 17.6 with their appropriate waveforms [7 – 9], and this principle can be extended to a combination of any of the previously shown topologies. Note that the two DC-side power supplies do not have any common point, or—in other words—the neutral point is not connected.

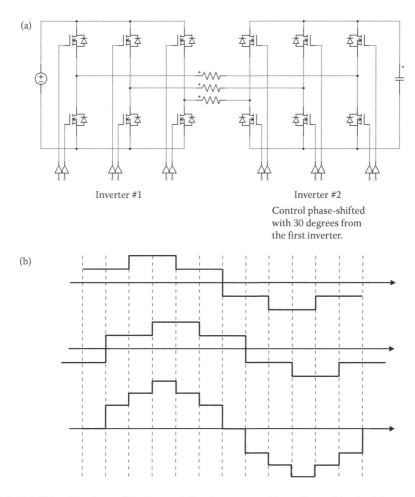

FIGURE 17.6 Simple combination multilevel converter: (a) topology; (b) principle.

17.2 DESIGN AND RATING CONSIDERATIONS

17.2.1 SEMICONDUCTOR RATINGS

The success of multilevel converter topologies comes from the reduction of voltage seen across each power semiconductor device during operation. This means we can use these topologies in Medium and High Voltage applications with power semiconductor devices rated in a lower class. First, the medium voltage IGBTs (up to 6 kV) are the highest rated monolithic devices and the use of multilevel converters opens up the active filtering possibilities in High Voltage applications (13 kV...50 kV). On another line of thought, the switching performance, voltage drop during conduction, and cost of low voltage IGBTs (rated 1200 V) are better when compared to the medium voltage IGBTs (up to 6 kV). Hence, the use of multilevel converters benefits from lower rating IGBTs to boost the overall system level performance.

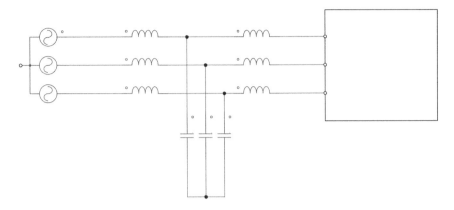

FIGURE 17.7   AC filter.

17.2.2 PASSIVE FILTERS

This Chapter has introduced the multilevel converters through their advantages in better approximating a sinusoidal waveform for an un-modulated operation. This helps reducing the requirements for the passive filters in AC power supplies or in grid-tied applications. This generic advantage is even more significant when PWM is used to control the power stage. Figure 17.7 shows the simplest filter structure possible. Alternatively, the trap filters can be associated to this *L-C-L* structure for trapping certain frequencies on the filter resonance.

It is also important to note that the operation of these filters in high-voltage applications implies the use of multiple series-connected (split) capacitors. Optimization of the way the capacitors are used within this type of connection becomes very important.

17.3   PWM ALGORITHMS

17.3.1 PRINCIPLE

The advantages of multilevel converters are not enough exploited when PWM operation is not utilized. This can be adjusted by synthesizing intermediary voltage level in average terms (as a moving average) by spending short intervals of time in between two adjacent voltage levels. This operation mode with pulse-width modulation is well-known from the conventional bridge-like 6-switch converters.

Since we need to work with adjacent voltage levels only, the PWM algorithms are simplified and easy to understand.

Figure 17.8 explains the PWM operation of a multilevel converter with the help of conventional PWM used for varying a voltage within the DC/DC converters. The goal of an inverter is to generate a sinusoidal like voltage, and this can be achieved with multilevel converters by depicting the portion of sinusoidal waveform that belongs in between two adjacent voltage levels (*level k*, *level k* + 1), followed by the adjustment of the average pulse value in accordance to that portion of sinusoidal waveform. Furthermore, generation of PWM within a three-phase system suggests

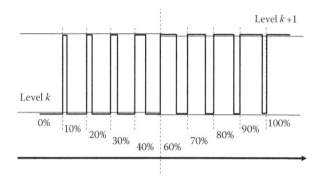

FIGURE 17.8   Principle of PWM applied to multilevel converters.

using the symmetries of a three-phase system within the PWM generation. There is a multitude of PWM algorithms reported in literature for multilevel converters and the most used methods are herein explained in detail.

17.3.2 SINUSOIDAL PWM

Let us first consider the generic multilevel converter shown in Figure 17.9. This structure makes abstraction of the actual converter schematic, allowing us to focus on the PWM generation algorithm without spending effort with the actual topology. Later on, we will define the particular aspects of PWM generation for each of the previously shown topologies.

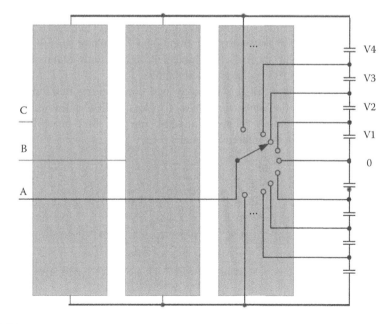

FIGURE 17.9   Schematic of an *m*-level 3-phase converter.

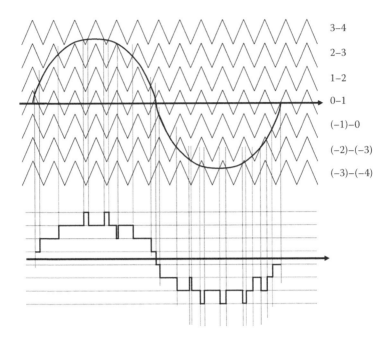

FIGURE 17.10   Sine-triangle PWM explained for a 5-level converter.

Generation of PWM voltage on each phase is illustrated in Figure 17.10 for a 4-level converter. The generic state switch from Figure 17.9 connects to the state depicted from Figure 17.10, based on the actual instantaneous position of the sinusoidal waveforms in respect to the triangular waveforms. The same algorithm is used for the other two phases with the conventional 120° phase shift. Figures 17.11–17.13 show the harmonic spectra when implementing this sinusoidal modulation to a 3-level converter with 10 kHz carrier modulation.

Despite the advantages of this topology with lower rating IGBT devices for a higher output voltage, the Sinusoidal PWM provides a somewhat limited output voltage. To increase the output voltage in a star-connected three-phase load, a third harmonic is injected in the reference signal.

$$
\begin{cases}
v_{a,\text{ref}} = \dfrac{v_{\text{dc}}}{2} \cdot \left[1 + m \cdot \cos(\theta) - \dfrac{1}{3} \cdot m \cdot \cos(3 \cdot \theta) \right] \\[2mm]
v_{b,\text{ref}} = \dfrac{v_{\text{dc}}}{2} \cdot \left[1 + m \cdot \cos\left(\theta - \dfrac{2 \cdot \pi}{3} \right) - \dfrac{1}{3} \cdot m \cdot \cos(3 \cdot \theta) \right] \\[2mm]
v_{c,\text{ref}} = \dfrac{v_{\text{dc}}}{2} \cdot \left[1 + m \cdot \cos\left(\theta - \dfrac{4 \cdot \pi}{3} \right) - \dfrac{1}{3} \cdot m \cdot \cos(3 \cdot \theta) \right]
\end{cases} \qquad (17.2)
$$

where m is the modulation index, now up to $2/sqrt(3) = 1.15$.

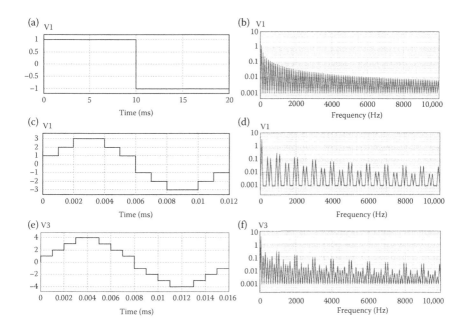

FIGURE 17.11 Harmonic spectra for un-modulated multilevel converters.

An alternative solution for the third harmonic injection, very well suited for analog implementation, is explained in [10]. The added signal is calculated from the three sinusoidal reference signals with

$$v_{\text{inj}} = \frac{\max(v_{a,\text{ref}}, v_{b,\text{ref}}, v_{c,\text{ref}}) + \min(v_{a,\text{ref}}, v_{b,\text{ref}}, v_{c,\text{ref}})}{2} \qquad (17.3)$$

and it can be seen as half of the distance between positive and negative envelopes of the three-phase waveforms.

If the calculation of the time intervals associated with each state is easy to be understood from Figure 17.10, the actual switching sequence is determined by properly selecting each state to provide the required voltage level.

The most common digital solution is to count the number of triangular waveforms below the reference waveform, and to enable turning-on switches on the same leg as the voltage is increasing.

At operation with a large modulation index of the diode-clamped inverter (Figure 17.12 g,h), the inner switches are switched less often than the outer switches. This shortcoming can be avoided with a proper PWM when using variable switching frequencies [11].

Advanced methods propose to improve the efficiency with

- Different frequencies for some of the carrier triangle waveforms [12];
- Shifting the boundary of the carrier triangle waveforms in order to balance the capacitor voltages [13].

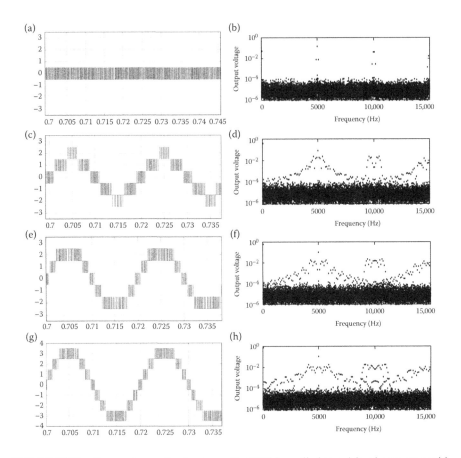

FIGURE 17.12 Harmonic spectra for sinusoidal PWM, applied to a 4-level converter with 5 kHz PWM, at different modulation indices (pole-neutral voltage shown).

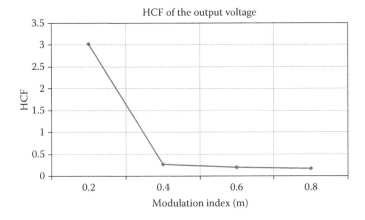

FIGURE 17.13 HCF of the output voltage in %, for the data shown on the previous figure.

17.3.3 SPACE VECTOR MODULATION

The advent in the early 1990 s of the *Space Vector Modulation* for conventional six-switch inverter has encouraged the use of this method to the analysis and PWM control of multilevel converters [14].

First, the possible vectorial positions within the complex plane have been identified. The multilevel converters are able to produce output voltage similar to conventional 6-switch inverters, at multiple voltage levels. For instance, the diode-clamped 3-level converter of Figure 17.5, can be operated as a conventional 2-level inverter with one capacitor voltage applied to the load through the inner IGBTs or can take advantage of switching both IGBTs on the same DC rail voltage as a series composed switch producing two times capacitor voltage on the load. This means the diagram of all available voltage space vectors will include two hexagons at 1× and 2× capacitor voltage as half of the diagonal (like radius). Moreover, the presence of all 12 switches in circuit allows some other combinations leading to other intermediary vectorial positions. A complete view on several multilevel converters is provided in Figure 17.14.

Design of a SVM method for a multilevel converter consists in identifying several vectorial locations in the neighborhood of the desired vector position, followed up by a vectorial decomposition of the desired vector into the adjacent vectorial positions. It can be observed from Figure 17.14, that any desired vectorial position will be inside a triangle formed in between three nodes. The calculation of the time intervals associated with each state is made by volt-second averaging similar to the conventional 2-level inverter.

$$t_a \cdot \vec{v}_a + t_b \cdot \vec{v}_b + t_c \cdot \vec{v}_c = T \cdot \vec{V} \tag{17.4}$$

and

$$t_a + t_b + t_c = T$$

Decomposing the vectorial equation on the two orthogonal axes of the complex plane leads to:

$$\begin{cases} t_a \cdot v_{a,\mathrm{Re}} + t_b \cdot v_{b,\mathrm{Re}} + t_c \cdot v_{c,\mathrm{Re}} = T \cdot V_{\mathrm{Re}} \\ t_a \cdot v_{a,\mathrm{Im}} + t_b \cdot v_{b,\mathrm{Im}} + t_c \cdot v_{c,\mathrm{Im}} = T \cdot V_{\mathrm{Im}} \\ \qquad\quad t_a + t_b + t_c = T \end{cases} \tag{17.5}$$

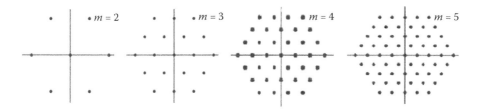

FIGURE 17.14 Vectorial positions in the complex plane of output voltage.

This system has three equations and three unknown variables, with an unique solution $[t_{a0}, t_{b0}, t_{c0}]$. This mathematical property is very important to defining the PWM control since it also says that there is no need for zero states when controlling a multilevel converter. Since the voltage steps (fast transitions in the waveform) are smaller, it guarantees a better waveform in terms of EMI and harmonic generation.

17.3.4 HARMONIC ELIMINATION

Similar to the problem exposed in Chapter 3, Section 3.6.1, the angular intervals for changing the waveform can be calculated by optimization criteria for any of the multilevel power converters described herein (Figure 17.15). Similar to Figure 3.20, the following is explaining the principle of waveform decomposition for computer optimization. The Fourier series development helps in the calculation of harmonics through simple addition of the component waveforms.

Considering equal voltage levels, the staircase waveform of Figure 17.15(a) is decomposed in the harmonic series:

$$V_n = V_{dc} \cdot \frac{1}{n} \cdot \left[\sin(n - \alpha_1) + \sin(n - \alpha_2) + \cdots + \sin(n - \alpha_n) \right] \qquad (17.6)$$

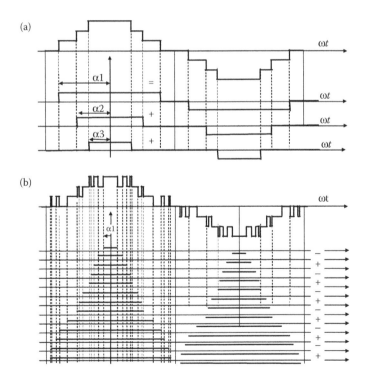

FIGURE 17.15 Decomposition in quasi-rectangular waveforms: (a) conventional staircase waveform; (b) multilevel PWM waveform.

The PWM waveform of Figure 17.15(b) is decomposed in the harmonic series:

$$V_n = V_{dc} \cdot \frac{1}{n} \cdot \left[\sin\left(n - \alpha_1\right) - \sin\left(n - \alpha_2\right) + \cdots - \sin\left(n - \alpha_n\right) \right] \qquad (17.7)$$

Since the calculation is very similar, we will not replicate it here [15]. Different analytical methods can be applied to solving the set of nonlinear equations derived from imposing different harmonic constraints for harmonic elimination and THD/HCF reduction. They can be implemented in mathematical programs like MathCAD.

A very interesting optimization is proposed in [16,17], where different voltage levels are considered as steps in the waveform. This allows for more degrees of freedom in harmonic optimization.

17.4   APPLICATION SPECIFICS

17.4.1   HVDC Lines

Transmission of energy on *High Voltage DC* lines was adopted for transmission of large amounts of energy at long distances. Examples include lines in between the United States and Quebec, or within Russia. The *"classic"* HVDC link works at 500 kV, under an installed power of 4000 MW [18]. Modern solutions are implemented for 800 kV lines, and a transmission power of 5000 – 7200 MW. The conventional solution consists in line-commutated power converters built of thyristors. These converters are slow in nature, yet reliable, and with a well-known operation. Recent efforts have encouraged the use of IGBT-like devices and power converters switching in the kHz range, with improved performance in the fast control of energy (more dynamics for better power quality). The drawback of this modern solution consists of additional loss in the range of 2–4% [2].

17.4.2   FACTS

The role of a *Flexible AC Transmission System* is to support the proper transmission of energy. There are multiple functions implemented with power converters under this name. The most known is the *Static Synchronous Compensator* (STATCOM), usually implemented with Voltage Source Inverters [18]. All of these functions are used to improve the quality of the power transfer.

17.4.3   Motor Drives

High voltage motor drives benefit from the multilevel converter technology with the benefits already mentioned:

- Lower ratings of the power semiconductors;
- Improved EMI radiation due to the lower steps in voltage waveform;
- Lower expectations for the input/output passive filters due to the more advantageous harmonic content when compared to conventional 6-switch bridge-like converters.

Considering these, several products have emerged on the market based on the multilevel converter technology, and are incorporating control software for motor drives applications. These target especially medium voltage, multimega-watt applications, like wind turbines, aviation power systems, and naval propulsion systems.

REFERENCES

1. Khomfoi, S. and Tolbert, L. 2011. *Chapter 31, Multilevel Power Converters, in Power Electronics Handbook*, 3rd Edition. Elsevier, Butterworth-Heinemann, Amsterdam/Boston/Heidelberg/London/New York/Oxford/Paris/San Diego/San Francisco/Singapore/Sydney/Tokyo.
2. Gemmell, B., Dorn, J., Retzmann, D., and Soerangr, D. 2008. Prospects of multilevel VSC technologies for power transmission. *IEEE PES Transmission and Distribution Conference and Exhibition*, pp.1–16.
3. Meynard, T.A. and Foch, H. 1992. Multi-level conversion: High voltage choppers and voltage-source inverters. *IEEE Power Electronics Specialists Conference*, pp. 397–403.
4. Nabae, A., Takahashi, I., and Akagi, H. 1981. A new neutral-point clamped PWM inverter. *IEEE Trans. Industry Appl.* IA-17, 518–523.
5. Baker, R.H. 1981. Bridge Converter Circuit, U.S. Patent 4 270 163.
6. Floricau, D., Gateau, G., and Leredde, A. 2010. New active stacked NPC multilevel converter: Operation and features. *IEEE Trans. Industrial Electron.* 57(7), 2272–2278.
7. Corzine, K. 2005. *Operation and Design of Multilevel Inverters—Report Developed for the Office of Naval Research.*
8. Kawabata, T. and Ejioglu, E. 1997. New open-winding configurations for high power inverters. *IEEE ISIE*, pp. 457–462.
9. Stemmler, H., and Guggenbach, P. 1993. Configurations of high-power voltage source inverter drives. *EPE'93*, pp. 7–14.
10. Menzies, R.W., Steimer, P., and Steinke, J.K. 1994. Five-level GTO inverters for large induction motor drives. *IEEE Trans. Industry Appl.* 30(4), 938–944.
11. Tolbert, L.M. and Habetler, T.G. 1999. Novel multilevel inverter carrier-based PWM method. *IEEE Trans. Industry Appl.* 25(5), 1098–1107.
12. Tolbert, L.M. and Habetler, T.G. 1999. Novel multilevel inverter carrier-based PWM method. *IEEE Trans. Industry Appl.* 35(5), 1098–1107.
13. Espinoza, J., Moran, L., and Sbarbaro, D. 2003. A systematic controller design approach for neutral-point-clamped three-level inverters. *Proceedings of the IEEE Industrial Electronics Conference*, pp. 2191–2196.
14. Rodriguez, J., Correa, P., and Moran, L. 2001. A vector control technique for medium voltage multilevel inverters. *Proceedings of the IEEE Applied Power Electronics Conference*, vol. 1, pp. 173–178.
15. Chiasson, J.N., Tolbert, L.M., McKenzie, K.J., and Du, Z. 2004. A unified approach to solving the harmonic elimination equations in multilevel converters. *IEEE Trans. Power Electron.* 19(2), 478–490.
16. Du, Z., Tolbert, L.M., Chiasson, J.N., and Li, H. 2005. Low switching frequency active harmonic elimination in multilevel converters with unequal DC voltages. *IEEE IAS Annual Meeting Conf. Rec.*, vol. 1, pp. 92–98, October.
17. Filho, F., Maia, H.Z., Mateus, T.H.A., Ozpineci, B., Tolbert, L.M., and Pinto, J.O.P. 2013. Adaptive selective harmonic minimization based on ANNs for cascade multilevel inverters with varying DC sources. *IEEE Trans. Industrial Electron.* 60(5), 1955–1962.
18. Retzmann, D. 2009. *Tutorial on VSC in Transmission Systems—HVDC and FACTS*, Siemens AG.

18 Use of IPM within a "Network of Switches" Concept

Apart from the conventional use of IPM devices shown in Chapter 10, there are other topologies that also benefit from using these devices. The main idea behind the topologies presented in this chapter is to split the switched mode control into multiple power switches instead of using a higher power rated device. This was also analyzed in Chapter 14. Multiple switches allow us to introduce more control of the output waveform with benefits in reducing loss and improve reliability. Eventually this would emerge into a new set of high-performance topologies assembling a "network of switches."

18.1 GRID INTERFACE FOR EXTENDED POWER RANGE

The first chapter of this book has presented standards for harmonics within the grid current that stand true independent of the power electronics topology used for energy conversion. The desire to reduce the packaging complexity of the power electronics equipment with the use of *Intelligent Power Modules* applies also to grid interfaces. Whilst the advent of IPM devices is noteworthy, their power levels tend to be limited. A novel power conversion principle is proposed in [1] to augment the power capability of an IPM device with a diode rectifier, or reversing the logic, to correct the harmonics of the input current within a diode rectifier by using an IPM device.

The conventional topology of an active power filter is considered as a starting point (Figure 18.1), and instead of dwelling into advanced control of current and power, the IPM hardware available on the market is herein used.

The power diodes within a diode rectifier conduct for 120° only, and there are 2 intervals of 60° on each phase without any current conduction. A current control algorithm is used to drive current into the grid from the three-phase power converter during such intervals only. The power to supply the three-phase power converter is taken from the load-side of the diode rectifier as a pure rectified voltage (Figure 18.2) and the references for the three phase currents follow Figure 18.3.

The current generated by the 6-switch power converter closes through the two conducting diodes and back to the power converter. This circulation will overload a little the diode rectifier. The current circulation though diodes improves the input (grid) current even if this is somewhat in an uncontrolled way, based on the offline current reference.

The current controller follows a feed-forward algorithm previously discussed in Sections 9.4 and 13.3.9.2 as well as in reference [2]. Synchronization with the grid

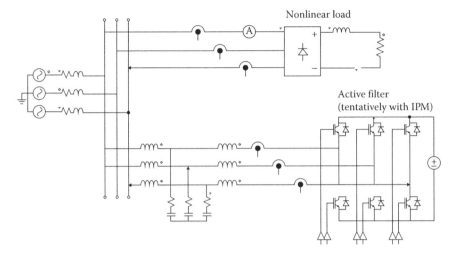

FIGURE 18.1   Conventional topology for an active filter.

voltage of the control system is very important herein for both the definition of the six operation modes, and the dead-time intervals shown in Figure 18.3. Each one of the six individual current controllers needs to achieve a fast locking to its triangular reference and to work after and/or before the others have started. Yet, their operation must start and finish after and respectively before the diodes have finished their switching of the current. Depending on the peculiar hardware, this may produce glitches in the resulting input current (see later on, Figure 18.6).

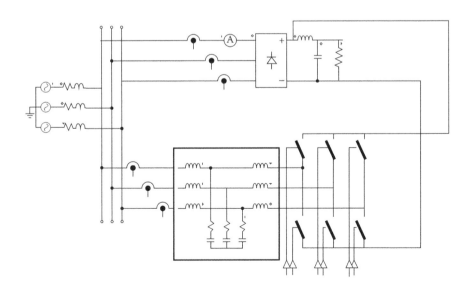

FIGURE 18.2   Novel hardware.

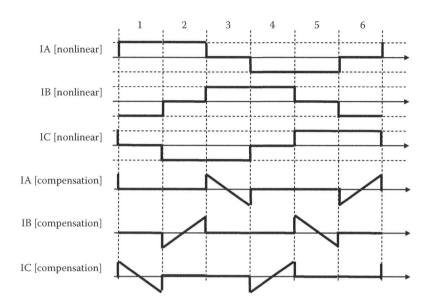

FIGURE 18.3   Theoretical current references for the novel method.

Figure 18.4 is showing individual current waveforms without passive filtering.

The two 60°-wide intervals when the 6-switch power converter influences the operation with the injection of triangular currents can be seen. During the remaining four 60°-wide intervals, the correction of the grid current yields from the circulation of the compensation current through the rectifier's conducting diode, and a pulsed current superimposes to the traditional square-wave current. The waveforms of Figure 18.4 have a high content in fundamental, hence satisfying the grid harmonic standards.

Using a triangular current reference induces a light third harmonic (Figure 18.6).

The compensation power converter is rated in respect to the load power. The maximum peak current for the compensation converter is at 0.5578 of the load current (Figure 18.3). The power loss and thermal requirements are extremely beneficial since each IGBT works in switched mode for 1/6 of the period, under the line-to-line voltage, at a current ranging from −0.5578 to 0.5578 of the load current.

Let us calculate the power loss distribution for each IPM package in different circuit configurations (Figure 18.5). For comparison we will use data provided by manufacturer along its own loss calculation formulas, for the example of a back-to-back motor drive.

A. *Conventional back-to-back motor drive (similar to data from manufacturer's example)*
   System Data:
   - Motor power rating 1 HP = 745 W A/C
   - DC bus voltage V_{dc} = 400 V
   - Modulation index of 0.8

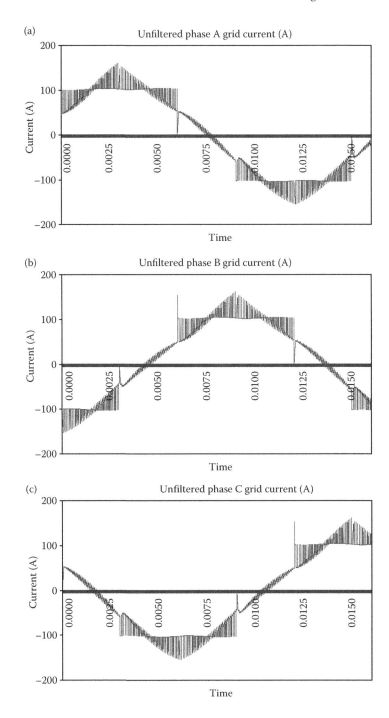

FIGURE 18.4 Grid phase currents (data samples printed with Excel).

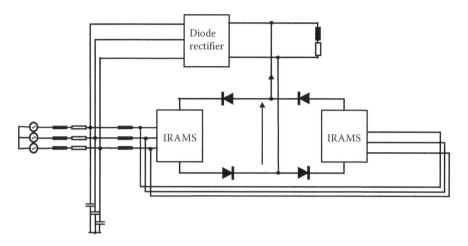

FIGURE 18.5 Implementation with IPM/IRAMS modules.

- Line to line voltage $V_{LL} = 113\ V_{rms}$
- Power factor PF = 0.6
- Phase current $I_{ph} = 4\ A_{rms}$
- Switching frequency $f_{sw} = 4\ kHz$

Power Loss Calculation:

- $P_{switching}\ (I + D) = 0.48\ W$
- P_{cond}, switch = 2.46 W
- P_{cond}, diode = 0.64 W
- Total P_{loss} per IPM module = 21.48 W

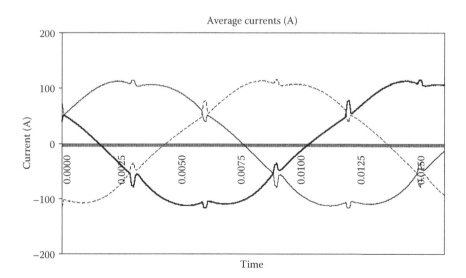

FIGURE 18.6 Currents calculated with moving average at 160 μs from samples in Excel.

B. *New converter in active filter role* (Figure 18.2), *used along 25 A rectifier load*
   System Data:
       Converter phase current $I_{ph} = 10$ A$_{rms}$
       Load current in rectifier $I_{rectifier} = 25$ A$_{dc}$
       Switching frequency $f_{sw} = 12.5$ kHz
       Power cycle = 0.33
   Power Loss Calculation:
       $P_{switching}$ $(I + D) = 1.16$ W
       P_{cond}, switch = 1.11 W
       P_{cond}, diode = 0.53 W
       Total P_{loss} per IPM module   = 16.80 W
C. *New converter in active filter role* (Figure 18.2), *used along 50 A rectifier load*
   System Data:
       Converter phase current $I_{ph} = 20$ A$_{rms}$
       Load current in rectifier $I_{rectifier} = 50$ A$_{dc}$
       Switching frequency $f_{sw} = 12.5$ kHz
       Power cycle = 0.33
   Power Loss Calculation:
       $P_{switching}$ $(I + D) = 2.42$ W
       P_{cond}, switch = 1.46 W
       P_{cond}, diode = 1.06 W
       Total P_{loss} per IPM module = 29.64 W

Another demonstration of capabilities for this new converter topology and its afferent control is made for a direct comparison of power losses when a certain rectifier load current (DC) is produced. Basically, we are seeking the method able to generate the DC current with lower amount of losses overall. The following examples are using numeric data especially selected to produce around 30W power loss for different converter designs.

A. *Direct three-phase IGBT 6-switch bridge producing a DC current of*
   $I_{rectifier} = 6.5$ A$_{dc}$
   System Data:
       $I_{ph} = 5.0$ A$_{rms}$,
       $m = 0.8$
       Unity Power Factor:
       $f_{sw} = 6.00$ kHz
   Power Loss Calculation
       $P_{switching}$ $(I + D) = 0.87$ W
       P_{cond}, switch = 3.42 W
       P_{cond}, diode = 0.80 W
       Total power loss P_{loss} per IPM module = 30.54 W

B. Conventional active filter for a rectifier load current of $I_{rectifier} = 20\,A_{dc}$

System Data:

$I_{ph} = 22\,A_{rms}$

Waveform = harmonic difference, conventional active filter definition

$f_{sw} = 12.5$ kHz

Power Loss Calculation:

$P_{switching}\,(I + D) = 1.68$ W

P_{cond}, switch = 2.99 W

P_{cond}, diode = 0.35 W

Total power loss P_{loss} per IPM module = 30.12 W

C. New method used for a rectifier load current of $I_{rectifier} = 50\,A_{dc}$

System Data:

$I_{ph} = 20\,A_{rms}$

Waveform = Figure 18.3

Power cycle = 0.33

$f_{sw} = 12.5$ kHz

Power Loss Calculation:

$P_{switching}\,(I + D) = 2.42$ W

P_{cond}, switch = 1.46 W

P_{cond}, diode = 1.06 W

Total power loss P_{loss} per IPM module = 29.64 W

In conclusion, this new control concept benefits from the intervals when diodes are not conducting current to inject a triangular waveform current that closes through the rectifying diodes, producing a natural compensation of the grid current. The results prove this method as an alternative to conventional grid interfaces as previously shown in Chapter 14. Using IPM devices allows a paradigm shift from conventional reasoning of saving or reducing the number of semiconductor components (Figure 18.5). Using IPM reduces the count of passive components, with advantages in size, weight, efficiency, and reliability.

Overall, one achieves a very low-cost solution, with a component minimized power stage, without additional power supplies or DC link capacitors, and a fully integrated electronics.

18.2 MATRIX CONVERTER MADE UP OF VSI POWER MODULES

18.2.1 Conventional Matrix Converter Packaged with VSI Modules

Considerable research effort has been dedicated to the three-phase AC/AC matrix converters [3–12] concerning both the realization of bidirectional switches and improving the PWM control. The operation and control of the conventional matrix converters was presented in Chapter 18.

One of the problems with AC/AC matrix converter relies in the realization of the bidirectional switches. Since the bi-directional switches are not available and they need special packaging, another possible approach for the hardware packaging

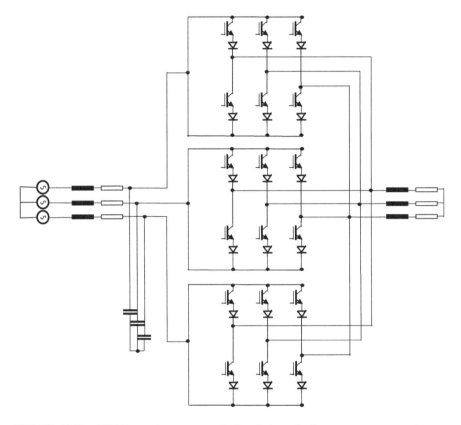

FIGURE 18.7 AC/AC matrix converter built of three 3-phase current source inverter modules.

considers three three-phase current source converter modules connected as in Figure 18.7. Realization of the matrix converter yields therefore based on the previous expertise and know-how on this topology and appropriate packaging.

Due to the symmetry of the power stage, there are two ways possible for the connection of the unidirectional switches. It is important to group the unidirectional switches as shown in Figure 18.7 with the DC side towards input. In this way, we avoid the need to short-circuit one converter leg when two output phases are connected to the same input. The generic SVM algorithm for matrix converters can be applied here. Figure 18.8 shows the actual implementation with IPM/IRAMS.

18.2.2 Dyadic Matrix Converter with VSI Modules

Given some hardware limitations of the conventional matrix converter, novel topologies employing unidirectional switches have been proposed in [13–16] with very promising results. The scheme proposed in [17,18] consists of three power

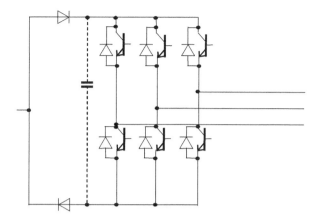

FIGURE 18.8 Realization of each CSI converter phase with conventional power modules.

modules working as Unity Power Factor SPWM Rectifiers that are employed to form the three-phase output voltage system. These topologies are reviewed here with respect to their implementation with IPM modules. Each power module ensures near sinusoidal, unity power factor current on the input side, while a reduced value of the output capacitor allows us to control the converter output to track a sinusoidal waveform.

The dyadic matrix converter theory was developed in [18] and it constitutes a possible base for the further development of PWM algorithms for any kind of direct AC-AC matrix converter. The output-input relationship is given by

$$[e] = [H] \cdot [v_2] \Rightarrow [v_2] = [H]^T \cdot [e] \tag{18.1}$$

It has demonstrated that $[H]$ can be decomposed into two matrices that are basically separating *the frequency changer* $[H_f]$ and the *static VAR compensator* $[H\varphi]$ components:

$$[H] = [H_f(t)] + [H_\varphi(t)] \tag{18.2}$$

Moreover,

$$[H_f(t)] = \begin{bmatrix} C_{12}(0) & C_{12}\left(-\dfrac{2\pi}{3}\right) & C_{12}\left(\dfrac{2\pi}{3}\right) \\ C_{12}\left(-\dfrac{2\pi}{3}\right) & C_{12}\left(\dfrac{2\pi}{3}\right) & C_{12}(0) \\ C_{12}\left(\dfrac{2\pi}{3}\right) & C_{12}(0) & C_{12}\left(-\dfrac{2\pi}{3}\right) \end{bmatrix} \tag{18.3}$$

$$[H_\varphi(t)] = \begin{bmatrix} C_{1\varphi}(0) & C_{1\varphi}(0) & C_{1\varphi}(0) \\ C_{1\varphi}\left(-\dfrac{2\pi}{3}\right) & C_{1\varphi}\left(-\dfrac{2\pi}{3}\right) & C_{1\varphi}(-\dfrac{2\pi}{3}) \\ C_{1\varphi}\left(\dfrac{2\pi}{3}\right) & C_{1\varphi}\left(\dfrac{2\pi}{3}\right) & C_{1\varphi}\left(\dfrac{2\pi}{3}\right) \end{bmatrix} \qquad (18.4)$$

Different forms for the above cell functions are considered in [19].

The schematic diagram of the matrix converter topology consisting of three 3-phase voltage-source converter modules is presented in Figure 18.9. Each converter module is controlled to produce, on its "*DC-side,*" an AC voltage superimposed on a DC voltage. The load does not see the DC components due to the 3-wire connection. The possibility to use three-phase voltage source converters packaged as power modules is the main advantage of this topology.

What concerns the control system, there are two possible ways of defining the control algorithm for a conventional matrix converter:

- Based on a scalar Sinusoidal PWM approach;
- Using a vector model of the AC/DC power converter.

While the original approach suggests the reference as being a sinusoidal waveform superimposed on a DC component (Figure 18.10a), a waveform with a higher

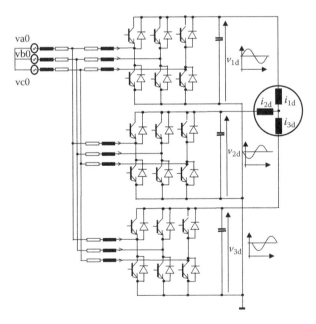

FIGURE 18.9 Schematic diagram of the power stage of the proposed three-phase voltage source matrix converter.

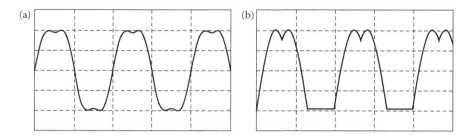

FIGURE 18.10   Load voltage: (a) sinusoidal reference; (b) novel reference.

content in third harmonic was proposed (Figure 18.10b). In this way, efficiency is improved as at any moment one converter does not work at all.

The converters behave as current sources pumping current into capacitors used on the output side for the required voltage source character. For this reason, the actual waveforms of the output voltages are strongly influenced by the passive components of the system, including the output filter. This is true for both the sinusoidal reference and the waveform with a third-harmonic injection reference [19]. Closed loop PWM control improves further the system performance.

The major advantage of the control method proposed in Figure 18.10b consists in a larger available output voltage. Reversing the design requirement: the power converter works with 15.4% less voltage across the output capacitor bank (filter) for a desired maximum output (phase) voltage. Moreover, the maximum dv/dt across the capacitor is less by 50%. This permits a lower switching frequency of the converter for the same harmonic performance in the phase voltage. In some cases, the power switches can be chosen at a lower maximum direct voltage for the same desired output voltage.

Additional harmonic aspects are discussed in [19].

18.3   MULTILEVEL CONVERTER MADE UP OF MULTIPLE POWER MODULES

Multilevel converters are used for higher voltages in the DC bus and consist of a network of switches able to produce more voltage levels in the output (AC) voltage. Some examples for the use of IPM devices to the definition of novel topologies worth be mentioned here for the completeness of information. The scheme is shown in Figure 18.11.

A second approach for using IPM modules to building multilevel converters has been recently the subject of a PhD thesis at University of Bologna [21]. The base scheme is shown in Figure 18.12.

18.4   NEW TOPOLOGY BUILT OF POWER MODULES AND ITS APPLICATIONS

18.4.1   Cyclo-Converters

Given the inherent circuit loss and reliability issues with passive components, engineers have explored power conversion topologies able to alleviate the need for passive

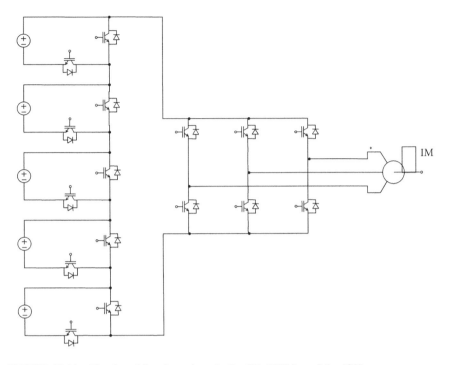

FIGURE 18.11   Novel multilevel topology built of Fuji IPM modules [20].

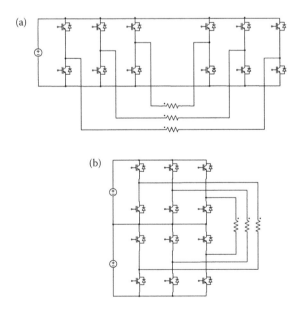

FIGURE 18.12   Series (a) and parallel (b) dual-IPM based multilevel converters (IRAMS).

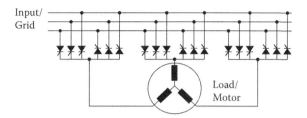

FIGURE 18.13   Principle of operation for a two-quadrant cyclo-converter.

components. The resulted converters are often named *direct converters*. Among such alternatives, the *cyclo-converters* are a very old solution [22–24].

Despite the advent of IGBTs, cyclo-converters are still used in high power applications, especially in applications such as cement mill drives, ship propulsion drives, rolling mill drives, Scherbius drives, grinding mills, mine winders, dynamometer equipment. The operation of a cyclo-converter is based on continuously changing the control angle of a SCR device in order to derive a waveform with a high content in fundamental at the desired load frequency. Figure 18.13 presents the operation principle for a three-phase to three-phase half-wave cyclo-converter. Each conventional rectifier can be controlled to vary its average output voltage between a positive maximum and a negative minimum, while the load current can have one direction only.

In order to achieve a full four-quadrant operation necessary for motor drives applications, two identical cyclo-converter systems of Figure 18.13 are connected in anti-parallel to form circulation paths for both negative and positive currents. Figure 18.14 shows the three-phase to three-phase full-wave cyclo-converter.

A modern and updated approach to the realization of direct conversion would build the four-quadrant cyclo-converter with IGBTs. Such converters can be seen as being similar to the matrix converters [25,26]. The topology is drawn in Figure 18.15 and the principle of operation is shown in Figure 18.16. The control method was changed from control of the firing angle to synchronized PWM algorithm, while the structure and operation of the system remains the same as for a cyclo-converter [27–30]. The advantage consists in achieving larger available voltages on the load with a simpler packaging strategy. These new converters are good competitors of the

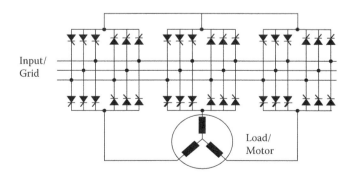

FIGURE 18.14   Full-wave four-quadrant cyclo-converters.

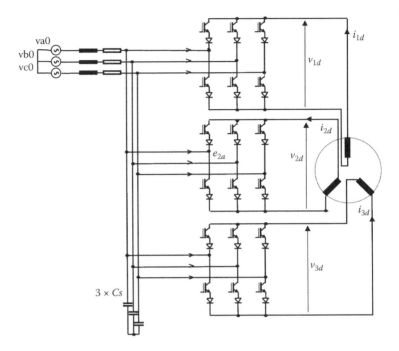

FIGURE 18.15 Implementation of the two-quadrant direct power converter.

well-known multilevel converters, matrix converters, or power electronics systems series-connected through the open-winding loads.

The AC/AC direct converter previously described was first reported in 1976 [31,45[1]] with the devices and control specific to that time. Details of operation were reported in [15–19] along with a *Space Vector Modulation* controller derived from the knowledge about *Current Source Converters*. Parallel efforts defined the control based on matrix representation of the modulating waveforms [20,21].

Each phase current is derived by controlling the DC side of the conventional grid-interface CSI. Later on, we have proved that the grid-side currents can be controlled with unity power factor and low harmonic content, and this control can

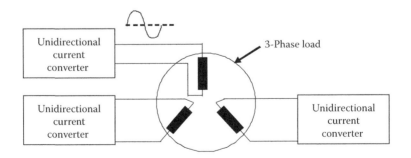

FIGURE 18.16 Two-quadrant IGBT-based cyclo-converter principle.

be extended even for unbalanced operation [27] or grid induced harmonics. The load-side current can be considered as a sinusoidal wave superimposed with a DC component. A special waveform (Figure 18.16b) has been considered in order to decrease the DC component to 82.8% and the current peak to 86.6% of the sinusoidal based solution.

A possible performance comparison can be achieved with a current source converter used commonly for medium and high power drives. If the rotor winding is built with large bars or in the double-cage configuration in order to repress the current (for direct start at nominal frequency with lower starting current and larger torque), the dependency on frequency: $K_s = 0.4 + 0.97 \cdot \sqrt{f}$ and can go up to 10. The proposed converter will dissipate less power inside the machine than the simple Current Source Inverter. Despite the harmonics reduction with PWM operation of the CSI, the machine loss is larger than that of the IGBT cyclo-converter and, more importantly, a large part of them are within the machine rotor.

The following waveforms explain the operation of the two-quadrant power converter, employing the control system described in the next section (Figure 18.17).

Figure 18.18 introduces the four quadrant direct AC/AC converter able to get closer to the industrial drives' case. Using open-winding load is more realistic in high-power applications where the load is typically with both terminals of each phase available, but similar operations can be achieved with star-connected load. This power electronics system is composed of 6 conventional CSI converters operated with a variable DC-side current. The control of each individual CSI is described in the next section. It includes a compensation for unbalance or grid harmonics while the DC-side currents are controlled in closed loop.

Figure 18.19 shows the complete schematic of the power stage for the 4-quadrant converter built up of Current Source Inverter modules. Each module can be implemented independent of the system and its control requirements can easily be depicted from the system requirements. Designing and building individual current source converters is the only design task for this system.

Table 18.1 provides a comparison with the back-to-back multilevel converter and with the series-connected converters following a rectification with power factor correction. One can observe the ability to obtain *sqrt*(3) times more voltage than a matrix converter or other topology without any special over-modulation algorithm [32], and with less passive components.

18.4.2 CONTROL SYSTEM

Let us review the operation of each individual CSI converter. As shown in Chapter 16, there are many control algorithms previously reported in literature [33–37].

It is important to understand that the converter used in the application from Figure 18.20 behaves as a voltage source on the output side. At any moment, there are two switches turned-on and the converter output (conventional DC voltage of the AC/DC converter) equals the difference of the voltages across two input filter capacitors. The voltage pulses obtained as output voltages will be filtered by the machine inductance to produce the desired currents. Observing this provides an analogy between the outputs of the proposed converter and the outputs

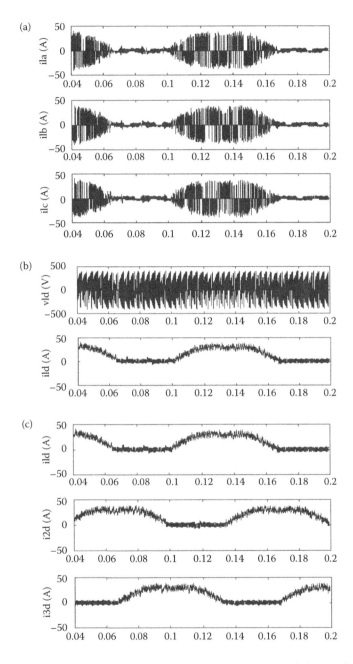

FIGURE 18.17 Waveforms for the first converter plus the command signals ($f_{\text{out}} = 10$ Hz, $I_m = 30$ A, $f_{\text{sw}} = 3.6$ kHz, $RL = 2$ V, $LL = 4$ mH). (a) Input phase currents for one of the converters; (b) output phase current and output voltage for one converter; (c) output phase currents for all three converters. (From Neacsu, D.O. 1999. *IEEE IECON 1999*, San Jose, CA, USA, November 29–December 3. With permission.)

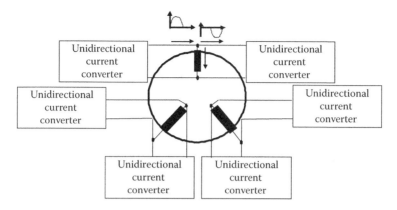

FIGURE 18.18 Four-quadrant direct AC/AC converter based on CSI modules.

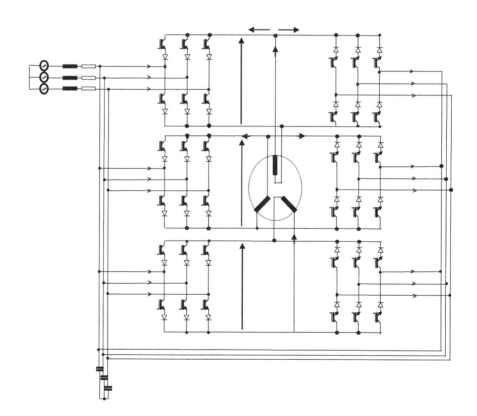

FIGURE 18.19 4-Q converter circuit diagram.

TABLE 18.1
Comparison Results for Four Quadrant IGBT Cyclo-converter

Hardware components (cost, size, weight)

	4Q Cyclo-conv.	Multilevel	2-Series
IGBT/diodes	36	24	24
Input filter	3× LC	3× LCL	3× LCL

Waveform characteristics

	4Q Cycloconv.	Multilevel	2-Series
Max. output voltage (RMS) versus input V_{ph} (RMS)	$0.866*sqrt(3)*V_{ph}$	$0.866*V_{ph}$	$2*V_{ph}$
Peak of the output current at the same fundamental of phase current	$Sqrt(6)*I_{ph}$	$Sqrt(2)*I_{ph}$	$Sqrt(2)*I_{ph}$

Operational and control characteristics

	4Q Cyclo-conv.	Multilevel	2-Series
Homopolar component in output current	NO	No	No
Smooth torque production	YES	Yes	Yes
Field-oriented vector control	YES	Yes	Yes
Input PF correction	YES	Yes	Yes

of a three-phase voltage source inverter. The control system uses a *Space Vector Modulation* based algorithm, and we are seeing the converter as a voltage source on the DC side. Hence, we can work with voltage vectors instead of the conventional current vectors (Figure 18.20).

Figure 18.20 shows a low power diode rectifier for the sensing of the input grid voltage. The circuitry also contains an input filter, current control, and the PWM generator. Designing the PWM circuitry based on voltage vectors allows us to neglect the load character and more importantly to define closed loop current control with the output (actuating measure) as voltage reference. Finally, a PLL loop is used as frequency multiplier locked to the supply frequency in order to produce the desired sampling frequency. The influence of the supply frequency variations can be reduced further.

18.4.3 PWM GENERATOR

The theory of PWM control of a Current Source Converter is briefly revisited herein. More details are available in Chapter 15. The load current I_d is considered as being constant during a PWM sampling interval, and it should be controlled to follow a variable reference over a fundamental frequency cycle.

Since one and only one switch in the upper and lower half bridge must be turned-on at a time, there are nine possible combinations for the "*ON*"-switches. Each pair of two switches turned-on determines a specific state of the converter and a *Space Vector* can be associated to each state. The output voltage can therefore coincide

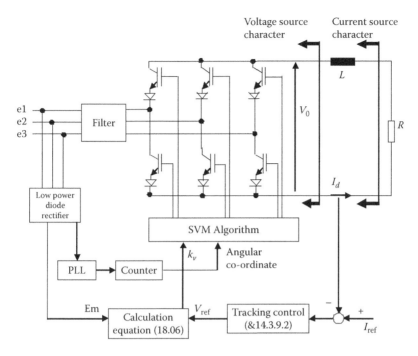

FIGURE 18.20 Grid interface with current source converter seen as voltage source on the DC side.

with any of the line-to-line voltages or can be zero. Only seven distinct positions of the Space Vector can be obtained: $I_1 - I_6$ and zero I_0, and they are shown in Figure 18.21. The synthesis of any virtual position of the current vector in the complex plane can be achieved with different active states being combined over a sampling interval.

As shown in Chapter 16, any desired position of the space vector I is always placed between two space vectors I_a and I_b, $(a,b = 1\ldots6)$ which represent the two

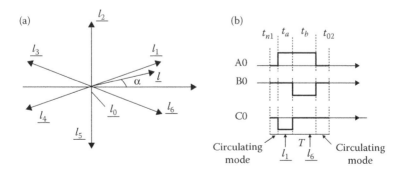

FIGURE 18.21 (a) Current space vector locations; (b) synthesis of a desired location of the space vector.

active states involved in the switching process. Writing the appropriate average relationship yields:

$$\bar{I}_a \cdot t_a + \bar{I}_b \cdot t_b + \bar{I}_0 \cdot t_0 = \bar{I} \cdot T \tag{18.5}$$

where T is the sampling period; t_a is the time assigned for the state I_a; t_b is the time assigned for the state I_b; t_0 - is the time assigned for the state I_0. The sampling interval is completed with a *zero* state that can be obtained by turning-on the switches on the same leg so that there is always a current path for the output inductive current.

Observing the circuitry from Figure 18.20 allows us to transform the current vector control into a voltage vector control. In this respect we need to consider the output voltage measured right after the power stage and before the load. This approach would further allow us to use the grid voltage as a feed-forward component able to compensate for harmonics or other grid distortion. Considering E_m the envelope of the rectified input voltage [36, 37] and V the desired voltage vector at the output of the converter, the space vector theory can be re-written in voltage terms:

$$
\begin{aligned}
t_a &= T \cdot k_v \cdot \sin(60 - \alpha) \\
t_b &= T \cdot k_v \cdot \sin(\alpha) \\
t_0 &= T - t_a - t_b \\
k_v &= \frac{R \cdot I_d}{E_m} = \frac{V}{E_m}
\end{aligned}
\tag{18.6}
$$

where V is one of the desired references for the three-phases (v_{1d}, v_{2d}, v_{3d}). The duration t_0 is shared by two zero states.

Any unbalance of the grid voltage is reflected in the real-time envelope of the grid voltage E_m, and hence it is compensated within the *Space Vector Modulation* algorithm (Figure 18.22). This compensation method benefits from the research results

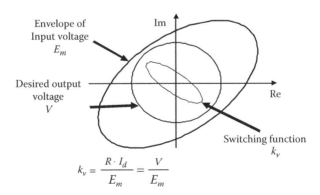

FIGURE 18.22   SVM for each power converter unit.

presented in [36]. This method has the drawback of limiting the maximum available voltage to the minimum value of E_m. The design of the controller follows the description from Section 13.3.9.2 [38,39].

18.5   GENERALIZED VECTOR TRANSFORM

Some of the previous converters have used a special shape of the phase currents or voltages. This shape of the converter waveform has the interesting property that the difference from one phase to another is always a pure sinusoidal waveform. However, a link between vector control methods and the proposed waveforms is necessary. Conventional vector control methods have orthogonal (d, q) or (x, y) coordinates as outputs.

Let us consider a general set of three voltages [40]:

$$[e] = \begin{bmatrix} e_a \\ e_b \\ e_c \end{bmatrix} \tag{18.7}$$

that needs to be transformed in another set of voltages

$$[v_2] = \begin{bmatrix} v_{1d} \\ v_{2d} \\ v_{3d} \end{bmatrix} \tag{18.8}$$

Usual vector control theory uses the time-invariant environment of $d - q - 0$ frame. We would like now to generalize this transform from a system with frequency ω_1 to a system with frequency ω_2 through an intermediary fixed gain DC controller. All possible control cases share the same general expression for the direct transfer function $[H_f]$ [15],

$$[H_f(t)] = \left[C(\omega_1) \right]_{3 \times 2} \cdot \left[P \right]_{2 \times 2} \cdot \left[C(\omega_2) \right]^T_{3 \times 2} \tag{18.9}$$

where

$$\left[C(\omega_i) \right] = \begin{bmatrix} b_1(\omega_i) & b_2(\omega_i) & b_3(\omega_i) \end{bmatrix} \tag{18.10}$$

represents a matrix composed of orthonormal base vectors of the input or output three-phase systems and $[P]$ represents a matrix of constant weights p_{ij}. For a three-phase system, the vector space has a dimension of three and only three terms will always be seen in the matrix defined with base vectors.

Symmetries within a three phase system contribute to a generalized form of Equation 18.09:

$$[H_f(t)] = \begin{bmatrix} C_{12}(0) & C_{12}\left(-\dfrac{2\pi}{3}\right) & C_{12}\left(\dfrac{2\pi}{3}\right) \\[3mm] C_{12}\left(-\dfrac{2\pi}{3}\right) & C_{12}\left(\dfrac{2\pi}{3}\right) & C_{12}(0) \\[3mm] C_{12}\left(\dfrac{2\pi}{3}\right) & C_{12}(0) & C_{12}\left(-\dfrac{2\pi}{3}\right) \end{bmatrix} \tag{18.11}$$

A *base* within a vector space consists in a system of vectors B that provides a unique representation of any member V of that Vector Space as a linear combination of vectors from B. For our 3-dimensional vector space, it yields:

$$\vec{V} = b_1(\omega_j) \cdot v_d + b_2(\omega_j) \cdot v_q + b_3(\omega_j) \cdot v_0 \tag{18.12}$$

or

$$\vec{V} = b_1(\omega_j) \cdot v_d + b_2(\omega_j) \cdot v_q \tag{18.13}$$

where v_d, v_q, v_0 are also called coordinates. If the vector space has a finite dimension, then all possible bases have the same number of elements. A base is orthonormalized if all its vectors have unitary magnitude and any two different vectors are orthogonals.

Generally, analysis of three-phase power converters is considering the orthonormalized base vectors as

$$[b_1(\omega_i)]^T = \sqrt{\frac{2}{3}} \cdot \left[\cos\omega_i t \quad \cos\left[\omega_i t - \frac{2\pi}{3}\right] \quad \cos\left[\omega_i t - \frac{4\pi}{3}\right]\right] \tag{18.14}$$

$$[b_2(\omega_i)]^T = \sqrt{\frac{2}{3}} \cdot \left[\sin\omega_i t \quad \sin\left[\omega_i t - \frac{2\pi}{3}\right] \quad \sin\left[\omega_i t - \frac{4\pi}{3}\right]\right] \tag{18.15}$$

$$[b_3(\omega_i)]^T = \sqrt{\frac{1}{3}} \cdot [1 \quad 1 \quad 1] \tag{18.16}$$

When the zero-sequence is omitted, $b_3(\omega_i)$ does not appear in the frequency changer term.

The intermediary factor, P_f plays the same role as the turn ratio of the primary to secondary windings of a magnetic transformer. For example, we provide three possible cases herein:

A. Choosing

$$[P]_{2\times2} = P_f \cdot \begin{bmatrix} 1 & 0 \\ 0 & -1 \end{bmatrix} \tag{18.17}$$

yields

$$C_{12}^A(x) = \frac{2}{3} \cdot P_f \cdot \cos\left[(\omega_1 + \omega_2)t + x\right] \tag{18.18}$$

B. Choosing

$$[P]_{2\times2} = P_f \cdot \begin{bmatrix} 1 & 0 \\ 0 & 1 \end{bmatrix} \tag{18.19}$$

yields

$$C_{12}^B(x) = \frac{2}{3} \cdot P_f \cdot \cos\left[(\omega_1 - \omega_2)t + x\right] \tag{18.20}$$

C. A general dependency including both type of terms yields

$$C_{12}^C(x) = P_f \cdot \left\{ \frac{1}{3} \cdot \cos\left[(\omega_1 - \omega_2)t + x\right] + \frac{1}{3} \cdot \cos\left[(\omega_1 + \omega_2)t + x\right] \right\} \tag{18.21}$$

Due to the definition of a base in a vector space, a base is not unique and new vector bases can be defined. This demonstrates that selection of Equations 18.17 through 18.19 is not the unique choice for a three-phase system. It opens up a new mathematical tool for working with references not sinusoidal but characterizing uniquely a three-phase system. The resulting waveforms have been also used in discontinuous PWM algorithms.

Figure 18.23 presents two possible waveforms considered as examples for this approach. Furthermore, to simplify the mathematics, the third term corresponding to the homopolar component is neglected in this analysis.

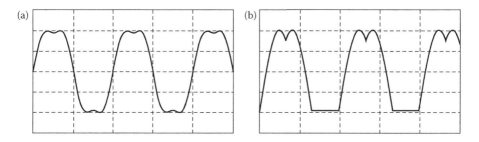

FIGURE 18.23 Different choices of output-side references functions.

Mathematical form of Figure 18.23a yields the following two functions that can be chosen as a base in a vector space.

$$
[b_1(\omega_2)]^T = \sqrt{\frac{2}{3}} \cdot \left[\cos\omega_2 t + \frac{1}{6} \cdot \cos(3 \cdot \omega_2 t) \quad \cos\left[\omega_2 t - \frac{2\pi}{3}\right] \right.
$$
$$
\left. + \frac{1}{6} \cdot \cos(3 \cdot \omega_2 t) \quad \cos\left[\omega_2 t - \frac{4\pi}{3}\right] + \frac{1}{6} \cdot \cos(3 \cdot \omega_2 t) \right]
$$

$$
[b_2(\omega_2)]^T = \sqrt{\frac{2}{3}} \cdot \left[\sin\omega_2 t + \frac{1}{6} \cdot \sin(3\omega_2 t) \quad \sin\left[\omega_i t - \frac{2\pi}{3}\right] \right.
$$
$$
\left. + \frac{1}{6} \cdot \sin(3\omega_2 t) \quad \sin\left[\omega_i t - \frac{4\pi}{3}\right] + \frac{1}{6} \cdot \sin(3 \cdot \omega_2 t) \right]
$$

$$(18.22)$$

Considering the same vector base from Equations 18.06 and 18.07 for $C(\omega 1)$ and the new base functions (Equations 18.15 and 18.16) on $C(\omega 2)$ yield the next form of the modulating signals.

$$
[H_f] = \begin{bmatrix}
C_{12}^D(0,0) & C_{12}^D\left(-\dfrac{2\pi}{3},0\right) & C_{12}^D\left(-\dfrac{4\pi}{3},0\right) \\[2mm]
C_{12}^D\left(-\dfrac{2\pi}{3},-\dfrac{2\pi}{3}\right) & C_{12}^D\left(-\dfrac{4\pi}{3},-\dfrac{2\pi}{3}\right) & C_{12}^D\left(0,-\dfrac{2\pi}{3}\right) \\[2mm]
C_{12}^D\left(-\dfrac{4\pi}{3},-\dfrac{4\pi}{3}\right) & C_{12}^D\left(0,-\dfrac{4\pi}{3}\right) & C_{12}^D\left(-\dfrac{2\pi}{3},-\dfrac{4\pi}{3}\right)
\end{bmatrix}
$$

$$(18.23)$$

where

$$
C_{12}^D(x,y) = \frac{2}{3} \cdot P_f \cdot \left\{ \cos\left[(\omega_1 + \omega_2)t + x\right] + \frac{1}{6} \cdot \cos\left[(\omega_1 + 3\omega_2)t + y\right] \right\} \quad (18.24)
$$

Analogously, we can consider the base vectors

$$
[b_1(\omega_2)]^T = \sqrt{\frac{2}{3}} \cdot \left[f[\omega_2 t] \quad f\left[\omega_2 t - \frac{2\pi}{3}\right] \quad f\left[\omega_2 t - \frac{4\pi}{3}\right] \right] \quad (18.25)
$$

$$
[b_2(\omega_2)]^T = \sqrt{\frac{2}{3}} \cdot \left[g[\omega_2 t] \quad g\left[\omega_2 t - \frac{2\pi}{3}\right] \quad g\left[\omega_2 t - \frac{4\pi}{3}\right] \right] \quad (18.26)
$$

where $f(\omega_2 t)$ represents the waveform presented in Figure 18.23b and $g(\omega_2 t)$ represents the same waveform with 90° out of phase. It yields:

$$f(\omega_2 t) = \left[u(\omega_2 t) - u\left(\omega_2 t - \frac{2\pi}{3}\right)\right] \cdot \cos \omega_2 t$$

$$+ \left[u\left(\omega_2 t - \frac{2\pi}{3}\right) - u\left(\omega_2 t - \frac{4\pi}{3}\right)\right] \cdot \cos\left(\omega_2 t - \frac{\pi}{3}\right) \quad (18.27)$$

where $u(x)$ represents the Heaviside function ($= 0$, for $x < 0$ and $= 1$ for $x > 0$).
Finally, denoting:

$$C_{12}^E(x,y) = \frac{2}{3} \cdot P_f \cdot \left\{\cos(\omega_1 t + x) \cdot f(\omega_2 t + y) - \cdot \sin(\omega_1 t + x) g(\omega_2 t + y)\right\} \quad (18.28)$$

yields

$$[H_f] = \begin{bmatrix} C_{12}^E(0,0) & C_{12}^E\left(0,-\frac{2\pi}{3}\right) & C_{12}^E\left(0,-\frac{4\pi}{3}\right) \\ C_{12}^E\left(-\frac{2\pi}{3},0\right) & C_{12}^E\left(-\frac{2\pi}{3},-\frac{2\pi}{3}\right) & C_{12}^E\left(-\frac{2\pi}{3},-\frac{4\pi}{3}\right) \\ C_{12}^E\left(-\frac{4\pi}{3},0\right) & C_{12}^E\left(-\frac{4\pi}{3},-\frac{2\pi}{3}\right) & C_{12}^E\left(-\frac{2\pi}{3},-\frac{4\pi}{3}\right) \end{bmatrix} \quad (18.29)$$

18.6 IPM IN IGBT-BASED AC/AC DIRECT CONVERTERS BUILT OF CURRENT SOURCE INVERTER MODULES

18.6.1 HARDWARE DEVELOPMENT

The previous sections have laid down the theoretical background for several direct AC/AC power converters, with reduced count of passive components, that can be implemented with circuitry based on conventional *Current Source Inverters* (CSI). Unfortunately, there is no CSI inverter packaged unitary for this purpose. Attentive to the developments on the market of power semiconductor devices, voltage source inverters (*intelligent power modules*) can be used for building CSI converters. Figure 18.24 illustrates this hardware implementation.

The PWM algorithms previously defined for current source converters cannot work with the hardware from Figure 18.24 since they assume shorting the DC bus during the zero-states. Such operation is prevented by the internal operation of the intelligent power module and a short dead-time is generally introduced by such module to prevent shoot-through. Additional requirements for the bootstrap power supply should be met, that is frequently enough operation of the low-side switch.

A new PWM algorithm is needed to use opposite active vectors during the zero-states in order to avoid shoot-through. The two opposite vectors are used to compensate each other within the average vector equation used for Space Vector Modulation generation. The details of this PWM algorithm are explained with Figure 18.25.

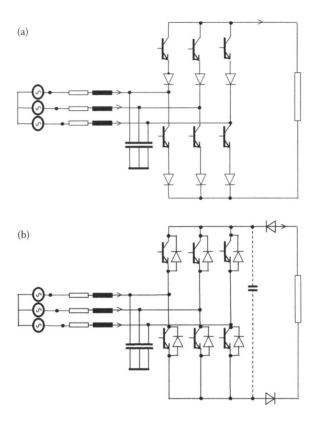

FIGURE 18.24   Conventional (a) and modified (b) CSI converters.

The vector applied during the first zero-state is selected to be the same as the nearby active vector, and the vector applied during the second zero-state is selected as the opposite of the vector applied during the first zero-state. Thus, the two vectors cancel each other over the average equation applied over the length of a sampling interval. Results for the current control over a single-phase are shown in Figure 18.26

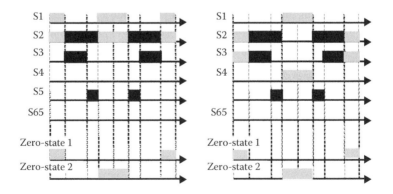

FIGURE 18.25   Conventional CSI algorithm (left) and VSI PWM algorithm (right).

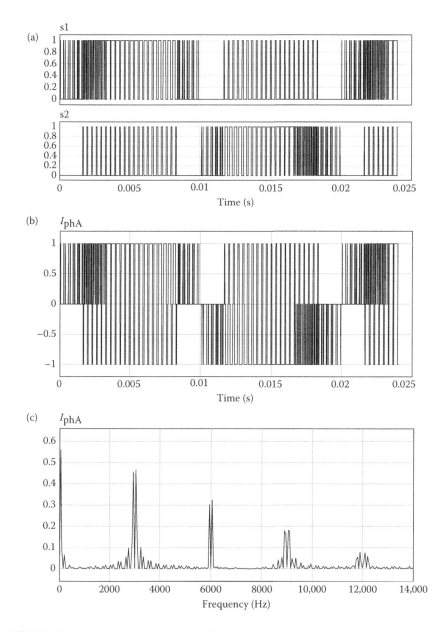

FIGURE 18.26 Converter operation with constant DC current at 3 kHz switching: (a) switching signals for top and bottom IGBTs, (b) phase current, (c) harmonic spectrum.

for switching at 3 kHz. The overall number of switching processes remains the same. The same design of the current control and PWM generator is applied for all six power modules.

Figure 18.27 illustrates the implementation with *Intelligent Power Modules* (IPM). A small LC input filter may still be used on the grid side and it can be omitted if the

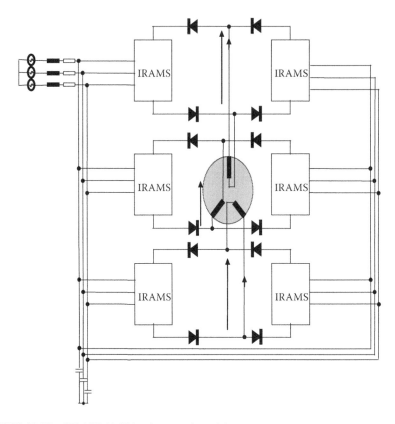

FIGURE 18.27   IPM/IRAMS implementation of the new converter.

grid is really stiff, satisfying the ideal voltage source requirement. On the grid side, each current is a summation of six currents, yielding a high content in fundamental.

18.6.2   PRODUCT REQUIREMENTS

The appliance market has recently seen a revolutionary change with the introduction of inverter-based motor drives. The appliance power system has the development structured into several levels:

- Conventional component level architecture
- Multiconverter subsystem level architecture
- Multiaxis and multiconverter system level architecture

The innovation of the appliance electronics is expected within the following subsystems:

- Semiconductors and packaging
- Circuit topology

- Communication and user interface
- Control algorithms

An important trend within the appliance power electronics relates to creation of integrated power modules based upon the merger of the device development with new circuit prototyping. The design criteria taken over from the system level design are Cost (product and maintenance), Energy Efficiency, Product lifetime and reliability, Influence on ambient medium (noise, temperature, grid harmonics, and so on).

The most important requirement for the appliance consumer market relates to system reliability. Different manufacturers show different customer expectations, with a high MTBF of 20 years lifetime with 90% reliability prevailing.

As discussed in Chapter 7, the electrolytic capacitor represents a critical component for the system lifetime since it has the lowest lifetime of all power conversion components. Capacitor lifetime is limited by the electrochemical degradation and accelerated by temperature and voltage stress. The lifetime of electrolytic capacitors has progressed from 1000 h at 65°C, 40 years ago, to up to 15,000 h at 105°C today. The lifetime is predicted by the capacitor vendors with the *Arrhenius* law equations. For instance, a capacitor rated with 10,000 at 105°C, could survive 160,000 at 65°C [39].

Hence, any solution that can eliminate passive components and/or reduce their importance in the calculation of the overall reliability is highly desirable. The AC/AC power converter subscribes to this concept, and the reliability is expected to be increased by the withdrawal of electrolytic capacitors.

It can be further proven that a power semiconductor module could contribute better performance to the system reliability than individual components. The IPM modules provide better thermal design and an excellent layout, both with effects on the system reliability. Using a power module supplied from the manufacturer rather than individual components is generally a better choice for the inverter application.

18.6.3 Performance

The previous sections have shown a series of novel direct AC/AC power converters derived from cyclo-converter concept, and intended to be implemented with standardized power modules. It is therefore attracting the use of this concept to the appliance market [39]. Since the most important advantage of using electronic control of motor drives consists of energy savings, the focus remains on cooling and thermal aspects of using this converter in the appliance application.

The thermal performance is presented herein in comparison with the conventional back-to-back solution. Thermal data is considered the same as in the manufacturer's example for IPM utilization. The intelligent power module is IRAMS10UP60, with a junction-case thermal resistance of 4.7°C/W. The thermal compound was Wakefield #120 characterized by a thermal resistance of 0.1°C/W. The suggested heat-sink is an *off-the-shelves* component, Aavid Thermalloy #66365, with a heat sinking area of 0.63 × 1.50 in, and a thermal resistance of 5.4°C/W.

Section 2.4 has presented the loss calculation for the IGBT based medium power converters [41,42]. Given the peculiar construction of the IPM module, an empirical calculation of the loss is presented in [43] and adopted herein. The results of this method used for a conventional motor drive application show 2.3 W power loss per switch, and 14.1 W per entire package, when operated in ambient temperature and trying to prevent a junction temperature close to 125°C.

This empirical model considers each switch individually and calculates the power loss with the following set of equations already presented in Chapter 2.4 as Equations 2.22 through 2.24 valid for International Rectifier's IRAMS devices.

$$E_{ON} = (h_1 + h_2 \cdot I^x) \cdot I^k = [(7.69e - 4) + (2.99e - 2) \cdot I^{-1.159}] \cdot I^2 \qquad (18.30)$$

$$E_{OFF} = (m_1 + m_2 \cdot I^y) \cdot I^n = [(1.76e - 2) + (4.34e - 2) \cdot I^{-0.492}] \cdot I^1 \qquad (18.31)$$

$$V_{CEON} = V_T + a \cdot I^b = 0.51 + 0.46 \cdot I^{0.649} \qquad (18.32)$$

The number of switching processes during each switching period is necessary in order to use these equations for the loss calculation of the entire system (Figures 18.28 and 18.29).

A. *Conventional back-to-back converter*

Switching Loss—IGBT = Switching processes are shown in the next figure.
Conduction Loss—IGBT = 180° total.
Switching Loss—Diode = Diodes are switched during the negative half waveform (reference), with a partial conduction.

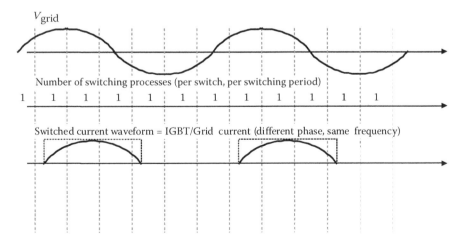

FIGURE 18.28 Distribution of switching processes used for switching loss calculation (conventional).

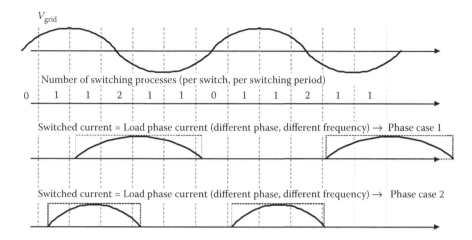

FIGURE 18.29   Distribution of switching processes used for switching loss calculation (IPM).

B. New topology based on multiple-IPM modules

  Switching Loss—IGBT = Switching processes are shown in the next figure.

  Conduction Loss—IGBT = 120° total.

  Switching Loss—Diode = Diodes are moved outside of the IPM module, with same conduction, and no switching loss.

The number of switching processes is very similar with the back-to-back converter. In any direct AC/AC converter, the switching loss cannot be expressed very easily since the switched current depends on both the load and grid frequencies.

Reference [43] discusses a practical case: a $1HP = 745W$ A/C compressor operated from $V_{dc} = 400$ V, modulation index of 0.8, $V_{LL} = 113$ V_{rms}, $PF = 0.6$, $I_{ph} = 6.35$ A_{rms}, $f_{sw} = 3$ kHz. Table 18.2 shows comparison of results between the conventional back-to-back converter and the IPM-based cyclo-converter. The load current is assumed to be at the same frequency. The power dissipated per module reduces at least 56% and this reduction can be even more important for certain combinations of input and output waveforms.

It is worthwhile to reverse the design reasoning and to observe that the new topology can deal with a load of $P = 2$ kW ($I_{ph} = 9.25$ A_{rms}, motor $PF = 0.6$, nominal frequency and voltage conditions of 120 V_{ac}/phase, 50 Hz, switching at 3 kHz), at the maximum thermal capacity of the setup. The system shown can therefore drive a 2 kW motor, in similar operation and thermal conditions as a back-to-back dual IPM module would drive a $1HP = 745$ W motor.

The results are very impressive for the system power density and packaging: the entire power stage (with straight-pin mounting, six individual heat-sinks, passive LC filtering and power connectors) accounts for ($2 \times 2 \times 7.5$ in =) 0.49l for 2 kW delivered power (that is 4.1 kW/l) [39]. This should compare to current industry goals of 4 kW/l, for this class of converters [44].

TABLE 18.2
(a) Power Loss Distribution Per IPM Package, for 1HP Load, with Voltage and Current Dictated by Operation of the Back-to-Back Converter

	Back-to-back converter	New converter
P_{sw}, per switch	0.32 W	Up to 0.32 W
P_{cond}, per switch	1.49 W	1.00 W
P_{diode}	0.53 W	0
P_{loss}, Entire IPM module	14.1 W	7.92 W

(b) Power Loss Distribution Per IPM Package, for 2 kW Load

	New converter
P_{sw}, per switch	0.75 W (variable)
P_{cond}, per switch	1.80 W
P_{diode}	0
P_{loss} Entire IPM module	15.30 W

18.7 USING MATLAB-BASED MULTIMILLION FFT FOR ANALYSIS OF DIRECT AC/AC CONVERTERS

18.7.1 INTRODUCTION TO HARMONIC ANALYSIS OF DIRECT OR MATRIX CONVERTERS

The previous sections have presented several direct converter topologies [45], each with different control methods. It is well-known that the PWM operation of power converters produces harmonics at both input and output [46,47]. In the conventional bridge-like topologies, we have to deal with two frequencies: the modulating waveform's fundamental frequency and the carrier/switching frequency. The challenges associated with the harmonic analysis of direct converters are multiple:

- The modulation signal contains both the input and output frequencies, typically independent from each other and without forming an integer ratio.
- The switching/carrier frequency is not a rational multiple of either the input frequency or the output frequency.

For a long time, the complex, vast, and time-consuming calculation required by the harmonic analysis of power switching converters encouraged the engineers to develop their own algorithms for a fast development in Fourier series [48–52]. The most used procedure consists of a double Fourier series expansion in two variables [48,52]. Later efforts presented the development for multiples frequencies [49,51]. Alternate results are given in [50,52]. All these solutions are providing a fast calculation of harmonics independent of the accuracy of the simulation model. However, they have a strong content in advanced mathematics and are each

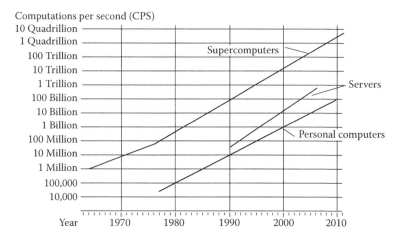

FIGURE 18.30   Evolution of processing power.

dedicated to a specific application. Hence, they are not very friendly for the practice engineers.

During this time, the Moore's law acted upon the computing hardware and today we have computers based on dual-, or quad-core processors, with 64-bits bus, huge Terra-Bytes memory capacities, and high clock frequencies. Figure 18.30 [53] provides a quick update on the development of computers. It says that the same simulation model we ran in 1 h 10 years ago can be run in under 30s now.

Another illustration of the contemporary computing power makes 2012 the year when the power electronics simulation waveforms could be achieved in real-time for the first time. *OPAL Corporation* reported at *Industry Forum of IECON2012* [54] the implementation of a complete motor-converter drive model on an advanced *Altera* FPGA with running at the system time scale. We consider this as one of the most important historic milestones for the simulations of power electronic systems.

These results should imply the advantages of using straightforward computer calculations for the analysis of power electronics waveforms. This is in par with technology development and it may diminish the academic struggle for harmonic calculation algorithms that unfortunately do not have too much practice value since the major manufacturers have settled already for their own algorithm.

The engineering practice of either multiconverter or multifrequency PWM converters provides signals not described with mathematical functions. Instead, these signals come from measurement acquired at a selected sampling interval from hardware or from simulation models (like in *Matlab-Simulink*, *PSIM*, *Plexim*, or alike), and hence they are not continuous but discrete signals. The duration of a signal is finite and in most cases it will not be the same as the period required by the *Fourier Theorem* or the signal may not be periodic at all. Due to these limitations, the frequency analysis devises a method to extract an estimate of frequency components which are not known *a priori*. The process is known as the *Discrete*

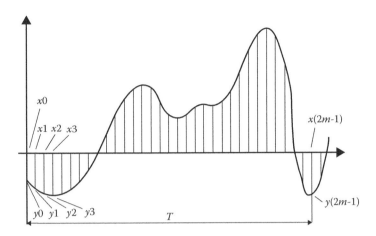

FIGURE 18.31 Sampling of the acquired time signal.

Fourier Transform (DFT). Section 3.5 in Chapter 3 have introduced us to harmonic analysis of PWM converters. The following are reconsidering this information for revealing other computer related aspects necessary to analysis of direct (matrix) converters.

The *Euler-Fourier integral* can be calculated by approximation when the analyzed time interval is sampled at a fixed rate leading to $2 \cdot m$ samples (Figure 18.31). Considering the Shannon Theorem confirms m different spectral lines.

It yields for each sample:

$$y_q = \frac{A_0}{2} + \sum_{n=1}^{m} \left(A_n \cdot \sin \frac{2 \cdot \pi \cdot q \cdot n}{2 \cdot m} + B_n \cdot \cos \frac{2 \cdot \pi \cdot q \cdot n}{2 \cdot m} \right) \qquad (18.33)$$

This represents an approximation of the Fourier series and also a discrete calculation of the Fourier integral. When we know all the samples y_q, each harmonic component can further be calculated with

$$A_0 = \frac{1}{2 \cdot m} \cdot \sum_{q=0}^{2m-1} y_q \qquad (18.34)$$

$$A_n = \frac{1}{m} \cdot \sum_{q=0}^{2m-1} y_q \cdot \cos \frac{n \cdot q \cdot \pi}{m} \qquad (18.35)$$

$$B_n = \frac{1}{m} \cdot \sum_{q=0}^{2m-1} y_q \cdot \sin \frac{n \cdot q \cdot \pi}{m} \qquad (18.36)$$

As shown in Section 3.5, the computer program yields very simple as structure:

- Calculate the argument $\beta = \pi \cdot q \cdot n / m$ and the harmonic functions for each sample y_q ($q = 1, \ldots, 2\,m - 1$);
- Calculate the individual products from A_n and B_n;
- Calculate A_n and B_n;
- Calculate the magnitude and phase for each harmonic.

For a reduced number of samples, performance can be improved when m is chosen as a multiple of 6 to speculate from the symmetries of the grid related waveforms. For an increased number of samples, the computer run time yields very large and some computer programs provide just calculation for the low frequency content. For instance, the instruction "FOUR" from SPICE provided a quick calculation for the first 9 harmonics only. SPICE after *Version 5* allows us to specify the number of desired low frequency harmonics. The computer run time can also be reduced within the *Fast Fourier Transform* that represents a peculiar case of DFT.

DFT is a discrete equivalent for the *Fourier Transform* and the sampling of the latter may hide some properties. This means we need to be careful with the selection of the resolution for the DFT calculation. Usual source of errors in calculation of DFT comes from the smoothing effect given by the interpolation of results. This acts as a low-pass filter with attenuation of the harmonic magnitudes as frequency increases. For better results we need increased number of samples which are equivalent to a higher bandwidth low-pass filter.

18.7.2 Parameter Selection

Once the data is acquired from the experimental setup or from simulation, a postprocessing tool is employed for harmonic analysis. It is vital to understand very well the settings of the parameters involved in the FFT/DFT analysis.

A. Resolution
The more the number of points in the transform, the better the frequency resolution.

Other than speed, resolution is the only other difference between a $2^9 = 512$-point transform and a $2^{20} = 1{,}048{,}576$-point transform. A power spectrum always ranges from the DC level (0 Hz) to one-half the sample rate being used. The number of points in the transform defines the power spectrum resolution (a 512-point Fourier transform would have 256 points in its power spectrum, while a 1,048,576-point Fourier transform would have 524,288 points in its power spectrum).

For example, if we want to see separate 20 and 21 Hz frequency components in the power spectrum of a power converter waveform, a 512-point Fourier transform might not show these individual components clearly since its entire power spectrum is only divided into 256 equally spaced points and the desired frequencies are so close together. For samples acquired at 1 MHz (1 μs sampling), the frequency bins would be spaced at 1 MHz/512 = 1.94 kHz = 1940 Hz.

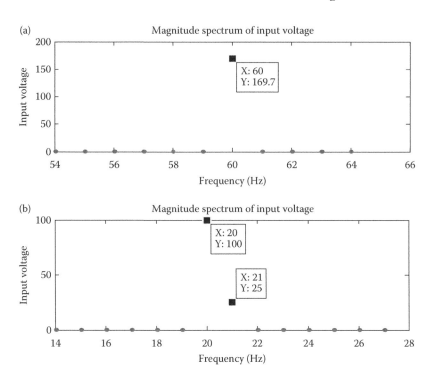

FIGURE 18.32 (a) A single 60 Hz grid voltage with 1 Hz frequency step in FFT representation; (b) a good selection of parameters allows to clearly distinguish between 100 V/20 Hz and 25 V/21 Hz components.

However, if the transform contains more points, it would be able to devote more points to the definition of closely spaced frequency components. In our example, a 1,048,576-point Fourier transform applied to a set of discrete signals depicted at a sampling rate of 1 μs, would space the frequency bins at 1 MHz/1,048,576 = 0.95 Hz. Re-arranging this selection yields $F_s = 1,048,576$ Hz = 1.048 MHz, and the frequency step 1 Hz. Thus, we are now able to see the difference between 20 and 21 Hz components (Figure 18.32b).

It is worthwhile here to make two comments:

- Most laboratory equipment sold as *Spectral Analyzers* offer a DFT with at most 32,768-points. In most cases they are not very suitable for analysis of power converters operated with both 50/60/400 Hz fundamental frequency and a carrier at 10–30 kHz. You can either benefit from the 0.1 Hz resolution through a windowing technique OR plot the entire range of DC-100 kHz with way lesser resolution.
- With the advent of computers, one can ask about the maximum number of DFT points we will ever need for analysis of power converters. Considering a maximum switching frequency of 50 kHz, a 1000-point PWM resolution (2^{10} counter, 50 ns acquisition rate), and a 0.1 Hz resolution in frequency definition yields: 20 MHz/0.1 = 200,000,000−134,217,728 = 2^{27}.

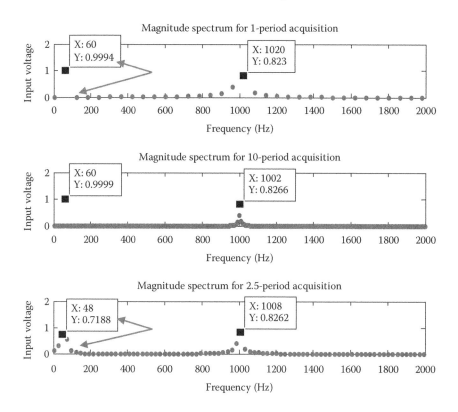

FIGURE 18.33 Spectral results when considering different time windows.

B. Accuracy

Figure 18.33 shows the spectrum of a signal composed of unity sine-wave components of 60 Hz and 1000 Hz, acquired over different time intervals. The spectrum appears somewhat *spread out* when we have a fractional number of periods (false harmonics in adjacent points) [51]. The spreading out (*leakage*) is due to energy being generated by the discontinuity at the end points of the waveform.

Solutions able to minimize *this leakage effect* consist of

- Use of conventional DFT instead of the optimized FFT;
- Multiplying the time series by a *window weighting function* before the FFT is performed.

Most window weighting functions attenuate the discontinuity by tapering the signal to zero at both ends of the window. With the window approach, the periodically incorrect signal will have a smooth transition at the end points which results in a more accurate power spectrum representation. However, if the waveform has important information appearing at the ends of the acquired time interval, it will be destroyed by this.

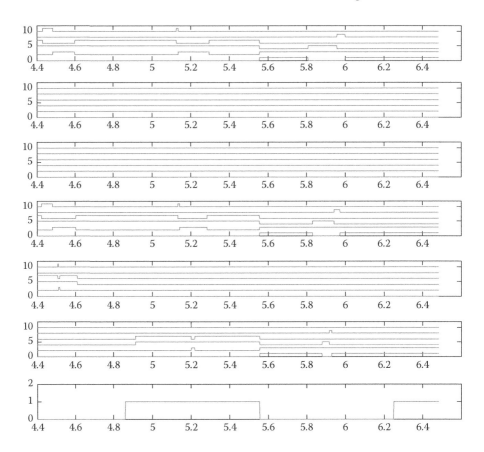

FIGURE 18.34 Example of PWM control signals for each of the six converters and the period flip-flop (twice the switching period).

Some popular windows are Hamming, Bartlett, Hanning, and Blackman. Each has different characteristics:

- The Hamming window offers the familiar bell-shaped weighting function, and produces a very good spectral peak with fair spectral leakage reduction only.
- The Bartlett window offers a triangular shaped weighting function and it produces a good, sharp spectral peak, good at reducing spectral leakage as well.
- The Hanning window offers a similar bell-shaped window and produces good spectral peak sharpness (as good as the Bartlett window), but the Hanning offers very good spectral leakage reduction (better than the Bartlett).
- The Blackman window offers a weighting function similar to the Hanning but narrower in shape. Because of the narrow shape, the Blackman window is the best at reducing spectral leakage, but the trade-off is only fair spectral peak sharpness.

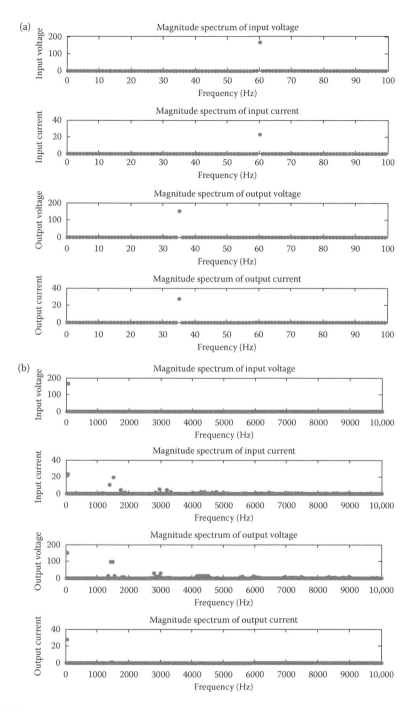

FIGURE 18.35 Sample of important waveforms for 1.44 kHz switching frequency on different zoom windows. (a) 100 Hz and (b) 10k Hz.

The choice of window function depends upon the trade-off between sharpness of peaks and decay of side-lobes. A Fourier analysis software package that offers a choice of several windows is desirable to eliminate spectral leakage distortion inherent with the FFT.

For the spectral-separation requirement of a direct power converter, the Blackman window is the best at bringing out the weaker term as a well defined peak.

18.7.3   FFT in MATLAB

MATLAB after Version 6.0 uses an implementation of the FFT called *FFTW*. The FFTW package was developed at MIT by Matteo Frigo and Steven G. Johnson (about 1997). Comparative studies performed on a variety of platforms show that its performance is typically superior to that of other publicly available FFT software, and it is even competitive with vendor-tuned codes (like in some hardware Spectral Analyzers). Contrary to vendor-tuned implementation, its performance is *portable*: the same program will perform well on most computer architectures without modification. Hence the name "FFTW" stands for the self-proclaimed title of "*Fastest Fourier Transform in the West.*"

No tuning or window selection is required in MATLAB when the dimension of the DFT is a power of 2. Moreover, for DFT dimensions that are powers of 2 between 2^{14} and 2^{22}, special preloaded information from MATLAB's internal database is used to optimize the computation.

The execution time for 1,048,576 real samples is less than a fraction of a second for a modern laptop [55].

18.7.4   ANALYSIS OF A DIRECT CONVERTER

A. Operation

We will use the above choice of a MATLAB DFT algorithm to the case of a direct converter. The operation of this multiconverter structure was explained in previous papers [15–19], and it follows the control of six *Current Source Converters* interconnected to provide a voltage source character on the load directly from the line-to-line grid voltages (Figure 18.34). The *Space Vector Modulation* is implemented herein with the sequence:

$$\textit{Zero-state} \rightarrow \textit{active 1} \rightarrow \textit{active 2} \rightarrow \textit{active 2} \rightarrow \textit{active 1} \rightarrow \textit{zero-state}$$

that is very advantageous in terms of minimizing the number of switching processes. The two *zero-states* within a sampling interval are equal with each other. A sample of the result is shown. We can see that three converters work at a given moment only. Moreover, pulses are distributed with the same sampling period.

A complete harmonic analysis is performed for a switching frequency of 1.44 kHz, for a converter with a highly inductive load, when a 35 Hz output waveform is created with a lower voltage magnitude (Figure 18.35).

B. Feed-Forward Compensation of the Grid Voltage

The operation of each converter produces a modulation of the output voltage with the envelope of the three-phase grid voltage. The *time·voltage* product corresponding to each pulse will thus depend on the instantaneous grid voltage. This yields a *time-varying spread-spectrum* of the output voltage and input current.

A compensation of the PWM generator based on the actual value of the grid voltage helps in providing the load with the desired *time·voltage* product for each PWM pulse. This has advantages in both stability of any closed–loop system and compensation of grid unbalances. Harmonic results for compensation of an 85% unbalance are shown for the first phase voltage (Figure 18.36). Differences can be seen in the spectra of the output voltages at under 400 Hz.

For a balanced grid, there should be no difference in between the spectra of the output voltage with or without compensation (also called adaptive PWM [25]) if the DFT parameters are selected properly. This is because there should be enough cycles of fundamental frequency to average the varying effect of the grid voltage envelope when the resolution is set to 1 Hz or under. This remark can also be used backwards, as a tool to detect proper use of the FFT/DFT analysis.

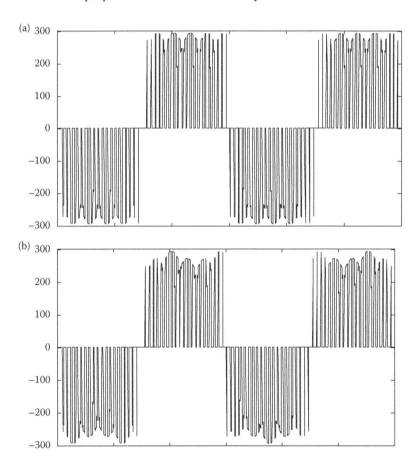

FIGURE 18.36 (a) Output voltage for the balanced input grid voltage; (b) Output voltage for 85% unbalance in the first phase of the grid voltage; (c) Spectra for PWM without compensation; (d) Compensation with input voltage measurement.

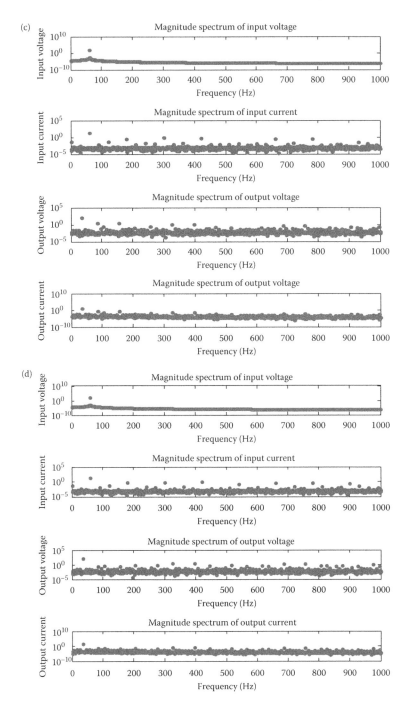

FIGURE 18.36 (continued) Stochastic ripple cancelation for interleaving the PWM controllers.

C. Synchronized PWM

The study of *Space Vector Modulation* algorithms has outlined the advantages of using an integer ratio between the sampling/switching frequency and fundamental frequency equal to a multiple of 6. This will benefit from all the symmetries in the complex plane representation and would avoid fluctuation of the *cycle-by-cycle* RMS (a *cycle* being defined at fundamental frequency).

However, spectral analysis with a fractional ratio between switching and fundamental frequencies shows that differences are minimal when the acquisition time is long enough (around 50 cycles). No important *leakage* is seen.

D. Using Interleaved Carriers

Interleaved operation of power converters is well-known for conventional bridge-type converters. Each dual (or bidirectional) converter corresponding to an output phase is controlled with a SVM carrier at 120° from each other. Contrary to interleaving conventional bridge-like converters, the instantaneous references for the three converters are different. Hence, the input current presents a stochastic ripple reduction of minimum $\sqrt{N} = \sqrt{3} = 1.73$ only [56]. All other waveforms and harmonics stay the same as in the case without interleaving (Figures 18.37 and 18.38).

18.7.5 AUTOMATION OF MULTIPOINT THD AND HCF ANALYSIS

Figures 18.39 through 41 show results for a back-to-back converter with a stepped-up 420 V DC bus, and open-neutral. Obviously, the number of actual switching processes differs from the direct converter despite the same PWM frequency of 1.44 kHz.

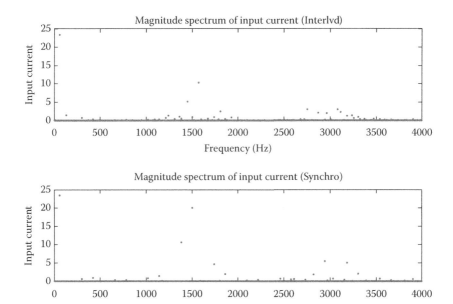

FIGURE 18.37 Stochastic ripple cancellation for interleaving the PWM controllers.

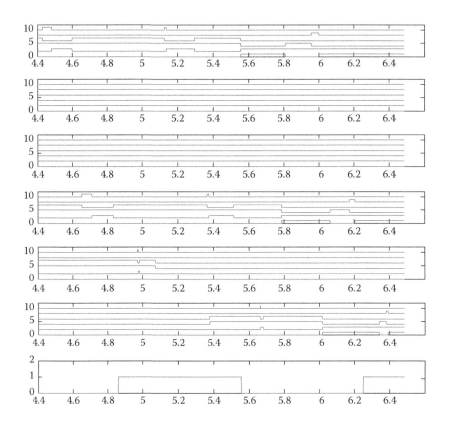

FIGURE 18.38 PWM control signals for each of the six converters, under interleaved operation (compare to Figure 18.34).

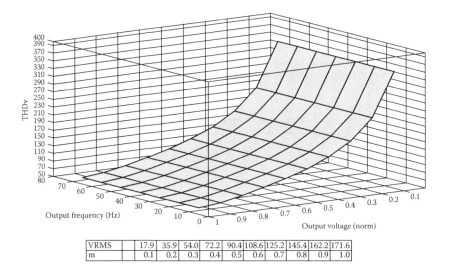

VRMS		17.9	35.9	54.0	72.2	90.4	108.6	125.2	145.4	162.2	171.6
m		0.1	0.2	0.3	0.4	0.5	0.6	0.7	0.8	0.9	1.0

FIGURE 18.39 THD for output voltage at different output magnitude and frequency.

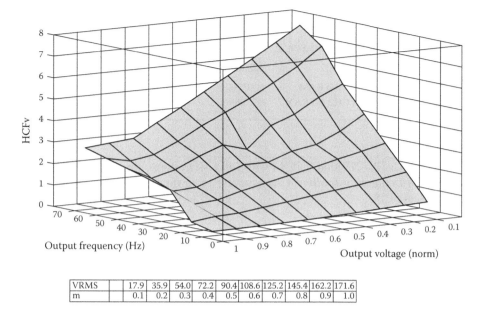

VRMS		17.9	35.9	54.0	72.2	90.4	108.6	125.2	145.4	162.2	171.6
m		0.1	0.2	0.3	0.4	0.5	0.6	0.7	0.8	0.9	1.0

FIGURE 18.40 HCF for output voltage at different output magnitude and frequency.

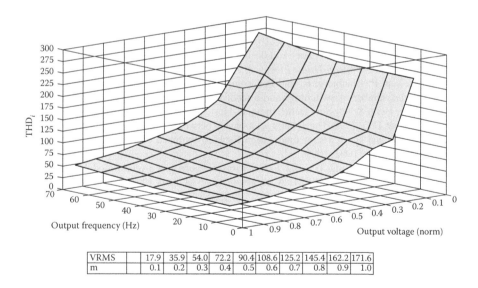

VRMS		17.9	35.9	54.0	72.2	90.4	108.6	125.2	145.4	162.2	171.6
m		0.1	0.2	0.3	0.4	0.5	0.6	0.7	0.8	0.9	1.0

FIGURE 18.41 THD of the input current at different output magnitude and frequency.

18.7.6 COMMENTS ON COMPUTER PERFORMANCE

Table 18.3 suggests running times for two conventional Windows-based computer architectures. The processing speed can be increased further by running the MATLAB code on the laptop's more powerful *Graphics Processing Unit*

TABLE 18.3
Computer Processing Power

Computer	Simulation	4-ch DFT	THD/HCF graph
Pentium 4, CPU 3.2 GHz, 1G RAM	10,000 s	~4 s	~120 s
Quad-Core i7, CPU 3.4 GHz, 8G RAM	~2900 s	~1 s	~40 s

card (GPUs like NVidiaGT555M with 144 CUDA cores) through the MATLAB's *Parallel Computing Toolbox*. Users suggest a four times speed-up compared to the quad-core CPU.

REFERENCES

1. Neacsu, D.O. 2012. Principle of a novel component minimized active power filter for high-power magnet supplies. *IEEE Industrial Electronics Conference IECON*, Montreal, Canada, October.
2. Neacsu, D.O. 2010. Analytical investigation of a novel solution to AC waveform tracking control. *IEEE International Symposium in Industrial Electronics*, Bari, Italy, July, pp. 2684–2689.
3. Huber, L., Borojevic, D., and Burany, N. 1992. Analysis, design and implementation of the space vector modulator for forced-commutated cycloconverters. *IEE Proceedings of the Series B*, 139(2), March, pp.103–113.
4. Neacsu, D.O. 1995. Theory and design of a space vector modulator for AC–AC matrix converter. *European Trans. Electrical Power Eng.* 4, 285–292.
5. Nielsen, P., Blaabjerg, F., and Pedersen, J.K. 1996. SVM matrix converter with minimized number of switchings and a feedforward compensation of input voltage unbalance. *PEDES '96*, New Delhi, India, 8–11.
6. Holmes D.G. 1992. The general relationship between regular-sampled pulse width modulation and space vector modulation for hard switched converters. IEEE-IAS Annual Meeting, pp. 1002–1009.
7. Zhou, K. and Wang, D. 2002. Relationship between space-vector modulation and three-phase carrier-based PWM: A comprehensive analysis [three-phase inverters]. *IEEE Trans. IE* 49(1), 186.
8. Casadei, D., Serra, G., and Tani, A. 1996. A general approach for the analysis of the input power quality in matrix converters. *IEEE PESC '96*, Baveno, Italy, vol. 2, pp. 1128–1134.
9. Nielsen, P., Casadei, D., Serra, G., and Tani A. 1996. Evaluation of the input current quality by three different modulation strategies for SVM controlled matrix converters with input voltage unbalance. *PEDES '96*, New Delhi, India, January 8–11.
10. Neft, C.L. and Shauder, C.D. 1988. Theory and design of a 30-HP matrix converter. *Proceedings of IEEE IAS '88*, pp. 248–253.
11. Halasz, S., Schmidt, I., and Molnar, T. 1995. Matrix converters for induction motor drive. *EPE'95*, Sevilla, Spain.
12. Watthanasarn, C., Zhang, L., and Liang, D.T.W. 1996. Analysis and DSP-based implementation of modulation algorithms for AC/AC matrix converters. *IEEE-PESC1996 Conference Record*, vol. 2, pp. 1053–1058.
13. Ziogas, P.D. 1986. Rectifier-inverter frequency changers with suppressed DC link component. *Trans. Ind. Appl.* IA-22(6), 1027–1036.

14. Kazerani, M. and Ooi, B.T. 1993. Direct AC-AC matrix converter based on three-phase voltage source converter modules. *Proceedings of IECON 1993*, Mavi Lahaina, Hawaii, November 15–19.

15. Ooi, B.T. and Kazerani, M. 1994. Application of dyadic matrix converter theory in conceptual design of dual field vector and displacement factor controls. *IAS Annual Meeting*, pp. 903–910.

16. Kim, S., Sul, S.K., and Lipo, T.A. 1998. AC to AC power conversion based on matrix converter topology with unidirectional swtches. *APEC*, Anaheim, CA, pp. 301–307.

17. Kazerani, M. and Ooi, B.T. 1993. Direct AC-AC matrix converter based on three-phase voltage source converter modules. *Proceedings of IECON*, Mavi Lahaina, Hawaii, November 15–19.

18. Ooi, B.T. and Kazerani, M. 1994. Application of dyadic matrix converter theory in conceptual design of dual field vector and displacement factor controls. *IAS Annual Meeting*, pp. 903–910.

19. Neacsu, D.O., Alistar, A., and Kazerani, M. 2002. Insightful analysis of carrier PWM algorithms for direct AC-AC matrix converters based on voltage-source converter module. *IEEE IAS 2002*, Pittsburg, USA, October 13–19, vol. 1, pp. 459–465.

20. Su, G.J., Adams, D., and Multilevel, D.C. 2001. Link inverter for brushless permanent magnet motors with very low inductance. *IEEE IAS Annual Meeting*.

21. Lega, A. 2009. Multilevel converters: Dual two-level inverter scheme. PhD Dissertation, University of Bologna.

22. Mohan, N., Undeland, T., and Robbins, W. 2002. *Power Electronics*. 3rd edition. John Wiley and Sons, New York.

23. Trzynadlowski, A. 2010. *Introduction to Power Electronics*. John Wiley and Sons, New York.

24. Trzynadlowski, A. 2010. *Introduction to Power Electronics*. John Wiley and Sons, New York.

25. Huber, L. and Borojevic, D. 1991. Space vector modulation with unity power factor for forced commutated cycloconverters. *IEEE IAS'91 Conference Record*, vol. I, pp. 1032–1041.

26. Chu, R.F. and Burns, J.J. 1989. Impact of cycloconverter harmonics. *IEEE Trans. Industry Appl.* 25(3), 427–435.

27. Neacsu, D.O. 1999. Current-controlled AC/AC voltage source matrix converter for open-winding induction machine drives. *IECON*, Denver, USA.

28. Kazerani, M. 2001. A direct AC/AC converter based on current-source converter modules. *IEEE PESC*, Vancouver, Canada.

29. Neacsu, D.O. 2003. IGBT-based "cycloconverters" built of conventional current source inverter modules. *IEEE International Symposium SCS 2003*, Iasi, Romania, July 10–11, vol.1, pp. 217–220.

30. Neacsu, D.O. 2004. Analysis and design of IGBT-based AC/AC direct converters built of conventional current source inverter module. *IEEE IAS*, Seattle, WA, October 2004, vol. 3, pp. 1824–1831.

31. Jones, J. and Bose, B.K. 1976. A frequency step-up cycloconverter using power transistors in inverse series mode. *Int. J. Electron.* 41(6), 573–587.

32. Neacsu, D.O., Rajashekara, K., and Gunawan, F. 2006. Overmodulation algorithm for zero-switching modulation. *IEEE ISIE 2006*, Montreal, Canada, July 9–12, pp. 1299–1304.

33. Weinhold, M. 1991. Appropriate pulse width modulation for a three-phase PWM AC to DC converter. *EPE J.* 1(2), 139–148.

34. Ciscato, D. 1992. PWM rectifier with low DC voltage ripple for magnet supply. *IEEE Trans. IA.* 28(2), 414–420.

35. Pan, C.-T. and Chen, T.-C. 1993. Modeling and analysis of a three phase PWM AC-DC converter without current sensor. *IEE Proc. B* 140(3), 201–208.

36. Enjeti, P. and Xie, B. 1992. A new real time space vector pwm strategy for high performance converters. *IEEE/IAS Annual Meeting Conference Record*, pp. 1018–1025.

37. Neacsu, D.O., Pastravanu, A., and Lucanu, M. 1993. Space vector based PWM AC/DC converter. *IEEE CAS Section SCS93 International Symposium*, Iasi, Romania, November 4–5, pp. 252–255.

38. Neacsu, D.O. 2010. Analytical investigation of a novel solution to AC waveform tracking control. *IEEE International Symposium in Industrial Electronics*, Bari, Italy, July, pp. 2684–2689.

39. Neacsu, D.O. 2010. Towards an all-semiconductor power converter solution for the appliance market. *IEEE Industrial Electronics Conference IECON*, Glendale, AZ, USA, November.

40. Pan, C.T. and Chen, T.C. 1993. Modeling and analysis of a three phase PWM AC–DC converter without current sensor. *IEE Proc. B* 140(3), 201–208.

41. Blaabjerg, F. and Pedersen, K. 1997. Optimized design of a complete three-phase PWM-VS inverter. *IEEE Trans. Power Electron.* 12(3), 567–577.

42. Neacsu, D.O. and Takahashi, T. 2000. Computer-aided design of a low-cost low-power snubberless three-phase inverter. *IEEE Workshop on Computers in Power Electronics*, Blacksburg, VA, June, pp. 204–210.

43. Wood, P., Battello, M., Keskar, N., and Guerra, A. 2002. IPM application overview— Integrated power module for appliance motor drives. *International Rectifier AN-1044*.

44. Staunton, R.H., Ozpineci, B., Theiss, T.J., and Tolbert, L.M. 2003. Review of the state-of-the-art in power electronics suitable for 10 kW military power systems. *ORNL/ TM-2003/209 Annual Report*, October.

45. Friedli, T. and Kolar, J.W. 2012. Milestones in matrix converters. *IEEJ J. Industry Appl.* 1(1), 2–14.

46. Holmes, G. and Lipo, T.A. 2003. *Pulse Width Modulation for Power Converters: Principles and Practice*. John Wiley and Sons, New York.

47. da Silva, E.R.C., dos Santos, E.C., and Jacobina, C.B. 2011. Pulsewidth modulation strategies. *IEEE Industrial Electron. Magazine* 5(2), 37–45.

48. Bowes, S.R. and Bird, B.M. 1975. Novel approach to the analysis and synthesis of modulation processes in power converters. *Proc. Inst. Electr. Eng.* 122, 507–513.

49. Odavic, M., Sumner, M., Zanchetta, P., and Clare, J.C. 2010. A theoretical analysis of the harmonic content of PWM waveforms for multiple-frequency modulators. *IEEE Trans. Power Electron.* 25(1), 131–141.

50. Fedele, G. and Frascino, D. 2010. Spectral analysis of a class of DC–AC PWM inverters by Kapteyn series. *IEEE Trans. Power Electron.* 5(4), 839–849.

51. Wang, B. and Sherif, E. 2013. Spectral analysis of matrix converters based on 3-D Fourier integral. *IEEE Trans. Power Electron.* 28(1), 19–25.

52. Gabriel, G.J. 2000. A general analytical theory of frequency conversion. *IEEE Trans. Circuits Syst. I Fundamental Theory Appl.* 47(2), 189–199.

53. Anon. 2012. The rise of the machines. *Popular Science* (www.popsci.com/content/ computing).

54. Lapointe, V. 2012. Using FPGA-based simulator to design and test power electronic controls used in automotive, industrial systems, aircrafts, micro-grid and MMC HVDC. *IEEE IECON Ind. Forum*, Montreal, Canada.

55. Anon. MATLAB DFT, *Internet R2012b Documentation*.

56. Perreault, D.J. and Kassakian, J. 1997. Analysis and control of a cellular converter system with stochastic ripple cancellation and minimal magnetics. *IEEE Trans. Power Electron.* 12(1), 145–152.

Index

For Product Safety Concerns and Information please contact our EU representative GPSR@taylorandfrancis.com Taylor & Francis Verlag GmbH, Kaufingerstraße 24, 80331 München, Germany

Printed and bound by CPI Group (UK) Ltd, Croydon, CR0 4YY

01/05/2025

01858484-0003